Developments in Mineral Processing, 8

Magnetic Methods for the Treatment of Minerals

Developments in Mineral Processing
D.W. Fuerstenau/advisory editor

1 A.J. Lynch
Mineral Crushing and Grinding Circuits.
Their Simulation, Optimisation, Design and Control

2 J. Laskowski (editor)
Mineral Processing.
Proceedings of the 13th International Mineral Processing Congress.
Warsaw, June 4–9, 1979

3 A.J. Lynch, N.W. Johnson, E.V. Manlaping and C.G. Thorne
Mineral and Coal Flotation Circuits.
Their Simulation and Control

4 H.J. Schulze
Physico-chemical Elementary Processes in Flotation.
An Analysis from the Point of View of Colloid Science
including Process Engineering Considerations

5 R.O. Burt
Gravitation Concentration Technology.

6 K.S.E. Forssberg (editor)
Flotation of Sulphide Minerals.

7 J. Srb and Z. Růžičková
Pelletization of Fines.

8 J. Svoboda
Magnetic Methods for the Treatment of Minerals.

Developments in Mineral Processing, 8

Magnetic Methods for the Treatment of Minerals

J. Svoboda

Ore-Dressing Division, Council for Mineral Technology, Randburg, South Africa

ELSEVIER
Amsterdam – Oxford – New York – Tokyo 1987

ELSEVIER SCIENCE PUBLISHERS B.V.
Sara Burgerhartstraat 25
P.O. Box 211, 1000 AE Amsterdam, The Netherlands

Distributors for the United States and Canada:

ELSEVIER SCIENCE PUBLISHING COMPANY INC.
52, Vanderbilt Avenue
New York, NY 10017, U.S.A.

ISBN 0-444-42811-9 (Vol. 8)
ISBN 0-444-41804-0 (Series)

Printed in The Netherlands

PREFACE

Magnetic separation embodies one of those branches of engineering and science where a wide gap exists between high and low technology.

Engineers and scientists can be found on remote mines who operate magnetic separators or who solve the equation of a motion of a particle in a magnetic separator in a university. Although both groups deal with the same technology, they very rarely communicate with each other and are seldom aware of each other's problems.

The aim of this book is to treat magnetic separation from both these viewpoints and to emphasize those points of magnetic separation in which lack of support in fundamental research is most evident. It is hoped that the book will help develop a greater overlap of the activities between the engineer and the scientist, promote the application of basic physical phenomena in engineering practice and gain the acceptance of the industry.

Magnetic separation is the technology where it is more than true that the novelty alone does not win out, unless the intrinsic physical and engineering potential is also present, and is sufficiently demonstrated to gain acceptance. In order to warrant such amalgamation of the activities of engineers and scientists the overlap of activities between these groups must be real. It will certainly not succeed by a mere passing over of the words. Magnetic separation thus remains an exciting branch of technology with a broad potential in diverse branches of engineering and with equally exciting prospects for fundamental research.

I am indebted to many organizations and individuals who have given me invaluable help during the preparation of this book.

The assistance of Mr J. Levin and Dr J. Kopp, who read the entire manuscript and made many valuable suggestions, is acknowledged with thanks. The first three chapters benefited from criticism by Dr L.A. Vermeulen.

Sincere thanks are due to the Council for Mineral Technology (Mintek) for partially sponsoring this book, to Mr G.G. Zwirn for his meticulous editing of the manuscript, to Dr H.W. Glen, Messrs G.G. Roberts, J.Z. Hoek, J. Welte and J.R. Northmore and Miss M. Craemer for their assistance with the preparation of equations and of some of the figures and photographs.

I am especially indebted to my wife Dr Olga Svoboda, who typed many versions of the manuscript, and helped with other aspects of the presentation of the book. Without her participation, this book would not exist.

J. Svoboda
Randburg, Transvaal
January 1987

ACKNOWLEDGEMENTS

I am very grateful to the following organizations which provided protographs or illustrations for reproduction or gave me permission to use the data from their publications:

Carpco, Inc., Aquafine Corp., Sala Magnetics, Inc., KHD Humboldt Wedag AG, Ore Sorters (Pty.) Ltd., Krupp Polysius AG, Jones Ferro-Magnetics Inc., Boxmag-Rapid Ltd., Biomagnetic Technologies, Inc., LDJ Inc., Dowty RFL Industries, Inc., F.W. Bell, Inc., S.G. Frantz, Inc., Eriez Magnetics (South Africa) (Pty.), Ltd., Bartles (Carn Brea) Ltd., Applied Magnetic Systems, Inc., Verlag Chemie GmbH, The Institute of Electrical and Electronics Engineers, Inc., Uplands Press, Ltd., Society of Mining Engineers, Inc., The Canadian Institute of Mining and Metallurgy, The Mining Journal, Ltd., American Institute of Physics, Marcel Dekker, Inc., The Institution of Mining and Metallurgy, Council for Mineral Technology, The South African Institute of Mining and Metallurgy.

CONTENTS

CHAPTER III

CHAPTER IV

INTRODUCTION

Although the ancient Greeks were familiar with some of the phenomena of magnetism as early as 550 B.C. and though the Chinese harnessed the directional properties of magnets as early as the first century B.C., it was not until 1845 that Faraday discovered that all substances were susceptible, in a varying degree, to the influence of a magnetic field. Ball, Norton, Edison and others applied this principle to the separation of iron ore from a gangue and carried the process even further in the successful separation of strongly magnetic minerals with the use of electromagnetic drum separators.

Up to 1895 Faraday's discovery that no substances were indifferent to the magnetic field had not brought the extension of magnetic methods of separation to materials whose magnetic susceptibilities were so small that they were classified as nonmagnetic [F1]. In 1895 Wetherill devised a separator that combined high magnetic field intensity with deflection of a grain stream. The machine was able to separate two components both of which were commonly considered nonmagnetic. In the ensuing time various types of disk, drum and roll dry magnetic separators were developed although the spectrum of minerals treatable by these machines was limited to sufficiently coarse and moderately strongly magnetic materials. In 1937 Frantz developed a magnetic separator consisting of an iron-bound solenoid packed with ferromagnetic steel ribbons [F2], and this proved to be an important milestone in the development of present high-intensity and high-gradient magnetic separators.

Nevertheless the genuine development of high-gradient magnetic separation (HGMS) began only in the sixties. The newly developed cyclic and continuous magnetic separators extended the range of the applicability of magnetic separation to many weakly paramagnetic and even to diamagnetic minerals of the micrometer size range.

The technique of high-intensity or high-gradient magnetic separation offers the solution to various complex problems encountered in chemical,biochemical, nuclear and other industries. The application of this method, however, is not limited just to the production phase of the industrial processes but may be used to treat solid wastes, waste waters and gases as well as to general problems of environmental protection. Jointly, with dramatic progress in the practical applications of HGMS, the last ten years have witnessed substantial changes in the formulation of the physical principles of magnetic separation.

Currently, the literature on magnetic separation is in a highly fragmented form. The theoretical and practical aspects of the classical magnetic separation, as developed and practised during the first half of this century, are described in an excellent treatise by Derkach [D1] which, however, quite understandably does not cover new dramatic developments in HGMS. A review by Gerber and Birss [G1] of the theoretical models of HGMS and of some applications was intended to fill this gap.

The objective of the present book is to present a fairly broad survey of magnetic separation as applied to and practised primarily by the mineral processing industries, although its use in other industries will be briefly reviewed. It includes information on the physical principles of magnetic separation, magnetic properties of minerals and their measurement, generation of magnetic field, theoretical and practical problems of magnetic separation and experience gained in the design and operation of magnetic separation systems.

The book should be of value to engineers and consulting metallurgists as well as to the students who want to learn more about this branch of technology. We have attempted to meet the needs of the growing number of engineers, technologists and applied physicists who are now engaged in the practical exploitation of magnetic separation.

Throughout the book the SI system of units is used, although the equivalents in the cgs system of units are given for most magnetic units.

Since italic typescript was not available on the word processor, all quantities, either vector or scalar, are indicated by bold-face type. The vectors are identified by means of an arrow over a symbol.

CHAPTER I

THE PHYSICAL PRINCIPLES OF MAGNETIC SEPARATION

I.1 PRINCIPLES OF MAGNETIC BENEFICIATION

Magnetic separation, which is an engineering practice of long standing, is used for the concentration of magnetic minerals and for the removal of suspended magnetizable particles from fluid streams. The separation is achieved by passing the suspensions or the mixtures of particles through non-homogeneous magnetic field, which leads to a preferential retention or deflection of the magnetizable particulates. The same objective is, however often achieved in a very different fashion, the common features being a competition between a wide spectrum of forces of various magnitudes and ranges. In magnetic separation the dominant external force is the magnetic force and the separation of one mineral from another or the removal of magnetizable particles from the streams depends upon their motion in response to the magnetic force and to other competing forces, namely gravitational, hydrodynamic, inertial and centrifugal forces. Interparticle forces of electromagnetic and electrostatic origin contribute to the overall scenario in a magnetic separator which is schematically depicted in Fig. 1.1 [O1].

It is thus clear that a necessary (but not sufficient) condition for a successful separation of more strongly magnetic and less strongly magnetic particles in a magnetic field of a separator, is that magnetic traction force $\mathbf{F}_{mag}^{(m)}$ acting on more magnetic particles must be greater than the sum of all the competing forces $\mathbf{F}_c^{i(m)}$. Simultaneously a magnetic force acting on less strongly magnetic particles, $\mathbf{F}_{mag}^{(n)}$ must be smaller than the sum of the corresponding competing forces. Therefore the following conditions have to be met in a magnetic separator:

$$F^{(m)}_{mag} \geqslant \sum_i F_c^{i\ (m)} \qquad (1.1)$$

$$F^{(n)}_{mag} \leqslant \sum_i F_c^{i\ (n)}$$

The slurry fed into the magnetic separator will thus be split into two or more components. In any real separation, both magnetic and nonmagnetic particles can be found in the magnetic fraction, nonmagnetic fraction and in middlings, and only in the limiting case can the separation be considered complete. The performance of a magnetic separator can be expressed by the recovery, the ratio of magnetic material in the mags relative to that in the feed, and the grade, the fraction of magnetic material in the mags. However, these definitions are not unique and they depend on whether the useful product is the magnetic or nonmagnetic fractions.

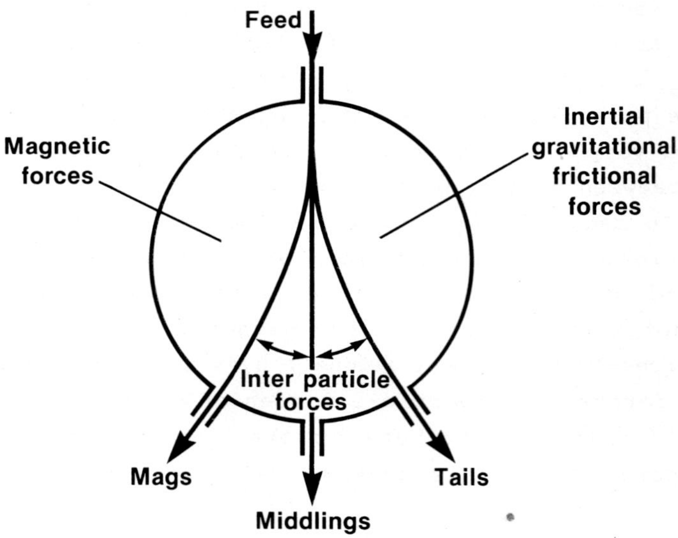

Fig 1.1 Schematic representation of magnetic separator (after Oberteuffer, IEEE Trans. Mag. [O1]).

We have thus seen that the magnetic, mechanical and interparticle forces determine the performance of a separator. These forces are dependent on both the nature of the feed to be separated as well as the character of the separation device.

In order to illustrate the role of fundamental magnetic quantities and of magnetic properties of minerals in magnetic separation, and before embarking upon a detailed analysis of the theory of magnetic separation, we shall briefly outline, in the following sections, the physical principles of magnetic separation.

I.1.1 Forces in magnetic separator

The traction magnetic dipole force acting on a paramagnetic particle in vacuum is given by

$$\vec{F}_m = \nabla \int_V (\vec{J} \cdot \vec{H}) dV \qquad (1.2)$$

where **J** is the magnetic polarization of a particle (in Tesla), **V** is the volume of a particle (in m^3) and **H** is the magnetic field strength (in $A\ m^{-1}$). ∇ is the operator of the gradient and acting on a scalar ϕ it can be written as

$$\nabla\phi \equiv \text{grad}\ \phi = \vec{i}\frac{\partial\phi}{\partial x} + \vec{j}\frac{\partial\phi}{\partial y} + \vec{k}\frac{\partial\phi}{\partial z}$$

where $\vec{i}$, $\vec{j}$, $\vec{k}$ are the unit vectors in the directions **x**, **y**, **z** respectively.

The integration is taken over the particle volume. If the particle is sufficiently small it can be reduced to a point dipole of moment $\vec{\mu}_m = \vec{J}V$ where **V** is the volume of the particle. The force on such point dipole is thus

$$\vec{F}_m = (\vec{\mu}_m \cdot \nabla)\vec{H} \qquad (1.3)$$

Permeability of a spherical paramagnetic or diamagnetic particle is given as

$$\mu = (1+\kappa)\mu_0 \qquad (1.4)$$

where κ is the volume magnetic susceptibility, and μ_0 is the permeability of free space numerically equal to $4\pi \cdot 10^{-7}$ H m^{-1}. Thus the magnetic polarization of the particle is given by

$$\vec{J} = \frac{\mu_0 \kappa \vec{H}}{1+\frac{\kappa}{3}} \qquad (1.5)$$

Combining eq. (1.5) and eq. (1.3) we can see that the force on a small spherical weakly magnetic particle placed in the external magnetic field can be written as

$$\vec{F}_m = \frac{\mu_0 \kappa V}{1+\frac{\kappa}{3}} (\vec{H}\nabla)\vec{H} \qquad (1.6)$$

or in a simplified form, assuming that $\kappa \ll 1$

$$\vec{F}_m = \frac{1}{2}\mu_0 \kappa V \nabla(H^2) \qquad (1.7)$$

For practical calculations it is sometimes advantageous to replace the magnetic field strength by the magnetic induction **B** and then eq.(1.7) reads as follows

$$\vec{F}_m = \frac{\kappa}{\mu_0} V|B|\text{grad}|\vec{B}| \qquad (1.8)$$

From eq. (1.6) we can see that the magnitude of the magnetic force is proportional to the product of the magnetic field strength and of the gradient of the magnetic field and this force is in direction of the field gradient, not of the magnetic field strength.

It is clear that an increase in the value of the magnetic field **H** will increase the magnetic force, both by increasing the magnetization of the particles and, in many cases, the value of the magnetic field gradient as well. To illustrate the role of various quantities contained in eq. (1.6) we shall consider the force on a spherical particle produced by a ferromagnetic filament in a uniform external magnetic field, as shown in Fig.1.2.

The diagram ilustrates the situation in a high-gradient magnetic separator.

Along the axis of symmetry, and for values of the external magnetic field $\mathbf{H}_o$ lower than the bulk saturation value $\mathbf{H}_s$ of the ferromagnetic filament, the magnetic field strength [S1] is given by the expression

$$H = H_0\left(1 + \frac{a^2}{r^2}\right) \qquad (1.9)$$

where **a** is the filament radius and **r** is the radius vector of a point at which the value of **H** is measured. For $\mathbf{H}_o > \mathbf{H}_s$, **H** is given by

$$H = H_0 + H_s \frac{a^2}{r^2} \qquad (1.10)$$

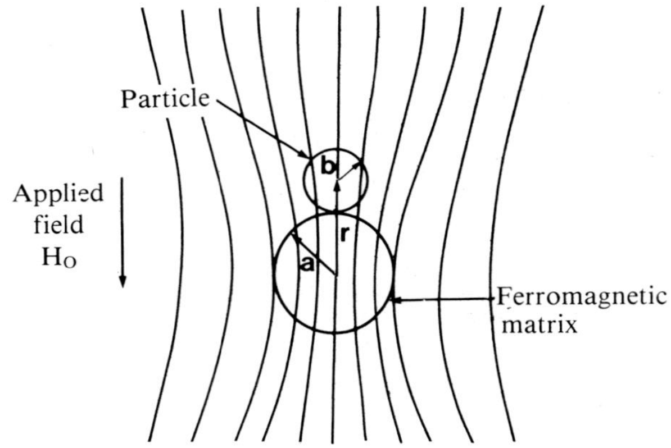

Fig. 1.2 Cross section of spherical particle, radius **b**, attached to ferromagnetic wire, radius **a**, magnetized by uniform magnetic field $\mathbf{H}_o$.

The magnetic field gradient for $\mathbf{H}_o < \mathbf{H}_s$ along the axis of the particle is given as

$$\frac{dH}{dx} = -2H_0 \frac{a^2}{r^3} \qquad (1.11)$$

when the magnetization of the particles is small. If we combine eq.(1.9) and eq. (1.11) with eq. (1.6) we arrive at the approximate expression for the magnetic force [O1] on a point-like spherical weakly magnetic particle

$$F_m = -\frac{8}{3}\pi\mu_0\kappa b^3\left(H_0 + H_0\frac{a^2}{r^2}\right)H_0\frac{a^2}{r^3} \qquad (1.12)$$

where **b** is the particle radius. Magnetic force as a function of the steel ball radius **a** for the uraninite particle of radius **b** and volume magnetic susceptibility $\kappa = 2.5 \times 10^{-7}$ (SI), is plotted in Fig.1.3.

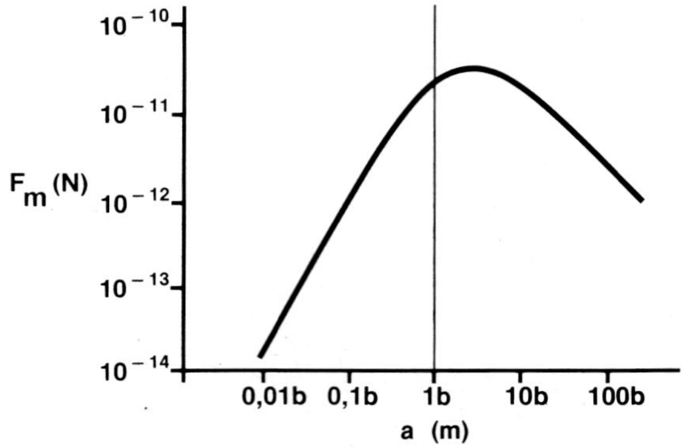

Fig. 1.3 Magnetic force $\mathbf{F}_m$ acting on a uraninite particle of radius **b** caused by a magnetized steel ball of radius **a** at **B** = 1 T, **v** = 0.1 m s^{-1}, as a function of a ball radius.

We can see that the magnetic force is maximized for a radius of a matrix collector of the same order of magnitude as the particle radius. If we take the derivative of $\mathbf{F}_m$ of eq. (1.12) with respect to the ratio **a/b**, the maximum is reached for a matrix collector radius 2.69 times that of the particle radius [O1]. A more rigorous treatment [A1] shows that the choice of the optimum collector diameter is less critical than that estimated by Oberteuffer [O1], since the maximum is flatter; however the position of this maximum is not significantly changed.

The variation of the total magnetic field strength, field gradient and magnetic force on a paramagnetic particle, for a matched system (i.e. a = 3b) is shown in Fig. 1.4 as a function of a distance from the surface of the collector along the axis of symmetry. This diagram indicates the importance of the magnetic force, as having its greatest rate of change across the diameter of the particle. The drop-off of magnetic field strength, field gradient and magnetic force away from a magnetized element, is a common feature of magnetic separators.

Eq. (1.12) also allows us to express the dependence of the magnetic force on the particle radius for a matched system (i.e. **a** = 3b). In this case eq. (1.12) can be rewritten as

$$F_m = -\frac{16}{27}\pi\mu_0\kappa b^2 H_0^2 \qquad (1.13)$$

Hence the magnetic force varies as the square of the radius of the particle in such a matched system.

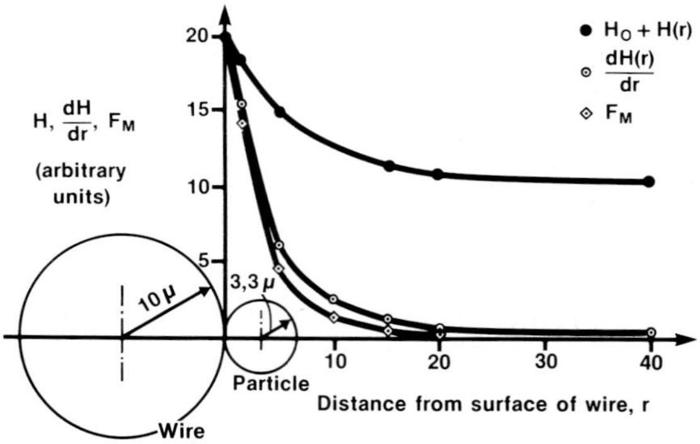

Fig. 1.4 Functional dependence of total magnetic field strength **H**(r), field gradient **dH/dr**, and magnetic force $\mathbf{F}_m$ on paramagnetic particle of radius 3.3 μm due to ferromagnetic collector of radius 10 μm in uniform magnetic field with distance from surface of collector (after Oberteuffer, IEEE Trans. Mag. [O1]).

I.1.2 Competing forces

As mentioned above, magnetic force in a magnetic separator competes with additional forces, namely forces of gravity, inertia and hydrodynamic drag. The importance of each force is a function of a type of magnetic separator. At this stage we shall limit our attention to the hydrodynamic drag and force of gravity, which is the usual situation in wet high-gradient magnetic separators.

For a spherical particle of density ρ_p the gravitational or buoyant force is given by

$$\vec{F}_g = \frac{4}{3}\pi(\rho_p - \rho_f)b^3\vec{g} \tag{1.14}$$

where ρ_f is the density of the fluid medium and **g** is the acceleration by gravity. Finally, one obtains for the hydrodynamic drag force

$$\vec{F}_d = 6\pi\eta b\left(\vec{v}(\vec{r}) - \frac{d\vec{r}}{dt}\right) \tag{1.15}$$

where η is the dynamic viscosity of the fluid, **dr/dt** is the velocity of the fluid and **v**(r) the velocity of the particle at position **r**.

It can be seen from the last three equations that because each force has a different dependence on the particle size **b**, the relative importance of the forces will vary with the particle size. The dependence of the gravitational force (eq. 1.14) on the third power of the particle radius means that the gravitational force will be significant for large particles. The hydrodynamic force, which in the Stokes region depends on the first power of the particle radius, will be important for small particles. When the field gradient is matched to the particle size the magnetic force varies as the square of the particle size, as follows from eq. (1.13). We can thus anticipate that the magnetic force will be greater than the competing forces and in a limited interval of particle sizes. The magnitudes of the forces that act on uraninite particles passing through the external magnetic field of the induction of 1 T, at the interstitial velocity 0.1 m s^{-1} with the field gradient generated by the matched steel ball matrix are shown in Fig. 1.5.

For large particles, the gravitational force clearly predominates while for very fine particles the hydrodynamic drag is dominant. Consequently the magnetic force is larger than either of the competing forces only over a limited size range. The fundamental condition (eq. 1.1) for separation to occur is thus met only for a restricted range of particle size and there is always only a narrow interval of particle size at which the separation conditions are optimum, as shown in Fig. 1.6.

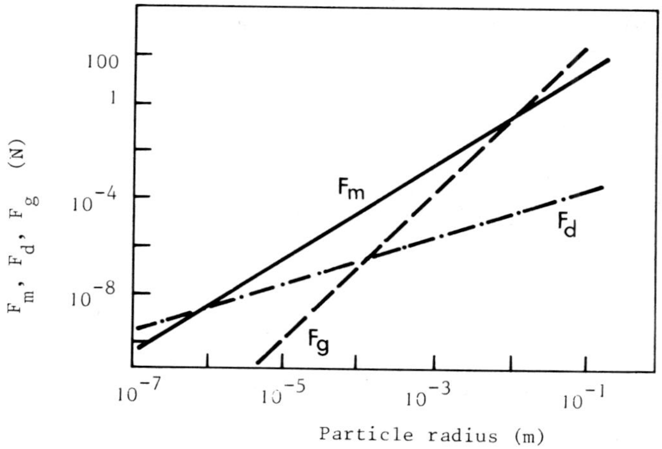

Fig. 1.5 Plot of magnetic ($\mathbf{F}_m$) and competing gravitational ($\mathbf{F}_g$) and drag forces ($\mathbf{F}_d$) versus particle size **b** for uraninite particle flowing through the ball matrix placed in the magnetic field of 1 T at interstitial velocity 0.1 m s^{-1}.

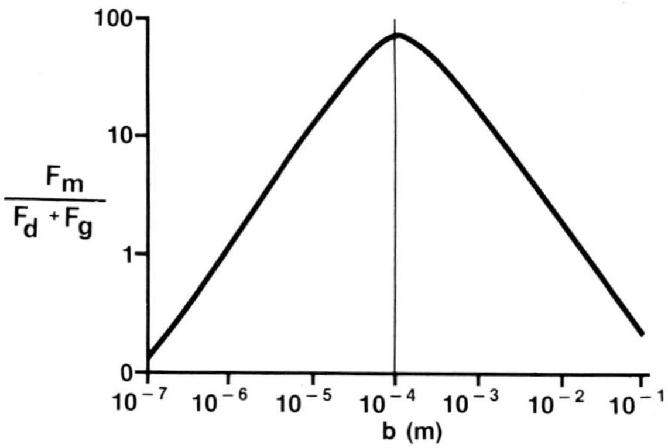

Fig. 1.6 Force ratio (magnetic force divided by competing gravitational and drag forces) versus particle size for a system of uraninite particles suspended in water and passed through a bed of steel balls of diameter 6 mm, placed in the magnetic induction 1 T. Flow velocity of the slurry 0.1 m s^{-1}.

We shall illustrate this model of the force balance by an example. Let us assume that a particle of siderite, of 100 μm diameter, volume magnetic susceptibility $\kappa = 5 \times 10^{-3}$ (SI) is passed through a wet high-gradient magnetic separator with the interstitial velocity of 0.1 m s^{-1}. The applied magnetic induction is 2 T and the field gradient is generated by steel balls of diameter 6 mm. As we shall see later $|\text{grad } B| \cong |B|/a$, so that in our case grad B $\cong$ 700 T m^{-1}. If we substitute these values into the eq. (1.8) we obtain $\mathbf{F}_m = 3 \times 10^{-6}$ N. The hydrodynamic drag can be evaluated from eq. (1.15), assuming that the carrying medium is water with the dynamic viscosity $\eta = 10^{-3}$ $kg^{-1}m^{-1}s^{-1}$. Thus $\mathbf{F}_d$ = 0.9×10^{-7} N. Likewise, the gravitational force can be calculated from eq. (1.14) and under our assumptions we obtain $\mathbf{F}_g = 2 \times 10^{-8}$ N. Under our experimental conditions the magnetic force is almost two orders of magnitude greater than the competing forces and the recovery of the siderite particles should be efficient. Successful separation of such siderite particles can be expected even at much lower magnetic field strengths. If the magnetic induction is reduced by a factor of three to 0.6 T, the resultant magnetic force ($\mathbf{F}_m = 3.2 \times 10^{-7}$ N) will be still larger than the sum of the competing forces, in accordance with the condition of eq.(1.1).

Similarly, a mineral with a magnetic susceptibility of an order of magnitude lower than siderite can still be recovered. On the other hand, the effect of the particle size on the overall balance of forces is much more dramatic. If we assume that the diameter of a siderite particle is 22 μm, then at the background magnetic induction of 2 T, and flow velocity 0.1 m s^{-1} magnetic force $\mathbf{F}_m$ is equal to 4.4×10^{-8} N. Similarly, the force of gravity $\mathbf{F}_g = 2.1 \times 10^{-10}$ N and the hydrodynamic drag $\mathbf{F}_d = 2.1 \times 10^{-8}$ N. Since the particle size was reduced from 100 μm to 22 μm, the magnetic force dropped by two orders of magnitude, while the hydrodynamic drag dropped just by an order of magnitude and we can expect a decrease in the recovery of fine siderite particles.

Although this simple model of the force balance does not take into account the kinetics of the particles in the slurry and their mutual interaction, it does give us practical information about the role of some important quantities affecting the performance of a magnetic separator. A rigorous theoretical treatment of the process of magnetic separation will be reviewed in Chapter III while the experimentally determined effects of the variables involved in magnetic separation will be discussed in Chapter IV.

From the above example it is obvious that the subject of

magnetic separation cannot be discussed without first defining and explaining the quantities usually encountered in magnetism, and the units in which these quantities are measured. The next section explains the system of units used in this book.

I.2 MAGNETIC UNITS

The units and symbols used are those of the Système International d'Unites, designated SI. The system originated in 1901 with the proposal made by the Italian engineer Giorgi that electrical units be based on the meter-kilogram-second system. The Giorgi system grew with the years and came to be called, first MKS system and later the MKSA system (A for ampère).

Most students today are taught in SI units and most scientific journals encourage their contributors to use SI units. Nevertheless, many papers on magnetism - and magnetic separation is no exception - are still published in cgs units. Those authors who follow the newly-established trend and convert to SI units find that the task of conversion is gigantic and that errors almost certainly creep in.

It is our wish that this book be understood by the many older technologists who have never used anything but the old cgs system, as well as by the rising generation of scientists and engineers who have been taught in the new SI system. Thus, when there is a choice, we tend to speak of magnetic induction **B** in Tesla (T), and the conversion to cgs unit of Gauss should present no difficulty if the values in T are multiplied by 10 000. When the magnetic field strength **H** is appropriate, both SI unit (A m^{-1}) and cgs unit (Oe), are given.

Magnetic susceptibility is rather troublesome. The quantity that enters most equations used in magnetic separation is invariably the volume magnetic susceptibility, which although dimensionless in both SI and cgs, differs by a factor 4π in the systems of units. Experimentally, it is easier to determine the specific (or mass) magnetic susceptibility and therefore most values of magnetic susceptibility to be found in the literature are those of the specific susceptibility, again invariably in cgs units. Various names are used for this unit, e.g. emu cm^{-1}, emu g^{-1} Oe^{-1} and cgs, but all of them represent the same unit, namely $cm^3\ g^{-1}$ in cgs system of units. Thus, in SI the unit is $m^3\ kg^{-1}$ and the conversion factor is $4\pi/1000$.

When we describe a magnetic field we employ two different entities: magnetic field strength **H** and magnetic induction (or flux density) **B**. **B** and **H** are both vector quantities having direction as well as magnitude. In free space **B** and **H** are not independent but are related by the equation

$$\vec{B} = \mu_0 \vec{H} \tag{1.16}$$

where μ_0 is the magnetic permeability of free space, numerically equal to $4\pi \times 10^{-7}$ and its dimension follows from eq. (1.16), namely $Vsm^{-1}\ A^{-1} = H\ m^{-1}$. In a magnetic material, **B** and **H** can vary independently. They are related by the equation

$$\vec{B} = \mu \vec{H} \tag{1.17}$$

where μ is the magnetic permeability; however they do not have to be in the same direction and μ is in general a tensor.

The magnetic induction **B** includes the contributions from external magnetizing forces and from the two quantities describing the intrinsic properties, namely the magnetization **M** measured in $A\ m^{-1}$ and the magnetic polarization **J** measured in Tesla. These quantities are related by the equation:

$$\vec{J} = \mu_0 \vec{M} \tag{1.18}$$

The magnetic induction, magnetic field strength and polarization are related, in the Kennelly system, by

$$\vec{B} = \mu_0 \vec{H} + \vec{J} \tag{1.19}$$

Given **H** and **B** or **J**, the curves can be plotted as shown in Fig. 1.7. Unlike **B** which continues to increase indefinitely with increasing magnetic field strength, the polarization of a material reaches a maximum, or the saturation value $\mathbf{J}_s$.

By combining eqs. (1.17) and (1.19), we can express the polarization either as a function of **H** or as a function of **B**

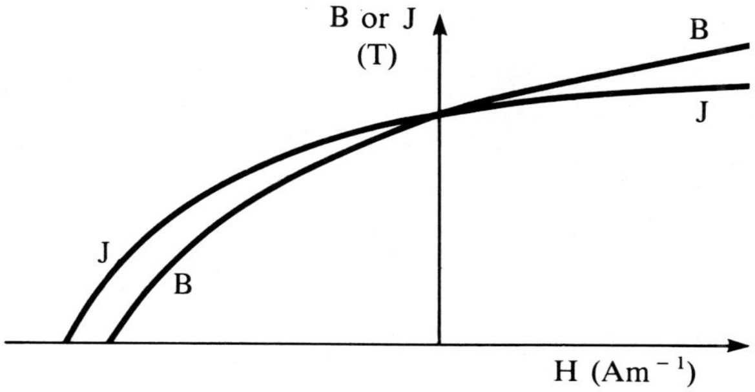

Fig. 1.7 **B** and **J** plotted against **H**.

$$\vec{J}=(\mu-\mu_0)\vec{H}=\left(1-\frac{\mu_0}{\mu}\right)\vec{B}=\frac{\mu_r-1}{\mu_r}\vec{B} \tag{1.20}$$

where $\mu_r = \mu/\mu_0$ is the relative permeability.

In the Sommerfeld system the magnetic induction is written as

$$\vec{B}=\mu_0(\vec{H}+\vec{M}) \tag{1.21}$$

as follows from eq. (1.18) and eq. (1.19).

In the cgs system, the unit of magnetic induction **B** is Gauss and permeability of free space μ_0 is dimensionless and by definition equal to unity. The cgs equivalent of eq. (1.21) is

$$\vec{B}=\vec{H}+4\pi\vec{M} \tag{1.22}$$

and for the magnetic field strength the name Oersted (Oe) was used. Conversion factors between SI and cgs units are given in Table 1.1 and comparison of SI and other units of magnetism can be found in [M1].

Although the conversion of the fundamental quantities of magnetism causes some difficulty, the situation with magnetic susceptibility is even more complicated, because volume, mass (specific) and molar susceptibilities are used in cgs units and there are some differences of opinion about the equivalent SI units. We shall briefly review the definitions of magnetic susceptibilities and shall attempt to indicate the conversion between the two systems of units.

An isotropic material in equilibrium in a magnetic field strength **H** has a magnetization **M** along the field direction. In the simplest case

$$\vec{M} = \kappa \vec{H} \tag{1.23}$$

or similarly

$$\vec{J} = \mu_0 \kappa \vec{H} \tag{1.24}$$

where the scalar quantity κ is the magnetization per unit field, or volume magnetic susceptiblity. In SI units, where polarization **J** and **B** are measured in Tesla, the susceptibility κ is dimensionless. Magnetic susceptibility can also be expressed with respect to the unit mass of material of density ρ, and then

$$\kappa = \rho \chi \tag{1.25}$$

where χ is the mass or specific magnetic susceptibility. As follows from eq. (1.25), in SI the unit of χ is $m^3\ kg^{-1}$. We have to point out that the magnetic polarization **J**, or the magnetization **M**, depend not only on the substance, but also on the shape of the particle placed in the external magnetic field. For a particle in a homogenous magnetic field **H**, shape-dependence of the magnetic polarization can be written

$$\vec{J} = \frac{\kappa}{1 + D\kappa} \mu_0 \vec{H} \tag{1.26}$$

where **D** is the shape-dependent demagnetization factor, and $0 < \mathbf{D} < 1$.

Conversion factors between volume and specific magnetic susceptibilities in SI and cgs systems of units are given in Table 1.1. Since the data on magnetic susceptibility in literature are

TABLE 1.1

Units and conversion factors

Quantity	SI unit	cgs unit	Conversion factor F+)
Magnetic induction B	Tesla (T)	Gauss (G)	10^{-4}
Magnetic field strength H	A m^{-1}	Oersted (Oe)	$10^{3}/4\pi$
Magnetic polarization J	Tesla (T)	emu cm^{-3}	$4\pi \cdot 10^{-4}$
Magnetization M	A m^{-1}	emu cm^{3}	10^{3}
Permeability of free space μ_0	H m^{-1}	1	$4\pi \cdot 10^{-7}$
Magnetic flux	Weber (Wb)	Maxwell (Mx)	10^{-8}
Volume magnetic susceptibility κ	-	-	$1/4\pi$
Specific (mass) magnetic susceptibility χ	m^{3} kg^{-1}	cm^{3} g^{-1} (or emu g^{-1} or emu $g^{-1}cm^{-3}$)	$1000/4\pi$
Magnetic moment μ_m	T m^{3}	emu	$4\pi \cdot 10^{-10}$
Specific polarization σ_m	T m^{3} kg^{-1}	emu g^{-1}	10^{-7}

+) The number by which the quantity in cgs units must be multiplied to obtain the quantity in SI units, i.e. SI = cgs x F. Thus 1 G = 10^{-4} Tesla.

often given in different systems of units, it will be useful to summarize the relationships between them, as follows

$$\kappa(\mathrm{SI}) = 4\pi\kappa(\mathrm{cgs}) \qquad (1.27)$$

$$\kappa(\mathrm{cgs}) = \rho\chi(\mathrm{cgs}) \qquad (1.28)$$

$$\kappa(\mathrm{SI}) = 4\pi\rho\chi(\mathrm{cgs}) \qquad (1.29)$$

$$\chi(\mathrm{SI}) = \frac{4\pi}{1000}\chi(\mathrm{cgs}) \qquad (1.30)$$

In these equations density is expressed in $g\ cm^{-3}$.

Another quantity sometimes encountered in magnetism is the specific polarization σ_m defined by:

$$\sigma_m = \frac{J}{4\pi\rho} \tag{1.31}$$

In the cgs system σ_m has units of $emu\ g^{-1}$, while in SI the unit is $T\ m^3\ kg^{-1}$.

I.3. MAGNETIC PROPERTIES OF MINERALS

I.3.1 The classification of minerals according to their magnetic properties.

All matter responds to the presence of the external magnetic field, although the degree of this response varies widely. The applied magnetic field interacts with the magnetic field of the atoms of the matter. The latter magnetic field exists as a consequence of the precessional motion of the charged nucleus, the orbital motion of electrons around the nucleus and the spin motion of electrons and nuclei.

According to their magnetic properties, minerals can be divided into three basic groups: ferromagnetic, paramagnetic and diamagnetic. Until recently the technological classification for magnetic separation differed from the fundamental physical categorization [Dl]. In the early days of magnetic separation, all minerals were divided into the following three groups:

Strongly magnetic minerals which can be recovered by a magnetic separator with the use of relatively weak magnetic induction, up to 0.15 T. The magnetic susceptibility of minerals responding to such a low magnetic field is usually greater than $5 \times 10^{-5}\ m^3\ kg^{-1}$ ($4 \times 10^{-3}\ cm^3\ g^{-1}$) and minerals like magnetite fall under this group.

Weakly magnetic minerals which can be recovered by a magnetic separator generating magnetic induction up to 0.8 T, with modest values of the gradient of the magnetic field. This broad group of minerals, which includes ferrimagnetic and paramagnetic minerals and paramagnetic minerals with parasitic ferromagnetism also comprises iron and manganese oxides and carbonates, ilmenite, wolframite, and biotite. Magnetic susceptibility ranges from 5 x 10^{-6} down to 1 x 10^{-7} m^3 kg^{-1} (4 x 10^{-4} to 8 x 10^{-6} cm^3 g^{-1}), although the lower limit of specific magnetic susceptibility is conditional and tends to decrease with the progress in development of new types of magnetic separators generating higher magnetic force.

Nonmagnetic minerals are those minerals that cannot be recovered by conventional magnetic separators. They include weakly paramagnetic minerals with specific magnetic susceptibility smaller than, say, 1 x 10^{-7} m^3 kg^{-1} (8 x 10^{-6} cm^3 g^{-1}) and diamagnetic minerals with negative magnetic susceptibility.

Although this classification is very rough (since it does not take into account, among other things, particle size, which is so important for the determination of the efficiency of magnetic separation), it well illustrates the classes of minerals according to their response to conventional magnetic separators, usually found in ore-dressing practice.

However, with the advent of HGMS the potential for the treatment of weakly magnetic and "nonmagnetic" minerals was extended in such a way that all magnetic classes of matter can be treated by magnetic separation, down to the submicrometre range, provided that proper device and experimental conditions are chosen.

I.3.1.1 Diamagnetism

Diamagnetism is associated with the tendency of electrical charges to partially shield the interior of the body from an applied magnetic field. The electron orbits the nucleus in the presence of the central electrostatic field. If an external magnetic field is applied, then, according to the Lenz's law the induced current produces the magnetic field that is opposite to the applied field. The magnetic moment associated with the current is a diamagnetic moment and the volume magnetic susceptibility, being the magnetic moment per unit field per m^3, can, for a monoatomic molecule, be written as

$$\kappa = -\frac{\mu_0 N_a e^2 z < r^2 >}{6m_0} \quad (1.32)$$

where $\mathbf{N}_a$ is the number of atoms per unit volume, **e** the electronic charge, **z** is the number of electrons in an atom, $\mathbf{m}_0$ the rest mass of electron and $< r^2 >$ the mean square distance of the electrons from the nucleus.

As follows from eq. (1.32), diamagnetism is present in all substances and exists in substances that have neither spin nor orbital magnetic moments. Diamagnetic susceptibilities are negative and small, and temperature-independent. For example the specific susceptibility of quartz is -0.57×10^{-8} m^3 kg^{-1} (-0.45×10^{-6} cm^3 g^{-1}). The values of selected diamagnetic metals and non-metallic compounds are given in Table 1.2. Table 1.3 summarizes the magnetic susceptibilities of certain diamagnetic minerals. The most complete list of the magnetic susceptibilities of minerals can be found in [A2].

TABLE 1.2

Specific magnetic susceptibilities of assorted diamagnetic elements and compounds

Element or compound	$(m^3kg^{-1}) \times 10^{-9}$	$(cm^3g^{-1}) \times 10^{-6}$
Copper	-1.081	-0.086
Lead	-1.38	-0.11
Silver	-2.39	-0.19
Zinc	-2.20	-0.18
Gold	-1.76	-0.14
Mercury	-2.14	-0.17
Tin	-3.9	-0.31
Arsenium	-0.9	-0.072
Germanium	-1.33	-0.11
Sulphur	-6.0	-0.48
Carbon	-6.2	-0.49
Nitrogen	-5.40	-0.43
Hydrogen	-24.96	-1.99
Helium	-5.91	-0.47
Sulphuric acid	-5.0	-0.40
Carbon dioxide	-5.91	-0.47
Water	-9.06	-0.72
Ethylalcohol	-9.3	-0.74

TABLE 1.3

Specific magnetic susceptibilities of assorted diamagnetic minerals

Mineral	Chemical formula	$(m^3kg^{-1}) \times 10^{-9}$	$(cm^3g^{-1}) \times 10^{-6}$
Baddeleyite	ZrO_2	-1.4	-0.112
Barite	$BaSO_4$	-3.8	-0.3
Calcite	$CaCO_3$	-45.2	-3.6
Fluorite	CaF_2	-7.9	-0.63
Litharge	PbO	-1.6	-0.13
Magnesite	$MgCO_3$	-6.4	-0.51
Molybdenite	MoS_2	-7.2	-0.57
Quartz	SiO_2	-5.7	-0.45
Ruby	Al_2O_3	-5.9	-0.47
Sphalerite	ZnS	-3.3	-0.26
Cerussite	$PbCO_2$	-23.9	-1.9
Mimetite	$Pb_5(Cl,AsO_4)_3$	-21.4	-1.7
Apatite	$Ca_5[PO_4]_3(F,Cl)$	-3.27	-0.26

I.3.1.2 Paramagnetism

In paramagnetic materials the magnetic atoms or ions have permanent intrinsic magnetic moments and occur in low concentrations. Susceptibility arises from the competition between the aligning effect of the applied magnetic field and the randomizing effect of thermal vibrations.

If the atom or molecule has a magnetic moment caused by an odd number of electrons (total spin of the system is non zero) or free atoms or ions with a partly filled inner electronic shell, then under an applied magnetic field the magnetic moment tends to orient along the direction of the field. In the absence of a magnetic field the magnetic moments are randomly oriented and the overall magnetic moment is zero. In the presence of the external magnetic field magnetization is given by the Langevin equation

$$M = N_a \frac{\mu_m}{\mu_0} L(x) \tag{1.33}$$

where μ_m is the magnetic moment of an atom and $\mathbf{L}(x)$ is the Langevin function

$$L(x) \equiv \coth x - \frac{1}{x} \tag{1.34}$$

where $x = \mu_m H/kT$, **k** the is Boltzmann constant and **T** the absolute temperature.

For $x \ll 1$ we have $L(x) \cong x/3$ so that the magnetization is

$$M \cong \frac{\mu_m^2 N_a H}{3\mu_0 kT} \tag{1.35}$$

Thus the dependence of the magnetization of a paramagnetic material on the field strength is expressed by a straight line. Therefore, magnetic susceptibility does not depend on the field strength, and in the limit $\mu_m H/kT \ll 1$ it becomes

$$\kappa = \frac{M}{H} = \frac{\mu_m^2 N_a}{3\mu_0 kT} = \frac{C}{T} \tag{1.36}$$

where **C** is the Curie constant and the equation (1.36) is known as the Curie law, which states that magnetic susceptibility of the paramagnetic substances is inversely proportional to the temperature.

In order to obtain the real paramagnetic contribution the diamagnetic susceptibility is subtracted from the paramagnetic one, i.e. $\kappa = \kappa_p - (-|\kappa_d|) = \kappa_p + |\kappa_d|$. The specific (mass) susceptibility of a paramagnetic material is typically 10^{-6} $m^3\ kg^{-1}$ ($80 \times 10^{-6}\ cm^3\ g^{-1}$). Selected values for metals and compounds are given in Table 1.4 and those for minerals in Table 1.5.

The magnetic susceptibilities of minerals represent the overall values for pure minerals and inclusions, and these usually cannot be eliminated. The composition and the level of impurities vary greatly and depend on the type of rock and on the effectiveness of the separation process. In fact, the very concept of a mineral lacks a strict definition in chemical and physical terms, not only because of iso- and polymorphysm but also because of micro- and macroinclusions of compounds differing from the mineral in question [A2].

I.3.1.3 Ferromagnetism

In ferromagnetic materials interaction between neighbouring atoms is so strong that the magnetic moments of all atoms are aligned parallel to each other against the disturbing force of thermal motion. The strong internal fields which align the spins are called molecular or Weiss fields and are mainly result of quantum mechanical process called exchange interactions. The order of magnitude of the molecular field 10^7 - 10^9 A m^{-1} (10^5 -10^7 Oe). Weiss assumed that there is a magnetic field proportional to the magnetization and that this field tends to keep the magnetic moments of neighbouring atoms parallel. The Weiss hypothesis predicts that a material is spontaneously magnetized even in the absence of an applied field at temperatures below the Curie point, at which temperature the thermal agitation overcomes the molecular

TABLE 1.4

Specific magnetic susceptibilities of assorted paramagnetic elements and compounds

Element or compound	$(m^3kg^{-1}) \times 10^{-9}$	$(cm^3g^{-1}) \times 10^{-6}$
Aluminium	7.67	0.61
Barium	1.88	0.15
Chromium	42.0	3.3
Lanthanum	8.68	0.69
Molybdenum	10.9	0.87
Neodymium	492	39.2
Niobium	28.9	2.3
Palladium	66.2	5.3
Platinum	12.3	0.98
Plutonium	27	2.2
Samarium	106	8.4
Sodium	8.53	0.68
Tantalum	10.6	0.84
Titanium	42.2	3.4
Uranium	20.6	1.6
Vanadium	73.0	5.8
Calcium	13.8	1.1
Zirconium	16.7	1.3
Oxygen	1355	107.8
$CuSO_4 \cdot 5H_2O$	76.7	6.1
$FeSO_4$	844	67.2
$NiSO_4 \cdot 7H_2O$	201	16.0
MnO	860	68.4

field. The Curie temperature T_C separates the disordered paramagnetic phase at $T > T_C$ from the ordered ferromagnetic phase at $T < T_C$.

Weiss also assumed that a ferromagnetic body can be subdivided into domains, each of which is spontaneously magnetized to saturation by virtue of a strong exchange magnetic field. The direction of domain magnetization is determined by crystallographic structure, but except for very fine grains, the domains are mutually oriented so that total magnetization of a ferromagnetic body may have any value between zero and saturation, depending upon the fields to which it had been and is exposed.

TABLE 1.5
Specific magnetic susceptibilities of selected paramagnetic minerals

Mineral	Chemical formula	$(m^3kg^{-1}) \times 10^{-9}$	$(cm^3g^{-1}) \times 10^{-6}$
Anatase	TiO_2	3.27	0.26
Asbestos	$Mg_3[Si_2O_5](OH_4)$	155.8	12.4
Bornite	Cu_5FeS_4	100.5	8.0
Cassiterite	SnO_2	2136.3	170
Chalcopyrite	$CuFeS_2$	1595.9	127
Dolomite	$CaMg(CO_3)_2$	15.1	1.2
Goethite	$HFeO_2$	250 - 380	20 - 30
Hausmanite	Mn_3O_4	500 - 760	40 - 60
Ilmenite	$(Fe,Mg,Mn)TiO_3$	200 - 1500	15 - 120
Limonite	$Fe_2O_3 \cdot H_2O$	250 - 760	20 - 60
Hematite	Fe_2O_3	500 - 3800	40 - 300
Malachite	$Cu_2(OH)_2CO_3$	100 - 200	8 - 15
Monazite	$(Ce,La,Dy)PO_4$	120 - 250	10 - 20
Rutile	TiO_2	12 - 50	1 - 5
Siderite	$FeCO_3$	380 - 1500	30 - 120
Wolframite	$(MnFe)WO_4$	380 - 1200	30 - 100

The exchange interaction between spins S_i and S_j can, on certain assumptions, be written as

$$U = -2J_eS_iS_j \tag{1.37}$$

where J_e is the exchange integral and is related to the overlap of the charge distributions of the atoms **i, j.** A positive value of the exchange integral describes the case of parallel spins (ferromagnetics) while a negative value describes the case of antiparallel spins (antiferromagnetics). The dependence of the exchange integral J_e on the interatomic distance is illustrated by the Slater curve shown in Fig. 1.8.

It can be seen, that α- Fe, Ni, Co are ferromagnetics ($J_e > 0$) while Mn, Cr, and γ- Fe are antiferromagnetics ($J_e < 0$).

The distinction between the magnetically disordered and ordered spin arrangements is illustrated in Fig. 1.9, which also summarizes some of the ordered magnetic spin structures.

The magnetic susceptibility of ferromagnetics is a complex function of the external magnetic field strength and of temperature. The temperature dependence of the magnetic susceptibility of the ferromagnetics is described by the Curie - Weiss law

$$\kappa = \frac{\mu_m^2}{3R} \frac{N_a^2}{\mu_0 (T - T_c)} \tag{1.38}$$

where **R** is the gas constant. Since magnetic polarization of a ferromagnetic material depends upon its magnetic history and not merely upon the field to which it is exposed at a particular

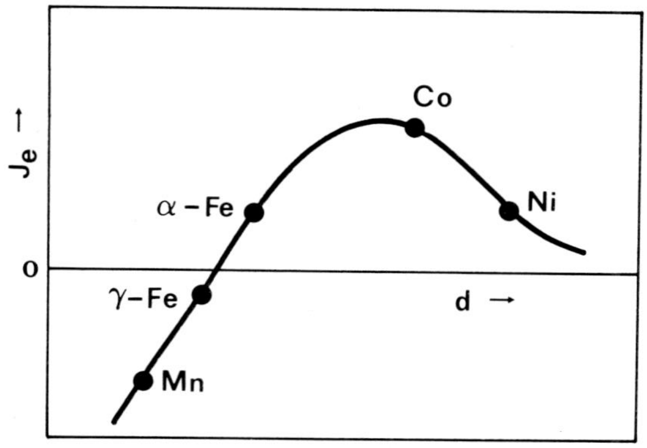

Fig. 1.8 Slater curve expressing the change in the sign of the exchange integral J_e with the relative interatomic distance **d.**

instant, it can remain magnetized after removal of a field. The ferromagnetics are thus characterized by hysteresis, or irreversibility of magnetization. The magnetization curve of hysteresis loop is the polarization **J** or induction **B** measured in the direction of the applied field **H** under static conditions. A hysteresis loop can be either a **J-H** diagram or a **B-H** diagram, as illustrated in Figs. 1.10a and 1.10b, respectively.

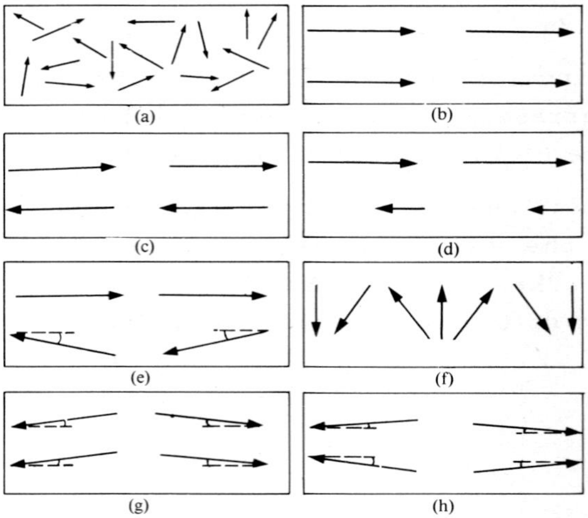

Fig. 1.9 Schematic representation of spin arrangements:
(a) disordered paramagnetic state
(b) parallel spins: ferromagnetism
(c) antiparallel spins: Neèl-type antiferromagnetism
(d) uncompensated antiparallel spins: Neèl-type ferrimagnetism
(e) triangular antiparallel spins: Yafet-Kittel-type ferrimagnetism
(f) helical spiral structures: compensated antiferromagnetic or uncompensated ferrimagnetic
(g) canted spin Dzialoshinskii-type weak ferromagnets
(h) canted spins compensated antiferromagnets: hidden canting
(after Schieber [S2])

Magnetic polarization attains the ceiling value known as saturation polarization, represented by the symbol $\mathbf{J}_S$. It is observed that if the magnetic field is decreased the original curve is not retraced. The value of polarization for $\mathbf{H} = 0$ is remanent polarization $\mathbf{J}_R$ and the value of **H**, for which $\mathbf{J} = 0$ is called the coercive force $\mathbf{H}_C$.

The polarization of a multidomain particle is [B1]

$$J = 3\mu_0 H \qquad -\frac{J_s}{3\mu_0} \leq H \leq \frac{J_s}{3\mu_0}$$

$$J = -J_s \qquad H < -\frac{J_s}{3\mu_0} \qquad (1.39)$$

$$J = J_s \qquad H > \frac{J_s}{3\mu_0}$$

if we assume that the magnetization in the internal magnetic field (the sum of external and demagnetizing fields) is perfectly easy.

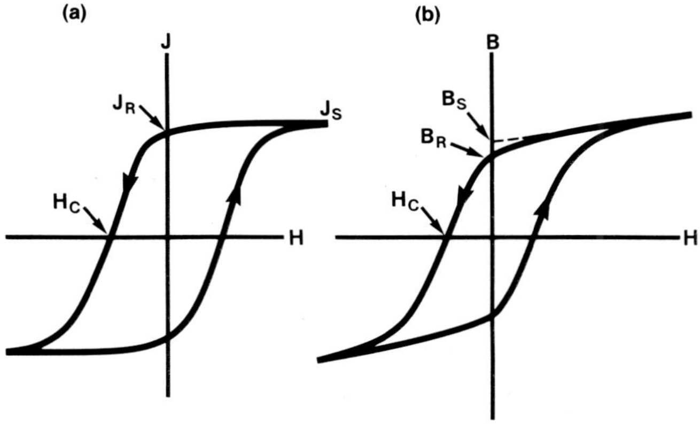

Fig. 1.10 The magnetization curves. The coercive force $\mathbf{H}_C$ is the reverse field necessary to bring the polarization $\mathbf{J}$ to zero; the remanence $\mathbf{J}_R$ is the value of $\mathbf{J}$ at $\mathbf{H} = 0$; the saturation polarization $\mathbf{J}_S$ is defined as the limiting value of $\mathbf{J}$ for large $\mathbf{H}$.

I.3.1.4 Antiferromagnetism

It has been mentioned that exchange interactions can be negative or positive. Néel suggested that there should be some substances with an antiferromagnetic order, i.e. with the moments in neighbouring atoms ordered in an antiparallel manner, as shown in Fig. 1.9(c). The magnetic behaviour of antiferromagnetic materials differs only slightly from that of paramagnetic materials. However, as in ferromagnetism, antiferromagnetic order decreases with increasing temperature and disappears at a critical temperature at which thermal energy is equal to the antiferromagnetic exchange energy. This temperature is known as the Néel point. Above the Néel point the material behaves like an ordinary paramagnetic material, but at the Néel point the magnetic susceptibility falls as the temperature is reduced. At temperatures below the Néel point the susceptibility of an antiferromagnet is dependent upon the direction of the field with respect to the axis of spin alignment. In a polycrystalline material it is the average susceptibility that is measured, but in a simple crystal we distinguish between $\kappa_{\parallel}$ and $\kappa_{\perp}$ measured parallel and perpendicular, respectively, to the spin axis. In a polycrystalline antiferromagnetic material the magnetic susceptibility [S3] is close to

$$\kappa = \frac{1}{3}\kappa_{\parallel} + \frac{2}{3}\kappa_{\perp} \tag{1.40}$$

Antiferromagnets have small (10^{-9} to 10^{-10} m^3 kg^{-1}, or 10^{-7} to 10^{-6} cm^3 g^{-1}) positive magnetic susceptibilities.

I.3.1.5 Ferrimagnetism

In a ferrimagnetic material atomic moments are ordered regularly in an antiparallel sense, but the sum of the moments pointing in one direction exceeds those pointing in the opposite direction. This situation is schematically illustrated in Fig. 1.9 (d). The word was coined by Néel to describe the magnetism of ferrites, of which the best known is the mineral magnetite.

The magnetic properties of ferromagnetics and ferrimagnetics are generally similar; both have saturation polarization within domains and Curie points at which saturation polarizations vanish. Ferrimagnetic ordering requires that the atoms that are magnetized in one direction have a greater effective magnetic moment than those magnetized in the opposite direction. They do not have to be atoms of a different element. In some common ferrimagnetic substances, iron atoms can be found in two different crystallographic sites, so that their nearest neighbours differ in number or type. This difference suffices to make the effective moments on the atoms in the two different sites differ in magnitude. With sublattice polarizations $\mathbf{J}_A$, $\mathbf{J}_B$, the net spontaneous polarization of the lattices is $(|J_B| - |J_A|)$; in magnetite $\mathbf{J}_B$ is the result of one ferric ion (Fe^{3+}) with a magnetic moment of 5 μ_B, where μ_B is the Bohr magneton, and one ferrous ion (Fe^{2+}), with 4 μ_B, per molecule of Fe_3O_4, while $\mathbf{J}_A$ is the result of one ferric ion (Fe^{3+}, 5 μ_B) only. The saturation polarization $(|J_B| - |J_A|)$ is thus 5 + 4 - 5 = 4 μ_B per molecule.

I.3.1.6 Superparamagnetism

Superparamagnetism is an intermediate state between behaviour as isolated paramagnetic atoms and behaviour as bulk ferromagnetic matter. A single domain particle has a magnetic moment μ_m which undergoes repeated thermal reorientation, but in the presence of the magnetic field $\mathbf{H}$ it has an average alignment determined by the Boltzmann factor ($\mu_m H/kT$), as for a classical paramagnetic material. Since it has no stable remanent polarization its behaviour is paramagnetic in nature, but because the magnetic moment is much larger than the atomic moment of a normal paramagnetic material, the alignment in a magnetic field is much stronger, i.e. the magnetic susceptibility may be much higher. Hence the use of the term superparamagnetism to describe its properties. The notion of superparamagnetism would therefore be based on two basic requirements: first, the magnetization curve must show no hysteresis, and second, the magnetization curve for an isotropic sample must be temperature-dependent [B6].

A small concentration of superparamagnetic particles can have a remarkable effect on the magnetic susceptibility, although simple paramagnetism of non-interacting single domains is not normally observed in minerals [S3].

Classes of magnetic materials discussed here display different behaviour as a function of temperature. The temperature dependence of magnetic susceptibility is schematically illustrated in Fig. 1.11, while Fig. 1.12 shows the shape of the magnetization curves of various magnetic materials.

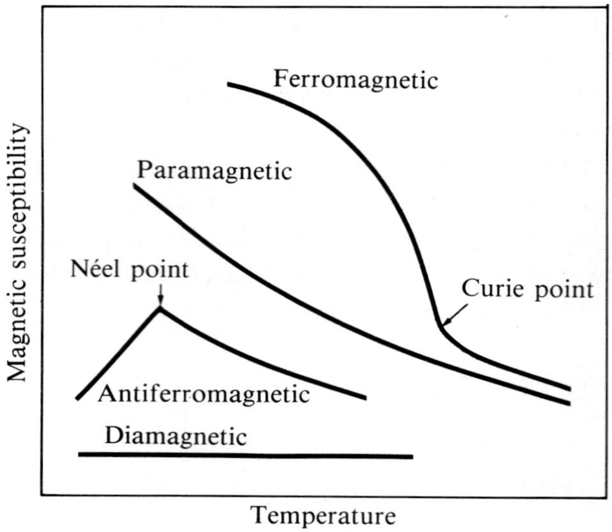

Fig. 1.11 Classes of magnetic compounds display different behaviour as a function of temperature.

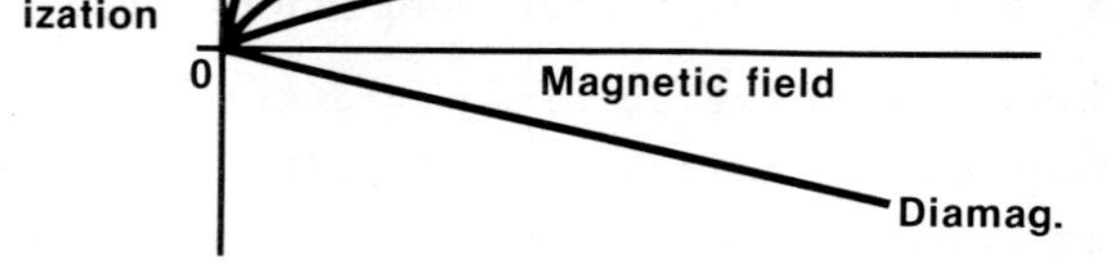

Fig. 1.12 Shapes of the magnetization curves of various classes of materials (not in scale).

I.3.2 Magnetic properties of strongly magnetic minerals

Magnetite iron ore is the basic mineral that was upgraded by magnetic separation because of its high value of magnetic polarization; the other frequently encountered strongly magnetic minerals are maghemite and pyrrhotite. The important characteristics of minerals with ordered structures are: the existence of the remanent polarization, coercive force and the effect of the particle size, shape and magnetic field on magnetic properties.

I.3.2.1 Magnetite

Magnetite (Fe_3O_4) is the most fundamental and important magnetic oxide in rocks. Its magnetic properties have long been studied and its crystal structure and contribution to magnetism have been classified on the basis of ferrimagnetism. Magnetite is therefore an inverse spinel, whose ferromagnetism fundamentally results from the superexchange interaction between two sublattices.

The saturation polarization at room temperature is 0.55 T (or specific polarization is 92-93 emu g^{-1}) and the Curie temperature is 578° C.

As already stated, the magnetic properties of magnetically-ordered mineral grains depend not only on composition, crystalline perfection, etc., but on the grain size. Although there is no sharp discontinuity in properties at any particular grain size, the most important distinction is that between very small grains, each of which is a single domain, and larger grains, which are multidomained. For any material there is a maximum size for spherical simple domain, referred to as the critical size. For magnetite this critical size is about 0.05 μm [S3].

The variation of the coercive force $\mathbf{H}_c$ (see Fig. 1.10), with grain diameter **d**, in granular magnetic materials is quite general and over a wide range of grain sizes is well represented by the relationship

$$H_c \sim d^{-n} \tag{1.41}$$

Measurements for different materials have given a value of $\mathbf{n}$ in the range $0.25 \leqslant \mathbf{n} \leqslant 1$. A dispersed powder with the particle size $\mathbf{d} > 100$ μm has a coercive force $\mathbf{H}_C = 800 - 1600$ A m^{-1} (10 - 20 Oe) and the ratio $\mathbf{J}_R/\mathbf{J}_S$ of remanent ($\mathbf{J}_R$) to saturation ($\mathbf{J}_S$) polarizations is $\mathbf{J}_R/\mathbf{J}_S = 0.1$ at room temperature [J1]. Both parameters increase with decreasing particle size (from 100 μm to 1 μm), as follows from eq. (1.41) and as shown in Fig. 1.13.

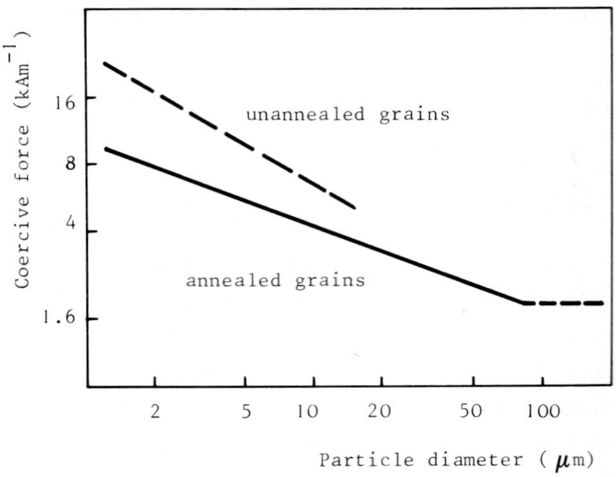

Fig. 1.13 Coercive force as a function of grain size for dispersed magnetite powders.For grains up to about 80 μm in diameter annealed grains follow eq.(1.41) with $\mathbf{n} = 0.4$; at larger diameters coercivity becomes less dependent upon grain size because domain structure is not determined primarily by grain size. (After Stacey and Banerjee [S3]).

Fine magnetite particles ($\mathbf{d} < 1$ μm) exhibit a maximum coercive force $\mathbf{H}_C = 1.4 \times 10^4$ A m^{-1} (180 Oe) and $\mathbf{J}_R/\mathbf{J}_S = 0.2$ while systems of very fine particles of diameter of several hundreds of angströms are characterized by much reduced values of $\mathbf{H}_C$ and $\mathbf{J}_R/\mathbf{J}_S$, as illustrated in Fig. 1.14.

This figure shows that at the transition between multidomain and single domain arrangements magnetic susceptibility achieves its minimum, while the coercive force is maximum. In a single domain system $\kappa \sim \mathbf{d}^{1/2}$ and for magnetite a decrease in magnetic susceptibility and an increase in coercive force occur for particle sizes 80 - 100 μm, while for hematite this critical size can be much higher (200 - 300 μm) [N1]. Dependence of the coercive force $\mathbf{H}_c$ and of specific magnetic susceptibility χ of pure magnetite on the particle diameter d is shown in Fig. 1.15. A reduction in the efficiency of magnetic separation and increased problems with demagnetization can be experienced in this particle size interval.

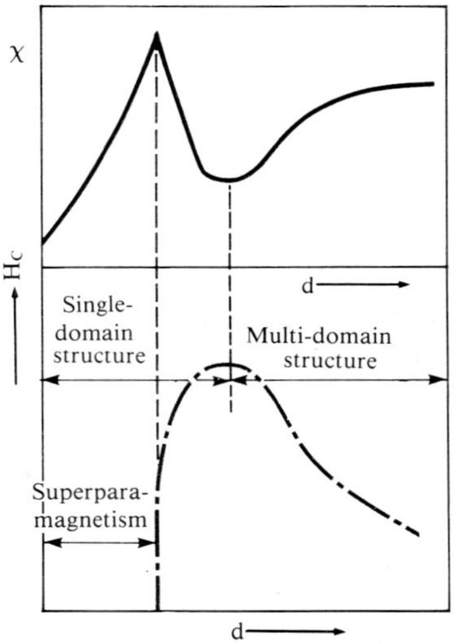

Fig. 1.14 Magnetic susceptibility χ and coercive force $\mathbf{H}_c$ as a function of a size of random clusters of ferromagnetic particles.

It was previously stated that the magnetic properties of ferromagnetic particles also depend on the shape of the body. The magnetic field inside a magnetized body ($\mathbf{H}_i$) is not equal to the field applied to it ($\mathbf{H}_o$). The difference usually known as the demagnetizing field ($\mathbf{H}_d$), is due to the fact that different parts of the body affect each other because of the long-range nature of the magnetic dipole interactions; in other words, in open magnetic circuits there is a return flux that goes in part through the sample itself. Consequently,

$$H_i = H_0 - H_d = H_0 - DM \tag{1.42}$$

where **D** is the geometrical demagnetizing factor that depends on the shape of the sample. With the increasing aspect ratio with the major axis parallel to the field, the demagnetization coefficient

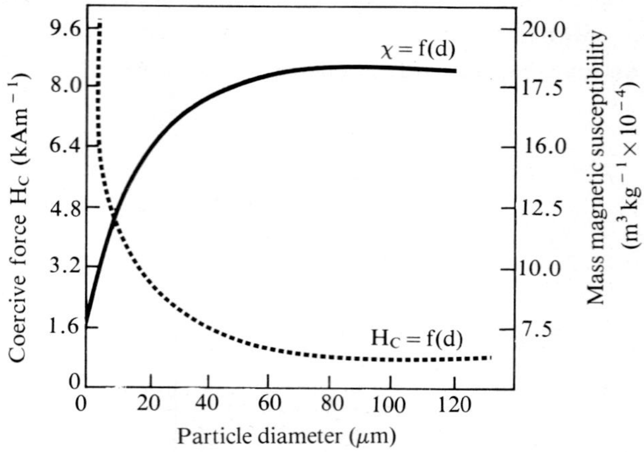

Fig. 1.15 Dependence of the coercive force $\mathbf{H}_c$ and specific magnetic susceptibility of magnetite on the particle size (after Derkach [D1]).

decreases. For the infinitely long cylinder whose axis coincides with the direction of the magnetic field $\mathbf{D} = 0$, and for a thin disc perpendicular to the field the demagnetization factor achieves its maximum value ($\mathbf{D} = 1$); for a sphere, $\mathbf{D} = 1/3$. The demagnetization factor thus ranges from 0 to 1 (in SI units) or from 0 to 4π (in cgs units). In general, $\mathbf{D}$ is a tensor except for directions along the principal axes of the spheroid.

The calculated values of the demagnetizing factor for an elipsoidal shape, together with the experimentally determined values for a cylindrical specimen, are given in Table 1.6.

For a mixture of minerals containing magnetite, the effective magnetic field $\mathbf{H}_i$ seen by the particles is

$$H_i = H_0 - (D + D_i)pM \tag{1.43}$$

where $\mathbf{D}_i$ is the internal demagnetizing factor and $\mathbf{p}$ the packing density of ferromagnetic particles.

The intrinsic magnetic susceptibility κ_i of the material relates **M** to $\mathbf{H}_i$, for small fields

$$M = \kappa_i H_i = \kappa_i (H_0 - DM) \quad (1.44)$$

so that the observed magnetic susceptibility of the grain which relates **M** to $\mathbf{H}_o$ is

$$\kappa = \frac{M}{H_0} = \frac{\kappa_i}{1 + D\kappa_i} \quad (1.45)$$

which is identical to eq. (1.26). For a mixture the observed magnetic susceptibility is again

$$\kappa = \frac{p\kappa_i}{1 + (D + D_i)p\kappa_i} \quad (1.46)$$

TABLE 1.6

Demagnetizing factors for rods and ellipsoids magnetized parallel to the major axis [B7]

Length-to-diameter ratio	Rod	Prolate ellipsoid	Oblate ellipsoid
1	0.27	0.3333	0.3333
2	0.14	0.1735	0.2364
5	0.040	0.0558	0.1248
10	0.0172	0.0203	0.0696
50	0.00129	0.00144	0.01472
100	0.00036	0.000430	0.00776
2000	0.0000009	0.0000019	0.000392

The intrinsic susceptibility κ_i is the susceptibility which in principle would be observed if it were possible to examine the material of the grain in bulk and to avoid self-demagnetizing effects. The susceptibility κ which is always observed in mineral magnetism provides a rather poor measure of κ_i for strongly magnetic minerals such as magnetite, because $\kappa_i > 1/D$, so that $\kappa = 1/D$. For small values of κ (<0.2) the intrinsic and bulk magnetic susceptibilities almost coincide. For large values of the intrinsic magnetic susceptibity κ_i, the bulk magnetic susceptibility κ remains constant, approximately equal to 1/D, as mentioned above. The situation is illustrated in Fig. 1.16.

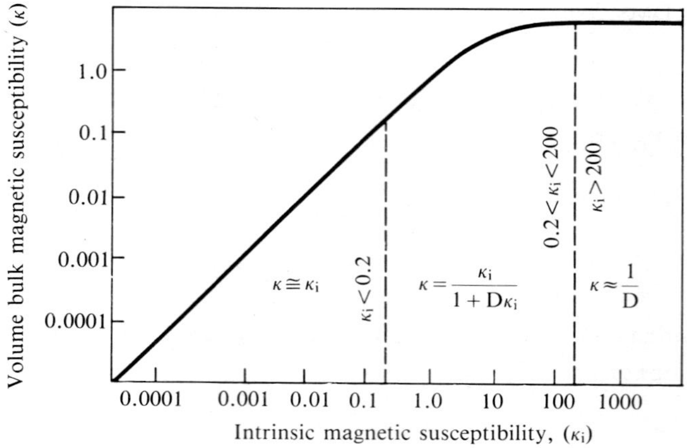

Fig. 1.16 Dependence of bulk magnetic susceptibility on the intrinsic magnetic susceptibility κ_i, for **D** = 0.16 (after Derkach [D1]).

We can thus see that in the interval of values of intrinsic magnetic susceptibility κ_i (0, 0.2) κ is equal to κ_i. For 0.20 < κ_i < 200 the bulk magnetic susceptibility κ is given by eq. (1.45). If we assume that the mean value of the demagnetizing factor **D** for elongated magnetite particles is equal to 0.16 [D1], κ ranges from 0.193 to 6.06.

Table 1.7 summarizes the values of specific magnetic susceptibility of magnetite placed in the external magnetic field **H** ranging from 0 to 240 kA m^{-1} (0 to 3 kOe).

It can be seen that the maximum values of specific magnetic susceptibility of magnetite are achieved at the magnetic field strength 8 - 16 kA m^{-1} (0.1 - 0.2 kOe). Magnetic separators used to recover magnetite usually generate magnetic field strength ranging from 60 - 120 kA m^{-1} (0.75 - 1.5 kOe) at which field the susceptibility is about five times lower than the maximum value. At these magnetic fields the coercive force of magnetite amounts to about 0.4 kA m^{-1} (5 Oe), the remanent polarization 0.11 T and the maximum polarization 0.48 T. [D1].

Results of the investigation of magnetic properties of artificial magnetite were reported in [G2]. The values of the maximum specific magnetic susceptibility (as a function of magnetic field) and those obtained at 80 kA m^{-1} (1 kOe) are summarized in

TABLE 1.7

Specific magnetic susceptibility of magnetite as a function of external magnetic field strength. (After Derkach [D1])

Magnetic field strength		Specific magnetic susceptibility after [P1]		after [D2]	
kAm^{-1}	kOe	$x\ 10^4 m^3 kg^{-1}$	$cm^3 g^{-1}$	$x\ 10^4 m^3 kg^{-1}$	$cm^3 g^{-1}$
240	3	.		3.46	0.028
160	2	.		5.00	0.040
96	1.2	.		7.59	0.060
64	0.8	9.53	0.076	10.00	0.080
48	0.6	11.13	0.089	11.72	0.093
32	0.4	15.25	0.121	14.03	0.112
24	0.3	18.00	0.143	15.31	0.122
16	0.2	22.45	0.179	16.88	0.134
8	0.1	27.48	0.219	18.73	0.149
4	0.05	16.50	0.131	18.00	0.143
2	0.025	14.00	0.111	18.00	0.143

The data in [P1] were obtained with pure magnetite containing 94.9 % Fe_3O_4 with a particle size distribution -74 + 53 μm.

TABLE1.8

Magnetic susceptibilities of artificial magnetite and of the original minerals [G2]

Reducing agent	Product	Maximum specific magnetic suscep. SI	Maximum specific magnetic suscep. cgs	Specific magnetic suscep. at 1 kOe SI	Specific magnetic suscep. at 1 kOe cgs
Hydrogen	Magnetite from				
	Hematite	1.11	88500	0.95	75500
	Martite	1.03	82000	0.87	69000
	Hydroxides	0.73	58000	0.68	54000
Natural gas	Magnetite from				
	Hematite	1.51	120000	1.12	89000
	Martite	1.46	110000	1.11	88500
	Hydroxides	0.60	48000	0.60	47500
Carbon	Magnetite from				
	Hematite			0.96	76500
	Martite	1.13	90000	0.84	66500
	Hydroxides	1.03	82000	0.86	68200
Original mineral					
	Magnetite	2.15	171000	1.36	108000
	Hematite	0.10	8300	0.09	7500
	Martite	0.03	2780	0.03	2720

The values of specific magnetic susceptibilities in Table 1.8 are given in $10^3\ m^3\ kg^{-1}$ (SI) and $10^6\ cm^3\ g^{-1}$ (cgs).

Table 1.8. It was concluded [G2] that the magnetic properties of artificial magnetite depend on the type of reducing agent. The largest values of χ were observed with magnetite with the use of a natural gas as reducing agent. Likewise, magnetic properties depend on the nature of the original material. Magnetite obtained by reducing the monomineral fractions of hematite exhibited the best magnetic properties. The reduction products have a large coercive force, dependent on the origin of the mineral and on the reducing agent.

Magnetic properties of magnetite can be affected by comminution as a consequence of the increase in free surface, rupture of chemical bonds and the formation of charged states and point and linear defects. Moroz and Ol'khovik [M20] found that in the course of comminution the saturation magnetization of magnetite decreases, a factor explained by the formation of α and γ-Fe_2O_3. On the other hand, the Curie temperature was found to remain constant (584° C) regardless of the comminution time. As magnetite comprises two sublattices with antiparallel magnetizations, a decrease in saturation magnetization in comminuted samples was associated with deviation of the sublattice magnetization from their antiparallel configuration. Annealing at 600° C helps to eliminate the point defects and restore magnetization.

I.3.2.2 Titanomagnetites

Titanomagnetites ($Fe_{3-x}Ti_xO_4$) are one of the most common magnetic minerals. The substitution of Ti ions for Fe ions in the magnetite lattice means that Ti^{4+} and Fe^{2+} replace $2Fe^{3+}$, i.e. substitution of the Ti^{4+} for Fe^{3+} also requires the conversion of a second Fe^{3+} ion to Fe^{2+}. Saturation magnetization, Curie point and coercive force vary in a regular manner with the composition parameter **x** from **x** = 0 (magnetite) to **x** = 1 (ulvöspinel). The magnetic properties of titanomagnetites are treated in detail in [D3] and Fig. 1.17 shows the particle size dependence of the initial volume magnetic susceptibility for various members of the titanomagnetite family.

The ratio of the remanent magnetization to saturation magnetization as a function of particle size is shown in Fig. 1.18.

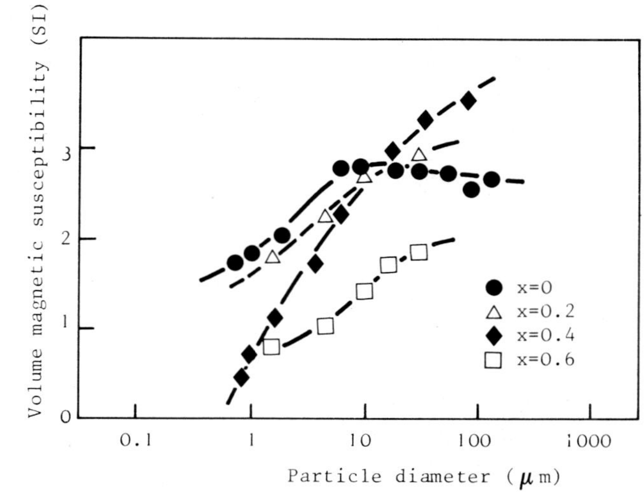

Fig. 1.17 Variation of initial volume susceptibility with particle size at room temperature for titanomagnetites. [D28, O16, T22]

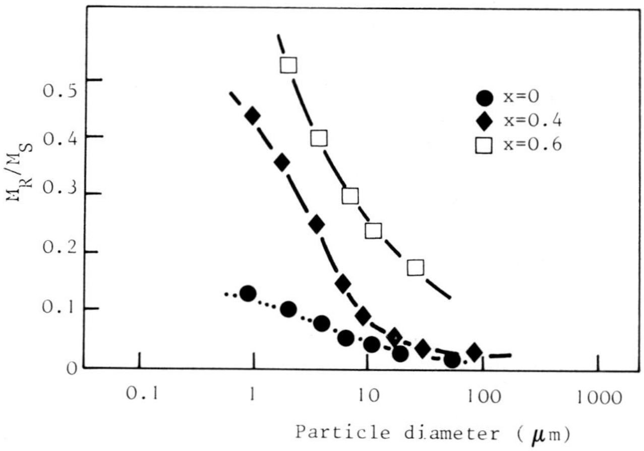

Fig. 1.18 Ratio of room temperature remanent magnetization to saturation magnetization as a function of particle size for titanomagnetites [D29, D28, O16, O17, T22].

I.3.2.3 Maghemite

Maghemite (γ - Fe_2O_3) has a chemical composition identical to that of hematite (α - Fe_2O_3). Unlike hematite, it is strongly magnetic (like magnetite) and also its crystal structure is similar to that of magnetite.

Maghemite is metastable and inverts to $\alpha - Fe_2O_3$ irreversibly when heated. The inversion temperature ranges from 275°C to 800°C [N1], depending on the previous history of the sample.

The saturation polarization J_S of maghemite is 0.53 T (or specific polarization 85 emu g^{-1}) at room temperature, quite close to that of magnetite. The similarity of J_S and of the magnetocrystalline anisotropy constant ensures that the rules applied to the magnetic properties of magnetite are also applicable to maghemite.

Table 1.9 summarizes some magnetic properties of natural and artificial magnetite and maghemite, as obtained from assorted Soviet ores [D1].

The results in Table 1.9 show that the specific magnetic susceptibility of natural magnetite, artificial magnetite and maghemite are not very different. The fundamental difference rests in the magnitude of the coercive force which attains the maximum values for artificial magnetite and the minimum values for natural

TABLE 1.9

Magnetic properties of natural and artificial magnetite and maghemite (after Derkach [D1])

Sample	Particle size (mm)	Magnetic polarization (T) max.	remanent	Coercive force H_c (kAm^{-1})	Spec.mag. susceptib. ($m^3kg^{-1}10^{-4}$)
Natural magnetite (Lebedinskii district)	-0.15	0.1	0.02	5.8	6.85
Artificial magnetite (Akkermanskii district)	-0.15	0.06	0.03	10.4	4.52
Artificial maghemite (Akkermanskii district)	-0.15	0.07	0.025	9.2	5.73
Natural magnetite (Vysokogorskii district)	-0.18 +0.12	0.25	0.03	3.2	5.67
Artificial magnetite (Krivoy Rog district)	-0.18 +0.12	0.24	0.08	10.3	5.57
Artificial maghemite (Vyatskii district)	-0.18 +0.12	0.15	0.05	9.6	4.12

magnetite. High value of coercive force of artificial magnetite complicates the demagnetization of the pulp of the roasted ore, induces the formation of the particle clusters and increases the entrainment of the gangue particles in the magnetic product [D1].

I.3.2.4 Pyrrhotite

Apart from iron-titanium oxides, sulphides with a range of composition between troilite FeS and pyrrhotite Fe_7S_8 possess intrinsic ferromagnetic properties.

The magnetic properties of iron sulphide FeS_{1+x} can be classified into three compositional ranges, i.e.

a)	$0 < x < 0.1$	antiferromagnetics at room temperature
b)	$0.1 < x < 1/7$	ferrimagnetics at room temperature
c)	$x \cong 0.1$	ferrimagnetics in a restricted temperature range

The saturation polarization of pyrrhotite is 0.08 T (or the specific polarization 13.5 emu g^{-1}) and the Curie point is 320° C. Magnetic susceptibility of pyrrhotite, as a function of the magnetic field strength, is given in Table 1.10.

The coercive force of pyrrhotite is quite high and amounts to 9.6 kA m^{-1}. Although the magnetic susceptibility is about ten times lower than that of magnetite permanent magnet separators with magnetic induction up to 0.15 T are normally used for the recovery of pyrrhotite.

TABLE 1.10

Specific magnetic susceptibility of pyrrhotite (after Derkach [D1])

Magnetic field strength		Specific magnetic susceptibility	
kA m^{-1}	kOe	(m^3kg^{-1} x 10^{-5})	(cm^2g^{-1} x 10^{-3})
1	0.012	3.8	3.0
8	0.1	5.0	4.0
16	0.2	6.2	4.9
24	0.3	6.8	5.4
32	0.4	6.2	4.9

I.3.3 Magnetic properties of weakly magnetic minerals

Magnetic minerals that are classified as weak from the view point of magnetic separation comprise a wide range of minerals. These minerals are not only paramagnetic but also include a variety of minerals with ordered structures (antiferromagnetic or ferrimagnetic). The information about magnetic properties of minerals is scarce, often contradictory and inconsistent. Although ample data are available on the magnetic susceptibility of pure chemical compounds, these values are usually of little value for the tasks solved by magnetic separation.

It is of prime importance that the presence of strongly magnetic, iron containing minerals in the samples used to measure magnetic susceptibility is determined, since these admixtures can cause serious errors.

A fundamental alteration in the magnetism of the material, such as the appearance of parasitic ferromagnetism, may be ascertained from changes in magnetic parameters such as susceptibility or coercive force. Because the magnetism of many important minerals is antiferromagnetic or ferrimagnetic (and therefore depending critically upon a balance between sublattice magnetizations), we should expect the generation of parasitic ferromagnetism to be relatively common in minerals. Similarly, in very fine grain particles of minerals departures from true antiferromagnetism are to be expected. Of course, such a parasitic magnetism may not be apparent at room temperature, since the Néel point of many minerals is below this temperature.

Weak or parasitic ferromagnetism can be caused by at least two defect mechanisms [F3]. In an antiferromagnet containing two perfectly compensating sublattices, the preferential occurrence of vacancies on one sublattice can give a bulk moment. A second type of mechanism that can generate weak ferromagnetism depends upon local imbalance. Local defect moments are oriented in an effectively self-compensating antiparallel alignment in virgin crystals of antiferromagnets if they are produced below their Néel point. However, when the material is heated to above its Néel point and subsequently cooled, the defects are preferentially aligned. It appears that this is a general property of antiferromagnets [F3].

Materials usually encountered in magnetic separation represent a wide spectrum of natural compounds, with large variations in the content of admixtures and impurities which can substantially affect the magnetic properties. The complexity of some minerals by the ambiguity of the chemical composition of many mineral compounds also contributes to the difficulties in applying the magnetic susceptibility values to the evaluation of the potential of magnetic separation. Although the values of the magnetic susceptibility of minerals, found in literature (the most comprehensive review is that given in a treatise by Andres [A2]) can serve as a guide in evaluating the potential of magnetic separation, the safest way to find out whether magnetic separation will be successful is to pass the sample through a magnetic separator.

I.3.3.1 Goethite

Goethite (α- FeOOH) is an antiferromagnetic mineral which ideally should have perfectly compensated spins and hence no net spontaneous magnetic moment. Samples of goethite also exhibit a weak ferromagnetism (or ferrimagnetism) which lies parallel to the antiferromagnetic spin axis [O2]. The disappearance of the weak ferromagnetism coincides with the Néel point of the antiferromagnetism. The origin of the weak ferromagnetism is unknown. The specific magnetic susceptibility of the monomineral and dressed fractions ranges from 2.5 x 10^{-7} m^3 kg^{-1} to 6 x 10^{-7} m^3 kg^{-1} (20 to 42 x 10^{-6} cm^3 g^{-1}) [A2].

The magnetic properties of natural goethite were also reported in [P2]; magnetic susceptibility data are shown in Fig. 1.19. Magnetic susceptibility is strongly field-dependent. Magnetic susceptibility χ_∞ extrapolated to infinite field strength is very weakly temperature dependent thus suggesting antiferromagnetic ordering. Particle size was also found to have a small effect on χ_∞, which ranged from 3.3 x 10^{-7} to 4 x 10^{-7} m^3 kg^{-1} (26 x 10^{-6} to 32 x 10^{-6} cm^3 g^{-1}).

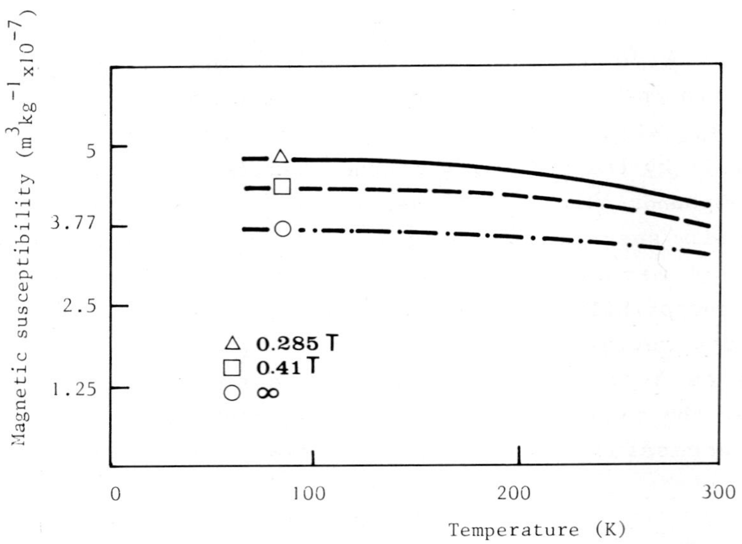

Fig. 1.19 Specific magnetic susceptibility χ of natural goethite as a function of temperature and magnetic field strength **H** (after Pastrana and Hopstock [P2]).

I.3.3.2 Hematite

Hematite (α - Fe_2O_3) is a much studied and little understood material. Ideal hematite is antiferromagnetic and has only a weak magnetization, about 0.5 per cent that of magnetite. The magnetic properties of hematite exhibit large variations and depend strongly on the sample preparation, concentration of impurities and other variables. Hematite exhibits weak or parasitic ferromagnetism (see Fig. 1.9), which develops as a consequence of a nonlinear or canted antiferromagnetic system [D3]. At temperatures below 263 K hematite becomes antiferromagnetic [M2]. The magnetization curves of the crushed fragments of a single crystal are strongly dependent upon grain size (Fig. 1.20).

The magnetic polarization fell from a value of 2 mT for a grain size of about 100 μm to effectively zero for 1 μm. The samples then exhibited a paramagnetic susceptibility of 2.5 x 10^{-7} $m^3 kg^{-1}$ (20 x 10^{-6} $cm^3 g^{-1}$).

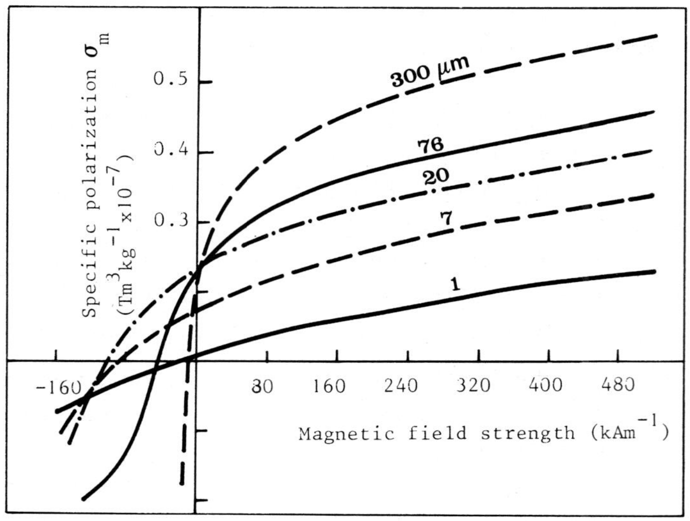

Fig. 1.20 Magnetization curves for hematite showing dependence upon grain size (after Chevallier and Mathieu [C1]).

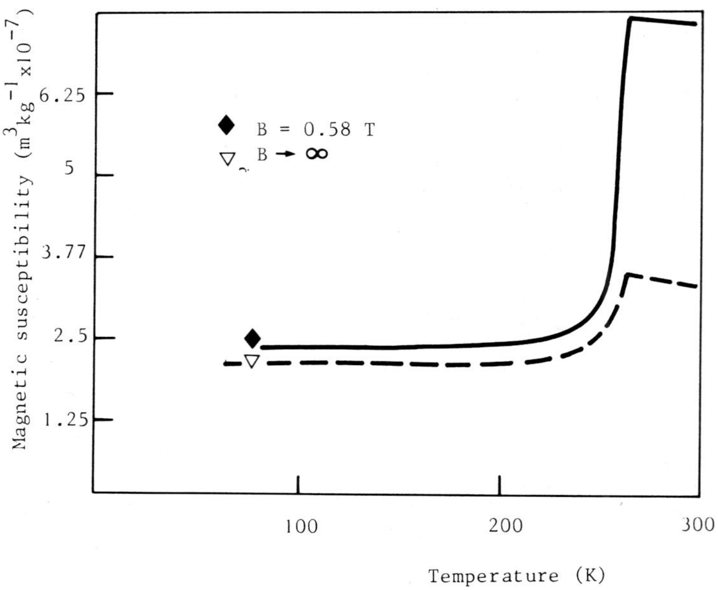

Fig. 1.21 Specific magnetic susceptibility χ vs. temperature for chemical reagent grade hematite at a magnetic induction of 0.58 T and extrapolated to infinite field (after Pastrana and Hopstock [P2]).

Fine powders of hematite have a very high coercive force, up to 2.5 x 10^5 A m^{-1} (3 kOe), in contrast with the low values of the order of 8 x 10^2 A m^{-1} (10 Oe) observed with crystals of both natural and synthetic hematite which are a few millimetres in size [S3].

The magnetic susceptibility of pure hematite grains is a complicated function of particle size, magnetic field and temperature, determined by the interplay between contributions from weak ferromagnetism and antiferromagnetism of hematite [D4]. The effects of temperature and magnetic field strength on specific magnetic susceptibility of pure hematite and Brazilian specular hematite [P2] are shown in Figs. 1.21 and 1.22, respectively.

It was observed [P2] that specific magnetic susceptibility decreased with increasing magnetic field strength. However at magnetic field strengths greater than 2.2 x 10^5 A m^{-1} (2.8 kOe) the magnetization curves assumed constant slopes. The effect of the field strength on magnetic susceptibility is discussed in paragraph I.3.4.

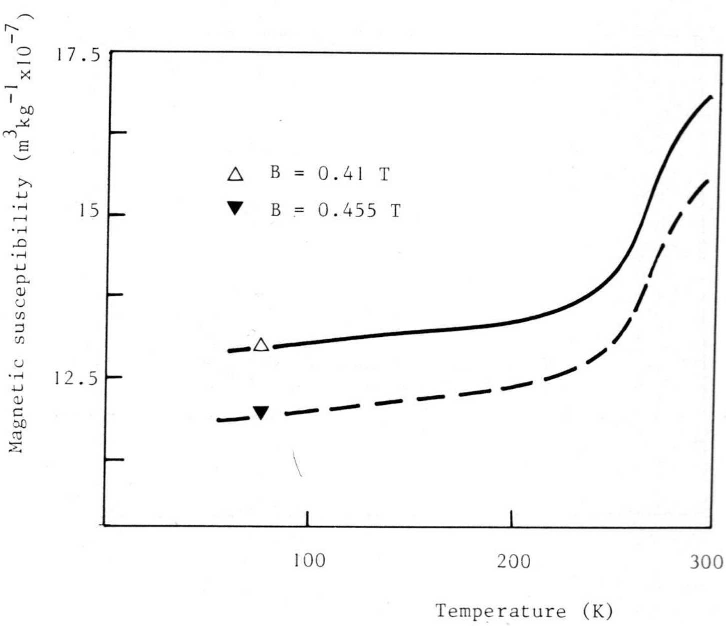

Fig. 1.22 The effect of temperature and magnetic field on the specific magnetic susceptibility of a sample of Brazilian specular hematite (after Pastrana and Hopstock [P2]).

For an accurate prediction of the results of magnetic separation, it is clear that variations in magnetic susceptibility with field strength, particle size, history and the origin of the sample must be taken into account. For example, in the case of hematite with a high degree of weak ferromagnetism, the assumption of constant magnetic susceptibility leads to an overestimation of the degree to which increasing magnetic induction increases iron recovery.

The variations in magnetic susceptibility of hematite and goethite with the origin of the ore are illustrated in Table 1.11.

TABLE 1.11
Specific magnetic susceptibility of hematite and goethite from various locations

Mineral	Specific magnetic susceptibility $m^3kg^{-1} \times 10^{-8}$	$cm^3g^{-1} \times 10^{-6}$	Ref.
Hematite (Malmberget, Sweden)	169	134	[H1]
Hematite (Gruvberget,Sweden)	3105	2471	[H1]
Specularite (Czechoslovakia)	92.3	73	[H1]
Specularite (Caue, Brazil)	225	179	[H1]
Hematite (Messina, South Africa)	143	114	[S4]
Goethite (Kepus, Romania)	62.3	49.6	[H1]
Goethite (Lengede, BRD)	54.5	43.4	[H1]

I.3.3.3 Siderite

Siderite ($FeCo_3$) is a naturally occurring form of ferrous carbonate having rhombohedral symmetry with a crystal structure of the $NaNO_3$ type. Siderite has an antiferromagnetic transition at T_N = 14 K and when heated under oxidizing conditions it is unstable. When siderite decomposes it can form such minerals as maghemite, magnetite and iron hydroxide. The final decomposition product is hematite [K18, B29].

Results of extensive measurements of magnetic susceptibility of siderite were reported by Powell and Miller [P17]. The measurements were made on powdered samples (-150 + 105 μm) at room temperature. It was found that susceptibility did not vary with the applied field, which indicated the absence of a ferromagnetic phase. The results of the measurements are shown in Table 1.12.

The magnetic susceptibility of the siderite crystals was also measured by Mookherji and Mathur [M24] as a function of temperature. The specific magnetic susceptibility at room temperature was found to vary between 7.9 x 10^{-7} m^3 kg^{-1} (63 x 10^{-6} cm^3 g^{-1}) and 9.3 x 10^{-7} m^3 kg^{-1} (74 x 10^{-6} cm^3 g^{-1}) as a consequence of the varying concentration of ferrous carbonate. The susceptibility-versus-temperature curves showed a broad maxima at around 300 K and an existence of the structural changes was inferred.

TABLE 1.12
Mass magnetic susceptibility of siderite ore [P17]

Source	Mass susceptibility (m^3kg^{-1}) x 10^{-7}	
	3.7 x 10^5 A m^{-1} (4.6 kOe)	7.9 x 10^5 A m^{-1} (9.9 kOe)
California, L.A.	10.7	10.5
Arkansas, Saline Co.	11.3	11.4
Arkansas, Pulashi Co.	9.4	9.5
Idaho, Kellogg	11.5	11.5
Texas, Cass Co.	9.8	9.7
Missouri, Bates Co.	8.4	8.3
Oklahoma, Rogers Co.	8.3	8.4
Connecticut,Roxbury	8.2	8.3
Louisiana,	9.1	9.1
Colorado	10.4	10.2
Germany, Rhine Prov.	11.0	11.0

I.3.3.4 Sphalerite

The magnetic properties of sphalerite (α - ZnS) are usually influenced by the presence of impurities. Sphalerite, as a rule, contains iron and manganese, the concentration of which ranges from 28 to 5 per cent [S5], and smaller amounts of cadmium (up to 5 per cent), indium, gallium, germanium, mercury, copper and lead (fractions of per cent).

Pure sphalerite is diamagnetic with specific magnetic susceptibility $\chi = -6.3 \times 10^{-9}\ m^3\ kg^{-1}$ ($-0.5 \times 10^{-6}\ cm^3\ g^{-1}$). However, sphalerite, free from admixtures of iron and manganese is very rare. Natural sphalerite is usually paramagnetic and magnetic susceptibility χ ranges from $3.8 \times 10^{-8}\ m^3\ kg^{-1}$ ($3 \times 10^{-6}\ cm^3\ g^{-1}$) to $5.9 \times 10^{-6}\ m^3\ kg^{-1}$ ($470 \times 10^{-6}\ cm^3\ g^{-1}$), with the iron concentration from 0.3 to 15 per cent. For most samples, however, χ assumes the values between $5 \times 10^{-8}\ m^3\ kg^{-1}$ ($4 \times 10^{-6}\ cm^3\ g^{-1}$) to $1.5 \times 10^{-7}\ m^3\ kg^{-1}$ ($12 \times 10^{-6}\ cm^3\ g^{-1}$). From a large number of samples only one was diamagnetic with $\chi = -3.6 \times 10^{-9}\ m^3\ kg^{-1}$ ($-0.29 \times 10^{-6}\ cm^3\ g^{-1}$) [S5].

It was observed that the magnetic susceptibility of the majority of samples did not depend on magnetic induction, while its value dramatically increased with the increase of iron concentration. It was also found that manganese did not decisively influence the magnetic properties of sphalerite, since its concentration in the mineral is low (from 0.015 to 0.9 per cent), at simultaneously high concentrations of iron. Cadmium, at relatively low concentrations (from 0.1 to 1.13 per cent) only negligibly reduces magnetic susceptibility of sphalerite.

The highest values of magnetic susceptibility were observed with sphalerite from high-temperature hydrothermal deposits, which are usually rich in iron.

I.3.3.5 Cassiterite

Cassiterite (SnO_2), as a rule, contains admixtures of niobium, tantalum, manganese, titanium, chromium and other elements. Investigation of magnetic properties of synthetic cassiterite has shown that the mineral was diamagnetic [S5] with specific magnetic susceptibility equal to $\chi = -3.3 \times 10^{-9}\ m^3\ kg^{-1}$ ($-0.265 \times 10^{-6}\ cm^3\ g^{-1}$). However, natural cassiterite, owing to the presence of iron, manganese and chromium is usually paramagnetic, although even strongly magnetic cassiterite can be found.

From a large number of the cassiterite samples from various deposits, only one was found to be diamagnetic [S5], with $\chi = -2.8 \times 10^{-9}\ m^3\ kg^{-1}$ ($-0.225 \times 10^{-6}\ cm^3\ g^{-1}$), which is close to the theoretical value. The remaining samples exhibited large values of magnetic susceptibility, which was found to be field-dependent. Iron in cassiterite can be present in a variety of forms. This is supported by the fact that only in a limited number of samples is magnetic susceptibility field-independent and it is consequently assumed that the iron in such samples is present as an isomorphic admixture. In the majority of samples, magnetic susceptibility was observed to be dependent on magnetic field strength and it is assumed that this behaviour can be explained by the presence of a strongly magnetic phase [S5]. It is argued that the upper limit of the concentration of isomorphous iron is 0.3 per cent, for concentrations above 0.3 per cent magnetic susceptibility becomes field-dependent.

I.3.3.6 Wolframite

Wolframite ($Fe_xMn_{1-x}WO_4$), the chief ore of tungsten, occurs in nature (with $0 \leqslant x \leqslant 1$) in the form of crystals commonly of the tabular form. This natural solid solution is the product of local reactions, i.e. its composition depends on the Mn/Fe ratio of the source material and on the formation temperature (between 250° C and 450° C) of the geological structure [G6]. The end-members of the wolframite series, ferberite $FeWO_4$ ($x = 1$, opaque black) and huebnerite $MnWO_4$ ($x = 0$, yellow red) show similar crystallographical structures but quite different magnetic structures. Quartz and hydrothermal sulphides of Fe, Zn, Pb, Cu (mainly pyrite FeS_2) are the principal minerals found in association with wolframite. They are often accompanied by other phases like cassiterite (SnO_2), scheelite ($CaWO_4$), hematite (Fe_2O_3) and fluorine (CaF_2), the trace elements currently found in wolframite are Nb, Ta, Sc, Mg and Sb.

Paramagnetic properties of wolframite are determined by the presence of iron and manganese. The role of divalent and trivalent iron differs in magnetic properties as the effective magnetic moment of Fe^{3+} is greater than that of Fe^{2+}. In most cases magnetic susceptibility increases with the increasing ratio Fe_2O_3/FeO [S5].

Spokes et al. [S6] reported a negative correlation between the iron content and the magnetic susceptibility. All members of the wolframite group were shown to be antiferromagnetic [B2], the antiferromagnetic transition temperature of ferberite being 207° C and that of huebnerite 257° C. A progressive decrease in mean magnetic susceptibility has been observed between 183 and 427° C [G3].

Results of the investigation of the magnetic pıoperties of different wolframites and wolframite concentrates were reported in [B2]. It was found that the concentrates and wolframite exhibited considerable variations in their magnetic properties and these were attributed to three possible major factors: the iron - manganese ratio, the presence of small quantities of hematite as inclusions and dilution effects due to non-magnetic mineral entrainment. When subjected to heat treatment all the samples showed a rapid drop in susceptibility.

I.3.3.7 Pyrite

Pyrite (FeS_2) is a semiconductor whose properties are strongly affected by impurities in natural crystals. The Mössbauer experiments established that there is no magnetic moment associated with iron atoms in FeS_2 [M3]. Nevertheless, the measurements of magnetic susceptibility indicate a small paramagnetic susceptibility ranging from 1 to 5×10^{-9} m^3 kg^{-1} (0.8 to 5.1×10^{-7} cm^3 g^{-1}). Measurements on pure pyrite crystals [B5] with well defined impurities show that in pure sample $\chi = 2.1 \times 10^{-9}$ m^3 kg^{-1} (1.7×10^{-7} cm^3 g^{-1}) at room temperature, with a small monotonic increase of χ with temperature. Magnetic susceptibility also increases with the increasing content of impurities.

Various methods of enhancing pyrite susceptibility have been proposed which include roasting and heating with microwaves. Hall et al. [H28] have shown that the oxidative alkaline pressure leaching of pyrite produces a large increase in volume magnetic susceptibility. The maximum increase observed was by a factor of 300 (from 0.0188×10^{-3} to 5.4×10^{-3} (SI)) while the increase was more pronounced for small particle sizes (<25 μm). The increase in susceptibility was attributed to the formation of a thin layer of disordered maghemite and magnetite, possibly with hematite.

I.3.3.8 Chromites

Natural chromium-bearing spinels of chromites (Fe, Mg) (Cr, Al, Fe)$_2$O$_4$ are of commercial interest both because of their role as refractory material in basic steelmaking (and as the only economically important ore for chromium) and because of their occurrence as difficult-to-separate contaminants in ores of economic potential, for example, in ilmenite sands [S47]. Furthermore, basic studies of magnetic properties of synthetic ferrochromites (Fe_3O_4 to $FeCr_2O_4$) and titanochromites (Fe_2TiO_4 to $FeCr_2O_4$) have shown that the presence of chromium in these spinels results in some unusual magnetic properties [M21], the most notable being the deviations from simple Néel-type ferromagnetism or antiferromagnetism, and the onset of viscous behaviour at temperatures below the temperature of magnetic ordering [R11, C29].

The magnetic susceptibility of chromite is a highly variable property, the variation being closely related to the variation in its chemical composition. Hawkes [H29] relates the susceptibility of chromite mineral to the percentage of end-member magnetite it contains. This implies that only those chromite ores having substantial amounts of $FeO.Fe_2O_3$ as an end-member can possess a positive magnetic susceptibility [H29, M22]. These views are supported by the experimental studies on the Indian chromite ores [M22]. Although it was found that the magnetic properties were close to those expected for rocks containing magnetite mineral, the specimens possessed no other ore mineral except chromite and it was concluded that magnetic susceptibility and the hysteresis behaviour of chromite were controlled by the end-member magnetite. Table 1.13 summarizes the values of mass magnetic susceptibility of four Indian chromite ores [M22, R12].

It was, however, reported by Kumar et al. [K19] that the chromite ores from Orissa, India, carry permanent magnetization, an observation that is in contrast to the assumption that the spinels of such composition are weakly magnetic. The possible influence of magnetite present as discrete ex-solutions in the chromite grains was ruled out on the grounds of widely different magnetic properties of chromites and magnetite.

TABLE 1.13
Mass magnetic susceptibility of Indian chromites [M22, R12]

Area	Mass magnetic susceptibility ($m^3\ kg^{-1} \times 10^{-6}$)
Lingampet	1.3
Sriramagiri	5.2 - 9.4
Gangineni	75.9 - 401
Kondapalli	105 - 787
Dendukury	0.5 - 0.7

The comparison of magnetic properties of natural and synthetic Fe-Cr spinels and magnetite is shown in Table 1.14.

TABLE 1.14
Magnetic properties of Fe-Cr spinels [K19]

Sample	σ_m (300 K) (emu g^{-1})	H_c (300 K) ($A\ m^{-1}$) x 10^3	T_c (^{o} C)
$FeCr_2O_3$	-	-	-185
$Mg_{0.6}Fe_{0.4}Cr_{1.4}Al_{0.6}O_4$	5.82	8 - 12	350
$Fe_{1.9}Cr_{1.1}O_4$	9.00	6 - 14	200
Fe_3O_4	92.00	16 - 24	580

Similar conclusions were drawn by Peoples and Eaton [P18] from their experimental investigation of magnetic properties of chromites from Montana, U.S.A. They found that the magnetic properties were those of the chromite itself and no correlation between magnetic susceptibility and Fe content was observed. The mass magnetic susceptibility at $5.2 \times 10^5\ A\ m^{-1}$ (6.4 kOe) ranged from 6.5×10^{-8} to $153 \times 10^{-8}\ m^3\ kg^{-1}$, although for an isolated sample the value of $635 \times 10^{-8}\ m^3\ kg^{-1}$ was measured.

The experimental study of the effects of temperature and mechanical and thermal treatments on the static magnetic properties of natural chromites [S47] has shown that crushing, or decreasing the particle size increases the polarization from the near zero value observed for unworked chromite (Fig. 1.23).

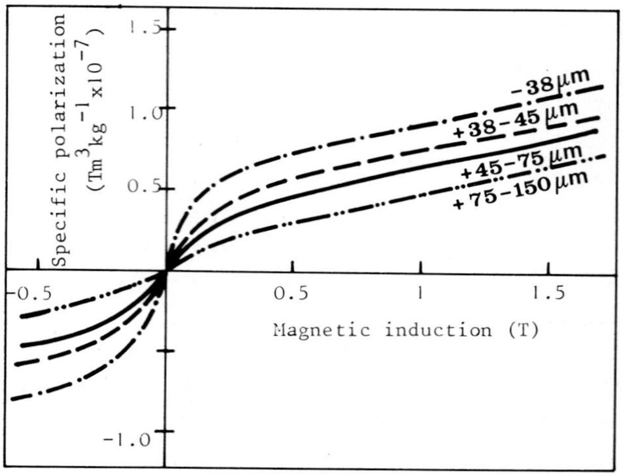

Fig. 1.23 The room temperature specific polarization as a function of applied magnetic field for size fractions of crushed Transvaal hard-lumpy-grade chromite (after Schwerer and Gundaker [S47]).

It was further observed that the high-field magnetic susceptibility (the slope of the magnetization curve in the high-field region) is almost independent of particle size. Upon cooling to 78 K the susceptibility increased and the saturation magnetization was temperature-independent. The magnetic parameters are summarized in Table 1.15.

The experimental results indicate [S47] that the ore-grade chromites order magnetically only at cryogenic temperatures ($T < 78$ K) and that the onset of long-range order occurs over an extended temperature range. At very low temperatures the magnetically ordered chromites exhibit viscous magnetic behaviour.

Mechanical work in the form of crushing produces an ordered magnetic phase in the chromites that is apparently magnetically ordered on a sufficiently large scale to produce multidomain-type **M**(H) characteristics. This induced magnetic phase has an ordering temperature of about 570° C, which is close to that for magnetite. This magnetic phase is probably associated either with Fe-enriched phases or with defect structures on the grain surfaces. This crushing-induced phase is almost fully recovered by annealing at temperatures in the range 500 to 600° C.

TABLE 1.15

Remanent specific polarization $\sigma_m(0)$, specific polarization at 7 kOe $\sigma_m(7)$ and the high-field mass magnetic susceptibility χ of chromites [S47]

Sample	T = 296 K			T = 78 K	
	$\sigma_m(0)$ emu g^{-1}	$\sigma_m(7)$ emu g^{-1}	$\chi \times 10^6$ cm^3 g^{-1}	$\sigma_m(0)$ emu g^{-1}	$\chi \times 10^6$ cm^3 g^{-1}
Transvaal, hard-lumpy					
as received	0.015	0.25	32	0.03	68
-150 + 75 μm	0.075	0.34	37		
-63 + 45 μm	0.35	0.61	37		
-38 μm	0.55	0.80	36	0.53	72
Transvaal, low SiO_2					
as received	<0.01	0.24	35	0.03	60
-150 + 75 μm	0.007	-	35	-	-
-38 μm	0.37	-	27	0.37	57
Masinloc, (Philippines)					
as received	<0.005	0.20	28	<0.01	49
-150 + 75 μm	0.006	0.19	26	-	-
-38 μm	0.13	0.31	26	0.14	48

Owada and Harada [O8] recently investigated the magnetic properties of various chromites and found that their paramagnetic susceptibilities have a positive correlation with the Fe^{2+} content of chromite. The authors speculated that the distribution of magnetic ions is not uniform in the crystal structure and so ferrimagnetism is created in the more concentrated sections. Magnetic susceptibilities, at room temperature and at 77 K, of the chromite investigated by Owada and Harada are summarized in Table 1.16.

I.3.3.9 Stannite

The recovery of tin from stannite (Cu_2FeSnS_4) and its oxidation products, hydrostannate and warlamovite, by flotation and gravity separation does not always give satisfactory results. On the other hand, recoveries of tin exceeding 80 per cent, with the concentration ratio of about 15 can be obtained [V1] by magnetic separation.

TABLE 1.16

Magnetic susceptibilities of chromites [O8]

Country, mine	Mass magnetic susceptibility			
	* 300 K	**	* 77 K	**
South Africa Montrose	33.8	4.25	54.5	6.85
Madagascar Adriamena Befandriana	31.0	3.90	71.4	8.97
India Sukrargi	26.5	3.33	45.3	5.69
Albania Kükes	28.6	3.59	48.9	6.14
Vietnam Cổdinh	33.5	4.21	85.2	10.71
Philippines Acoje	31.6	3.97	50.2	6.31
Philippines Masinloc	25.8	3.24	54.3	6.82

* $cm^3\ g^{-1} \times 10^{-6}$ ** $m^3\ kg^{-1} \times 10^{-7}$

The specific magnetic susceptibility of stannite, hydrostannate and warlamovite, as a function of the applied magnetic field is given in Table 1.17 [C3]. In the natural state, the minerals are weakly magnetic substances, not containing ferromagnetic inclusions, as indicated by the constant value of magnetic susceptibility for various values of the magnetic field strength.

Voevodin et al. [V1] have calculated the theoretical values of magnetic susceptibility for given minerals from their chemical composition using the expression (in SI units)

$$\kappa = 423.6 \times 10^{-6} \sum_i \frac{\mu_m c_i}{m_i} \tag{1.47}$$

where $\mathbf{m}_i$ is the atomic mass of the paramagnetic ion of type **i**, $\mathbf{c}_i$ is its concentration and μ_m its magnetic moment. This equation is valid under the assumption that the paramagnetic ions are magnetically diluted, i.e. they do not mutually interact. The calculated magnetic susceptibilities and the contributions from assorted ions are summarized in Table 1.18.

TABLE 1.17

The specific magnetic susceptibility ($m^3\ kg^{-1} \times 10^{-9}$) of tin minerals as a function of magnetic field (after Chichernikov [C3])

Mineral	Magnetic field strength **H** ($A\ m^{-1} \times 10^5$)				Mean
	4.4	0.7	10.4	11.8	
Stannite	9.0	9.08	9.02	9.03	9.03
Hydrostannate	8.21	8.33	8.33	8.29	8.29
Warlamovite	8.20	8.25	8.15	8.20	8.20

TABLE 1.18

The calculated specific magnetic susceptibilities of tin minerals ($m^3\ kg^{-1} \times 10^{-9}$) (after Voevodin et al.[V1])

Paramagnetic ions	Effective magnetic moment (in Bohr magnetons	Content of paramagnetic ions		
		Stannite	Hydrostannate	Warlamomovite
Fe^{2+}	4.9	6.53	-	-
Fe^{3+}	5.92		6.36	7.56
Cu^{2+}	2.73	16.77	14.7	12.78

Paramagnetic ions	Contribution from ion to magnetic susceptibility		
	Stannite	Hydrostannate	Warlamovite
Fe^{2+}	11.86	-	-
Fe^{3+}	-	16.9	20.33
Cu^{2+}	8.34	7.31	6.35
TOTAL	20.20	24.21	26.68

It can be seen from Table 1.18 that, apart from iron, Cu^{2+} ions have a considerable effect on magnetic susceptibility, and that the calculated values are greater than the experimental values (Table 1.17). This can be explained by the nature of the magnetic ordering of the paramanetic ions in the crystal lattice of the minerals.

Owing to low values of magnetic susceptibility the selective recovery of tin by magnetic separation is not easy. On the other hand, preliminary chemical treatment and calcination of the minerals can appreciably change magnetic properties. As a result of the treatment with a 10 per cent solution of sodium hydroxide at 80° C for 15 minutes the magnetic susceptibility at $\mathbf{H} = 4.4 \times 10^{5}$ A m^{-1} increased about 30 times for hydrostannate and warlamovite, and dependence on the magnetic field was observed [V1]. The field dependence indicates the presence of a ferromagnetic phase in the composition of these minerals, which appears during the treatment with alkali. For stannite the magnetic susceptibility remains unchanged.

The magnetic properties were found to be affected by thermal treatment [D13]. It was determined that thermal treatment both in oxidizing and reducing media at temperatures greater than 400° C produced new strongly magnetic phases as a consequence of the transition of the paramagnetic ions of iron into a ferromagnetic mineral phase. The magnetic susceptibility was found to depend strongly on the applied magnetic field and the concentration of a strongly magnetic phase was found to increase with the duration of roasting, as shown in Fig. 1.24.

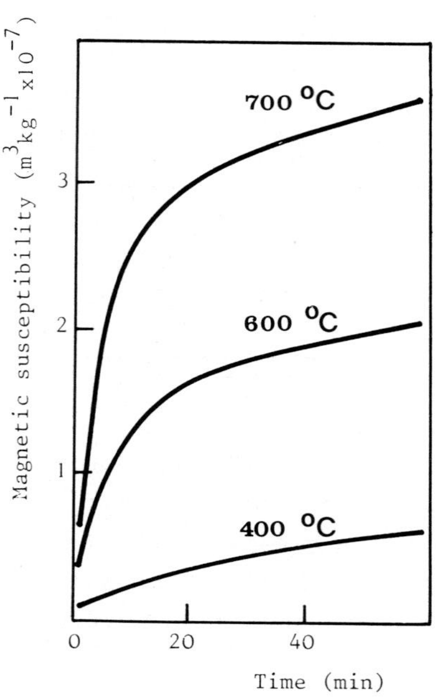

Fig. 1.24 The effect of duration of the oxidizing roasting on the specific magnetic susceptibility of stannite at different temperatures (after Dobrovolskaya et al. [D13]).

Dobrovolskaya et al. [D13] also determined the optimum conditions of thermal treatment that result in the maximum increase in magnetic susceptibility. It was found that for stannite the best results were obtained with oxidizing roasting at 700° C for 1 hour while for hydrostannate reducing roasting at 800° C for 15 minutes was recommended. Similarly for warlamovite, roasting in the air at 800° C for 15 minutes produced a mineral, the magnetic susceptibility of which increased by two orders of magnitude.

The effect of temperature of the oxidizing roasting on magnetic susceptibility is depicted in Fig. 1.25 and selected results are summarized in Table 1.19.

As a result of the above analysis on the magnetic properties of tin minerals, two methods of tin recovery can be proposed. The ore can be either ground below 200 μm and beneficiated by wet high-gradient magnetic separation, or it can be heat-treated in the air, coarse fraction separated by low-intensity magnetic separation and the slimes treated by HGMS [V1]. Thermal treatment reduces the magnetic induction needed to recover tin from 1.2 T for untreated ore to 0.3 T for roasted ore [D13].

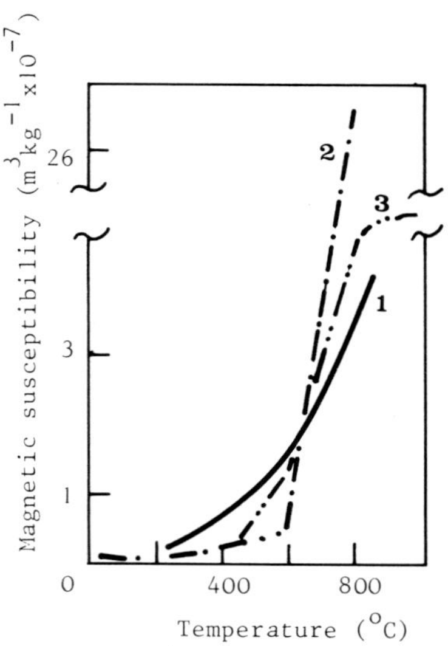

Fig. 1.25 The effect of temperature of the oxidizing roasting on the magnetic susceptibility of stannite (1), warlamovite (2) and hydrostannate (3), at magnetic induction 0.55 T (after Dobrovolskaya et al. [D13]).

TABLE 1.19

The effect of the chemical and thermal treatment on the magnetic susceptibility of tin minerals ($m^3\ kg^{-1} \times 10^{-9}$) (after Voevodin et al.[V1])

Mineral	Treatment	Magnetic field strength, H $Am^{-1} \times 10^5$			
		4.4	7.0	10.4	11.8
Stannite	Alkali	9.12	9.06	9.1	9.07
Hydrostannate	treatment	230.6	197.2	179.0	174.0
Warlamovite		260.4	225.6	210.1	104.5
Stannite	Heat treat-	9.68	8.59	7.98	7.72
Hydrostannate	ment in	544.5	346.2	245.2	207.0
Warlamovite	reducing medium *	784.8	459.0	305.5	-
Stannate	Heat treat-	368.6	247.5	183.6	162.0
Hydrostannate	ment in	245.7	186.0	154.1	146.0
Warlamovite	air *	2680.5	1639.9	1075.8	-

* T = 800° C, 15 minutes

I.3.3.10 Magnesite

Magnesium carbonate ($MgCO_3$) is a diamagnetic substance and the presence of a paramagnetic or ferromagnetic impurity in naturally occurring magnesites usually renders magnesite feebly magnetic. The nature of the changes in magnetic properties is likely to be different, depending upon the state in which the ferruginous minerals may occur.

Magnesite can be broadly classified into two types:

a) generally free from ferruginous contamination,

b) containing about 2 to 4 per cent of ferruginous mineral.

Whereas the ferruginous materials in the former group constitute impurities in trace amounts, those in the latter group occur in major proportions as essential minerals, presumably in the form of a solid solution [B30]. The overall range of composition of different Indian magnesites was as follows: SiO_2 0.6 to 8.0, MgO 36.4 to 46.7, CaO 0.1 to 9.7, and Fe_2O_3 0.2 to 4.4 per cent. Traces of limonite, hematite, pyrrhotite or magnetite contribute to the Fe_2O_3 and act as carriers of the magnetic properties [M23]. The mass magnetic susceptibilities of the Indian magnesites, as measured by Banerjee et al. [B30] are given in Table 1.20.

TABLE 1.20

Magnetic properties of Indian magnesites (after Banerjee et al. [B30])

Sample	Ferruginous constituents (%) as:			χ (m^3 kg^{-1} x 10^{-7})	
	FeO	Fe_2O_3	Total Fe	Magnesite	Iron
Pure $MgCO_3$				-0.05	
Salem		0.19	0.13	-0.03	19.2
Mysore		0.25	0.18	0.08	77.5
Ajmer		0.23	0.16	0.11	103.1
Almura	1.57	1.06	1.96	0.43	25.2
	1.21	1.06	1.68	0.43	29.2
	1.76	0.59	1.78	0.47	31.3
	1.98	0.92	2.18	0.56	28.8
	2.77	0.83	2.74	0.71	28.8
	0.89	1.25	1.58	0.39	28.0
	1.67	0.26	1.48	0.27	22.3

The weakly paramagnetic susceptibility of the Salem magnesite may be due to the diluted ferruginous phases. In contrast, magnesites from Mysore and Ajmer exhibit high magnetic susceptibility, in spite of low amounts of iron. Furthermore, magnetic susceptibility was found to decrease with the increase in the applied field. The impurities are probably of the ferromagnetic type and it was suggested that pyrrhotite and magnetite were present in those samples.

The measurements suggest that the Almura magnesites contain ferrous iron in a magnetically dilute system and it was suggested that siderite in a solid solution was present.

I.3.3.11 Manganese oxides

According to Slater [S48], five of the ten 3d states of the manganese atom are occupied and all five 3d electrons have parallel spins. In manganese oxides (depending on the degree of oxidation) the number of electrons in d-orbitals varies; nevertheless the electrons remain unpaired thus creating a permanent magnetic moment and paramagnetism of manganese oxides.

In magnetically diluted oxides (e.g. pyrochroite $Mn(OH)_2$) the 3d electrons are shielded and the orbital component of the magnetic moment is quenched. In magnetically concentrated compounds the exchange interaction causes an oxide to be ferromagnetic or antiferromagnetic. Manganosite MnO, braunite β - Mn_2O_3 and pyrolusite β - MnO_2 are typical antiferromagnetics. Their magnetic susceptibilities obey the Curie-Weiss law (eq.(1.38)) and are summarized in Table 1.21.

TABLE 1.21
Magnetic properties of the manganese oxides [Y2]

Mineral	χ ($m^3 kg^{-1}$ x 10^{-7})	T_N (K)	μ_m (exp.) (Bohr magneton)
Manganosite MnO	8.5	122	5.93
Pyrochroite $Mn(OH)_2$	18.7		5.8
Hausmanite Mn_3O_4	7.3-78	42.5	
Manganite γ - MnOOH	3.5-5.3		4.9
β- Braunite β - Mn_2O_3	11		4.9
Pyrolusite β - MnO_2	3.4	84	3.8

Wide variations in the values of magnetic susceptibilities in natural ores are caused by the presence of magnetic impurities.

I.3.3.12 Titanium dioxides

Titanium dioxides exist in three crystalline forms, viz, tetragonal rutile and anatase and orthorhombic brookite. Although many experimental values of magnetic susceptibility of rutile have been reported in the literature, they exhibit considerable disagreement.

It was found by Senftle et al. [S49] that above 77 K rutile exhibits a feeble temperature-independent paramagnetism and the mean mass magnetic susceptibility was found to be equal to 8.4 x 10^{-10} m^3 kg^{-1} (0.067 x 10^{-6} cm^3 g^{-1}) at room temperature while the values obtained by other investigators, and listed in [S49] range from -3.8×10^{-9} to 4.6×10^{-9} m^3 kg^{-1} (-0.3×10^{-6} to 0.37×10^{-6} cm^3 g^{-1}).

Magnetic susceptibility of anatase exhibited a strong dependence on temperature and this behaviour was ascribed to the presence of trivalent titanium sites. The values of magnetic susceptibility were affected by the presence of impurities and a value of 2.5×10^{-10} m^3 kg^{-1} (0.02×10^{-6} cm^3 g^{-1}) was obtained as a maximum value of anatase powder.

I.3.3.13 Cu-Fe-S Minerals

Bornite Cu_5FeS_4

Bornite is a common mineral occurring in many copper deposits and its magnetic properties are in close relation to the crystal transformation. The first transformation appears at about 180° C, where magnetic susceptibility abruptly decreases and exhibits a thermal hysteresis [T8]. The transition to the cubic form occurs at 300° C. Magnetic susceptibility decreases paramagnetically with increasing temperature and shows anomalies at the temperatures mentioned above. The value of specific magnetic susceptibility at room temperature was measured to be between 8.8×10^{-9} m^3 kg^{-1} (0.7×10^{-6} cm^3 g^{-1}) and 1.6×10^{-7} m^3 kg^{-1} (13×10^{-6} cm^3 g^{-1}) and was found to be independent of the applied magnetic field [K23].

Cubanite $CuFe_2S_3$

Cubanite is weakly ferromagnetic in the natural orthorhombic form, but above 245° C it transforms to a cubic modification and behaves paramagnetically [T8]. On cooling, the magnetic susceptibility of cubic cubanite shows no anomaly at 245° C, but increases abruptly at liquid nitrogen temperature and reveals the ferrimagnetic properties.

I.3.4 The effect of ferromagnetic admixtures on magnetic susceptibility of paramagnetic minerals

Magnetic susceptibility of a paramagnetic mineral containing magnetite depends on the concentration of magnetite. Susceptibility of such a mixture does not rise proportionally to the increase in the percentage concentration of magnetite but much faster.

Laurila [L1] has shown that a mixture of ferromagnetic and nonmagnetic particles has a magnetic susceptibility

$$\kappa \sim \beta c \tag{1.48}$$

where **c** is the volume concentration of ferromagnetic particles and β is given by

$$\beta = \frac{1}{3}\left(\frac{1}{D_a} + \frac{2}{D_b}\right) \tag{1.49}$$

where $\mathbf{D_a}$, $\mathbf{D_b}$ are the demagnetizing factors in the direction of the long and short axes, respectively.
For spherical particles this gives

$$\kappa \sim 3c \tag{1.50}$$

If the proportion between the long and short axes of the ovoid particles is **a/b** = 10, the susceptibility is

$$\kappa \sim 18c \tag{1.51}$$

This clearly shows the effect of the form of the particles, even when they are randomly oriented.

The volume magnetic susceptibility κ_M of such a mixture can then be calculated from the expression

$$\kappa_M = \alpha^2 \kappa_0 \times 10^{-4} \tag{1.52}$$

where κ_o is the volume magnetic susceptibility of pure magnetite and α is the magnetite content in per cent. In calculating the specific magnetic susceptibility one has to realize that the density of a mixture increases with increasing concentration of magnetite. In most magnetite ores the intergrown grains consist of magnetite (density ≃ 5000 kg m^{-3}) and of gangue minerals (density 2800 kg m^{-3}) [D1].

Karmazin [K1] has shown that the specific magnetic susceptibility (in m^3 kg^{-1}) of a mixture of magnetite and a gangue mineral can be written as

$$\chi_M = 2.4 \times 10^{-11}(\alpha + 27)^3 \qquad (1.53)$$

Magnetic susceptibility of mixtures of ferromagnetic and weakly magnetic minerals can be affected by the application of ultrasound [G2]. It was shown that for some minerals the ultrasound irradiation causes a reduction of magnetic susceptibility. The concentration of ferromagnetic impurities decreases because of their stripping from the grain. In other cases (e.g. ilmenite) the effect of ultrasound is to increase the value of magnetic susceptibility. The effect was explained by a selective removal of diamagnetic particles from the sample and by an increase of a relative content of strongly magnetic ones. Ultrasound can thus be used to change the magnetic properties of a mineral with the aim to improve the efficiency of magnetic separation.

Although the magnetic susceptibility of paramagnetic material is independent of the magnetic field, minerals displaying various forms of magnetic order (parasitic ferromagnetism, canted spin ferromagnetism or antiferromagnetism) or paramagnetic minerals containing the admixtures of the magnetically ordered materials exhibit a field-dependent magnetic susceptibility. This phenomenon can seriously alter the predictions of the results of magnetic separation based on the field-independent magnetic susceptibility. The effects of paramagnetic and ferromagnetic components can be separated by a method devised by Honda and Owen [O3,H2].

The measured magnetic susceptibility, in a low applied magnetic field ($\mathbf{H} < 5 \times 10^5$ A m^{-1}, or 6 kOe) can be represented by the Owen-Honda equation

$$\kappa = \kappa_p + \kappa_d + \frac{M}{H} = \kappa_\infty + \frac{M}{H} \qquad (1.54)$$

where κ_p is the paramagnetic susceptibility, κ_d the diamagnetic susceptibility, **M** the ferromagnetic saturation magnetization and **H** the applied magnetic field. κ_∞ is the magnetic susceptibility extrapolated to the infinite magnetic field strength and can be obtained as an intercept on the susceptibility axis of a κ versus 1/**H** plot. If all the ferromagnetic minerals or fragments in the sample are saturated, this type of plot yields a straight line whose slope is saturation magnetization **M**.

Equation (1.54) assumes that all ferromagnetic mineral particles are saturated; at low magnetic fields ($< 5 \times 10^5$ A m^{-1}) this condition is strictly true only for needle-like crystal inclusions. A spherical ferromagnetic particle is completely unsaturated and acts like a paramagnetic particle. In between needle-like and spherical particles there may be irregularly shaped particles that are only partially saturated. Therefore, the Owen-Honda equation that applies to all shapes of ferromagnetic inclusions can be rewritten [A3] as

$$\kappa = \kappa_p + \kappa_d + \frac{M}{H} + \left[\frac{f(H)}{H} + \kappa_s\right] \qquad (1.55)$$

where κ_s is the apparent paramagnetic susceptibility of those mineral particles that have a large demagnetization factor and therefore have not started to saturate at low magnetic fields; **f**(**H**)/**H** is a term describing the susceptibility of partially saturated, irregularly shaped ferromagnetic particles and **M**/**H** is the susceptibility of the completely saturated ferromagnetic particles.

The variation in magnetic susceptibility of hematite [N2] with a magnetic field is shown in Fig. 1.26 and a significant change can be seen for magnetic fields less than 8×10^5 A m^{-1} (10 kOe). The field-independent magnetic susceptibility κ_∞ was found to be equal to 1.43×10^{-3} (SI) (1.14×10^{-4} (cgs)) for hematite and 4.0 x 10^{-3} (SI) (3.18×10^{-4} (cgs)) for ilmenite. Saturation magnetization **M** was equal to 1.62 kA m^{-1} for hematite and 1.17 kA m^{-1} for ilmenite.

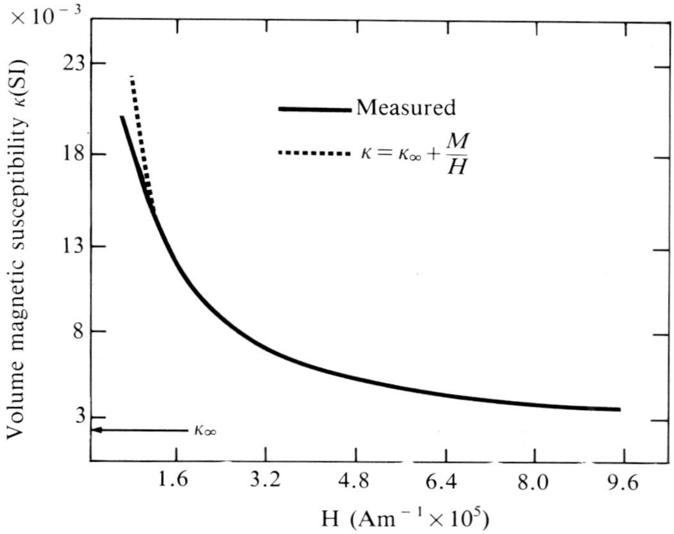

Fig. 1.26 Volume magnetic susceptibility of hematite as a function of applied magnetic field: $\kappa_\infty = 0.14 \times 10^{-2}$ (SI), **M** = 1.62 kA m^{-1} (after Nesset and Finch [N4]).

Pastrana and Hopstock [P2] investigated the effect of temperature and magnetic field on magnetic susceptibility in terms of eq. (1.54). They found that κ_∞ increased only slightly with temperature but the effect of temperature on κ was dramatic. The results indicate that most of the increase in κ results from an increase in **M**. Although in impure samples **M** may merely be an indication of the content of ferromagnetic impurities, measurements on pure samples of hematite indicated that the quantity M represents the contribution from an intrinsic weak ferromagnetism of hematite. From the plot χ versus **H** Pastrana and Hopstock obtained that χ_∞ decreased from about 3.8×10^{-7} m^3 kg^{-1} (30×10^{-6} cm^3 g^{-1}) above 280 K to about 2.5×10^{-7} m^3 kg^{-1} (20×10^{-6} cm^3 g^{-1}) below 250 K. It was also observed that κ_∞ was independent of particle size while **M** showed significant variations between size fractions, although no consistent pattern of variations could be found.

Thus, the quantity κ_∞, i.e. the magnetic susceptibility extrapolated to infinite field strength, is characteristic of each mineral, and is nearly constant for samples of the same mineral from different localities. On the other hand, saturation magnetization **M**, which indicates the degree of weak ferromagnetism, varies widely between samples from differing localities and between varying size fractions of the same sample.

It is thus clear that ferromagnetic admixtures to paramagnetic minerals, or a certain degree of magnetic order in weakly magnetic minerals, can significantly influence the value of magnetic susceptibility as a consequence of its dependence on magnetic field, particle size and temperature. A correct evaluation of magnetic susceptibility in terms of eq. (1.54), under given experimental conditions can only be effected with sophisticated measuring equipment such as a vibrating sample magnetometer, which is rarely available to an engineer. Although some information can be obtained with the use of Frantz isodynamic separator [N3], serious errors in the evaluations of the potential of magnetic separation can develop when using the literature values, which rarely reflect the complicated character of the magnetic susceptibility of natural minerals.

I.4 GENERATION OF MAGNETIC FIELD IN MAGNETIC SEPARATOR

The aim of magnetic separation is to recover or remove particles with sizes ranging from several millimetres down to a fraction of a micrometre, with a wide spectrum of magnetic susceptibilities, from ferromagnetic to diamagnetic. Furthermore, it is frequently necessary in mineral processing to separate two or more different mineral species, the magnetic susceptibilities of which do not differ very much, with a wide range of particle sizes. The magnetic field strength needed to recover such particles thus varies substantially from one application to another and different ways of generating the magnetic field are employed.

As can be seen from eq. (1.8) the magnetic force required to separate a magnetizable particle depends not only on the physical properties of a mineral and on the magnetic induction, but also on the value of the gradient of the magnetic field. Therefore, to achieve a selective separation of a valuable mineral from unwanted ones, a carefully selected combination of the magnetic induction and of the field gradient must be used to warrant a high recovery and simultaneously a high grade of the concentrate. Magnetic separation can be considered successful only if it produces the right grade of the concentrate with the right recovery.

In some of the magnetic filtration applications the "brute force" approach is often used to warrant a high degree of removal of solids from suspensions. Consequently an effort is usually made to maximize the external magnetic field or field gradient, or both. This approach is also employed in certain isolated applications in mineral processing (e.g. brightening of kaolin) where the useful product is the nonmagnetic fraction. The combination of a maximum magnetic force with a long retention time leads to a high quality nonmagnetic product, although sometimes at the expense of losses of the nonmagnetic component into the magnetic fraction.

When applied to mineral benefication this "brute force" approach usually results in a magnetic concentrate of very low quality, which in many cases differs little from the grade of the feed, although the researches are sometimes misled by a high mass yield, a consequence of a high nonselective capture of particles onto the matrix of a magnetic separator. Such a simplified generalization on magnetic separation caused the term "high gradient magnetic separation" (HGMS) to be used [G1] for those applications of magnetic separation in which a fine steel wool was used as a matrix to generate the field gradient. As shown in Chapter VI, "lower-gradient" and "high-gradient" matrices perform equally well even in many magnetic filtration applications and a frequent effort to maximize the field gradient and magnetic induction is often self-defeating and costly while the results do not necessarily correspond to the effort exerted.

The desired magnetic induction and the field gradient is thus a function of the type of application, required metallurgical performance and economic criteria. In the following section we shall review the basic techniques of the generation of magnetic field and the gradient of magnetic field, as used in magnetic separation.

I.4.1 Permanent magnets

Permanent magnets embrace the oldest magnetic material known to man, i.e. natural magnetite or lodestone, as well as commercially significant fine-particle magnets developed during the past 30 years.

The early history of magnetic separation is closely associated with permanent magnets which were used to remove strongly magnetic matter from wastes or to recover strongly magnetic (mostly iron) minerals from ores. With the development of powerful electromagnets the interest in permanent magnets as applied to magnetic separation diminished, but in recent years we have witnessed a remarkable renaissance of interest in permanent magnets, a phenomenon associated with the impressive development of new permanent magnet materials, based on rare-earth metals.

Although the magnetic induction generated by permanent magnets rarely exceeds 1 T, i.e. the value that can be easily produced by electromagnetic coils, permanent magnets offer the following advantages over the coils.

a) Permanent magnets do not require any source of energy or any cooling fluid.
b) Magnetic induction is very stable over time; it fluctuates in an interval of 10^{-8} and decreases at a rate of the order of 10^{-3} per year.

On the other hand permanent magnets possess several disadvantages:

a) Magnetic induction is difficult to vary.
b) The magnitude of the magnetic induction is sensitive to temperature, the rate of variation being about 10^{-4} per degree Celsius.

The permanent magnet provides useful field energy in an air gap and thus requires a high remanent polarization J_R and a sufficiently high coercive force $\mathbf{H}_c$ (see Fig. 1.10) to preserve adequate magnetic induction in the presence of the air gap which tends to demagnetize it. In addition to remanent polarization $\mathbf{J}_R$ (or remanent induction $\mathbf{B}_R$, related to $\mathbf{J}_R$ through eq.(1.19)), a third property of permanent magnets is used, namely, $(\mathbf{BH})_{max}$, the maximum value of the product **B** . **H** in the second quadrant, i.e. where **B** is positive and **H** is negative. This energy product indicates essentially the "strength" of the magnet and is related to the total energy stored in the magnetic field. The product $(\mathbf{BH})_{max}$ clearly depends on $\mathbf{B}_R$ and $\mathbf{H}_C$, as well as the shape of the hysteresis curve. For a given $\mathbf{B}_R$ the theoretical maximum value of $(\mathbf{BH})_{max}$ is $\mathbf{B}_R{}^2/4\mu_0$, which can be obtained when the **J** vs **H** hysteresis loop approaches a rectangular shape.

Fig. 1.27 shows how the $(\mathbf{BH})_{max}$ obtainable from permanent magnets material has improved during the last 100 years. When a new class of material is discovered, a very rapid improvement in

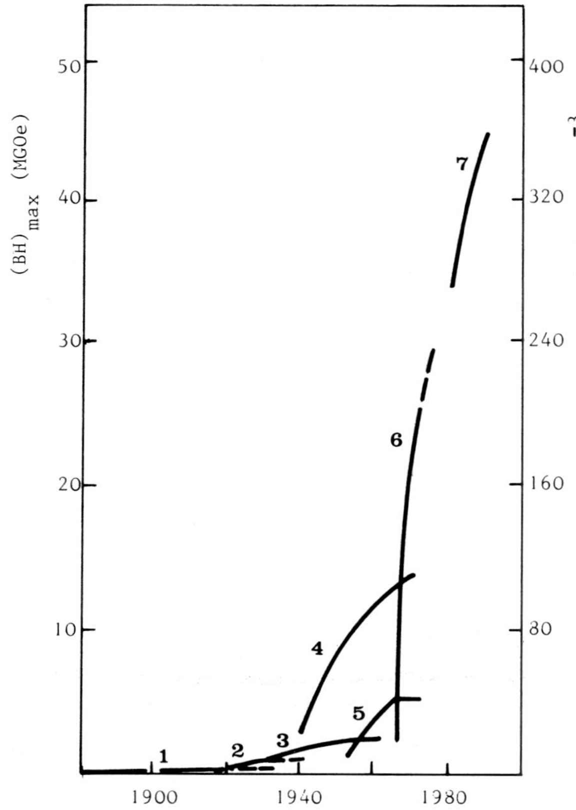

Fig. 1.27 Improvement of $(\mathbf{BH})_{max}$ of different classes of permanent magnets during the last hundred years. (1) Pre-cobalt steels, (2) cobalt steels, (3) isotropic Al-Ni-Co, (4) anisotropic and columnar Al-Ni-Co, (5) barium and strontium ferrite, (6) rare-earth cobalt, (7) Nd-Fe-B alloys.

$(\mathbf{BH})_{max}$ takes place for a few years followed by an asymptotic approach to a maximum, until a breakthrough is made with a new class of material [M1].

Useful permanent magnets may be based on either ferromagnetic or ferrimagnetic materials. Because the bulk material generally consists of many domains orientated randomly, the net magnetization can be negligible. However, the exposure of such a material to a magnetic field causes the domains to reorient along the applied field, leading to the increase in polarization. The preferred orientation of the domains does not fully disappear when the applied field is removed that leads to the residual polarization, and a negative magnetic field must be applied to reduce the polarization of zero.

Thus, the understanding why a material is a good permanent magnet corresponds to understanding why the magnetic domains do not readily reorient in response to even small applied magnetic fields. Many different mechanisms may restrain domain reorientation, and only a brief comment will be presented here; further discussion is available, e.g. in [M1].

The coercivity of the presently available permanent magnets is due to an anisotropy arising primarily from one of two effects: magnetocrystalline anisotropy or particle shape and size effects. The magnetoscrystalline anisotropy occurs because the quantum mechanical interactions among the electrons lead to a lower energy when the spontaneous magnetization lies along certain preferred easy directions in a crystal. In cubic crystals, symmetry produces easy directions that are $< 90^{\circ}$ apart in rotation. Crystal structures with less symmetry will generally have fewer easy axes, thus hindering rotation further [P3]. Most desirably, the crystal structure will have only one easy axis requiring a 180° rotation.

Shape anisotropy occurs because in an elongated particle, an external field can more readily saturate parallel to the long dimension. Further, in the absence of other factors such as crystalline anisotropy, a single domain elongated particle will have its spontaneous polarization aligned along its long dimension, so that very fine particles or grain sizes are desirable. Enhanced coercivity can be achieved by combining the shape anisotropy with a crystal phase having at least moderate crystal anisotropy and partial alignment. Fig. 1.28 shows the advance in $\mathbf{H}_c$ with time.

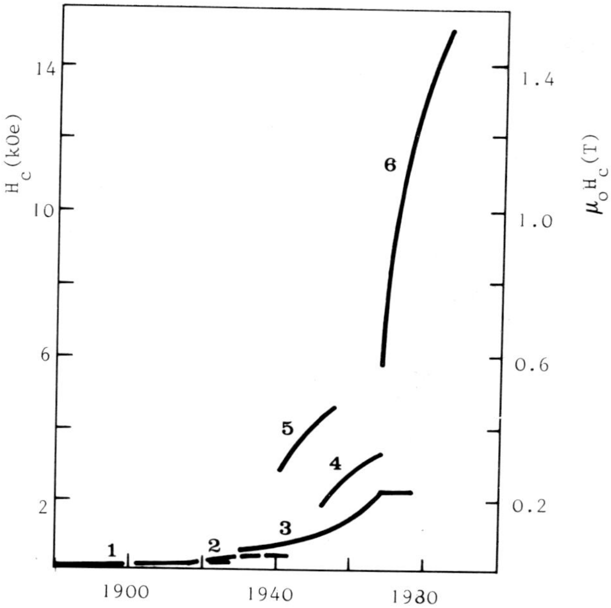

Fig. 1.28 Improvement of coercivity of different classes of permanent magnets during the last hundred years. (1) Pre-cobalt steels, (2) cobalt steels, (3) all Al-Ni-Co alloys, (4) barium and strontium ferrite, (5) platinum-cobalt, (6) rare-earth-cobalt. (After Mc Caigh [M1]).

Commercial permanent magnets

Table 1.22 shows representative examples of the types of standard permanent magnets commercially available [P3,S7,H3].

Ferrites are possibly the most important commercial permanent magnets at the present time, their success lies partly in their cheapness and partly in their high coercivity.

Ferrites are produced from iron oxide calcined with barium or strontium carbonate - cheap and non-strategic raw materials. They are mechanically hard and brittle, their magnetic properties are characterized by relatively low values of remanent polarization, and the coercive force is satisfactory. The maximum energy product $(\mathbf{BH})_{max}$ ranges from 7 to 30 kJ m^{-3} (1 to 4 MGOe).

The next most commonly used permanent magnets are a class

known as Alnicos. They are expensive when compared to ferrite magnets since they contain 20 to 40 per cent cobalt with iron, nickel, aluminium and various other additions. The various combinations of $\mathbf{H}_C$ and $\mathbf{B}_R$ which are available in different Alnico alloys result from differences in composition and processing. The Alnicos have desirably high $\mathbf{B}_R$ values, good thermal stability, but comparatively low coercivities. The energy product ranges from 10 to 80 kJ m^{-3} (1 to 10 MGOe).

Magnets based on the $SmCo_5$ phase have both high remanence and high coercivity and thus a high energy product. The coercivity of these alloys arises from the strong magnetic crystalline anisotropy of this hexagonal phase. Bulk magnets are made by sintering magnetically aligned fine powders (~2 to 3 μm particles). While clearly superior to other materials, the high cost of these Sm-Co-based magnets has limited their use although recently marketed magnetic separators that use this magnet material are finding a wide-spread application in a variety of demanding applications.

As shown in Table 1.22, Alnico and Sm-Co alloys contain significant amounts of cobalt. Cobalt is one of the most important strategic elements and its potential for future supply - as a result of interruptions caused by political instability in Africa - has led to a continued effort to develop high performance magnets containing no cobalt. In so far as cost plays a role, neither of the main components (samarium and cobalt) is particularly cheap and it would therefore be desirable to use iron-based compounds rather then cobalt based compounds. Sagawa et al. [S7] have succeeded in manufacturing Nd-Fe-B compound in which the energy product in excess of 36 MGOe was achieved. Rare-earth iron (R-Fe) intermetallics especially those containing the abundant light rare earths have long been studied as candidate materials for high performance permanent magnets. It was noticed that there exists a variety of stable compounds in the ternary systems composed of light rare earths, Fe and small quantities of third elements. From a number of ternary compounds that had actually been synthetized, a remarkable compound consisting of Nd, Fe and B was found [S7].

This phase has an approximate composition $Nd_2F_{14}B$, the Curie temperature is 585 K and the saturation polarization at room temperature is 1.57 to 1.61 T. This gives a theoretical limit to $(\mathbf{BH})_{max}$ = 509 kJ m^{-3} (64 MGOe). In practice, a $(\mathbf{BH})_{max}$ value of

TABLE 1.22

Examples of commercial permanent magnets [P3,S7,H3,V7,A18,N6]

Material	B_R(T)	H_c		$(BH)_{max}$	
		(A m^{-1})	(kOe)	(MGOe)	kJ m^{-3}
Carbon steel $Fe_{98}CMn$	1.0	4 x 10^3	0.05	0.2	1.6
Alnico 5 $Fe_{50}Ni_{15}Al_8Co_{24}Cu_3$	1.31	5.6 x 10^4	0.7	6.5	52
Alnico 8 $Fe_{36}Ni_{15}Al_7Co_{36}Ti_6$	0.85	1.3 x 10^5	1.6	5.1	41
Ferrites e.g. BaO . $6Fe_2O_3$	0.39	1.6 x 10^5	2.0	3.5	28
$SmCo_5$	0.85	1.3 x 10^6	16	18-22	143-175
Cu-Ni-Fe $Cu_{60}Ni_{20}Fe_{20}$	0.57	4.7 x 10^4	0.59	1.85	15
$Nd_{15}F_{77}B_8$	1.40	9.5 x 10^5	12	35-45	280-360
Au-Fe-Ni	0.53	4.0 x 10^3	0.5	1.1	9.1
Au-Fe-Co	0.34	8.1 x 10^3	1.0	1.1	8.2
(Sm,Pr,Nd)-Co	1.05	1.2 x 10^6	15	27	216
$Nd_{16}Fe_{66}Co_{11}B_7$	1.32	6.5 x 10^5	8	42	336
$Nd_{15}Fe_{62.5}B_{5.5}Co_{16}Al$	1.32	8.8 x 10^5	11	41	328

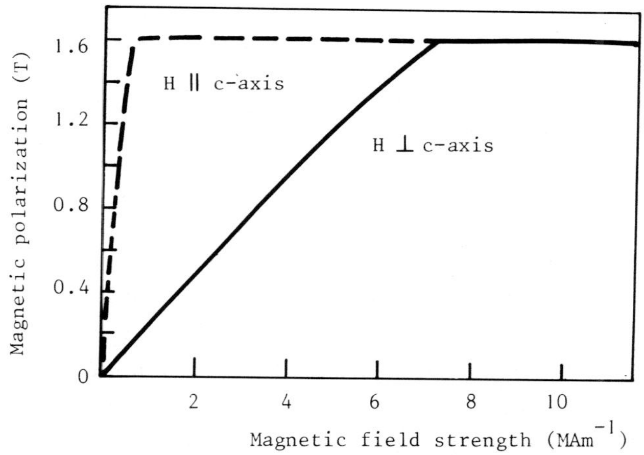

Fig. 1.29 Magnetization vs magnetic field curves of a single crystal of $Nd_2Fe_{14}B$ at room temperature (after Sagawa et al. [S8]).

435 kJ m^{-3} (55 MGOe) should be achieved under production conditions [O7]. Magnetization curves for this compound are shown in Fig. 1.29 [S8].

Based on this compound a very high energy product magnet was developed using a powder metallurgy technique [S9] or by melt spinning [H4]. This magnet, which has an approximate composition $Nd_{15}Fe_{77}B_8$, shows a remanent polarization 1.23 T, a coercive force of 960 kA m^{-1} (12 kOe) and a maximum energy product as high as 360 kJ m^{-3} (45 MGOe). These parameters were further increased by improvements to the technology [N6]. The effect of the additions of heavy rare-earth on the coercive force is illustrated in Fig. 1.30, which shows the demagnetization curve for two magnets. Partial substitution of Nd by Dy increases the coercive field.

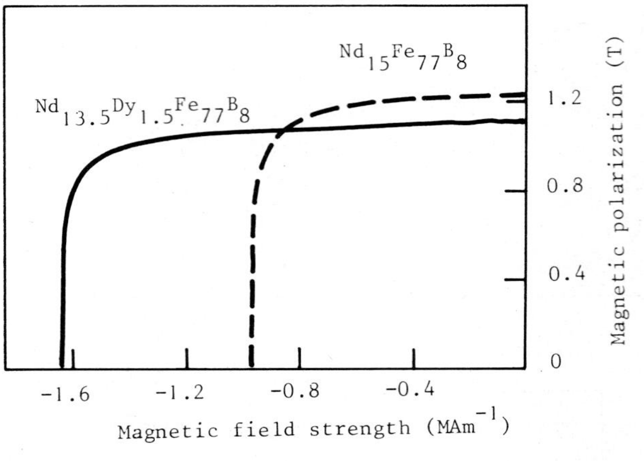

Fig. 1.30 Demagnetization curve of $Nd_{13.5}Dy_{1.5}Fe_{77}B_8$ sintered magnet (after Sagawa et al. [S7]).

The great interest in these alloys is for three reasons. Firstly, the samarium-cobalt alloys have shown there is a market for relatively expensive high-energy magnets. Secondly, the best energy product produced so far for Nd-Fe-B magnet is about 50 per cent more than the best Sm-Co alloys. Thirdly, neodymium is much more abundant in rare-earth ores than samarium, consequently, given demand, it should be cheaper. Being already cheaper than the Sm-Co alloys in terms of cost of equivalent magnetic power, Nd-based magnets are widely expected to have a substantial commercial impact - it is considered that they might even largely supersede Sm-Co magnets and possibly replace Alnico magnets in some applications [S10].

Moreover, it is assumed that the price of the neodymium-based magnets could fall as production increases from the present very low level to not only that of Alnico, but also below that of the hard ferrites [J7]. The size of the existing Western world market for permanent magnets is given in Table 1.23 in terms of both simple tonnages and also gross energy product.

If it is assumed that the appropriate basis for comparing the value of different magnetic material is the cost per unit of magnetic energy they supply, then it can be seen from Table 1.24 that the Nd-Fe-B alloys are already cheaper than Sm-Co, albeit more expensive on a mass basis [J7].

It was shown by Jacobson and Evans [J7] that should the Nd-Fe-B alloy entirely replace the rival magnetic materials, about 18 000 tonnes of alloy would be required per annum. It is estimated that the price would fall from the current of \$ 200/kg to \$ 18/kg.

The Curie temperature of the Nd-Fe-B alloy is about 300°C. This proximity to the room temperature causes an appreciable variation in the flux density as the temperature varies. It also

TABLE 1.23

The size of the present market for permanent magnets (U.S.A., Japan and Western Europe) [J7]

Magnet type	Production,1981 (t/annum)	Specific energy J kg^{-1}	Energy kJ	Energy % of total
Sm-Co	240	19.5	4700	0.7
Alnico	8100	6.3	51000	7.3
Ferrite	98000	6.6	647000	92.0

TABLE 1.24

The relative value of magnetic materials [J7]

Magnet type	Cost, 1983 (\$/kg)	Magnetic energy density (kJ m^{-3})	Density (kg m^{-3})	Cost/Joule (\$/J)
Ferrite	8	29	4400	1.2
Alnico	20	45	7200	3.2
$SmCo_5$	150	160	8200	7.7
$Nd_{15}B_8Fe_{77}$	200	290	7400	5.1

means that such magnets do not have useful properties a temperatures greater than 150°C. The thermal coefficient of Nd-Fe- magnets is almost three times as strong as that for $SmCo_5$ magnets The thermal aging confines the maximum operating temperature t about 80° C for Nd_{15} Fe_{77} B_8 magnets and to about 130° C fc $Nd_{14}DyFe_{77}B_8$ magnets [R13]. The coercivity of the materia decreases drastically with increasing temperature and is very smal at temperatures considerably lower than the Curie temperatur [H31].

Nd-Fe-B alloy is tougher and less brittle than the othe magnet materials. On the other hand, it is reported that the allo has a strong tendency to react with normal environments and must therefore be protected by a paint or plating system. It is believe that this tendency to corrode is due to the presence of a phase rich in Nd (which oxidizes very readily) at grain boundaries.

However, these problems appear to be tractable in the short term. By substituting cobalt for some of the iron a sizeable increase in the Curie temperature can be obtained. The Curie temperature of the system $Nd_{15}Fe_{77-2x}Co_{2x}B_8$ increases almost linearly from 312° C (for x = 0) to 525° C (for x=10), while B_R

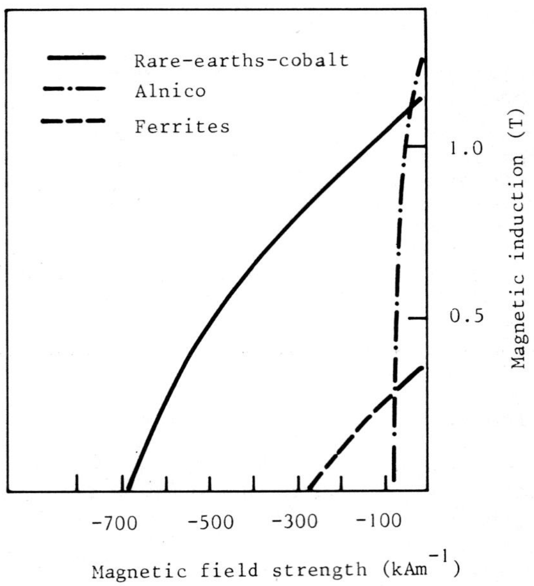

Fig. 1.31 Typical demagnetization curves (2nd quadrant of hysteresis loop) for ferrites, AlNiCo and RECo magnets.

remains unchanged with increasing Co content [A18]. On the other hand, the oxidation problem might be overcome by either addition of cobalt or by the introduction of heavier rare-earth compounds in place of the residual neodymium. Such modifications to the composition of the alloy are designed to simultaneously enhance their magnetic properties. Mizoguchi et al. [M44] have found that the combined addition of cobalt and aluminium in the Nd-Fe-B system improves the magnetic properties and temperature characteristics.

Comparison of the demagnetization curves of the commercial magnet materials is shown in Fig. 1.31. The improvement in primary magnetic properties of the Nd-Fe-B magnets over existing Sm-Co magnets is illustrated in Table 1.23.

I.4.2 Electromagnets

Of the several ways of generating a magnetic field in magnetic separation, the choice is dictated mainly by the required field magnitude and by the volume in which a given magnetic field is to be generated. An effective separation of weakly magnetic particles of a limited size requires an increase of both the magnetic field and of grad **H** considerably beyond the values available with permanent magnets. Furthermore, as the magnetic field falls very quickly from the surface of a permanent magnet, the effective volume in which a sufficiently high magnetic force can be created is limited. In order to generate a sufficiently high magnetic field in a large volume that would guarantee a large throughput of a magnetic separator an electromagnet must be used, in which the electric current passing through the windings magnetizes a great mass of iron which in turn produces a field in the working volume. Judicious location of the iron and of the windings on it may increase considerably the efficiency of the magnet. Although the field strength obtained in this way is limited by the saturation magnetization of the iron, for most large-scale applications of magnetic separation for the recovery of weakly magnetic minerals of the micrometer size, the magnetic force generated by such electromagnets is usually sufficient.

We shall demonstrate the laws governing the magnetic circuit by a toroid, wound homogenously with **N** turns, carrying a current **I**

(Fig. 1.32). For a closed curve going completely through the ring (the dotted line), Ampère's law gives

$$\oint Hdl = NI \qquad (1.56)$$

where **H** is expressed in A m^{-1}, **l** in meters, **I** in amperes. The left-hand side of eq. (1.56) is called the magnetomotive force (mmf).

For a fraction of the toroid of length **l** we have

$$\int Hdl = \frac{Bl}{\mu_0\mu_r} = \frac{\phi l}{\mu_0\mu_r S} \qquad (1.57)$$

where ϕ is the total flux and **S** is the area of cross-section of the ring. If the ring consists of several materials, of length l_j, cross-sectional area S_j, and relative permeability μ_{rj}, then the total magnetomotive force is given by

$$NI = \phi \sum_j \frac{l_j}{\mu_0\mu_{rj}S_j} \qquad (1.58)$$

where

$$\Lambda = \sum_j \frac{l_j}{\mu_0\mu_{rj}S_j}$$

is the magnetic reluctance of the circuit. It follows from eq. (1.58) that if all linear dimensions of the magnetic circuit, and a number of ampère-turns are multiplied by a factor **q**, then the total flux is multiplied by q^2 . However, since this flux is spread out over an area which is q^2 times larger, the magnetic induction remains constant. As a result the current density in the winding space goes down by a factor **q** and likewise the heat dissipation per unit volume by a factor q^2. This makes cooling much easier. Lord Kelvin's theorem makes it possible, before constructing a large magnet, to build a model on a reduced scale and to measure magnetic field and stray field as a function of ampère-turns.

Eq. (1.58) can be applied to an iron ring with a slit (Fig. 1.32)[D5]. Suppose l_j is the circumference of the iron, and l_g the width of the slit ($l_j \gg l_g$), then the magnetic reluctance of this circuit is:

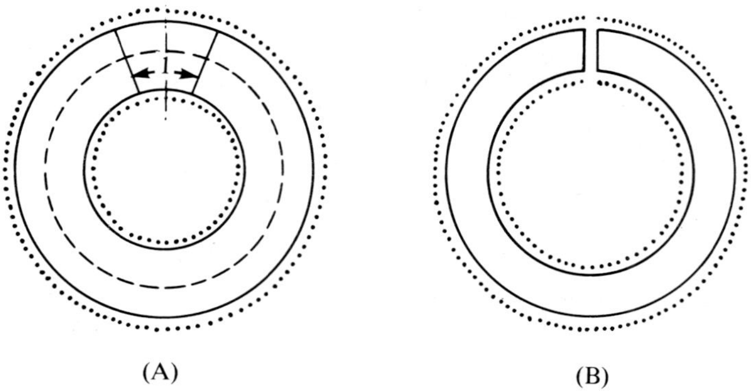

Fig. 1.32 (a) Toroid wound homogeneously. (b) Iron toroid with a slit. (After De Klerk [D5]).

$$\Lambda = \frac{l_j}{\mu_0 \mu_r S_j} + \frac{l_g}{\mu_0 S_g} \tag{1.59}$$

If we neglect leakage flux, the flux of B must be the same over any cross-section of the magnetic circuit, so

$$B_i S_i = B_g S_g \tag{1.60}$$

Combining eqs. (1.58),(1.59) and (1.60),

$$B_g S_g \left[\frac{l_j}{\mu_0 \mu_r S_j} + \frac{l_g}{\mu_0 S_g} \right] = NI \tag{1.61}$$

the magnetic flux is

$$\phi = B_g S_g = \frac{NI}{\Lambda} \tag{1.62}$$

Magnetic flux is therefore equal to the magnetomotive force divided by the sum of the reluctances of the iron and of the air gap. The situation in the iron-yoke electromagnet is shown schematically in Fig. 1.33.

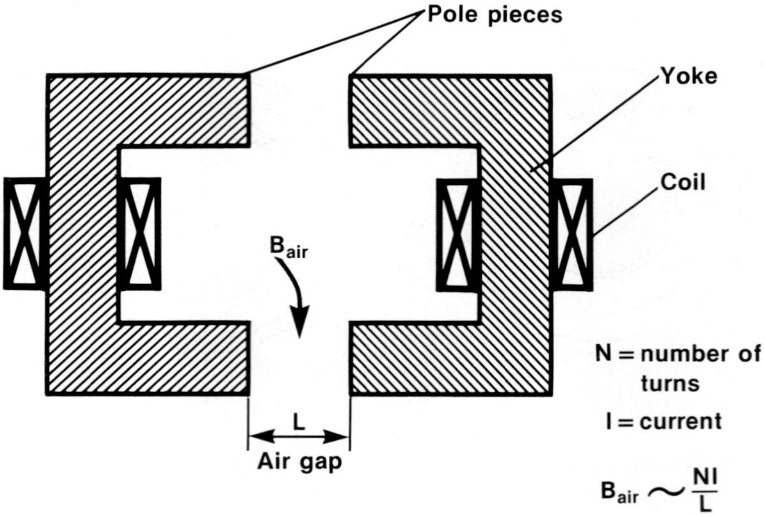

Fig. 1.33 Schematic diagram of the electromagnet with iron yoke.

The magnetic induction obtained in practice proves to be smaller than the values according to eq. (1.62). There are two effects which decrease the field [D5]. In the first place the lines of induction do not cross the air slit perpendicularly: they spread out (Fig.1.34). The imperfections of the circuit can be described by the so-called "leakage factor" **q** and eq. (1.62) then reads

$$\phi = \frac{NI}{\frac{l_j}{\mu_0 \mu_r S_j} + \frac{l_g}{\mu_0 q S_g}} \tag{1.63}$$

The second effect, closely correlated with the first one, is that the iron opposite the field slit is saturated sooner than that near the slit. Because of the spreading out of the induction lines near the slit the flux density in the middle of the circuit is larger than at the ends. The permeability therefore is lower and the magnetic reluctance is higher. The influence of this effect on $\mathbf{B}_g$ is considerable. If we have a magnet with a saturated middle part and we increase the magnetomotive force somewhat, then a small increase in **B** causes an appreciable decrease of μ_r. So the magnetic reluctance increases and the flux remains constant. An

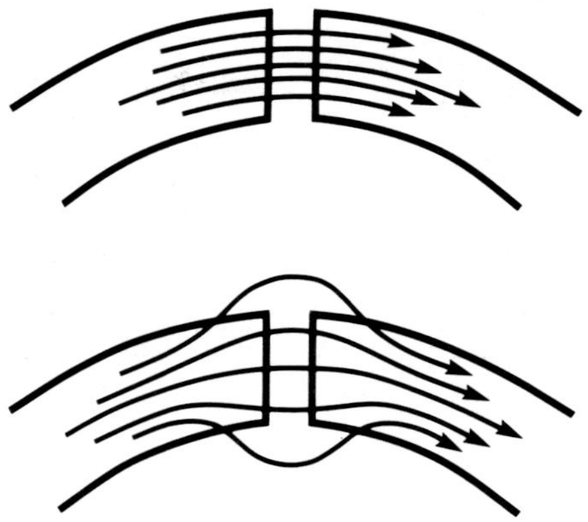

Fig. 1.34 Field pattern in the slit (after De Klerk [D5]).

increase in the number of ampère-turns increases only the length of the saturated part of the iron, but the polarization near the slit hardly becomes any stronger.

A magnet is most effective if, with increasing ampère-turns, the saturation of the iron starts at the slit and penetrates from there into the iron. In order to satisfy this requirement, it is important not to distribute the turns over the iron, but to concentrate them near the pole-faces [D5, D20]. Moreover, the yoke and the poles should be made of material with an initial permeability as high as possible, e.g. steel with a very low carbon concentration. Figure 1.35 summarizes several patterns of the winding and pole arrangements. It can be seen that pole cores of conical shape are better than cylinders, and the steeper and shorter they are built, the higher the efficiency and uniformity of the flux.

The field of a magnet can be concentrated by introducing pole-tips - truncated cones between the poles which concentrate the flux. They increase the field at the cost of a reduction of the working space. The material of the pole-tips has to satisfy requirements different from those of the yoke and the pole pieces. For the yoke and the pole pieces a very high initial permeability is required, while for the pole-tips a high saturation polarization is important.

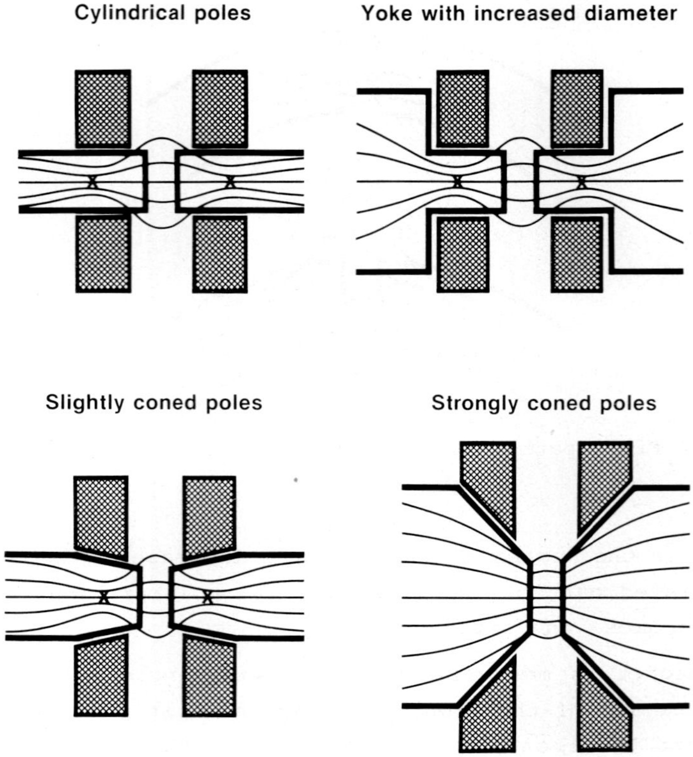

Fig. 1.35 Field pattern in the gap for various shapes of the pole-pieces. (a) Cylindrical poles. (b) Yoke with increased diameter. (c) Slightly coned poles. (d) Strongly coned poles. (After De Klerk [D5]).

Calculation of a field in a gap

We shall assume that the magnetic field is produced by the effective poles at the ends of two uniformly and axially magnetized pole pieces. If the pole-pieces are flat and cylindrical, as shown in Fig. 1.36 then the magnetic induction in the gap is given by

$$B = J_s \gamma_c \left[1 - \frac{x}{(r_2^2 + x^2)^{\frac{1}{2}}} \right] \tag{1.64}$$

where $\mathbf{J}_s$ is the saturation polarization of the pole-piece material (2.14 T for Armco and 2.37 T for iron-cobalt) and γ_c is the correction factor to be discussed later.

For tapered poles with coinciding apexes (Fig. 1.37) the magnetic induction consists of two terms

$$B = J_s \gamma_c \left\{ \left[1 - \frac{x}{(r_1^2 + x^2)^{\frac{1}{2}}} \right] + \frac{x r_1^2}{(r_1^2 + x^2)^{\frac{3}{2}}} \ln \frac{r_2}{r_1} \right\} \quad (1.65)$$

The first term represents the contribution of the flat face and the last term the contribution of the tapered section.

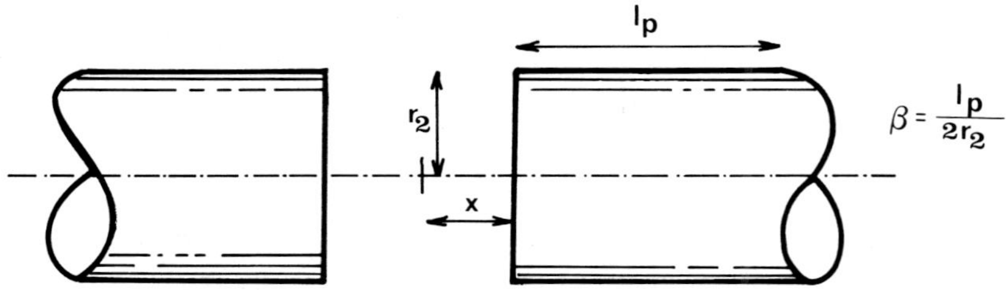

Fig. 1.36 Cylindrical flat poles.

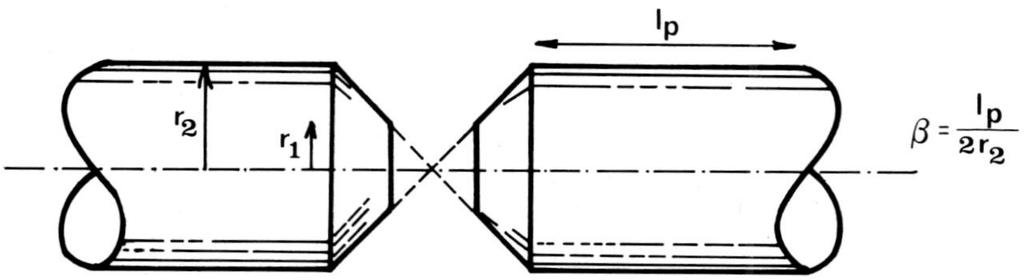

Fig. 1.37 Tapered poles with coinciding apexes.

It has been shown by Montgomery [M33] that eq. (1.64) gives good prediction results if $\gamma_c \cong 0.7$ for $\beta \geq 1$, while for $\beta \cong 0.5$ the correction factor γ_c equal to unity gives a good agreement between calculated and measured values of magnetic induction. In magnets with tapered pole-pieces eq. (1.65) has a very good predictive capability for $\gamma_c \cong 1$.

A wide range of magnetic separators was built with the use of the electromagnets with iron poles. Among these separators, the carousel high-gradient magnetic separators were able to extend the range of applicability of magnetic separation to weakly magnetic, finely ground minerals, and although the maximum available magnetic induction rarely exceeds 0.8 T, large working space guarantees high throughputs and therefore cost-effective operation. A typical example of a magnetic separator using an electromagnet with an iron yoke is the Jones separator which is widely used to beneficiate weakly magnetic iron ores. Other companies (e.g. Eriez Magnetics, Boxmag Rapid) market similar units based on the same principle of the magnetic field generation. A detailed description of such units will be given in the following chapter.

I.4.3 Solenoid magnets

As has been already mentioned, the magnetic induction obtained with electromagnets with an iron yoke is limited by the saturation of the iron. To obtain still stronger fields, even at larger volume, one has to return to the original technique of the magnetic field generation, the construction of a solenoid. Instead of making use of the magnetization of iron, the field should be obtained directly from the current in the electrical conductor. Hence the magnet must be constructed in such a way that the windings are as close as possible to the field space, and the current through them must be as high as possible. Thus we come to a magnet consisting of a short thick coil, usually clad with a steel return frame which facilitates the passage of the magnetic flux.

An electromagnet with an iron yoke (shown schematically in Fig. 1.38) is compared with an iron-clad solenoid (Fig. 1.39). Although electromagnets with iron yokes are frequently used by magnet designers, such an arrangement is inefficient. Although because of high permeability of iron the magnetic flux will preferably pass through the iron, it is rather difficult to convey magnetic flux to distant places owing to strong stray magnetic fields.

The total magnetic induction in the working volume of a magnet is equal to the sum of the contribution from the coil ($\mathbf{B}_c$) and of the contribution from the magnetized iron ($\mathbf{B}_i$)

$$B_{TOT} = B_c + B_i \qquad (1.66)$$

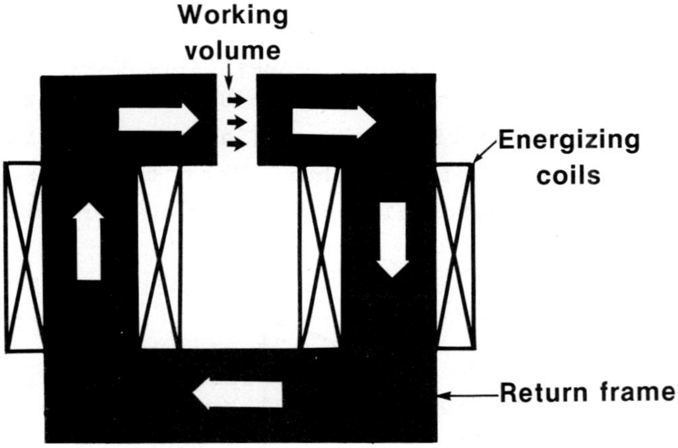

Fig. 1.38 Schematic diagram of electromagnet with iron yoke.

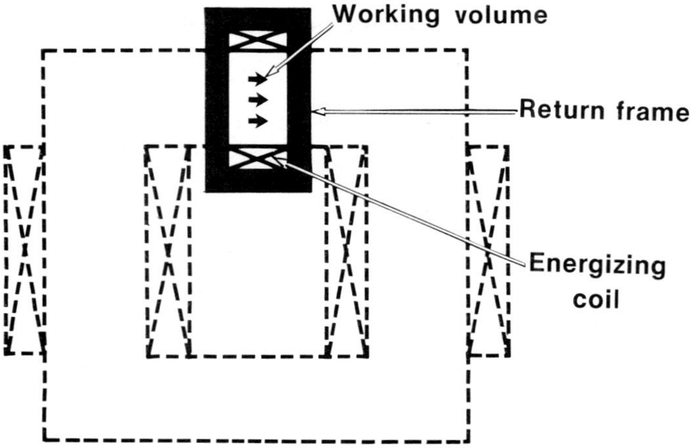

Fig. 1.39 Solenoid magnet (solid) compared to electromagnet with iron yoke (dotted) (after Marston [M4]).

Approximately 90 per cent of $\mathbf{B}_i$ comes from the iron in the proximity of the working volume. Therefore, approximately the same portion of iron of the yoke could have been eliminated without reducing $\mathbf{B}_i$. Furthermore, the contribution from the coils in the classical configuration (Fig. 1.38) in the direction of the field in the working space is very small. The coils are used just to magnetize the iron, and to generate high $\mathbf{B}_i$ [M4]. If the iron-clad solenoid is used (Fig. 1.39) the contribution of the coil to the flux density $\mathbf{B}_c$ is much higher than in an electromagnet, while $\mathbf{B}_i$ remains the same. Comparison of the magnetization curves for these two magnetic circuits is shown in Fig. 1.40. It can be seen that $\mathbf{B}_i$ (electromagnet) $\cong$ $\mathbf{B}_i$ (iron-clad solenoid) and $\mathbf{B}_c$ (electromagnet) << $\mathbf{B}_c$ (iron-clad solenoid). It is thus clear that the iron-clad solenoid uses much less material and exhibits better parameters.

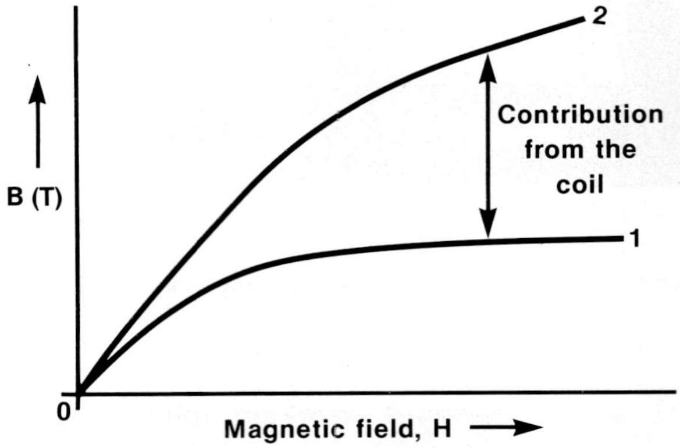

Fig. 1.40 Contribution from iron cladding (1) and from the coil to the total magnetic induction (2) in the working volume as a function of the magnetic field strength

The idealized circuit of the iron-clad solenoid is shown in Fig. 1.41. It follows from Ampère's law (eq. 1.56) that

$$BL = \mu_0 NI \tag{1.67}$$

under the assumption that the magnetic reluctance of the iron frame can be neglected. The actual number of ampère-turns **NI** that are necessary to generate **B** can only be calculated when the leakage of the flux is included, i.e.

$$BL = \mu_0 NI(1 + q) \tag{1.68}$$

It is accepted that the leakage factor **q** $\cong$ 0.2 [M4,T1].

It becomes obvious from eq. (1.68) that the number of ampère-turns required to generate a given magnetic induction is determined solely by the length of the field (**L**), and is independent of the radius of the working volume. Solenoids shown in Fig. 1.42 have a constant value of **L**, while their radii **r** differ and the number of

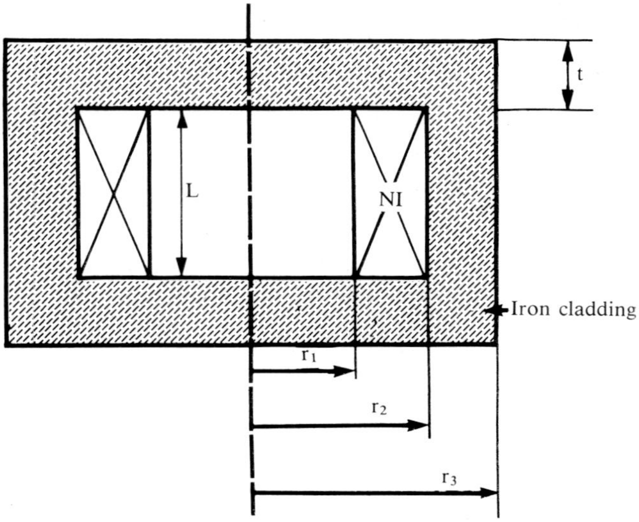

Fig. 1.41 Schematic diagram of iron-clad solenoid: **L** = length of the coil, $\mathbf{r}_1$ = inner radius of the coil, $\mathbf{r}_2$ = outer radius of the coil, $\mathbf{r}_3$ = outer radius of the return frame, and **t** = thickness of the iron cladding.

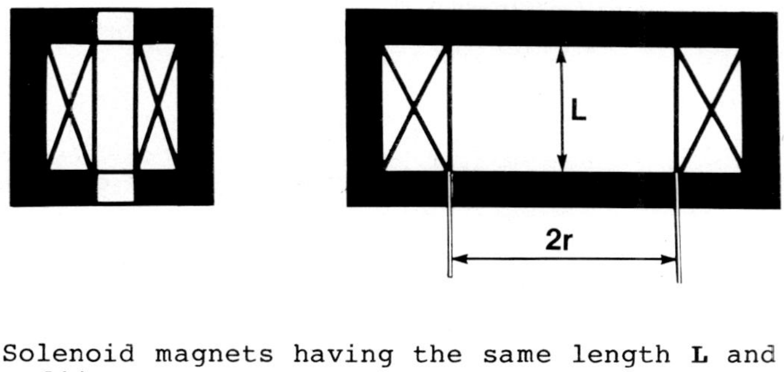

Fig. 1.42 Solenoid magnets having the same length **L** and different radii **r**.

ampère-turns needed to create a required magnetic induction is the same; only the length of a conductor, and therefore the electric resistance of the windings are different.

The Ampère's law can also be used to calculate the magnetic induction generated directly by the coil itself and we obtain [M5]

$$B_c = \frac{NI}{2\beta(\alpha-1)r_1} F(\alpha,\beta) \times 10^{-4} \tag{1.69}$$

where $\beta = L/2r_1$, $\alpha = r_2/r_1$ and **F** is the so-called "field factor" defined as

$$F(\alpha,\beta) = \frac{4\pi\beta}{10}\left(\operatorname{arcsinh}\frac{\alpha}{\beta} - \operatorname{arcsinh}\frac{1}{\beta}\right) \tag{1.70}$$

In eq. (1.69) $\mathbf{B_c}$ is in Tesla and $\mathbf{r_1}$ in cm.

The elemental current per unit cross-section is

$$j\lambda = \frac{NI}{L(r_2 - r_1)} \tag{1.71}$$

which defines the overall current density **j**λ ; **j** is the current density in the conductor and λ the filling factor defined as

$$\lambda = \frac{\text{active section of the winding}}{\text{total section of the winding}}$$

If we assume that the current density **j** and the resistivity ρ_R are constant over the volume of the winding, then the power is given by

$$W = 2\pi\beta\rho_R\lambda(\alpha^2 - 1)j^2 r_1^3 \tag{1.72}$$

The power can be expressed in terms of number of ampère-turns as

$$W = \frac{16\pi\rho_R}{\lambda L}\cdot\frac{r_2 + r_1}{r_2 - r_1}\cdot(NI)^2 \tag{1.73}$$

or as a function of magnetic induction

$$W = \frac{\pi\rho_R B^2 L(1+q)^2}{\mu_o^2\lambda}\cdot\frac{r_2 + r_1}{r_2 - r_1} \tag{1.74}$$

The power can be delivered at high current, few turns and high voltage, or at low current, many turns and high voltage.

The number of ampère-turns, that are necessary for a desired field can be found from eq. (1.68) or (1.69) and the necessary voltage can then be found from the relation **V** = IR, where the resistance of the winding **R** is computed from **R** = $\rho_R l_T/s$, where l_T is the total length of a conductor and **s** its active cross-sectional area. From Fig. 1.41 we have

$$l_T = N\pi d_{mean}$$

where $\mathbf{d}_{mean}$ is the mean diameter of a turn and is given as

$$d_{mean} = r_1 + r_2$$

The active cross-sectional area of the conductor **s** = λ**S/N**, where **S** is the overall cross-sectional area of the winding

$$S = L(r_2 - r_1)$$

Consequently, the resistance of the winding can be written as

$$R = \frac{\pi\rho_R N^2(r_1 + r_2)}{\lambda L(r_2 - r_1)} \tag{1.75}$$

Thickness of the iron cladding

The basic requirement for the iron cladding is that it must not be saturated. As follows from Fig. 1.41 the magnetic flux in the air gap is equal to $\phi = J_s S_c$ where $\mathbf{J}_s$ is the saturation polarization of iron and $\mathbf{S}_c$ the cross-sectional area of the cladding. In order to avoid oversaturation the following inequality must be satisfied

$$BS_g \leqq J_s S_c \tag{1.76}$$

from which the thickness **t** of the iron cladding can be found. The choice of the value of $\mathbf{J}_s$ is very important for a correct size of the iron frame, and in most cases the values of $\mathbf{J}_s$ are not available for the commercial steels. Although the maximum safe value is about 1.5 T, lower values would be a more appropriate choice, a factor which necessarily leads to the overdesign of the cladding and an increase in the mass of the magnet.

Low carbon steel that is frequently used for the magnet cladding has chemical composition shown in Table 1.25, while Table 1.26 gives the equivalents of low carbon steel.

TABLE 1.25
Composition of low carbon steel used for magnet cladding

	Composition (%)			
AISI	C	Mn	P	S
1010	0.08 - 0.13	0.30 - 0.60	0.04	0.05

TABLE 1.26
Comparison of international standards

AISI/SAE	1010
German St. No.	1.1121
DIN	Ck10
AFNOR (France)	XC10
B.S. (UK)	045 M10
JIS (Japan)	S10 C
GOST (USSR)	08; 10
CSN (ČSSR)	12010

It can be shown that the magnets for which $\mathbf{L} > 2r_1$ do not require the iron pole above and below the working volume. Where $\mathbf{L} < 2r_1$, the presence of iron cladding in the vicinity of the working space is necessary. Also, as follows from eq. (1.67) the number of turns needed to generate a given magnetic induction is determined by the length of the flux path through the air. It was shown by Cioffi [C2] that

$$B \sim \frac{NI}{l_g} \qquad (1.77)$$

where $\mathbf{l}_g$ is the so-called effective length of field path. A plot of l_g vs. solenoid axial length (Fig. 1.43) shows that l_g initially increases faster than axial length and for long solenoids approaches the asymptotic value $\mathbf{L}$. The external field path, given by the difference in ordinates between curves $\mathbf{l}_g$ and $\mathbf{L}$

is negligibly small for long solenoids but for short solenoids is greater than the axial length. Consequently, the most significant decrease in l_g, or gain in **B** can be obtained in solenoids of relatively small axial length by replacing the external air path with a magnetic circuit of low reluctance. Therefore, the contribution of the iron is the largest if the coil has a considerable stray field, so that the appreciable fraction of $\oint \mathbf{H}d\mathbf{l}$ is in a path outside the coil. In the case of a long thin solenoid in which the stray field is small, the influence of an external iron cladding is practically negligible.

In order to increase the throughput of such a solenoid magnet the diameter must be increased. The length of the air gap, an important determinant of magnet cost and energy consumption remains constant. Consequently, the slight increase in the cost of material cost and energy input is compensated by the increased diameter of a solenoid.

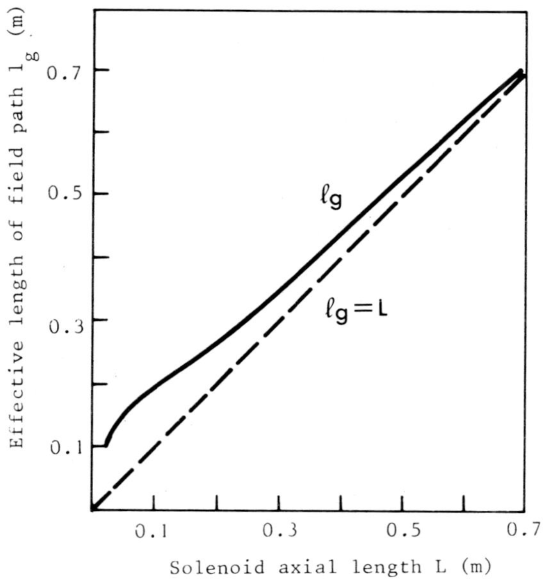

Fig. 1.43 Effective length l_g of the field path of the solenoid, as a function of the axial length **L**. (After Cioffi [C2]).

The cooling of the magnet

One of the major problems in the design of magnets which generate steady high magnetic fields is that of the removal of heat from the winding. High-power solenoids are cooled by passing fluid through many small passages in the windings. To achieve the best possible heat transfer, the coolant is brought into direct contact with the conductors where heat is generated. In modern high-intensity, high-gradient separators the magnets are wound with hollow copper conductors which greatly facilitate cooling. A photograph of such a hollow conductor is shown in Fig. 1.44. A detailed discussion of the cooling of magnets can be found in [D5, M5, P4].

A solenoid coil enclosed in a return steel frame that eliminates a heavy iron yoke was proposed by Bitter [B3] in 1936 and this concept has since been widely used, particularly in high energy physics and in other branches of experimental physics. The patent claiming the application of iron-clad solenoids to magnetic separation was filed by Frantz [F2]. Workers from Magnetic Engineering Associates (now Sala Magnetics), and from J.M. Huber Corporation, using Frantz's concept, filed a patent application in 1971 claiming a ferromagnetic bound solenoid surrounding a canister that contained a matrix.

At the present time several companies in various parts of the world are marketing cyclic and continuous magnetic separators that employ iron-clad solenoids, a development that produced a substantial increase of the available magnetic field in a much greater working space. For moderate magnetic fields at moderate working volumes, electromagnets with iron yoke offer the best solution thanks to their simplicity and low energy consumption; for large-scale applications of magnetic separation to finely ground weakly magnetic minerals, iron-bound, water-cooled solenoids are the best.

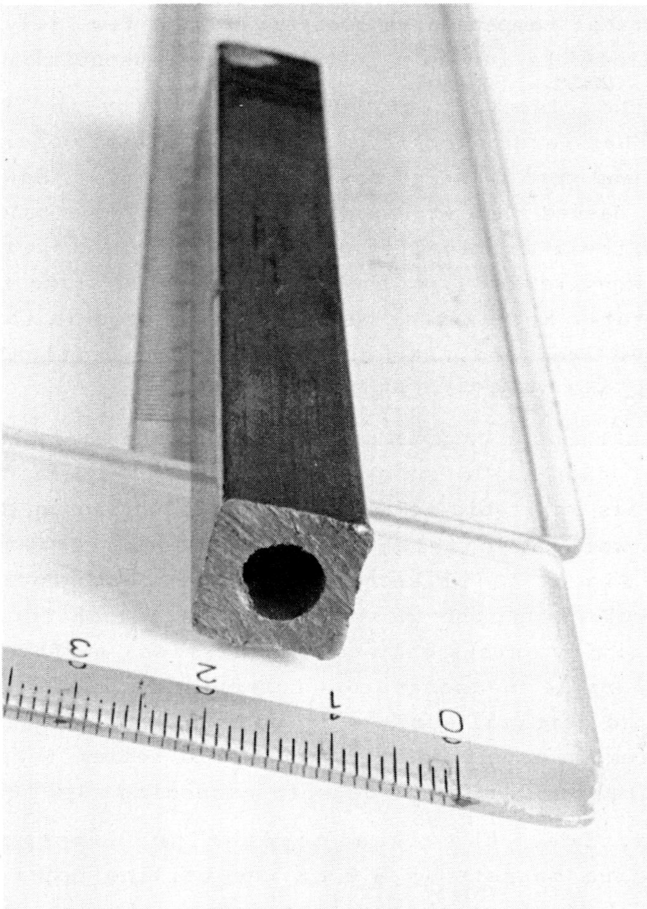

Fig. 1.44 Hollow copper conductor used to wind powerful solenoid magnets.

1.4.4 Superconducting magnets

Superconductivity is a remarkable phenomenon whereby certain metals, when cooled to very low temperatures, become perfect conductors of electricity. Unlike the gradual changes in electrical resistance shown by all metals at more familiar temperatures, the superconducting state appears quite abruptly at a critical temperature T_C, which is characteristic of the metal in question.

Below this temperature the resistance is not just very small; according to the results from some very sensitive experiments, it is zero. Critical temperatures are very low, a few Kelvin (K), and the use of liquid helium as a refrigerant is essential.

After the discovery of superconductivity in 1911 it was recognized that large electromagnets, capable of generating very high fields and consuming no power, could be built. However hopes were quickly dashed when it was discovered that superconductors, in addition to their critical temperature, have a critical field, above which they revert from the superconducting state to a normal resistive state. Metals that were investigated in those times, i.e. mercury, lead and tin, showed that the critical magnetic induction $\mathbf{B}_C$ was small (~0.05 **T**).

The breakthrough came in the late 1950s and early 1960s when a new class of high-field superconducting alloys was discovered. These materials were able to remain superconducting up to very high fields and were also able to carry extremely high current densities. Fig. 1.45 illustrates the properties of niobium titanium, currently the most popular superconducting alloy. Superconductivity exists only below a critical surface, defining maximum tolerable temperature, current density and magnetic induction. The critical surface shown in Fig. 1.45 indicates that Nb-Ti will remain superconducting at liquid helium temperature (4 K) in a 5 T field at a current density exceeding 3×10^9 A m^{-2}.

In practice, the usual operating temperature for superconducting magnets is 4.2 K, the boiling point of liquid helium. Fig. 1.46 shows the critical curves for the two common technological superconductors at 4.2 K; some properties of assorted technical superconductors are summarized in Table 1.27.

Although more than 2000 superconductors are known to exist, very few have been produced on an industrial scale. This is because the industrial materials have to carry large currents in the presence of high magnetic fields; they have to be prepared in long lengths in various geometries and they should be flexible enough to be wound into coils. The simplest form of a superconductor is one in which many filaments with diameters of the order of 10 μm are embedded in a high conductivity copper matrix. Only three superconducting stabilized materials are available commercially [S11]: Nb-Ti ductile alloy and the Nb_3Sn and V_3Ga conductors. The earlier alloys Nb-Zr and Nb-Ti-Zr, had become

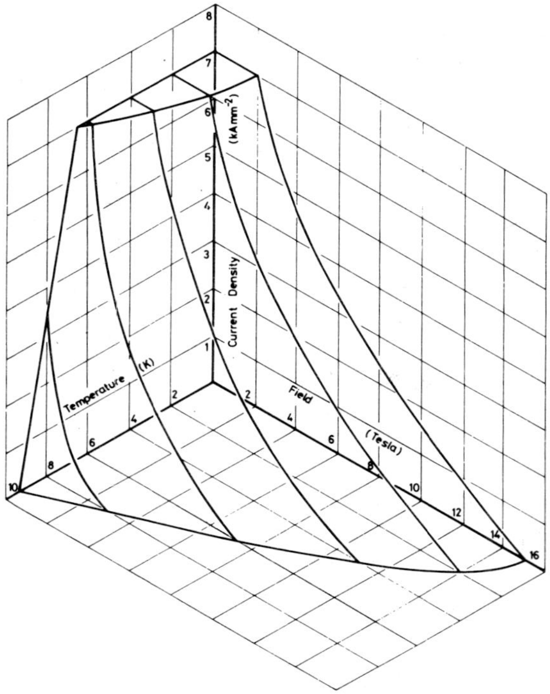

Fig. 1.45 Critical-current surface for a commercial superconducting alloy of niobium-titanium (after Wilson [W1]).

obsolete with the arrival of the superior material Nb-Ti. Cu(Nb-Ti) composite wires have been widely used for constructing large and small superconducting magnets. However, their inherent limitation to 8.8 T led to the development of still higher field and higher T_c materials, viz. Nb_3Sn and V_3Ga. Nb_3Sn is a good material for use in the production of magnetic fields up to 15 T; beyond 15 T V_3Ga should be more useful as in this field range it has a higher critical current density than Nb_3Sn. Because Nb_3Sn and V_3Ga are brittle compounds, they could not be drawn with copper and new techniques were developed [S11]. No new high T_c materials were found in the last decade and the development of advanced superconductors is now concentrated on the achievement of high values of critical current densities at high magnetic fields on materials already known. It appears that superconducting magnets with fields above 25 T can be expected in the future [F17].

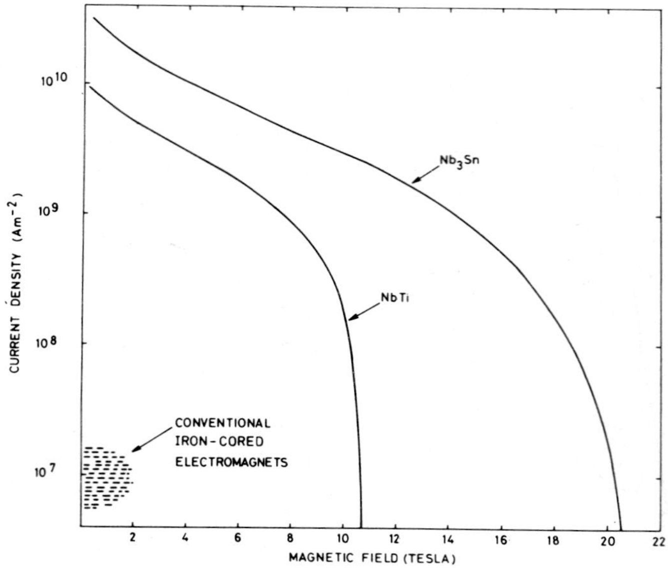

Fig. 1.46 Critical curves for the two common technological superconductors, niobium titanium and niobium tin, at a constant temperature of 4.2 K. Superconductivity prevails below the curves and normal resistivity above. Shaded area at bottom left illustrates the usual operating range for conventional electromagnets (after Wilson [W1]).

The current densities shown in Fig. 1.46 are quite extraordinary by everyday standards. Normal house wiring, for example, is rated at about 10^7 A m^{-3} and fuses blow at about 10^8 A m^{-3}. Air-cooled conventional electromagnets run at about 2 x 10^6 A m^{-3} and water-cooled copper windings at about 10^7 A m^{-3}.

The only power required by a superconducting magnet is the refrigeration power that is needed to cool it to low temperatures and a small current supply to initiate the flow of current round the superconducting circuit [W1].

Following the discovery of high-field superconductors it became clear, that superconducting magnets have the potential for much better performance than their conventional counterparts. They should be able to produce large volumes of high field for a small consumption of electrical energy. Furthermore they should be able to operate at high overall current densities, allowing a given number of ampère-turns, i.e. a given field, to be provided by a

TABLE 1.27

Some properties of technical superconductors [B4, H24, F17, L25]

Material	Critical temperature T_c (K)	Critical induction at 4.2 K (T)
Pb	7.3	-
Nb	9.24 + 0.1	0.1 - 0.5
$Mo_{0.5}$-$Re_{0.5}$	12.6	1.5
$Pb_{0.5}$-$Bi_{0.5}$	8.4	1.5
Nb_3Sn	18.0	19 - 25
V_3Ga	15.0	22.0
$Nb_{0.75}Zr_{0.25}$	10.8	>8.0
$Nb_{0.6}Ti_{0.4}$	10.6	11.8
$Nb_3Al_{0.8}Ge_{0.4}$	20.7 + 0.2	41.0
Nb_3Al	16 - 20	20 - 35
NbN	16 - 18	20 - 35
V_3Si	16.9	14
Nb_3Ge	22.5	37.0
$PbMo_6S_8$	14.5	40 - 55

rather small volume of winding. Apart from reducing the capital cost by making the magnet very compact and light, this method should also make it possible to produce high gradients of magnetic field by a judicious arrangement of the windings.

With the use of a superconducting magnet in mineral separation it should be possible to generate very high magnetic fields at low input power, in large volumes. The need for these very high magnetic fields has not yet been demonstrated; although simple theoretical models indicate that high slurry velocities, matched to high magnetic fields could be used, thereby increasing the throughput. Furthermore, by using the multipolar superconducting magnets, it is possible to generate sufficiently high gradients of magnetic field that would allow us to eliminate a matrix that still remains a weak spot in current high-intensity, high-gradient magnetic separators.

Although it has been envisaged [K2] that magnetic separation will be the first industrial application of superconductivity, superconducting magnets have not yet made a considerable impact in mineral processing. In order to make progress in this direction, it is necessary that the superconducting system performance be realistically evaluated in the laboratory and on the pilot scale. Also, for most large scale applications of magnetic separation the magnetic device is a small part of a complex (and highly energy-consuming) overall flowsheet. Although the cost incentives for reducing energy consumption are changing in a direction that favours superconductivity and although the cost advantages of superconductivity are real, these advantages and incentives are often not as substantial as the promoters of superconductivity would like to see [M6].

In order to introduce such a new technology as superconductivity into the conservative mining industry two fundamental requirements must be met. The first requirement is that the use of a superconducting magnet must be technologically justified. The vast majority of magnetic separation processes can be carried out at magnetic inductions not exceeding 2 T and therefore the need for very high fields (over 2 T) has not yet been proved. The trade-off between the field and flowrate, which means that higher slurry velocity is possible at high magnetic fields, can perhaps be used in exceptional applications, but it probably does not apply to most mineral processes and is not justified from the first principles. On the other hand, large separation volumes, which are available with superconducting magnets at low energy consumption, are an attractive prospect. Although open-gradient superconducting separators solve the problem of a matrix blockage, the versatility of such separators is limited.

The second requirement is that the superconducting separator be economically viable. Superconducting coils are generally regarded as expensive. During small-scale operation the cost of refrigeration becomes a significant factor and its reliability must be taken into account. However, for large systems the superconducting system could be less expensive in initial costs and significantly less expensive to operate that a conventional system [S12]. The initial costs of small systems, both conventional and superconducting, are compared in Fig. 1.47 for various field levels and are shown as a function of magnetic volume.

The slow implementation of superconducting magnets in the minerals industry is due to several reasons:

- Superconductivity is an expensive technology that pays only above a certain power level or size of the device
- The need for magnetic induction above 2 T has not been demonstrated
- Superconductivity is a new technology and has to compete with well proven and well-established technology.

To benefit from low running costs, superconducting devices have to be extremely reliable. Continuous operation under plant conditions without any disturbance is essential to compete with established conventional magnetic separators. High capital cost prevents simple changes of the system which then limits the flexibility of the machine. It is obvious that superconductivity is best applied where it offers a unique solution, i.e. where no other competitive solution exists [H5]. At this stage it appears that the areas where superconducting magnetic separators are economically justifiable will remain few [M7].

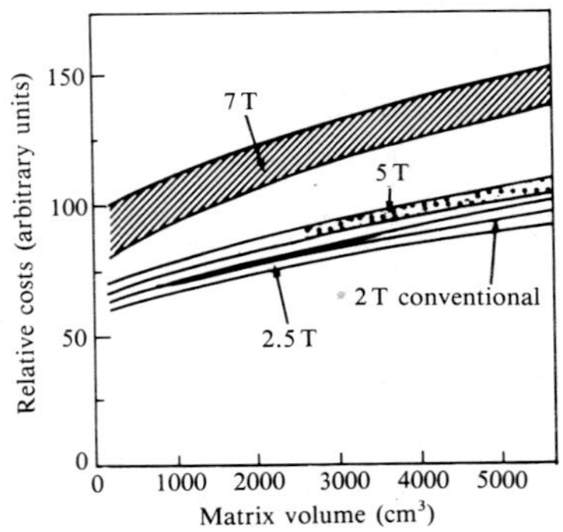

Fig. 1.47 Economic comparison of conventional and superconducting magnets (after Stekly [S12]).

I.4.5 Generation of the field gradient

A large variety of designs have evolved over the years to produce efficient magnetic field gradients. These designs can be divided into three basic categories:

1. Open-gradient separators
2. Closed-gradient separators
3. Matrix separators.

In open-gradient separators the field gradient is generated either by a set of permanent magnets or electromagnets of alternating polarity or by multipolar windings. Such a multipolar system is usually enclosed by a rotating drum on whose surface magnetic force is generated.

Separation in such open-gradient devices is achieved by a pick-up mechanism, in which the magnetic particles are captured onto a surface of the drum, or by a selective deflection of particles according to their magnetic susceptibility and size.

In closed-gradient separators a high field gradient is achieved either by a suitable shaping of the pole pieces of opposing magnetic poles or by closing an open pole system by an appropriate drum or roller. Magnetic induction is induced on the surface of this element and gradient is created by a suitable shaping of the drum. A typical example of this system is induced magnetic roll.

In matrix separators, the gradient of the magnetic field is generated by means of induced magnetic moments in the ferromagnetic bodies (matrix), placed in the applied external magnetic field. Magnetizable particles can be attached to the surface of the matrix elements and thus separated from the rest of the pulp. Ferromagnetic bodies of various shapes have been used in high intensity magnetic separators (balls, mesh, expanded metal, rods, steel wool, grooved plates etc.) and a proper choice of the matrix is one of the crucial prerequisites for a successful separation. Since the field gradient is inversely proportional to diameter **a** of the matrix elements, $\text{grad}B \sim J/a$, the magnetic force acting on the magnetizable particles is maximized by using as fine a matrix as possible. On the other hand, the dimensions of the matrix elements must be matched with the mineral particle size to avoid straining and to ensure the selectivity of separation. Discussion of the matrix selection can be found in Section IV.5. Typical values of the gradient of magnetic induction for various magnetic separators are shown in Table 1.28.

In order to produce large surface areas of a very high magnetic field gradient, one must apply a strong uniform magnetic field. This is because the ferromagnetic matrix is often of relatively low density and therefore does not conduct much magnetic flux. The magnetic field needed to magnetize such a fine matrix must be produced by a magnet capable of sustaining a strong magnetic field in what is essentially empty space [O4]. This situation is depicted in Fig. 1.48. It can be seen that a background magnetic induction of at least 0.7 T is needed to saturate the steel wool matrix, compared with an induction of about 0.2 T that is necessary to saturate the bulk material [K3]. The large background induction needed to saturate above that expected on the basis of bulk sample figures is due to demagnetization effects in the steel wool. Furthermore, average field gradients and peak field gradient shown in Fig. 1.48 increase dramatically above the 0.7 T level up to about 1.2 T, where substantial saturation occurs of the steel wool fibres.

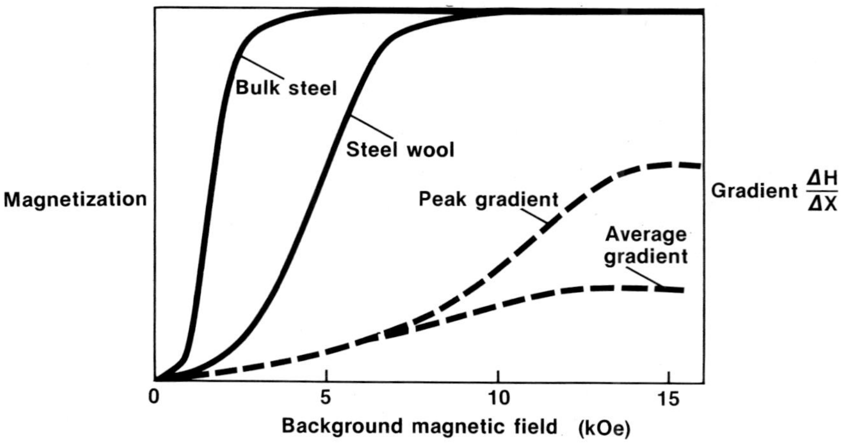

Fig. 1.48 Comparison of the magnetization curves and of the field gradient for steel wool and bulk steel (after Kolm [K3]).

The behaviour of the field gradient at high background magnetic inductions has been the subject of numerous discussions. It is generally accepted that field gradient remains constant with increasing magnetic induction and this argument supports the notion that magnetic force increases with increasing background magnetic induction. However, there are some indications that magnetic field gradient actually decreases with increasing background magnetic field. This fact raises doubts concerning the validity of increasing the efficiency of magnetic separation by increasing the background magnetic induction. The effect of the applied magnetic field on the spatial distribution of the field gradient was treated by Kulda [K4].

TABLE 1.28

Average values of magnetic induction gradient for assorted matrices

Separator or matrix	Gradient of magnetic induction	
	$T\ m^{-1}$	$kG\ cm^{-1}$
Steel balls	2000	200
Grooved plates	2000	200
Wire mesh 1 mm	6000	600
Steel wool 100 μm	4×10^4	4×10^3
Permanent magnet	5	0.5
Split pair (OGMS)	40 - 100	4 - 10
Quadrupole magnet	30 - 160	3 - 16
Belt separator	1 - 10	0.1 - 1
Davis tube	40	4
Frantz isodynamic	20	2
Permroll	300	30

Kulda assumed that a steel sphere of permeability μ_1 and radius **a** is placed into the homogeneous magnetic field $\mathbf{H}_o$, the surrounding medium having a permeability μ_2 (Fig. 1.49). The background magnetic field generates the homogeneous field in the sphere given by

$$H_1 = \frac{3\mu_2}{\mu_1 + 2\mu_2} H_0 \qquad (1.78)$$

and the corresponding magnetization can be written as

$$M_1 = \kappa_1 H_1 = \frac{3\mu_2(\mu_1 - \mu_0)}{\mu_0(\mu_1 + 2\mu_2)} H_0 \qquad (1.79)$$

where κ_1 is the magnetic susceptibility of material of the sphere. Since the effect of the demagnetization factor **D** can be described as

$$H_1 = H_0 - DM_1 \qquad (1.80)$$

it follows from eqs. (1.78) and (1.79) that

$$D = \frac{\mu_0(\mu_1 - \mu_2)}{3\mu_2(\mu_1 - \mu_0)} \qquad (1.81)$$

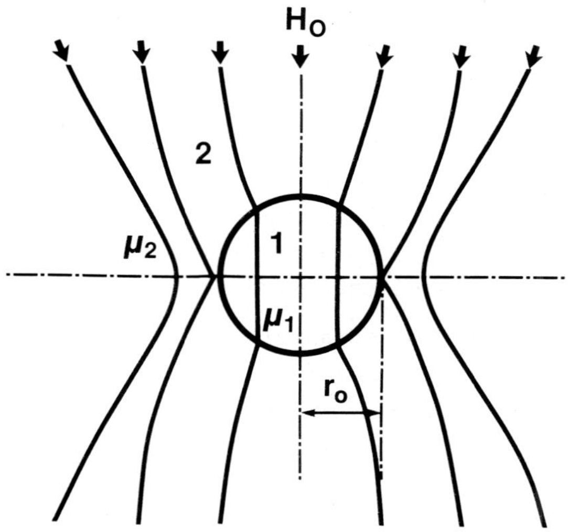

Fig. 1.49 Magnetic field pattern in the proximity of a steel sphere placed in a homogeneous background magnetic field.

If the ferromagnetic sphere is placed in the air ($\mu_2 = \mu_0$), then $\mu_1 >> \mu_2$ and the field strength in the sphere (eq. (1.78)) will drop to a very low value and a corresponding magnetic induction $\mathbf{B}_1 = \mu_1 \mathbf{H}_1$ will be three times greater than the induction of the backgroung homogeneous field. This induction, however, determines the magnitude of the permeability of the ferromagnetic sphere μ_1. This permeability exhibits nonlinear characteristics whereby μ_1 decreases with increasing $\mathbf{B}_1$. The inside of a sphere will be oversaturated in a strong background field $\mathbf{H}_o$, μ_1 will drop to a value comparable to the permeability of medium μ_2; and thus the magnetic induction and field strength inside the sphere will approach the value of the background magnetic field.

The radial component of a derivative of both components of the field strength (in spherical coordinates) can be written

$$\frac{\partial H_{r2}}{\partial r} = 6\frac{\mu_1 - \mu_2}{\mu_1 + 2\mu_2} \cdot \frac{a^3}{r^4} H_0 \sin \varphi \qquad (1.82)$$

$$\frac{\partial H_{\rho 2}}{\partial r} = -3\frac{\mu_1 - \mu_2}{\mu_1 + 2\mu_2} \cdot \frac{a^3}{r^4} H_0 \cos\varphi \tag{1.83}$$

It can be seen that the value of the gradient components at a point **r** is dependent on the ratio of permeabilities μ_1 and μ_2. These components increase with increasing value of the background field $\mathbf{H}_0$, but at very large field strengths the saturation of the sphere will become important and after reaching the maximum the magnitude of the gradient will start to decrease.

It is thus clear that a straightforward application of simple theoretical models, to increase the efficiency of magnetic separation by increasing either the magnetic force or the magnetic field gradient can lead to erroneous results. The choice of the magnetic field and the type of matrix must be separately determined for every application. Factors like the number of minerals, their particle size distribution, magnetic susceptibilities, and desired selectivity must be evaluated before the operating conditions can be chosen. Furthermore, such questions, as whether the magnetic fraction or nonmagnetic fraction is a valuable fraction, and whether a greater stress is placed on the recovery of a valuable mineral or on the grade of the valuable fraction, must be answered before a selection of the conditions of separation can be made.

I.5 MEASUREMENT OF MAGNETIC PROPERTIES

This section briefly describes experimental techniques used for the measurement of magnetic quantities. Extractive metallurgists using magnetic separation as a tool for mineral benefication will be interested primarily in magnetic induction and magnetic susceptibility. Although there is a wide range of measuring instruments available commercially, very few are accesible to a metallurgist. However, in view of the fact that the required precision of measurement of the magnetic induction in magnetic separation is not very high, several affordable devices can be acquired. Commercial instruments for the measurement of magnetic susceptibility of a wide spectrum of minerals, ranging from ferromagnetic to diamagnetic are much less numerous. Moreover they are, as a rule, costly, and , in most cases, too complicated and tedious to be used for routine measurements. In fundamental research in magnetic separation knowledge of magnetization is often desirable, as well as basic parameters of the hysteresis loop.

I.5.1 Measurement of magnetic field

I.5.1.1 Ballistic galvanometer

The most direct laboratory method for measuring magnetic induction is to use a search coil in series with the secondary coil of a standard mutual inductance and a ballistic galvanometer. According to Faraday's induction law, a change in magnetic flux in a coil gives rise to a voltage across the coil terminals. If this voltage is a momentary one induced in a search coil that is quickly removed from a static magnetic field, or in a coil rotated 90° from the initial orientation of the coil axis parallel to the field, then the voltage pulse can be measured with a ballistic galvanometer. The search coil should consist of a number of turns of fine insulated wire wound upon a non-conducting frame, whose area of cross-section is accurately known. The search coil is placed with its plane perpendicular to the magnetic field: on removing it suddenly from the field, the galvanometer is deflected.

The ballistic galvanometer differs from an ordinary galvanometer in that the moving element has a large amount of inertia. Damping of the moving system is controlled by resistance in the galvanometer circuit and this is an important consideration when the galvanometer is in use. For critical damping, the galvanometer coil will return to zero without oscillation in a minimum time. If the galvanometer is underdamped, it will exhibit, in returning to zero, an oscillator conditions, while the overdamped galvanometer will return slowly to zero without oscillation. Generally, critical damping or overdamping of the galvanometer is used in testing. [B35,Z1,P5].

I.5.1.2 The fluxmeter

In practice the ballistic galvanometer is replaced by a fluxmeter. This is virtually a moving-coil ballistic galvanometer in which torsional control is extremely small and heavy damping is produced almost entirely by electromagnetic couples which are caused by currents induced in the coil by its motion.

A search coil is connected to the fluxmeter and the change in flux that produces the induced electromagnetic force occurs within the search coil. This change may be caused by a change in flux that links a stationary coil, or by a motion of the search coil in the field, so that there is a change in the flux linkage of the coil.

The flux-density sensivity of the fluxmeter depends upon the value of the flux density within the air gap of the galvanometer. Sensitivity will be greater for weaker fields in the gap [Z1, B35, B8, S13] and depends on the effective coil surface and the quality of the integrator. A photograph of a fluxmeter is shown in Fig. 1.50.

I.5.1.3 The Hall effect probe

For steady readings in a steady field, the Hall probe is very useful. The small size of the probe sensing area (10 mm^2 and less), and its static physical characteristics - no motion of the probe is required, and there are no moving parts, enable it to be used in areas that previously precluded measurement of magnetic fields.

The basic premise of the Hall probe is that when a current **I** is flowing along one axis of a conductor or semiconductor and when the magnetic field **H** is applied perpendicular to the plane of the probe (Fig. 1.51), then an electromagnetic force **E** is generated perpendicular to both current **I** and magnetic field **H**. By a suitable choice of the parameters of the external circuit the emf **E** can be made proportional to **H**, over at least a limited range of **H** values:

$$E = K\frac{HI}{d}$$

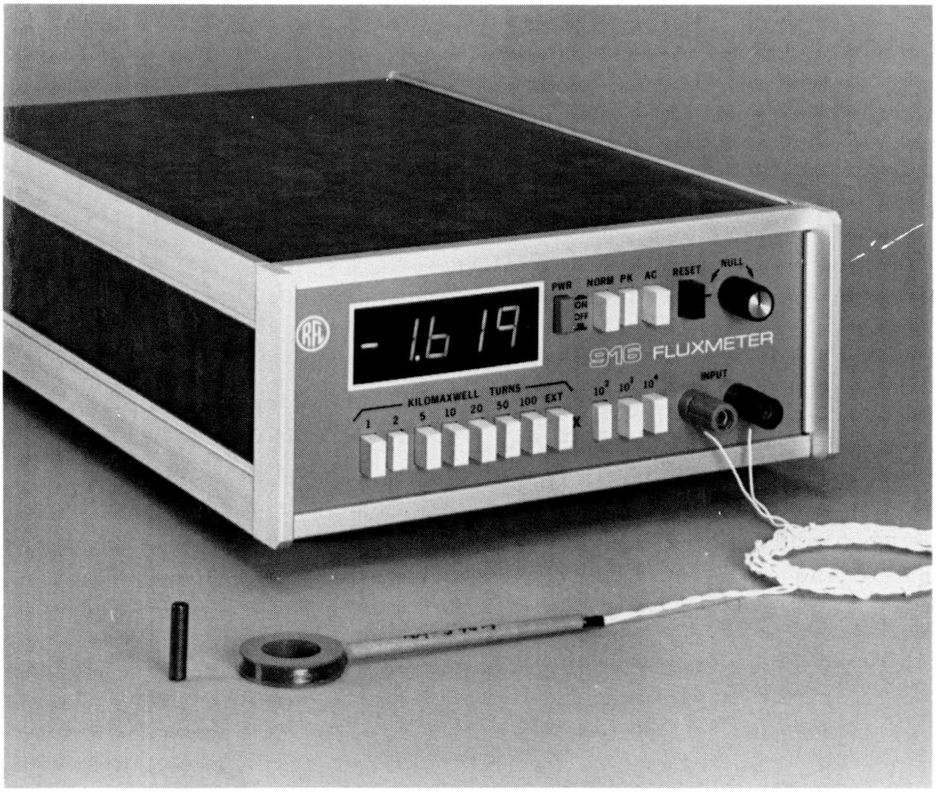

Fig. 1.50 Fluxmeter (courtesy of Dowty RFL Ind. Inc.)

where **K** is the Hall constant and **d** is the thickness of the Hall element.

Germanium, indium antimonide and indium arsenide are materials that have been used in Hall probes. Indium antimonide is very sensitive, but indium arsenide is now often preferred because it has the smallest temperature coefficient.

Several factors set limits on the obtainable accuracy, the most serious being the temperature coefficients of the Hall voltage. In order to overcome this problem, temperature stabilization can be employed, or it may be taken into account in the probe calibration by monitoring the temperature during

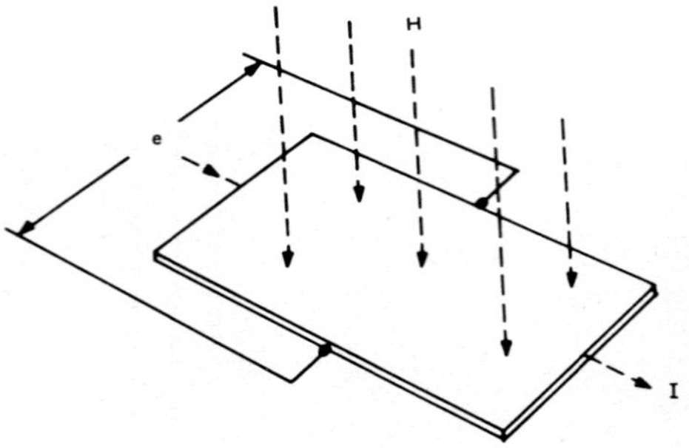

Fig. 1.51 Magnetic flux, drive current and Hall effect voltage relationship.

measurements. Since temperature coefficient depends on the level of magnetic field, relatively complex calibration is needed. Another complication can be that of the planar Hall effect, which makes problematic the measurement of a weak field component normal to the plane of the Hall plate if a strong field component is present and parallel to this plane. The measurement of the Hall voltage sets a limit of about 0.1 mT (1 G) on the sensitivity [H6].

The Hall probe can be used to measure dc fields, ac fields and pulsed fields. In practice the Hall element, usually crystalline in form, is mounted on a substrate of inert material such as aluminium oxide. The element is generally only a few hundredths of a milimetre thick and thus needs a mechanical support. The element and substrate, plus some form of protective encapsulation usually results in a probe having a thickness of 0.5 to 1 mm. The Hall element is extremely fragile and requires as much protection as is commensurate with usability. A cracked element is not repairable [M8].

Hall probes are available in various shapes, and their use depends on the direction of the magnetic field to be measured. The transverse (flat) probe (Fig. 1.52) is used to measure a field in electromagnets with pole pieces, while the axial probes (Fig. 1.53) are used to measure the fields parallel to the probe handle, such as solenoidal fields. The overall view of a typical gaussmeter is shown in Fig. 1.54.

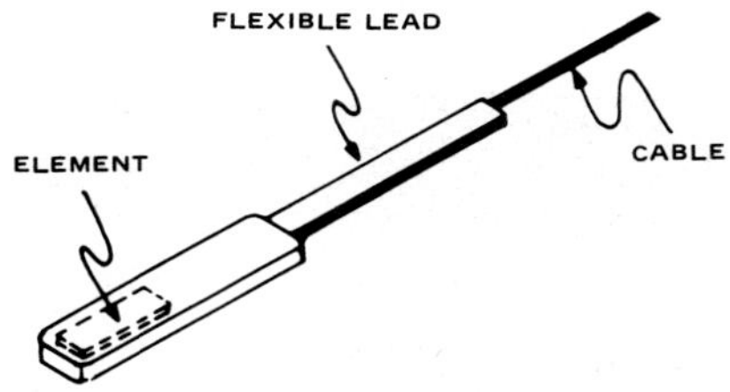

Fig. 1.52 Typical flat (transverse) probe (courtesy of Dowty RFL Ind. Inc.)

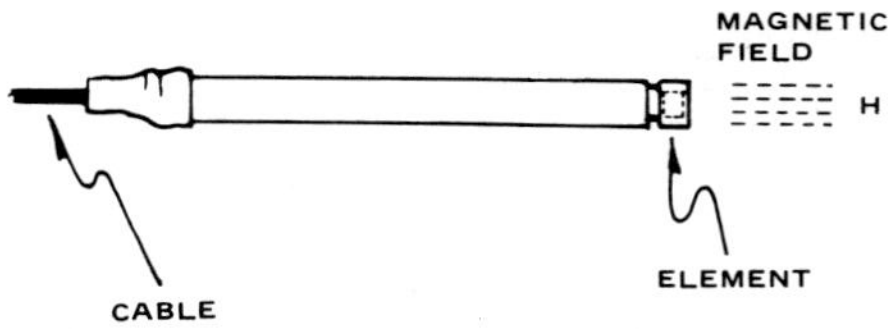

Fig. 1.53 Axial probe configuration (courtesy of Dowty RFL Ind., Inc.)

A comparison of measuring methods by fluxmeter and Hall probe is shown in Table 1.29.

TABLE 1.29

Comparison of fluxmeter and Hall probe

Property	Fluxmeter	Hall probe
Simplicity	Moderate	High
Speed	Moderate	High
Precision	High	Limited
Flux measurement	Yes	No
Measurement in air gap	Yes	Yes
Measurement in material	Yes	No
Temperature influence on the measurement	No	Yes
Sensitivity	High	Limited
Linearity	Good	Moderate
Search coil	Yes	No
Relative motion between field and coil	Yes	No

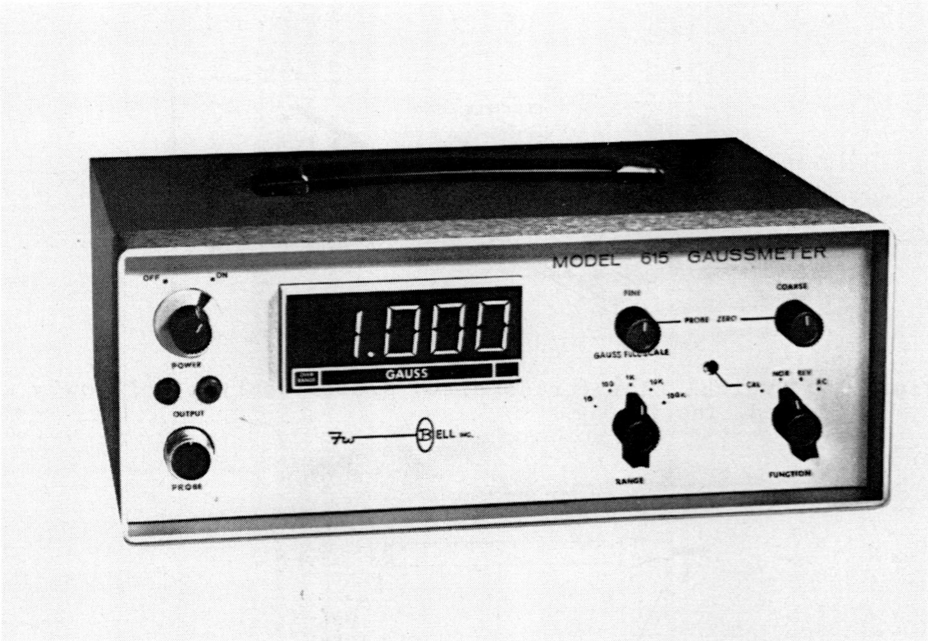

Fig. 1.54 Hall probe gaussmeter. Measurement range: from 0 to 10 T (courtesy of F.W. Bell, Inc.)

I.5.2 Measurement of magnetic susceptibility

All techniques for the determination of magnetic susceptibility or magnetization fall into two principal categories. The first is based on the measurement of force that acts on a sample when it is placed in a non-uniform magnetic field. The other method makes use of the emf induced in a coil when the magnetic field changes, or when there is a relative motion of the sample and coil. Two conditions can exist, viz., a steady state emf when alternating fields or vibrating samples are used, and a ballistic effect produced by a single motion of the sample or a change in field [L2].

None of these techniques is superior to the other and all possess their distinctive merits and limitations.

A comprehensive review of the magnetic measurements can be found in various monographs [B35, Z1, L2, B8] and in review articles [D6, F4, H7, F5].

I.5.2.1 Force method

In the simplest case, the magnetic force on a specimen of volume **V** in an nonhomogeneous field can be written, as follows from eq. (1.8), as

$$F_m = V\frac{\kappa_1 - \kappa_2}{\mu_0}B\frac{dB}{dz} \quad (1.84)$$

where the field gradient is confined to the Z-direction, either by the shape of the pole piece or by the position of the gradient coils. κ_1 is the susceptibility of the sample and κ_2 that of the surrounding medium; κ_2 can usually be neglected, unless an oxygen atmosphere surrounds the sample. For a 1 cm^3 sample, in a magnetic induction of 1 T and with a gradient of 0.01 T over its length, the force will range from approximately 10^{-1} N to 10^{-6} N for the values of magnetic susceptibility of interest in magnetic separation. Gradients of 10 T m^{-1} can be obtained easily. Measurements of the body force on small samples in a nonhomogeneous field, as given in eq. (1.84) are often referred to as the Faraday method.

When the sample has a uniform cross-section and extends from a region of strong magnetic induction $\mathbf{B}_1$ to a region of small induction $\mathbf{B}_2$ above the pole pieces, the force becomes

$$F_m = \frac{\kappa_1 - \kappa_2}{2\mu_0}S(B_1^2 - B_2^2) \quad (1.85)$$

where **S** is the area of cross-section of the rod. The measurement of samples of uniform cross-section such that the force is given by eq. (1.85) is referred to as the Gouy method.

The Gouy method gives absolute values of susceptibility since **F**, **S** and **B** can all be measured independently, while it is difficult to determine **dB/dZ** in the Faraday method. The larger force on the Guoy sample often facilitates measurements of greater accuracy. It is necessary, however to use 5 to 10 g of material in some shape of uniform cross-section for the Gouy technique, while 0.1 to 1 g often suffices for the Faraday method. Finally, the field gradient of several tenths of tesla over the Gouy sample length complicates the interpretation of field dependence studies [L2].

A schematic diagram of magnetic susceptibility balance is shown in Fig. 1.55.

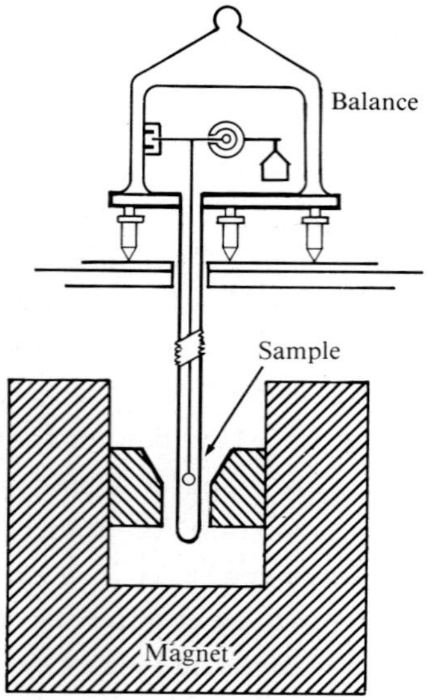

Fig. 1.55 Magnetic susceptibility instrument.

Certain precautions must be taken when using the above methods. The magnetic force on the sample holder must be measured independently and subtracted from the total force measurement. In the Gouy method it is a common practice to make a dummy sample holder long enough to extend into the region below the pole pieces to give a compensating force.

I.5.2.2 Alternating current method

The alternating current method is a convenient one and with a commercially available inductivity bridge the duration of measurement can be reduced to a minimum. Samples of any shape and form can be measured and the mass of the specimens can range from a few grams to several tens of grams.

A schematic diagram of the inductivity bridge is shown in Fig.

1.56. The primary of the differential transformer **Tr** is supplied with an ac voltage from the generator **G.** Two voltages of the same amplitude and the opposite phases are generated on the secondary side. Two measuring coils are connected to the secondary winding of the transformer: the pick-up coil **L**, into which a specimen is inserted during the measurement and the identically designed balancing coil **L'**, equipped with a small ferrite plug **F** for manual zeroing of the bridge.

The terminals **L** and **L'**, that are not connected to the transformer, are connected to the bridge network output point **P**, from which the signal of unbalance is taken for further processing. By inserting a specimen into the measuring coil, the bridge being zeroed, the inductivity of the coil will increase. The induced signal of unbalance will be compensated immediately by the effect of the feedback loop. A voltage proportional to the inductivity change and thus to the magnetic susceptibility of the sample will appear on the voltmeter. A general view of such a commercially available bridge, manufactured by Geofyzika Brno (Czechoslovakia) is shown in Fig. 1.57.

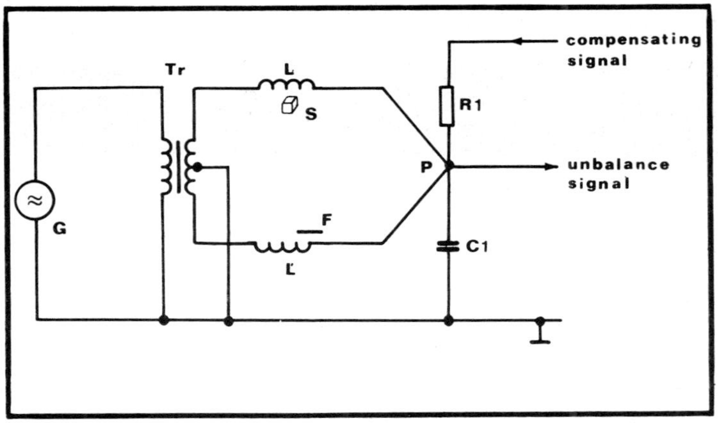

Fig. 1.56 Simplified circuit diagram of the inductivity bridge.

The advantages of the bridge are high accuracy, fast measuring rate and the ability to measure samples of various sizes and forms. High sensivity enables one to measure the magnetic susceptibility of minerals ranging from diamagnetic to ferromagnetic. The main disadvantage is that it is impossible to measure magnetic susceptibility as a function of magnetic field. As shown in Section I.3.4 the measurements of susceptibility in the case of weakly magnetic materials are generally of restricted use unless they are made over a sufficient range of magnetic field strengths (designed to eliminate the effect of ferromagnetic impurities and to investigate their behaviour at high magnetic field), usually employed in magnetic separation.

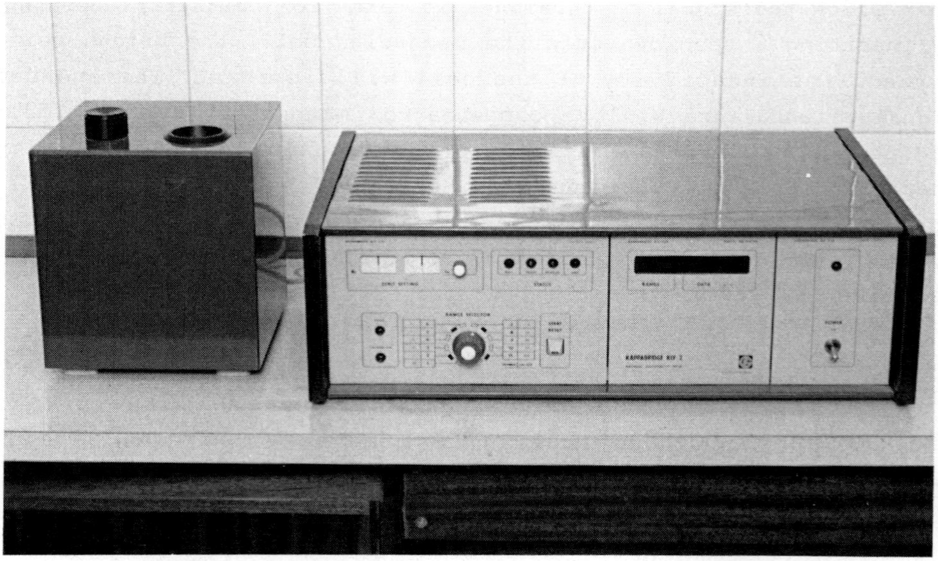

Fig. 1.57 Kappa - bridge.

I.5.2.3 Vibrating sample magnetometer

By detecting the electric potential produced, magnetic susceptibility can be measured by vibrating the magnetic material in a homogeneous magnetic field. The magnetometer which uses such a method is shown in Fig. 1.58 and was described by Foner [F6].

Material under study in the vibrating sample magnetometer (VSM) is inserted in the sample holder in such a way that it rests in a pair of sensing coils between the poles of a magnet. The sample holder is mounted in a transducer assembly which passes through the centre of the driving coil. The transducer is driven by

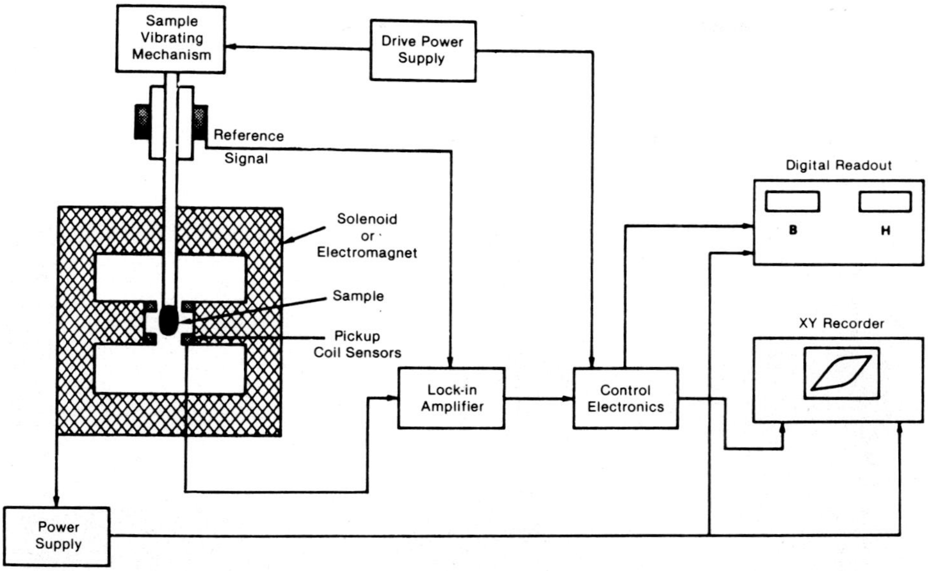

Fig. 1.58 LDJ Model 9000 vibrating sample magnetometer block diagram (courtesy of LDJ Inc.)

a power amplifier which itself is driven by an oscillator. The specimen is constrained to vibrate at small amplitude with known frequency and phase, along one axis only. As it does so, it induces a signal in the pick-up coil. The magnitude of the signal is dependent on the magnetic properties of the sample.

Originally, the VSM was developed for use with electromagnets for which the field was transverse to the vibration direction. Axial fields of solenoid magnets offer a greater freedom of coil arrangements and improved sensitivity [F5].

The magnetometer has become a widely used instrument for determining the magnetic properties of a wide range of materials, from diamagnetic to ferromagnetic. Hysteresis and magnetization curves, magnetic moments and susceptibilities can be measured as a function of magnetic field and temperature. VSM thus offers a opportunity to measure a full spectrum of magnetic parameters, required in magnetic separation research and magnetic separator design. A general view of the VSM, manufactured by LDJ Inc., Troy, Mich., U.S.A., is shown in Fig. 1.59.

An alternative to VSM is a vibrating coil magnetometer: the sample is stationary and the detection coil is vibrated. The main problem here is that the coils must move precisely along the applied field lines to avoid a large back-streamed pickup produced by the applied field [F5].

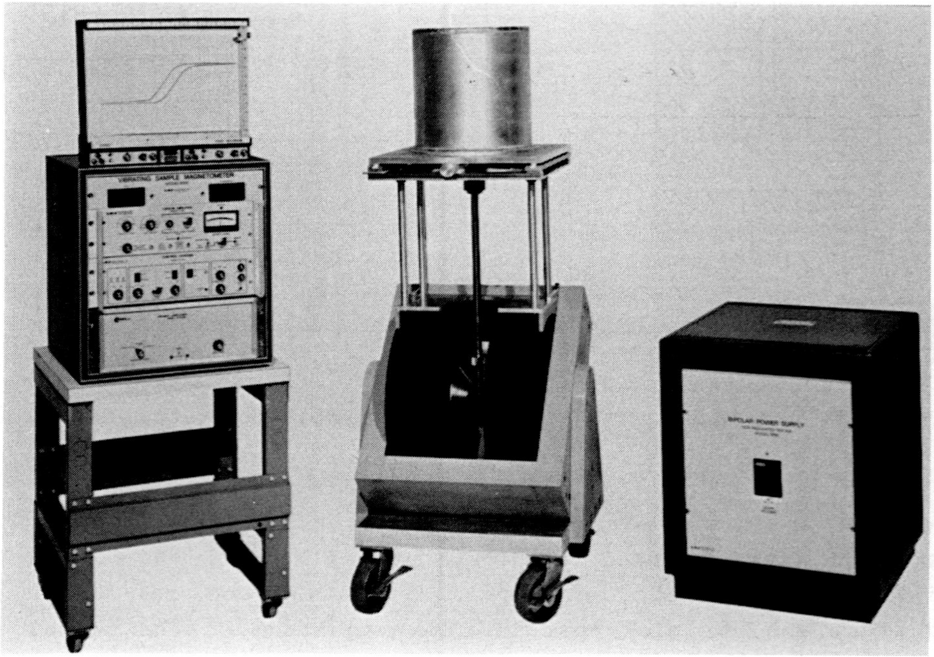

Fig. 1.59 Vibrating sample magnetometer LDJ Model 9000 (courtesy of LDJ Inc.).

I.5.2.4 <u>Superconducting magnetometer</u>

Superconducting quantum interference devices - SQUIDS - are today the most sensitive detectors of magnetic flux. Their energetic sensitivity is almost comparable with the Planck constant. SQUIDS have been playing an important role not only in physics, but also in geology, medicine and other branches of human endeavour. The principles of the operation of SQUID have been reviewed in a number of monographies and review papers [L3, B9, G7, O5].

Recently, variable-temperature superconducting magnetometers - susceptometers which use the SQUID magnetic flux detector became commercially available. In such a susceptometer, a uniform magnetic field is applied along the axis of a first derivative gradiometer and a sample is moved back and forth between the two pick-up coils. The resulting peak-to-peak signal from the SQUID is proportional to the magnetization of the sample. Measurements of magnetic moment and susceptibility can be made over a temperature range from 2 K to 400 K and in the magnetic fields from 0 to 5 T. The instrument can detect changes in specific magnetic susceptibility of only 10^{-13} m^3 kg^{-1} (10^{-11} cm^3 g^{-1}) in fields as low as 0.1 T. These limits improve at higher fields, giving the ultimate sensitivity of 10^{-14} m^3 kg^{-1} (10^{-12} cm^3 g^{-1}) from 1 T to 5 T (1 g sample).

The SQUID magnetometer developed by S.H.E. Corp. comprises all required subsystems, including those needed to control the magnetic field and temperature. The necessary cryogenic environment is maintained by a liquid helium reservoir having sufficient capacity to operate continuously for one week; the refilling of the reservoir takes 15 minutes.

The cross-section through the magnetic and cryogenic system is shown in Fig. 1.60 and the overall view of the magnetometer in Fig. 1.61.

I.5.2.5 Frantz isodynamic magnetic separator

This device is perhaps best known for its geological and mineralogical applications in performing precise separations of minerals of different magnetic susceptibilities. McAndrew [M9] demonstrated that the susceptibility of paramagnetic minerals could be determined on the Frantz separator. Magnetic response curves as discussed by Dobby et al. [D7] allow the distribution of magnetic susceptibility in mineral mixtures to be determined. It was also shown [N3] that the Frantz separator can be used for the determination of the magnetic parameters of the magnetic field-dependent susceptibility minerals.

Although the precision of magnetic susceptibility curves determined from the magnetic response curve is much lower than that determined by the specialized techniques discussed above, the accessibility, versatility and simple operation make the Frantz

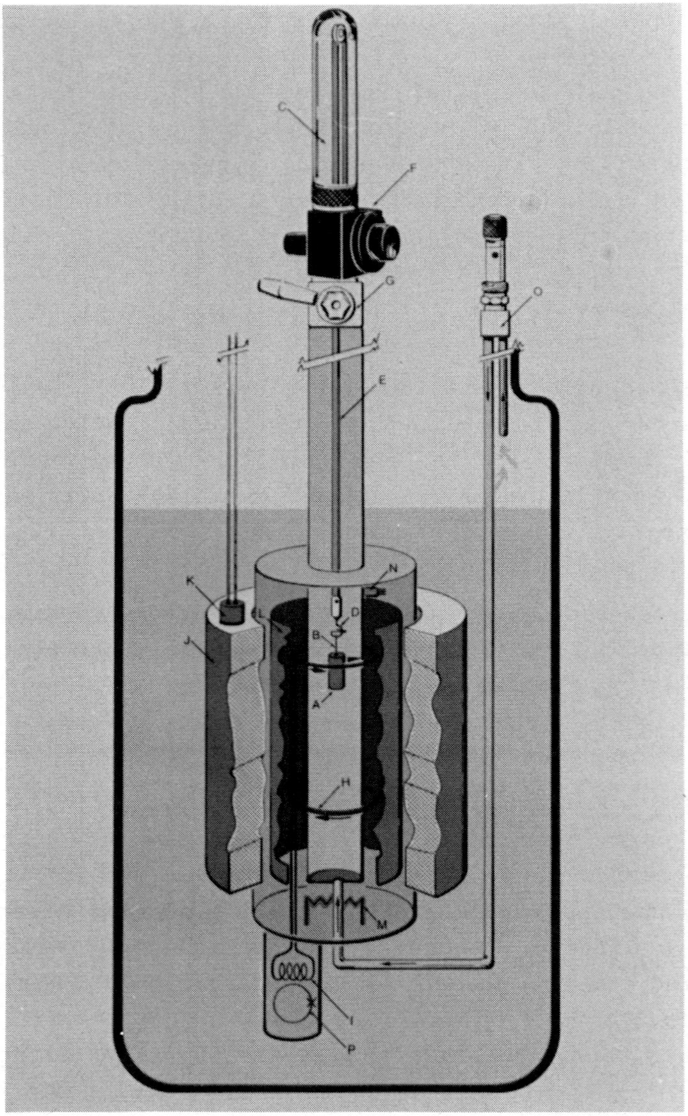

Fig. 1.60 Cross-section through the magnetic and cryogenic section of the S.H.E. SQUID magnetometer. A - sample, B - cotton thread, C - air lock chamber, D - sample support hook, E - metal tape, F - sample transport mechanism, G - air lock valve, H - pick-up coils, I - SQUID input coil, J - superconducting magnet, K - superconducting switch, L - superconducting magnetic shield, M - heater, N - thermometer, O - regulating valve, P - SQUID sensor. (Courtesy of BTi).

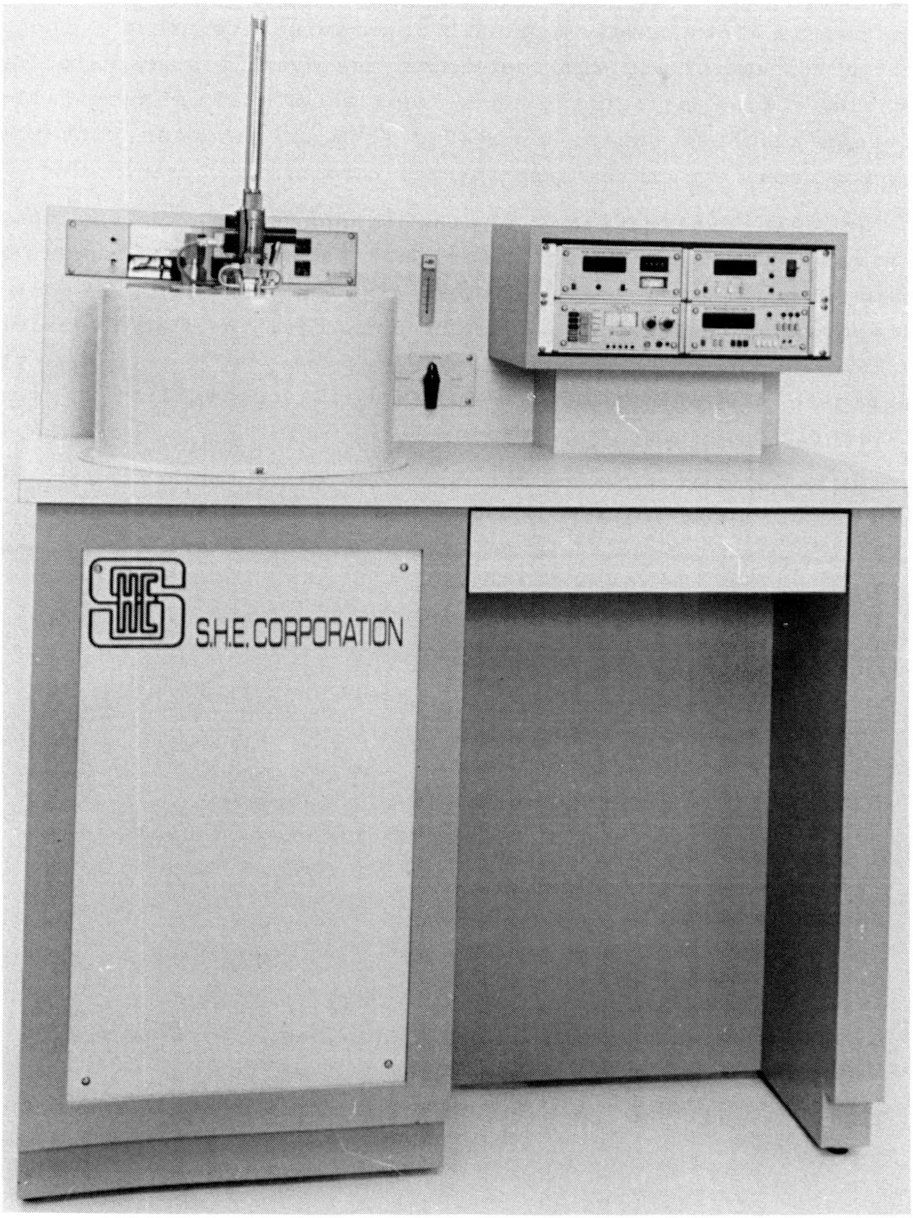

Fig. 1.61 S.H.E. superconducting magnetometer (courtesy of BTi).

separator an ideal instrument for a metallurgist faced with the problem of predicting the magnetic separability of a mineral. Moreover it determines separately the magnetic susceptibility for various groups of minerals present in the ore, while all other techniques measure just the mean magnetic susceptibility of the whole sample. If strongly magnetic impurities are present their magnetic susceptibility will determine the overall susceptibility of the sample. With the Frantz separator each magnetically different fraction can be separated from one another with the precision dependent on the experimental conditions chosen.

The main disadvantages of the Frantz separator are that relatively large quantities of samples must be used, ground to a certain particle size. Although very fine samples (say - 25 μm) are impossible to treat on a Frantz separator, the "wet" version [G8, P6, S50] can be used to process very fine samples and also to investigate the effect of the magnetic susceptibility of the carrying fluid.

A detailed discussion of the operation of the Frantz separator can be found in Section II.6.2.

CHAPTER II

REVIEW OF MAGNETIC SEPARATION TECHNIQUES

In reviewing magnetic separators one has to examine hundreds of patents and papers, consequently it is not possible to give adequate coverage to all of them. Moreover there appear to be more models and modifications of magnetic separators then there are applications, and since each embodiment of a magnetic separation technique possesses distinct merits as well as limitations, no single technique is universally applicable to all types of magnetic separation.

Various classification schemes exist according to which magnetic separators can be divided into categories, such schemes being determined by the dominant physical or technological feature. Magnetic separators can thus be grouped as follows:

a) according to the medium carrying the ore:
   1) dry
   2) wet

b) according to the purpose of the system [D2]:
   1) removal of iron and similar materials for the protection of expensive machinery subjected to damage or wear
   2) extraction of valuable magnetic constituents
   3) removal of deleterious substances, such as discolourations from clay

c) according to the way the magnetic field is generated:
   1) permanent magnet
   2) electromagnet with iron yoke
   3) solenoid
   4) superconducting magnet

d) according to how magnetic particles are separated from the feed sample:
   1) settling
   2) deflecting

e) according to the magnitude of the magnetic field:
   1) low-intensity magnetic separation
   2) high-intensity magnetic separation

Although there are many classification schemes [D1], some of them lose their relevance with new developments in magnetic separation and the current status in magnetic separation technology can be best described by combining groupings a) and e) as discussed above. Low-intensity separators are primarily used for the treatment of ferromagnetic minerals or paramagnetic minerals of high magnetic susceptibility, and of sufficiently large particle size. A wide variety of relatively simple devices exists in this category and photographs and sketches of many of the early commercial designs can be found in literature [D2, T2, D1] and in manufacturers' catalogues. Nevertheless several typical low-intensity magnetic separators will be reviewed here, to illustrate their principles of operation.

The main emphasis is on high-intensity high-gradient magnetic separators (HGMS). These have undergone a remarkable development in the last twenty years. Machines with improved performance have been developed and a theoretical description of the separation process in HGMS has been established.

A current trend in the design of magnetic separators is towards higher capacities per unit so as to obtain lower capital and operating costs per tonne of material treated. Two other parameters for evaluating magnetic separators are the energy consumption per tonne of material treated and the mass per tonne of a machine.

Almost without exception, the changes in the design of magnetic separators, have come about for economic reasons. The old adage still prevails that "necessity is the mother of invention".

Mechanical and process changes that reduce the capital and/or operating costs at magnetic separators usually take the following forms:

a) An increase in the size of units, which reduces the unit capital cost per tonne of capacity
b) Changes in design to make machine of the same size more efficient in capacity or metal recovery
c) The introduction of a known process previously regarded as uneconomical
d) The development of an entirely new process previously not in commercial use.

II.1 DRY LOW-INTENSITY MAGNETIC SEPARATORS

Dry low-intensity magnetic separation is confined mainly to the concentration of coarse sands that are strongly magnetic, the process being known as "cobbing". The cobbing of magnetite, the recovery of iron values from blast furnace and steel making slags and purification of a great variety of materials all use similar equipment.

II.1.1 Dry drum separator

Drum-type separators are the most common low-intensity machines in current use. The version, shown schematically in Fig. 2.1 is based on the original design of Ball and Norton and dates back to the beginning of this century. The separator consists of a rotating non-magnetic drum containing three to six stationary magnets of alternating polarity. These were initially electromagnets, but lately they were replaced by permanent magnets.

The material to be separated is fed at the top of the rotating drum surface; the non-magnetic fraction leaves the drum (Fig. 2.1), while the magnetic fraction is retained by the magnetic force at the drum surface until carried outside the effective magnetic field. By selecting of the proper separator configuration and

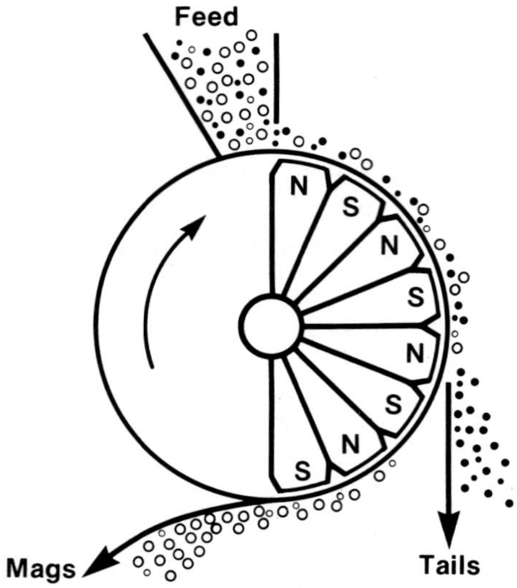

Fig. 2.1 Dry drum separator.

operating conditions, various products can be obtained which may be acceptable as final tailings, as final concentrate or as middlings requiring grinding and separation. Dry separators allow a better control over the grade of the products as compared with wet separators.

The theoretical description of low-intensity magnetic separation has attracted only limited attention as the process is generally considered a simple operation. As in other types of separators, whether or not a particle is detached from the drum depends upon the relative magnitudes of the forces attracting the particle to the drum and those tending to detach it.

An early investigation [K6] of the particle movement on the drum revealed that the majority of magnetizable particles perform their own rotation, the actual velocity of the particle being greater than that of the drum. Those particles that exceeded in size one-fifth of the distance between the centre lines of the poles did not acquire any rotation of their own when passing over the poles; instead, they merely turned over or bounced when approaching the poles. Some particles, smaller than one-fifth of the distance between the poles, did not perform any rotation of their own at a slow rate of feed, but remained over one pole, slipping on the surface of the drum. These slipping particles easily collect other particles with which they form chains. However, when large amounts of material are fed onto the drum surface, they tend to pile up and form chains that follow almost exactly the movements of the drum, without rotation of their own.

Laurilla [L4] made a fundamental investigation of the periodic magnetic field of the cylindrical array of magnets with radial poles used in magnetic drum systems. Brown [B10] has shown that the radial and azimuthal fields of such a system are best described in terms of a Fourier series comprising only cosine functions related to the pole periodicity. Brown has also shown that the azimuthal component of the field and its radial derivative contribute appreciably to the radial force attracting particles to the drum surface. Birss et al. [B11] developed a simpler Fourier series and constructed equations of motion for small magnetizable particles in drum magnetic separators.

The consistent analysis of forces acting on the particles that move through the magnetic field of a separator with top feed was performed by Derkach [D1] and later by Hopstock [H8].

II.1.2 Force considerations

The geometrical arrangement of the drum separator with a radius **R** is shown in Fig. 2.2. The z-axis is the axis of rotation of the drum and the radial and azimuthal components of the magnetic force acting on a ferromagnetic particle are given by

$$F_r = \mu_0 V \left(M_r \frac{\partial H_r}{\partial r} + M_\Theta \frac{\partial H_\Theta}{\partial r} \right)$$

$$F_\Theta = \mu_0 \frac{V}{r} \left(M_r \frac{\partial H_r}{\partial \Theta} + M_\Theta \frac{\partial H_\Theta}{\partial \Theta} \right) \qquad (2.1)$$

The subscripts **r** and θ indicate the radial and azimuthal components, **V** the volume of particle, **M** the magnetization and **r** the distance from the centre of the drum. By taking into account the demagnetizing factor of a particle, and the fact that ore

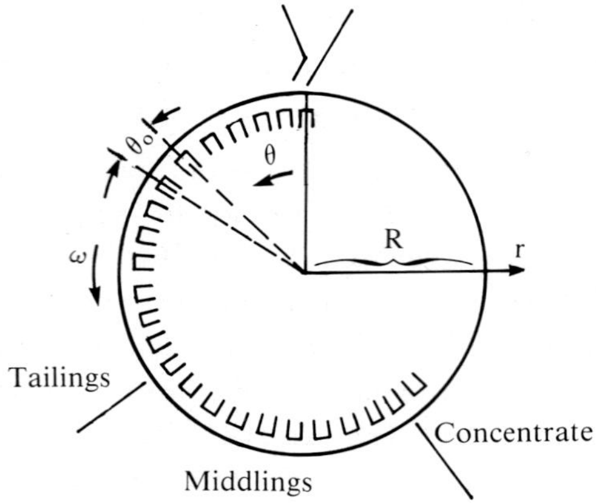

Fig. 2.2 Geometrical arrangement of a drum.

particles are frequently locked particles consisting partially of magnetite and partially of weakly magnetic gangue minerals, eqs. (2.1) can be, for a spherical particle, rewritten as

$$F_r = \frac{3}{2}\mu_0 \nu_m V \frac{\partial}{\partial r}(H^2) \tag{2.2}$$

$$F_\Theta = \frac{3}{2}\mu_0 \nu_m \frac{V}{r} \frac{\partial}{\partial \Theta}(H^2) \tag{2.3}$$

where ν_m is the fraction of the particle volume consisting of magnetite.

To make use of eqs. (2.2) and (2.3) an expression is necessary for the magnetizing field produced by the magnet array as a function of **r** and θ. For **r** > **R** these components are given by

$$H_r = \sum_{n=1}^{\infty} (a_n \cos n\Theta + b_n \sin n\Theta)\left(\frac{R}{r}\right)^{n+1} \tag{2.4}$$

$$H_\Theta = \sum_{n=1}^{\infty} (a_n \sin n\Theta - b_n \cos n\Theta)\left(\frac{R}{r}\right)^{n+1} \tag{2.5}$$

where $\mathbf{a}_n$ and $\mathbf{b}_n$ are coefficients to be determined experimentally. The series expressions are not in a form suitable for general analysis of the magnetic force and the following approximate equations by Hopstock [H8] were found to give a good fit to experimental data

$$H_r = H_0 \cos\left(\pi \frac{\Theta}{\Theta_0}\right) \exp\left[\frac{-\pi(r-R)}{\Theta_0 R}\right] \tag{2.6}$$

$$H_\Theta = H_0 \sin\left(\pi \frac{\Theta}{\Theta_0}\right) \exp\left[\frac{-\pi(r-R)}{\Theta_o R}\right] \tag{2.7}$$

$$H = H_0 \exp\left[\frac{-\pi(r-R)}{\Theta_0 R}\right] \tag{2.8}$$

where $\mathbf{H}_o$ is the magnetic field at the drum surface and θ_o is defined in Fig. 2.2. The approximation is most accurate for $\mathbf{r} - \mathbf{R} > \theta_o \mathbf{R}/\pi$. By inserting eq. (2.8) in eqs. (2.2) and (2.3), we obtain the magnetic force components

$$F_r = -3\mu_0 \nu_m V H_0^2 \frac{\pi}{\Theta_0 R} \exp\left[\frac{-2\pi(r-R)}{\Theta_0 R}\right] \tag{2.9}$$

$$F_\Theta = 0 \tag{2.10}$$

The radial force decays exponentially with $\mathbf{r}$ and at the drum surface eq. (2.9) reduces to

$$F_r = -\frac{3\pi\mu_0 \nu_m V H_0^2}{\Theta_0 R} \tag{2.11}$$

The most important force that acts to detach particles from the drum is the centrifugal force, given by

$$F_c = \rho_p V \omega^2 R \tag{2.12}$$

where ρ_p is the density of particles and ω the angular velocity of the drum. A particle will be detached from the drum when the magnitude of the centrifugal force exceeds that of the magnetic attractive force. The critical angular velocity ω_c at which the centrifugal and magnetic forces are just in balance is found from eqs. (2.11) and (2.12) to be

$$\omega_c = \left(\frac{3\pi\mu_0\nu_m H_0^2}{\rho_p\Theta_0 R^2}\right)^{\frac{1}{2}} \qquad (2.13)$$

Since the particle size does not appear in eq. (2.13) the dry drum separator can be applied to a wide range of particle sizes. If the drum separator is used to recover magnetite, eq. (2.13) can be converted to

$$f_c = 5.4\times10^{-5}\,\frac{H_0}{R}\left(\frac{G}{\Theta_0}\right)^{\frac{1}{2}} \qquad (2.14)$$

where f_c is the critical revolution rate in revolutions per minute for a particle containing **G** % iron.

II.1.3 Elements of design

The feed rate **Q** per unit width of drum can be written

$$Q = \omega R\rho_b d \qquad (2.15)$$

where ρ_b is the bulk density of the feed and **d** the depth of bed of feed. For ores with a median size greater than 1 mm the bed depth is generally chosen to be equal to the median size, while with finer feeds multilayers of particles are required to obtain economically acceptable feed rates. By inserting eq. (2.13) in eq. (2.15) we obtain

$$Q = \left(\frac{3\pi\mu_0\nu_m}{\rho_p\Theta_0}\right)^{\frac{1}{2}}\rho_b H_0 d \qquad (2.16)$$

The throughput can be increased by increasing H_o or by reducing the angle θ_o between adjacent poles. The former method decreases the selectivity of separation while the latter, in contrast, increases the degree of agitation applied to a cluster of particles and thus improves the selectivity of separation. The optimum distance **p** between adjacent poles at the drum surface can be computed from the relation

$$p = 2\pi(\delta + b) \cong \Theta_0 R \tag{2.17}$$

where δ is the distance between the magnet faces and the drum surface and **b** the particle radius. For fixed **p**, eq. (2.16) can be rewritten

$$Q = \left(\frac{3\pi\mu_0\nu_m R}{\rho_p p}\right)^{\frac{1}{2}} \rho_b H_0 d \tag{2.18}$$

Eq. (2.18) indicates that for a fixed pole pitch, other factors being equal, drum capacity increases as the square root of drum diameter. This fact suggests the choice of a relatively small drum diameter as an optimum. On the other hand by increasing the drum diameter the number of poles can be increased and the selectivity of separation improved.

The performance of a dry drum magnetic separator can thus be predicted from a knowledge of four easily determined parameters: magnetic field at drum surface ($\mathbf{H}_0$), angular spacing between magnets (θ_0), drum radius (**R**) and revolution rate (**f**). By proper choice of these parameters a high degree of control can be maintained over the grade of concentrate.

Commercially available models have a wide range of sizes: diameters from 300 to 1500 mm and lengths from 300 to 4000 mm. Coarsely crushed material as large as 8 mm in diameter can be treated economically at feed rates in excess of 150 t/h per metre of length. Typical radial magnetic induction values for current drum separators are of the order of 0.1 to 0.15 T at the drum surface.

II.2 WET LOW-INTENSITY MAGNETIC SEPARATORS

Magnetic separators of the wet low-intensity type are by far the most common in use today. They can only be used for the concentration of very strongly magnetic minerals, and in fact the iron mineral so concentrated in any quantity is magnetite [D8].

Most modern wet separators are based on the drum separator, shown in Fig. 2.1. This machine consisted of a number of fixed electromagnets with poles inside a non-magnetic rotating shell which carried the minerals to be separated across the face of the magnets; the drum itself was submerged in a tank of water. This separator is also used extensively for recovery of heavy media in the sink-float technique of gravity concentration.

Drum-type separators generally consist of from three to seven fixed poles which alternate in polarity and which are either electromagnets or permanent magnets, the latter being used in most machines. The non-magnetic cylinder which encloses them rotates round the fixed poles, carrying the material to be separated across the face of the magnets and the concentrate out of the submerged tank. The size of feed treated in these separators rarely exceeds 5 mm because of the problem of removing tailings from the spigots without causing blockage. Sizes coarser than 5 mm are usually treated in dry magnetic processes or with belt-type machines.

II.2.1 Types of drum separator tanks

The design of the magnetic separator tank is very important. It must take into account [D8]:

(a) presentation of the feed to the drum
(b) removal of the non-magnetic fraction
(c) nature of the washing action to be adopted.

While many variations exist in tank design, there are three basic groups, based on the pulp flow, as follows:

(i) concurrent
(ii) counter-rotation
(iii) counter current.

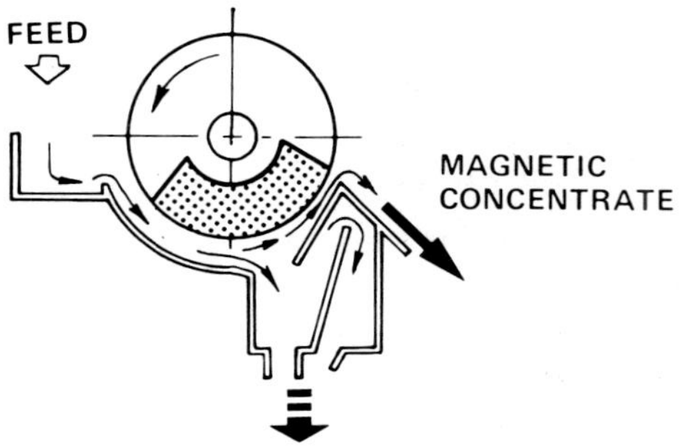

Fig. 2.3 The concurrent tank (courtesy of Eriez Magnetics, Inc.).

(i) <u>Concurrent tank</u>

The concurrent tank design is shown schematically in Fig. 2.3. The feed is introduced at one end of the separator through the feed box and flows in the direction of drum rotation. Magnetic material is picked up by the magnet and the non-magnetic material is discharged at the bottom through the tailing discharge opening. Magnetic particles continue to cling to the surface of the drum and pass through an opening between the drum shell and the tank wall. Then the magnetic fraction is discharged down the discharge lip.

With concurrent separators it is important that proper pulp level be maintained. Simple overflow control is provided by adjustable weir-bars on the side of the tank. Overflow volume should be approximately 10 to 20 per cent of the total tailing flow [S14]. For best results, feed must remain constant so that proper water level is maintained. Insufficient pulp level or excess overflow will result in excessive losses or an inadequately cleaned concentrate.

The concurrent design of the tank is used almost exclusively for treating coarser material in the size range 5 mm and below. Although high capacity rates are obtainable the arrangement neither results in the achievement of the maximum recovery nor the grade of the concentrate.

(ii) <u>Counter-rotation tank</u>

In the counter-rotation design of the tank (Fig. 2.4) the feed is introduced through a special feed-box to the drum which rotates in the opposite direction of pulp flow. Magnetic material is picked up by the drum and discharged almost immediately. Since tailings flow along the entire magnetic arc of the drum before being discharged, losses are minimal. Very high recoveries can thus be achieved with this design and the grade of the concentrate is only of secondary importance.

Feed solids for the counter-rotation tank should be smaller than 1 mm for the best results. As with the concurrent type tank, proper water level must be maintained, with overflow volume held within 10 to 20 per cent of the total tailing flow.

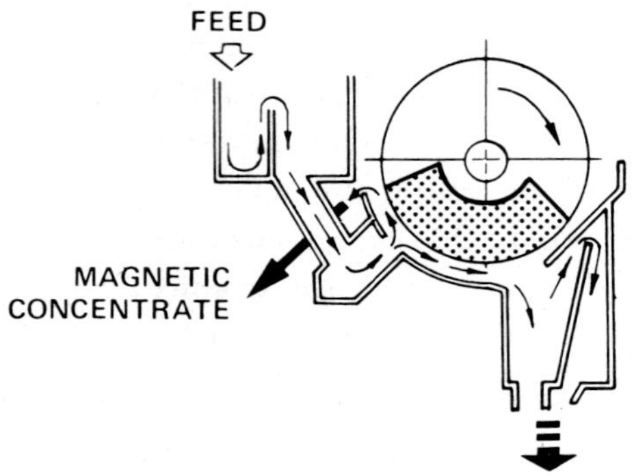

Fig. 2.4 The counter-rotation tank (courtesy of Eriez Magnetics, Inc.)

(iii) <u>Counter-current tank</u>

The counter-current tank, sometimes called a Steffenson tank, is shown in Fig. 2.5; it is mostly used for finishing where very thorough cleaning is required. The term "counter-current" derives from the fact that the tailings must flow counter to the rotation of the drum when leaving the machine. Feed is introduced near the bottom of the drum, from where the magnetic particles are picked up by the drum and agitated as they are carried through wash water jets. Tailings flow out from the opposite end of the tank through a tailing discharge overflow, which also acts as the pulp level control. This ensures a high recovery of magnetic material and a high quality concentrate. These machines are most suited to finely ground material of 100 μm and less. Coarse ores could produce settling problems.

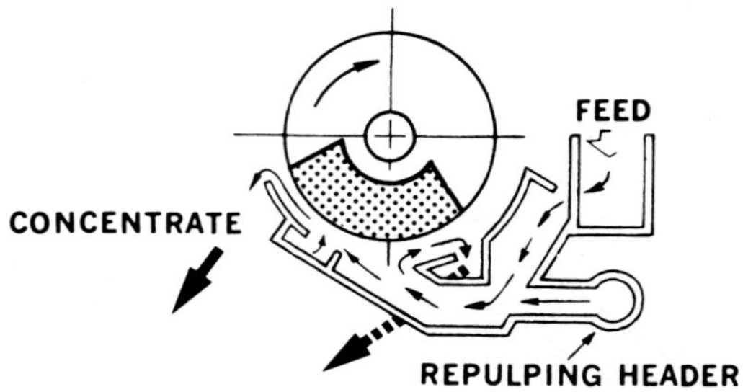

Fig. 2.5 The counter-current or Steffenson tank (courtesy of Eriez Magnetics, Inc.)

II.2.2 Design of the drum

The separating drum consists of two main parts: a stationary magnetic circuit and a shell, made of non-magnetic material, which freely rotates round the magnets. The capacity of such a separator is determined by the magnetic induction on the surface of the drum, by the gradient of the magnetic field in the pulp and by the size of the drum.

Nowadays, design of the drum is closely associated with the development of the permanent magnets used to generate the magnetic field. With barium ferrite permanent magnets, of the energy product of 20 - 22 kJ m^{-3} (2.6 - 2.8 MGOe) the thickness of the wall of the drum can be a maximum of 2.0 - 2.5 mm [K7]. Therefore, the length of the drum is limited to about 1500 mm. With the increased quality of the barium ferrite with the energy product of 2.8 - 3.2 MGOe (22 - 25 kJ m^{-3}) the wall thickness could have been increased to 3.0 - 3.5 mm, and therefore, the diameter and the length of the drum could have been increased to 900 and 3000 mm, respectively.

By combining the strontium and barium ferrites, the wall thickness can be increased to 4 - 5 mm and drum diameter and length to 1500 mm and 4000 mm, respectively [K7].

The capacity of the three types of drum magnetic separators discussed above, depending on drum diameter, is shown in Fig. 2.6 and Table 2.1.

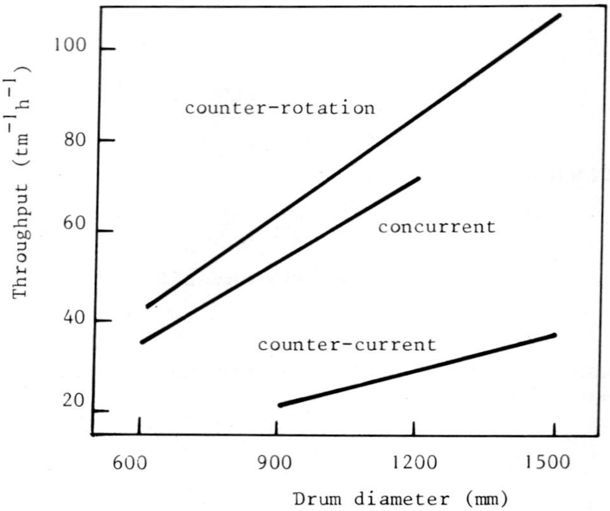

Fig. 2.6 Throughput as a function of the drum diameter (after Kolář et al. [K7]).

TABLE 2.1

Capacity of drum magnetic separators [K7]

Drum diameter	Concurrent drum	Counter rotation drum	Counter current drum
(mm)	Capacity (t/m h)	Capacity (t/m h)	Capacity (t/m h)
600	35	40	
900	50	60	20
1200		85	30
1500		100	35

The development of new, permanent-magnet materials has increased the magnetic induction available on the drum surface and inside the separation tank. For instance, the earlier system generated 0.16 T on the drum surface and 0.06 T at 40 mm from the drum surface, while new magnetic circuits create 0.22 T on the surface and 0.09 T at 40 mm [C4].

The improvement in parameters of the magnetic circuit and in the design of the separation tanks produced an increase in:

1) the diameter and length of the drum and therefore, the capacity of the machine
2) the gradient of the magnetic field and thus an increase in the efficiency of recovery of fine ferromagnetic particles
3) the depth of the magnetic field and easier recovery of larger grains.

II.3 HIGH-INTENSITY MAGNETIC SEPARATORS

Weakly magnetic minerals can only be removed from an ore feed in magnetic fields much greater than those available in low intensity magnetic separators. Dry high-intensity magnetic separation has been used commercially since the beginning of the century, while wet, high-intensity and high-gradient magnetic separators were developed only recently.

Three types of dry high-intensity magnetic separators are in common use, viz;

(1) the cross-belt type
(2) the induced magnetic roll type
(3) the permanent magnetic roll type.

II.3.1 Cross-belt magnetic separator

This is one of the oldest types of separator used to concentrate moderately magnetic ores (Fig. 2.8).

Dry material is fed in a monolayer onto the conveyor belt and is carried between the poles of the magnetic system. The belt, with its load, passes between the poles of two or more electromagnets. The lower pole-pieces are flat and immovable, but the

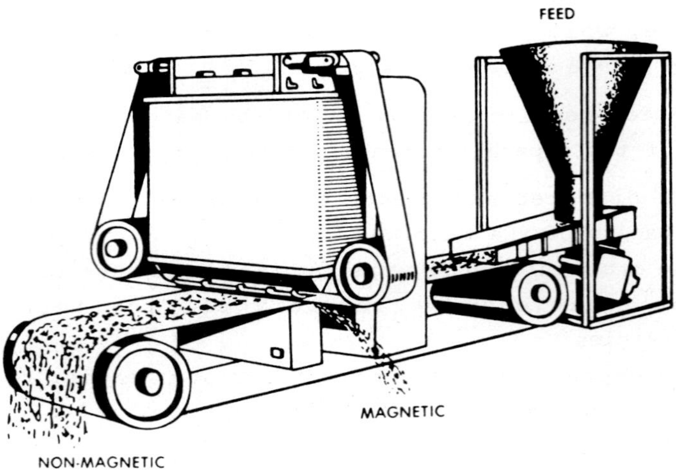

Fig. 2.8 Cross-belt magnetic separator.

corresponding upper pole-pieces are shaped and may be raised or lowered as required for the material undergoing treatment. An endless cross-belt runs around a series of pulleys and across each upper pole-face surface at right angles to the conveyor belt. As the material passes, the magnetic particles are attracted to the cross-belt, are carried with it towards the discharge side and fall into a suitably placed bin. The material may be passed successively between several pairs of magnet poles when it is desired to make several products in one pass. A non-magnetic tailing discharges as the conveyor belt passes over its end pulley. Machines with two to eight pairs of poles are available, and each pair can be adjusted independently by regulating the position of the upper poles and by altering the current through the coils. The height of the upper pole-pieces above the belt should be set at least 2.5 times the size of the largest particle in the feed. The conveyor belt width ranges from 100 to 600 mm, the distance between the conveyor belt surface and the upper belt surface being from 3 to 25 mm, depending on the size and magnetic properties of the feed. Particles ranging from 4 mm to 75 μm may be treated successfully, but they should preferably be sized into several fractions [D2].

Typical values of magnetic induction in a five-magnet cross-belt separator are given in Table 2.2.

The main advantage of the cross-belt separator is that several types of magnetic products can be recovered in one pass at increasing magnetic field strength while several separate passes are needed in other dry magnetic separators. Since the electromagnets are independently adjustable the selective removal of different magnetic minerals is possible. Low-intensity magnet can be installed to scalp off any ferromagnetic particles present. A five-magnet cross-belt separator is shown in Fig. 2.9.

TABLE 2.2
Cross-belt magnetic induction (T)

Gap(mm)	1st magnet	2nd magnet	3rd magnet	4th magnet	5th magnet
2	1.0	1.2	1.7	1.8	2.0
5	0.6	0.8	1.4	1.4	1.4
7	0.5	0.7	1.2	1.2	1.2
10	0.4	0.6	0.9	1.0	1.0

Fig. 2.9 Cross-belt separator.

II.3.2 Induced magnetic roll separator

The induced magnetic roll (IMR) separator consists of a revolving laminated roll formed of alternate magnetizable and non-magnetic discs. The roll is placed between specially shaped poles (Fig. 2.10) of an electromagnet. The electromagnet induces a magnetic field in the magnetic laminations of a roll forming local regions of high magnetic field gradients, as shown in Fig. 2.11.

Material to be treated is fed in a controlled thin stream by a vibratory feeder to the top of the roll. As the roll revolves, the material passes through a narrow gap between the pole of the magnet

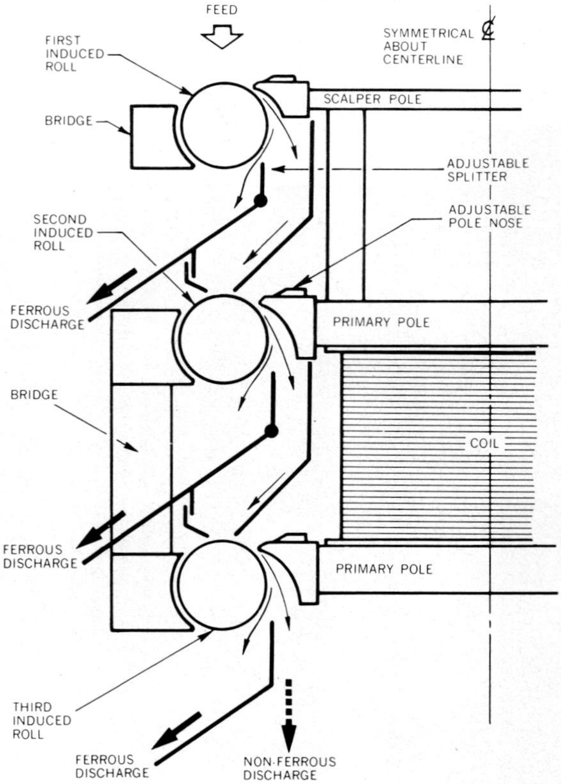

Fig. 2.10 The schematic diagram of an induced magnetic roll separator (courtesy of Eriez Magnetics Inc.).

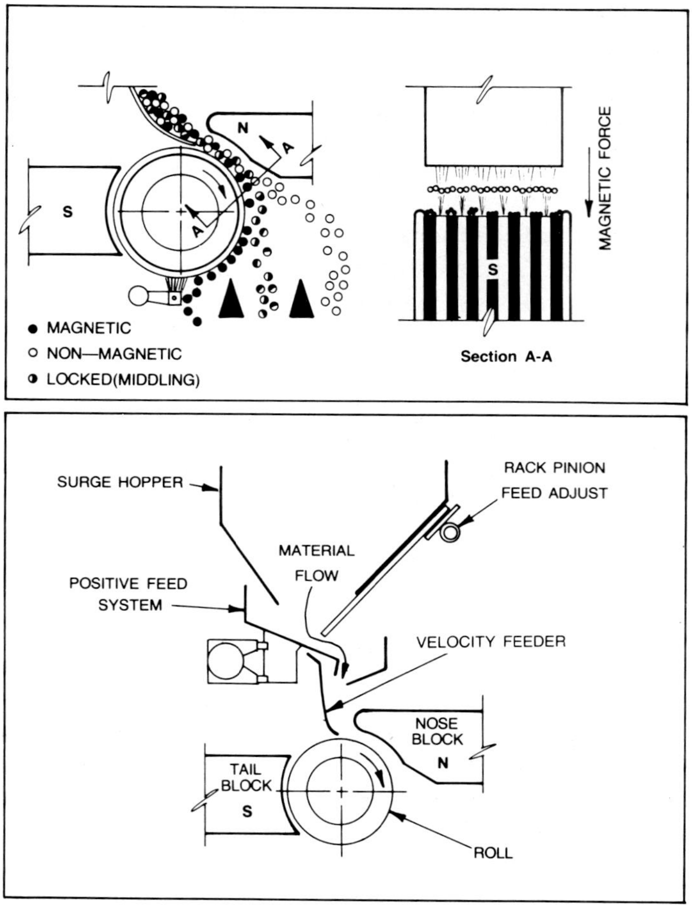

Fig. 2.11 The schematic diagram of the operation of an induced magnetic roll separator (courtesy of Carpco, Inc.).

and the roll and the non-magnetic particles are discharged from the roll. The magnetic particles are attracted to the roll, and discharged into a separate chute when they enter a non-magnetic region. Ferromagnetic material can be removed with a separate magnetic scalper before feeding to IMR, to prevent plugging the gap.

The gap between the feed pole and roll is adjustable and also the setting of the splitter is of great importance. The typical values of magnetic induction on the surface of a laboratory IMR as a function of the gap width, are given in Table 2.3 and in Fig. 2.12.

It is a general practice to ensure that each successive roll is of greater magnetic induction than the preceding roll. This is achieved by adjusting the poles so that a smaller gap between the pole and the roll is produced on each successive roll.

Because of variations in the character and size of the materials it is difficult to express capacity in terms of mass. However, in most applications capacity ranges from 2 to 3 t h^{-1} per metre of roll length. Material to be treated must be dry, free-flowing and for best results should be within the size range

TABLE 2.3
Magnetic induction on the IMR roll [M10].

Gap (mm)	Magnetic induction (T)
2	1.7
5	1.2
7	1.1
10	0.8
12	0.7

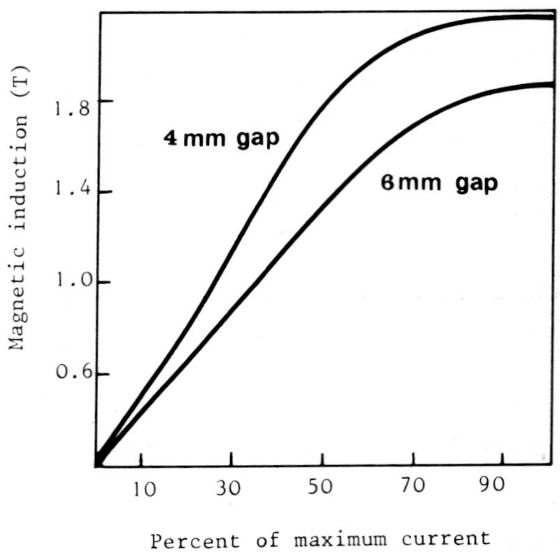

Fig. 2.12 Magnetic induction in the air gap of IMR as a function of the electric current (courtesy of Carpco, Inc.).

of 2 mm to 100 μm. The gap should be adjusted in such a way that it is approximately equal to 2.5 times the average grain size.

Induced magnetic roll separators are marketed by several manufacturers, e.g. Carpco, Eriez and Reading. The maximum length of the roll is 0.75 m, except for the Carpco separator which is 1 m in length. This separator uses hollow rolls which reduce the eddy currents during rotation, (most other IMR use solid rolls). Fig. 2.13 shows a six-roll IMR.

Although the recovery of ilmenite from beach sands is the biggest application, IMR separators have been successfully applied in numerous other branches of minerals treatment, viz. the production of high-quality (i.e. low-iron) materials for the glass and ceramics industry. They often constitute the final finishing part in the flowsheet, and the removal of magnetic impurities from andalusite, feldspar, wollastonite and other minerals is well proven.

IMRs are also used for concentration of the industrial minerals (chromite, iron ores, monazite, wolframite etc.).

There are two principal technical limitations for IMRs, namely: relatively low capacity per unit and limited particle size range. The limitations are related to the maximum allowable air-gap and to the need to feed a thin stream of particles. To permit unobstructed flow through the separator it is desirable to increase the air gap for larger material flow and for increasing particle size. It should be noted, however, that increasing the air-gap means reducing the magnetic field, thereby limiting the thickness of the material layer fed onto the roll, and reducing the capacity.

Fig. 2.13 Induced magnetic roll separator (courtesy of Carpco, Inc.).

II.3.3 Permanent magnet separators

II.3.3.1 The Laurilla high-intensity permanent magnet separator

Although the maximum achievable magnetic induction using ceramic permanent magnets is of the order of 0.2 T, Laurilla [L5, L6] developed a drum type wet magnetic separator with permanent magnets which generated higher gradient in narrow grooves on a drum surface. The separator shown in Fig. 2.14 consists of a large number of magnetic circuits incorporating ceramic permanent magnets, the magnetic field of which is concentrated in narrow slots on the surface of the drum. By adjusting the width of the slots the magnetic induction and the gradient can be matched to the magnetic properties of the ore. A secondary magnet system inside the drum cancels the field at a certain point, where magnetic particles caught in the slot can be removed. The capacity of this machine is low and care must be taken that the grooves do not become clogged.

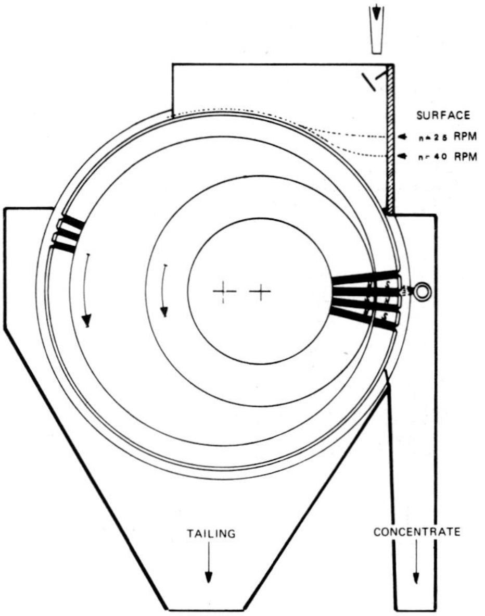

Fig. 2.14 The Laurilla high-intensity permanent magnet separator.

II.3.3.2 Permanent magnetic roll separator

The most recent development in the field of permanent magnet separators, conducted by E.L. Bateman Ltd. resulted in the design and manufacture of the Permroll separator. The active part of this separator is a roll consisting of disks of Sm-Co, or Nb-Fe-B permanent magnets interleaved with mild steel disks. The most favourable ratio of the widths of the magnet and of the steel is 4 : 1. Mild steel gives the most satisfactory results and special steels usually do not improve the performance of the separator. The magnet in this configuration generates magnetic induction up to 1.6 T on the surface of the roll and field gradients of the order of 300 T m^{-1}. For an easy removal of magnetic particles the roll is covered by a thin belt supported by a second (idler) roll, as shown schematically in Fig. 2.15. Below the conveyor is a hopper which collects the discharging material while adjustable splitters divert the different fractions into collection pans placed beneath the hopper.

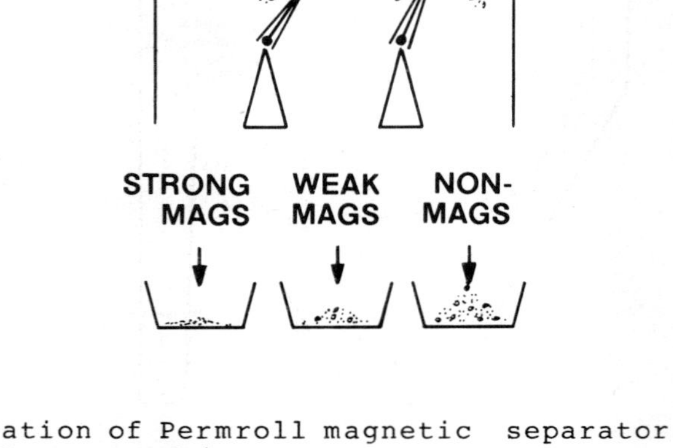

Fig. 2.15 Operation of Permroll magnetic separator (courtesy of Ore Sorters Ltd.)

Fig. 2.16 Separation by Permroll (courtesy of Ore Sorters Ltd.).

The process of separation by Permroll is shown in Fig. 2.16; a function of the splitters is illustrated in Fig. 2.17, while Fig. 2.18 shows magnetic and nonmagnetic fractions obtained by Permroll from a beach sand.

The magnetic force generated by the roll is of the same order of magnitude as that in IMR. A spectrum of minerals that can be successfully treated by either magnetic separator is practically identical. Several advantages exist, however, for Permroll separators, viz. low energy consumption. Since permanent magnets are used, no eddy current is present. It is estimated that only about 10 per cent of the electric energy required by IMR would be used by the Permroll for identical applications [A4]. Another advantage is that there are no mechanical problems like bearing failure and roll breakage. The overall size and mass are reduced and in a typical case 60 per cent of volume space and 10 to 20 per cent of mass is required compared to IMR installation [A4]. The

Fig. 2.17 The adjustment of splitters allows the collection of a number of fractions (courtesy of Ore Sorters Ltd.)

lack of an air gap means that large particles (up to 25 mm) can be separated and the rolls cannot be damaged by oversize material. A separate scalping stage is not needed since the belt detaches and easily discharges any ferromagnetic particles. On the other hand, the wear rate of the magnets, which is still an unknown factor, may well be the major cost item. In addition belt wear can be quite high, while the presence of the belt reduces the effective magnetic induction and the field gradient. The effect of the various belts used with Permroll on the magnetic induction on the roll surface at the centre and at the end of the roll, is summarized in Table 2.4.

One disadvantage is that the conditions of separation can be controlled only by the roll speed. The treatment of fine particles is hampered by their adhesion to the belt by electrostatic interactions.

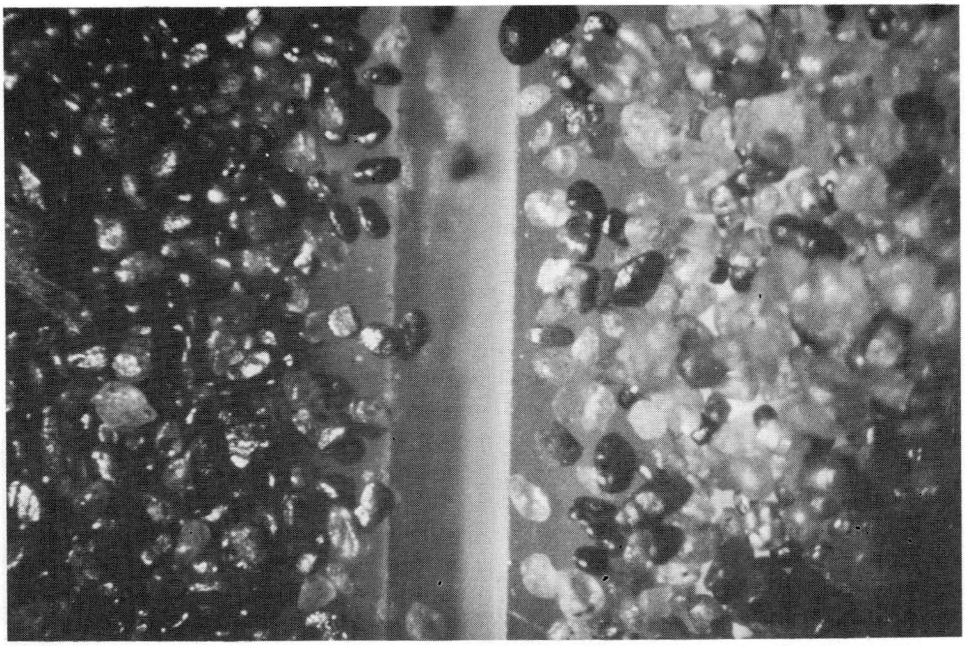

Fig. 2.18 Magnetic fraction (ilmenite) and non-magnetic fraction obtained by a permanent roll separator from a beach sand (courtesy of Ore Sorters Ltd.)

A photograph of a laboratory type Permroll separator is shown in Fig. 2.19.

Production-size machines have rolls 1 m long, rolls 1.5 m long are under development. The standard diameter of a roll is 75 mm, although 100 mm rolls are available. Rolls of greater diameter exhibit a larger depth of the magnetic field and a larger retention time of particles in the field. Capacity ranges from 1.5 t h^{-1} per metre of length for fine materials, to 15 t h^{-1} per metre for 12 to 15 mm particles. The modular design enables a number of passes to be arranged into a compact unit. A photograph of the production-size unit is shown in Fig. 2.20.

TABLE 2.4

Magnetic induction on the surface of Permroll (Sm-Co magnet) measured in the centre and at the end of the roll by a Hall probe

	Magnetic induction (T)	
	Centre	End
No belt	0.81	0.45
0.13 mm belt	0.79	0.42
0.2 mm belt	0.70	0.38
0.7 mm belt	0.46	0.24

Fig. 2.19 Laboratory Permroll (courtesy of Ore Sorters, Ltd.).

Fig. 2.20 The production-size Permroll separator (courtesy of Ore Sorters Ltd).

II.3.3.3 Motion of a particle in a permanent magnetic disk separator

The pattern of the magnetic field generated by a permanent magnetic roll separator, and the dynamics of the particles on the roll, have never been studied thoroughly. Although Kopp [K8] developed a simple model of the particle motion in such a disk separator, the steel disks were assumed to be replaced by air gaps. It was assumed that the magnets were disk-shaped and faced others with the same polarity (Fig. 2.21).

The following forces act on a magnetizable particle:

(a) a radially-inward magnetic force

(b) a radially-outward centrifugal force

(c) a vertically-downward gravitational force.

The exact force balance determines the point (or angle, as measured from the vertical) where the particles separate from the belt.

Magnetic force is given by eq. (1.8) and can be rewritten as

$$F_m = \mu_0 \chi m H \text{ grad } H \tag{2.19}$$

where **m** is the mass of a particle of specific magnetic susceptibility χ. For the elimination of mass dependence, magnetic acceleration due to magnetic force is defined as

$$a_m = \mu_0 \chi H \text{ grad } H \tag{2.20}$$

If $\chi = 10^{-6}$ m^3 kg^{-1} ($\cong 100 \times 10^{-6}$ cm^3 g^{-1}), $\mathbf{H} = 10^6$ A m^{-1} ($\cong 10$ kOe) and grad $\mathbf{H} = 10^7$ A m^{-2} (1 kOe cm^{-1}), then $\mathbf{a}_m \cong 10$ m s^{-2}, or roughly the acceleration of gravity. This is the right order of magnitude: if $\mathbf{a}_m \ll \mathbf{g}$ the magnetic force is ineffectual, while for $\mathbf{a}_m \gg \mathbf{g}$ the separation will be non-selective.

An exact calculation of the magnetic field of a permanent magnet assembly is difficult, since it depends on the distribution of the stray magnetic field and on the presence of the inleaving ferromagnetic steel disks. Under a simplified assumption that the disks of a magnet are separated by an air gap of thickness **t** and that the radius of a magnet disk is large in comparison with its thickness, the upper limit for the field in the air gap is given by

$$H_0 = \left[\frac{(BH)_{max}}{\mu_0}\right]^{\frac{1}{2}} \frac{w}{t} \tag{2.21}$$

where **w** is the thickness of the magnetic disk. The variation of the magnetic field in the midplane is exponential and thus

$$H = H_0 \exp\left(-\frac{z}{t}\right) \tag{2.22}$$

where $\mathbf{H}_o$ is the field at the disk edge (z = o), and **z** is the radial distance from the edge. There is also a variation of the magnetic field in an axial direction, but for the particles large in comparison with the gap width **t** the axial variation can be neglected.

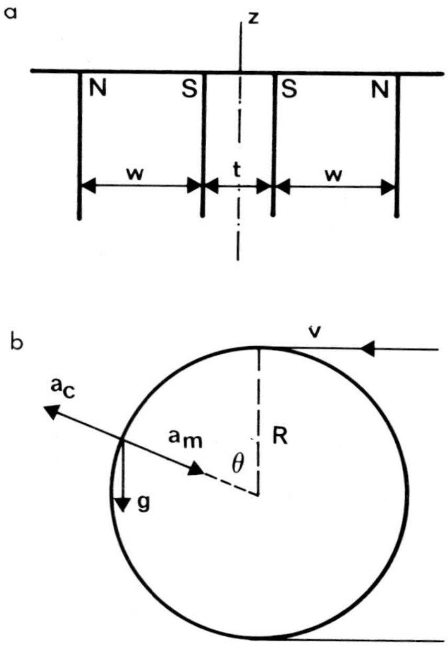

Fig. 2.21 The geometry of the disk separator: (a) radial view, showing the polarity of the magnets which are disks of width **w**, the air gap **t**, and the radial distance from the surface **z**; (b) axial view, showing the three accelerations of the particle at an angle θ from the vertical, **v** is the belt speed and **R** the disk radius (after Kopp [K8]).

As follows from eq.(2.22) the field gradient becomes

$$\text{grad H} = \frac{dH}{dz} = -\frac{H}{t} \tag{2.23}$$

so that the magnetic force can be written as

$$F_m = -\frac{\mu_0 \chi m H_0^2}{t} \exp\left(\frac{-2z}{t}\right) \tag{2.24}$$

From eqs. (2.21) and (2.24) it follows that the magnetic force is directly proportional to the maximum energy product of the magnet material. The magnetic force is the maximum for **t** = $(2z)^{1/2}$. The average particle diameter should thus be matched to this value, or conversely, the separator should be designed for the most abundant particle size.

The angle of separation

On the non-rotating belt a non-magnetic particle will fall off at a slip angle θ_s given by

$$f_s = \tan \Theta_s \tag{2.25}$$

where f_s is the coefficient of friction between the particle and the belt. Observed slip angles range from 35° to 45°.

When the belt is moving with velocity **v**, the magnetic acceleration is given by eq. (2.20), while the centrifugal acceleration is

$$a_c = \frac{v^2}{R} \tag{2.26}$$

A balance of forces shows that the particle will separate from the belt when

$$g \sin \Theta = f_s(a_m - a_c + g \cos \Theta) \tag{2.27}$$

or

$$\frac{\Delta a}{g} = \frac{\sin \Theta}{f_s} - \cos \Theta \tag{2.28}$$

where $\mathbf{\Delta a} = a_m - a_c$ is the effective inward acceleration. In the case $f_s = 1$ eq.(2.28) can be solved for θ

$$\sin 2\Theta = 1 - \left(\frac{\Delta a}{g}\right)^2 \tag{2.29}$$

but in other cases θ must be evaluated numerically. The results are shown in Fig. 2.22. The separation angle can be read off this graph for given values of $\mathbf{\Delta a}$ and $\mathbf{f_s}$. Only the values $\theta < 135^o$ are important; if $\theta > 135^o$ the particles will not fall off untill they leave the magnetic region.

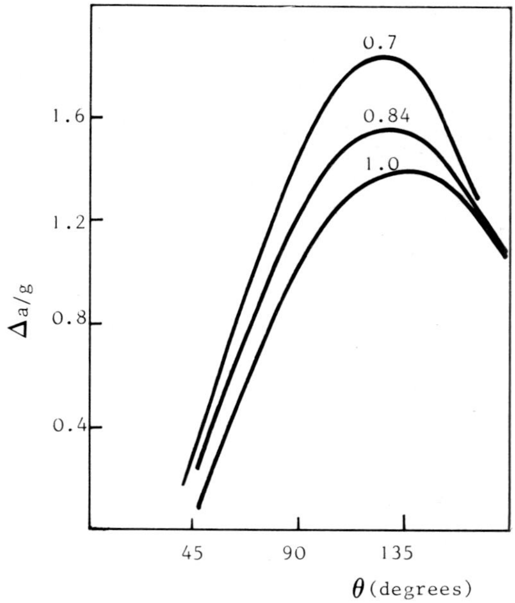

Fig. 2.22 The solution of the separation angle equation for three different values of the coefficient of friction . The vertical axis measures the normalized differential magnetic acceleration $(a_m - a_c)/g$ (after Kopp [K8]).

If all input parameters (magnetic and centrifugal accelerations and slip angle) are known, the separation angle can be computed as a function of particle size and magnetic susceptibility. This has been done in Fig. 2.23 for the case of ferrite permanent magnets under the following conditions: air gap **t** = 6 mm, force term $\mu_o \mathbf{H} \mathrm{grad} \mathbf{H} = 64 \times 10^{11}\ A^2\ m^{-3}$, centrifugal term: $\mathbf{a}_c/\mathbf{g} = 0.23$.

Since one can obtain the same separation angle for different combinations of size and susceptibility as in other types of magnetic separators, this implies that a careful size classification of the feed material improves the selectivity of separation.

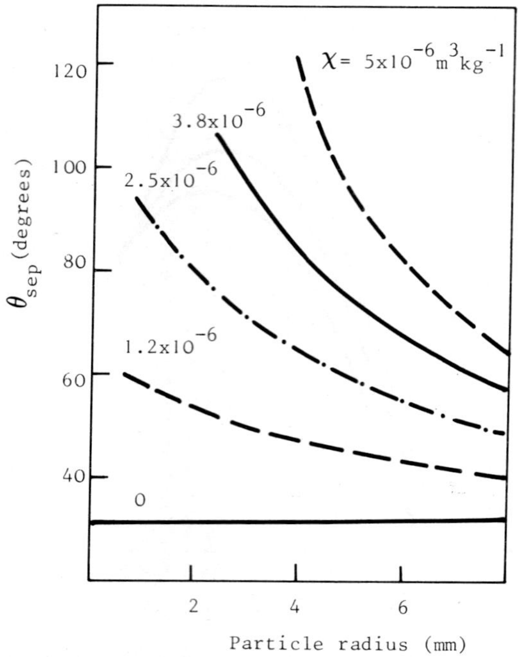

Fig. 2.23 The separation angle as a function of the particle radius for different values of the magnetic susceptibility (after Kopp [K8]).

<u>Capacity</u>

The capacity of a belt separator can be expressed as

$$Q = Wvfb\rho_b \tag{2.30}$$

where **W** is the belt width (in m)
v is the belt speed (in m s^{-1})
f is the fraction of the belt covered with particles
b is the mean particle radius (in m)
ρ is the density of the feed (in kg m^{-3})

For **W** = 0.5 m, **v** = 0.40 m s^{-1}, **f** = 0.3, **b** = 5 x 10^{-3} m and ρ = 3000 kg m^{-3} we obtain: **Q** = 0.9 kg s^{-1} or 3.2 t hr^{-1}.

Capacity can be increased by:

a) increasing the belt width, with 1.5 m as the practical upper limit

b) increasing the belt speed, which leads to a higher centrifugal force, that must be compensated by a higher magnetic force
c) increasing the roll radius
d) increasing **f**, which is limited by particles beginning to overlap and thus experiencing weaker magnetic force.

The maximum attainable throughput seems to be about 10 t/h per metre of roll [K8] for particles in the millimetre size range.

It is observed that the magnetic force is directly proportional to the maximum energy product of the material used; hence the use of Sm-Co or Nd-Fe-B alloys is desirable. Furthermore, the wider the gap between the disks, the higher the magnetic force. Because the first few millimetres from the surface of the roll determine the efficiency of the separation the use of a thin belt is advantageous.

The model was developed under the assumption that the magnet's disks are separated by an air gap. In a real separator, use is made of soft-iron spacers between the disks. Although the distribution of the magnetic field in such an arrangement will differ from that in a roll with air slots, the measurements show that the above model is generally valid [K8].

II.3.3.4 Submerged roll permanent magnetic separator

Although the development of permanent magnetic roll separator typifies the recent dramatic progress in magnetic separation, its use is limited by two considerations. Firstly, since a permanent magnetic roll separator is a "dry" machine, it can be used for the treatment of sufficiently coarse feeds only, its capability to treat the fines being is limited. Secondly, the permanent magnetic roll separator can be used effectively mainly in flowsheets where dry concentration methods are decisive. In the opposite case, an expensive drying step has to be included which, in view of relatively low market values of the majority of the materials beneficiated by permanent magnetic roll separators is a prohibitively expensive solution. The need for a "wet" permanent magnetic roll separator thus becomes apparent and it is paradoxical that the design and development of the first "wet" permanent magnetic roll separator was motivated by the limitation of a dry permanent magnetic roll separator in a situation where the product of separation has a high market value.

De Beers Industrial Diamond Division Ltd. (South Africa) [H32, H33] established that the use of a dry Sm-Co magnetic roll creates certain benefits in the diamond recovery process. However, as diamond recovery is mainly confined to wet concentration methods, it was evident that the greatest benefit to the industry would be a wet version of a permanent magnetic roll separator, thereby obviating the need for an intermediate drying operation.

The feed system

Initial tests were conducted with the use of a magnetic roll which was totally submerged in a tank of water. The diamondiferous gravel was fed onto the submerged belt via a fluidized feeder and drop chute. Owing to numerous problems experienced with this feed system [H33] the wet-chute feeder was designed. The chute, set at an angle of 15°, incorporates a water-distribution box which delivers a film of water at a controlled velocity down the width of the chute to match the peripheral speed of the roll. The situation is schematically depicted in Fig. 2.24.

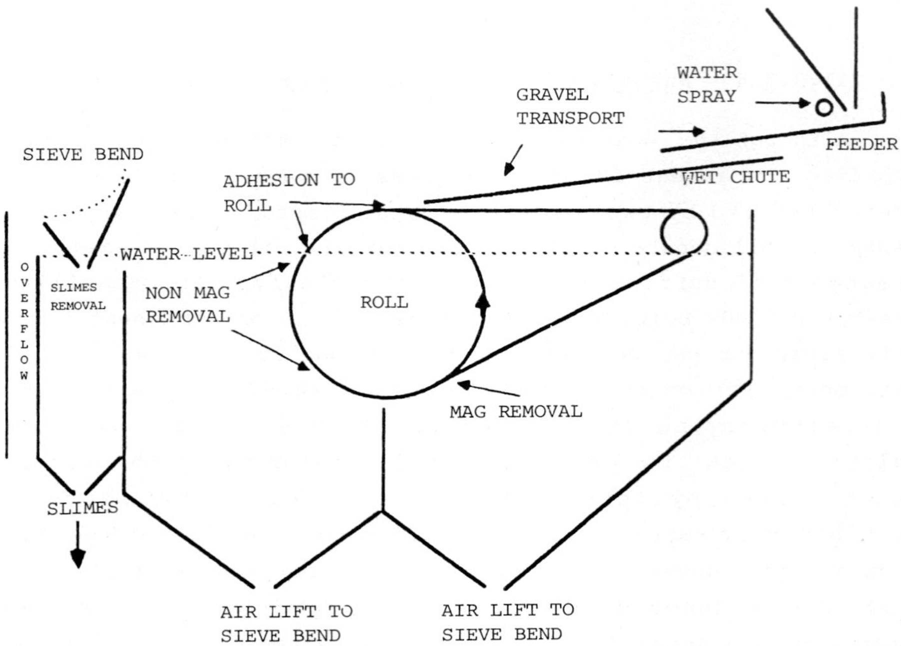

Fig. 2.24 The schematic diagram of a wet submerged roll permanent magnetic separator (after Hyland [H33]).

The diamondiferous gravel is fed from a fluidized feeder on the chute from where the particles disperse in a monolayer across the width of the chute and are then accelerated to the same peripheral speed as the roll before being discharged onto the roll. It was found [H33] that this feeding technique offered a fourfold increase in throughput over the drop chute arrangement.

When the chute water and belt are travelling at the same speed the feed particles adhere to the roll until the particles enter the water. At this point the non-magnetic particles leave the roll under gravity while the sufficiently strongly magnetic particles are carried by the belt past the splitter.

If the feed is not travelling at the same velocity as the roll then either of two conditions will prevail: at high feed velocities the feed particles will crowd the working area on the roll, causing a non-selective separation by entraining non-magnetic particles in clusters of magnetic particles, or low feed velocity merely reduces throughput.

The water level

It has been established that the optimum water level in the separator tank is between 20 and 30 mm below the top of a 308 mm diameter roll [H33]. If the water level is too high, the weakly magnetic particles have insufficient opportunity to adhere to the roll and usually report to the non-magnetic fraction. Conversely, a low water level can cause the non-magnetic particles to be carried by the belt into the zone of the magnetic product.

Process of separation

In wet-roll permanent magnet separators only strongly magnetic particles adhere to the roll while paramagnetic and diamagnetic particles sink in the water in a rather indistinct manner. The diamagnetic particles are also influenced by water currents caused by the rotation of the roll. These currents cause the particles to move towards the roll. The strongly magnetic particles adhering to the belt are pulled off the roll by the belt at the point where the belt leaves the magnetic roll. A region exists at the bottom of the roll where no particles come off, and it is here where all weakly magnetic particles have been removed and strongly magnetic fraction still adheres, that the splitter is positioned, approximately 20 mm below the roll. This narrow gap prevents eddy currents from carrying non-magnetic particles over the splitter. The comparison of the splitter positions in dry and wet roll separators is shown in Fig. 2.25.

The development work has subsequently led to the installation of four production units (BaFe magnets, roll diameter 308 mm, width 1 m) to treat old tailings dumps material after dense media separation [H33].

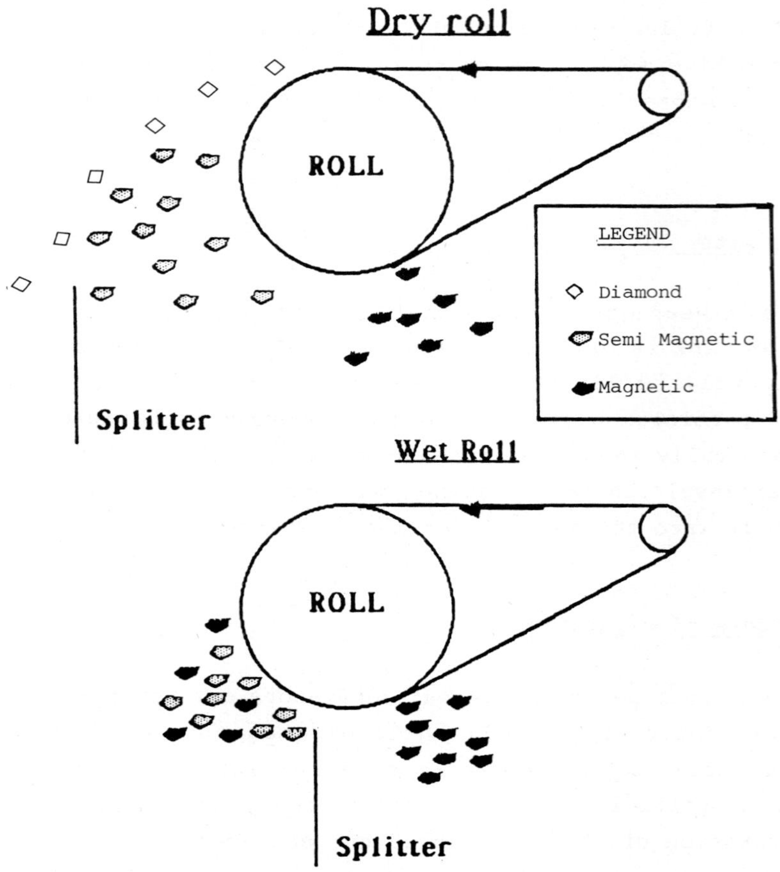

Fig. 2.25 The splitter positions in dry and wet-roll permanent magnet separator (after Hyland [H33]).

II.3.3.5 Ferrous wheel separator

A ferrous wheel magnetic separator combines the basic ideas of a high-gradient matrix separator with the low-cost permanent magnet circuit. The first version of the ferrous wheel separator was constructed in the mid-seventies with the aim of developing an inexpensive dry benefication machine for iron ore. The separator was tried for wet separation as well.

The principles of design and operation are shown in Figs. 2.26 and 2.27. A vertical matrix carrying a ring (wheel) rotates through a magnetized zone where the pulp is fed by gravity into the inner side of the ring. Magnetic particles adhere to the matrix elements and non-magnetic particles pass through the matrix and are discharged at the bottom of the ring. Permanent magnets generate the horizontal magnetic field over an air-gap that can be adjusted for a particular application.

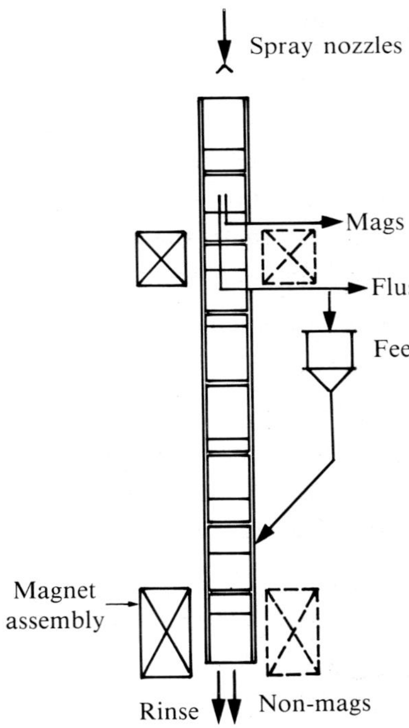

Fig. 2.26 End view of a ferrous wheel magnetic separator showing principle of material and water flows (after Arvidson and Fritz [A6]).

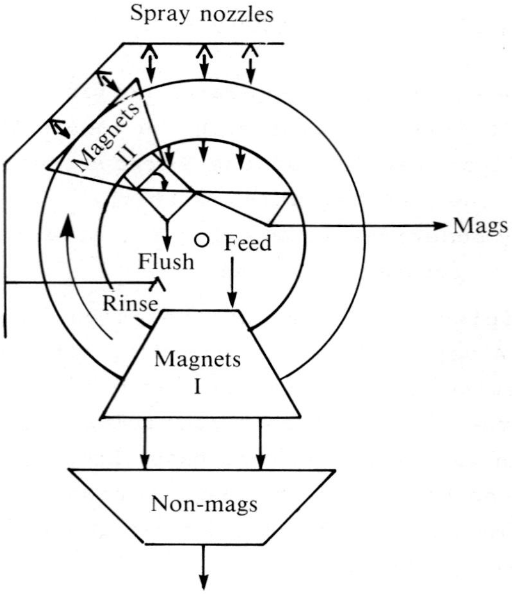

Fig. 2.27 Side view of a ferrous wheel magnetic separator showing principle of material and water flows (after Arvidson and Fritz [A6]).

The matrix with the collected magnetic product may be rinsed while it is in the magnetic zone to displace non-magnetic particles. The magnetic product is washed outside the magnetic zone in direction opposite to the feed-flow direction.

The matrix consists of woven-wire screen fabric, folded to make up a stack. This is inserted in every pocket in the ring and is held only there by friction.

The capacity of the separator is relatively low, varying between 1 to 5 t h^{-1} per wheel. Power is consumed solely by the drive unit; for the largest size considered this is about 5 kW [A6].

A direct full scale comparison is yet to be made but reportedly the laboratory and pilot tests show performance similar to high-gradient magnetic separators [A6]. Encouraging results were obtained on hematite ores, illmenite, clay, glass sand, fly ash, wollastonite and other minerals.

Low capital and operating costs make this machine an

interesting alternative to a high-gradient magnetic separator in those applications where a not too-large magnetic field is needed, or where a high grade magnetic product is desired.

II.4 WET HIGH-INTENSITY HIGH-GRADIENT MAGNETIC SEPARATORS

II.4.1 Dry versus wet separation

Although dry high-intensity magnetic separators have been successfuly applied to the beneficiation of a large spectrum of minerals, several basic disadvantages exist which prohibit a wider use of a dry process. For dry magnetic separation to be successful the ore must be completely (i.e. not partially) dry. Wet grinding is technically and economically well developed and thus expensive heating is not required. Dry magnetic separation almost invariably requires the ore to be sized into several fractions each of which must be spread in monolayers over the separator. With dry methods, care must also be taken to ensure control of possible dust hazard, an expensive precaution both in capital and operating costs. Furthermore, the dry separators have considerably lower capacities than wet machines. While dry separators frequently yield an excellent separation on material coarser than 75 μm, on unsized material containing a large portion of the fines the wet process is the only acceptable one. Considerable advantages must therefore be offered by dry separation to justify its use as against wet separation.

In order to extend the process of magnetic separation to fine weakly magnetic minerals one must use wet magnetic separators generating sufficiently high magnetic force in the working space. A major development was achieved in this direction in 1955 by Jones, who combined a high magnetic field strength with Frantz's idea of a magnetized matrix and who increased the magnetic force by several orders of magnitude compared to dry high-intensity magnetic separators. His wet high-gradient magnetic separator became the starting point of a remarkable progress in the field and resulted in the development of a wide range of high-gradient magnetic separators, which extended the application of magnetic separation to such classes of materials otherwise considered untreatable by magnetic means.

II.4.2 Separators based on electromagnets with iron yoke

II.4.2.1 Jones magnetic separator

The Jones batch high-intensity separator is shown in Fig. 2.28. The magnetic field in the working gap is generated by a conventional electromagnet and the field gradient is enhanced by a matrix consisting of grooved plates of a salient pole or of triangular groove types, placed in the air gap. The slurry is gravity-fed onto the matrix where magnetic particles are captured onto the plates while the non-magnetic fraction passes through and is collected in a trough below the magnet.

Magnetic fraction is removed from the matrix by switching the magnetic field off and flushing the matrix.

For treatment of large quantities of ore the batch separator is impractical. Consequently a continuous machine was developed [S15,S16], a schematic diagram of which is shown in Fig. 2.29. Magnetic yokes (2) fitted with coils (3) are enclosed in air-cooled casings. A rotor shaft (4) supported in roller bearings carries (depending on the type of unit) one or two rotors (6). Round the periphery of the rotors are the plate boxes (7) where separation takes place. The drive (5) is located directly on the rotor shaft and comprises a worm gearing driven by an electric motor through V-belts. The feed pulp flows through the separator via fitted pipes and launders.

The feed required is a slurry with particles smaller than 1 mm. The slurry flows through the feed pipes (8) and feed-boxes into a trough and then into the plate-boxes (7). The feed points are at the leading edges of the magnetic field. Feeding is continuous thanks to the rotation of the plate-boxes on the rotors. Each rotor has two symmetrically disposed feed points.

Magnetizable particles are captured onto the grooved plates while the non-magnetic fraction passes straight through the plate

Fig. 2.28 Jones batch magnetic separator (courtesy of Humboldt Wedag AG).

boxes and is collected in a launder underneath (11). Before leaving the magnetic field any entrained non-magnetic particles are washed out by low-pressure rinse water (13) and these middlings are collected in a separate launder (12) under the rotor.

When the plate-boxes reach the demagnetized zone midway between the two magnetic poles, where the magnetic induction changes the direction, and magnetic force is essentially zero, the adhering magnetic particles are washed out with high-pressure water sprays (14) normally operating at 500 kPa. Below the rotors are the collecting launders (9) with product discharges. A photograph of a flush device for the magnetics is shown in Fig. 2.30.

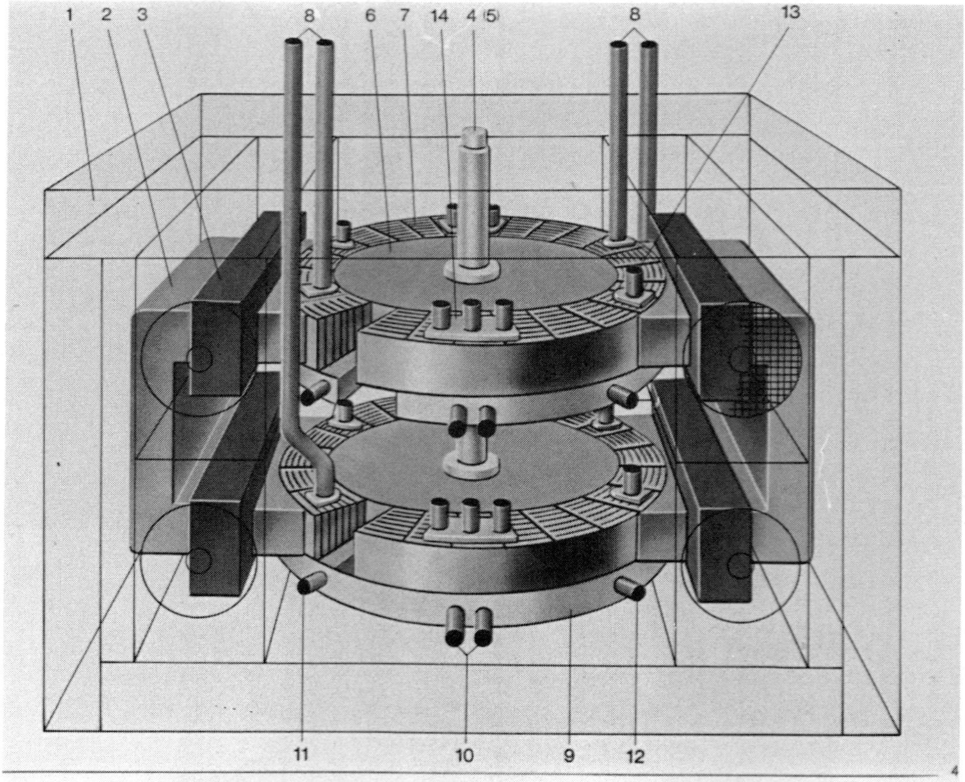

Fig. 2.29 Perspective view of a Jones continuous high-gradient magnetic separator (courtesy of Humboldt Wedag AG).

Matrix plates

Two types of plates can be used in the Jones separator: salient pole type and triangular groove type. When the salient pole type is used it is comparatively easy to wash off the entrained non-magnetic particles hereby producing a reasonably clean magnetic fraction. This type of plate also gives a high recovery and thus is suitable for most purposes, but when the grade of the magnetic fraction is less important and a high recovery is desired, the triangular groove type is often best.

Fig. 2.30 Flush device for magnetics (courtesy of Humboldt Wedag AG).

The plates are arranged vertically in the plate-boxes, as is shown in Fig. 2.31.

The gap between the grooved plates is dependent upon the grain size and magnetic susceptibility of the magnetizable particles to be processed. As a rule of thumb, the gap width, measured between two grooved plate tips, should be from 1.5 to 2 times the maximum particles size in the feed [Bl2].

Fig. 2.31 View of a plate box (courtesy of Humboldt Wedag AG).

Optimum separation depends not only on matching the space between the grooved plates but also their design which must be matched to the particle size distribution and magnetic susceptibility of the ore. For this purpose three different types of plates are normally used:

4R plates with 4 grooves per inch of plate width

8R plates with 8 grooves

12R plates with 12 grooves.

Of these, the 8R plate is a standard one, being used mostly in operation with feed particle sizes smaller than 0.5 mm. 12R plates are used for fine material with particle sizes smaller than 100 μm. 4R plates can be used for treatment of coarse material. The three types of grooved plates are shown in Fig. 2.32.

The plates are usually made of low carbon steels that are protected by a hard chrome layer. The height of the plates, as a rule, is 190 mm, and the filling factor of the plates (defined as a ratio between the volume of the plates and the volume of the plate-box) ranges between 50 to 80 per cent, depending on the type of plates and the gap between them.

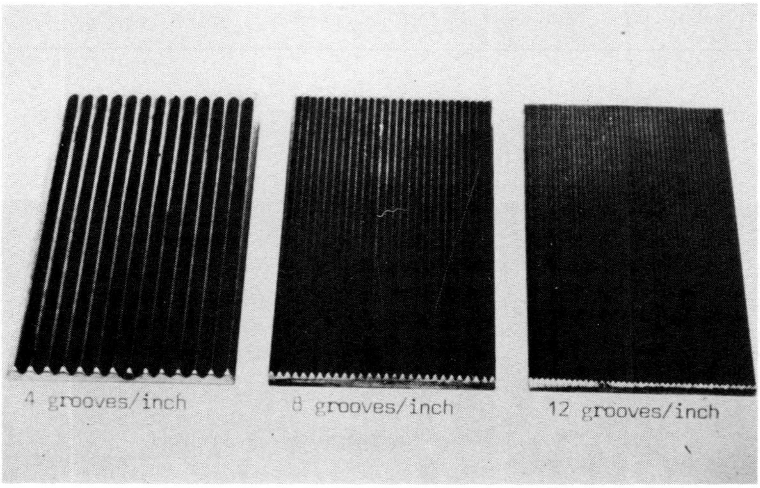

Fig. 2.32 Different grooved plates (courtesy of Humboldt Wedag AG).

Magnetic induction

The maximum background magnetic induction in the gap (in the free space, when the plates are removed) is usually 0.8 T to 1 T. Owing to the limited working space width and to a modest background induction energy consumption is low and on average does not exceed 1kW t^{-1}.

The Jones separator is manufactured by Humboldt Wedag AG, Köln, BRD and is available in a variety of sizes, as summarized in Table 2.6

A photograph of a DP 317 separator is shown in Fig. 2.33 and the erection of the steel frame and iron yoke is illustrated in Fig. 2.34.

TABLE 2.6
Types of Jones double-rotor high-gradient magnetic separators

Type	Capacity (t/h)	Mass (t)	Rotor diameter (m)
DP 335	180	114	3.35
DP 317	120	98	3.17
DP 250	75	70	2.50
DP 180	40	41.7	1.80
DP 140	25	29.2	1.40
DP 112	15	22.4	1.12
DP 90	10	16.2	0.9
DP 71	5	13.4	0.7

Fig. 2.33 A Jones high-gradient wet magnetic separator DP 317 (courtesy of Humboldt Wedag AG).

Fig. 2.34 The erection of the steel frame and iron pole-pieces of a Jones DP 317 separator (courtesy of Humboldt Wedag AG).

Applications

The Jones magnetic separator represents one of the largest applications of magnetic separation in the mining industry. Thanks to a sound mechanical design and reasonably trouble-free performance, the separator can be used to recover a wide spectrum of weakly magnetic minerals. Although the Jones separator has been applied to beneficiate glass sand, feldspar, bauxite, chromium and other ores, the main emphasis is on the upgrading of weakly magnetic hematite ores. A detailed discussion of the applications of the Jones separator in the mining industry is given in Section V.

II.4.2.2 Carpco high-gradient magnetic separator

A design of the magnetic circuit, different from that of the Jones separator, was employed by Carpco, Inc. in their Carpco-Amax high-gradient magnetic separator [C5,H9]. As shown schematically in Fig. 2.35 the carousel (A) containing the matrix rotates between two poles of the electromagnet, which generates a maximum background magnetic induction of 0.35 T. The unit operates in a continuous mode by feeding the pulp into the high-induction areas (B) at each pole pair. Although steel balls were used as a matrix, other shapes were used as well [M11].

The separator was successfully used in a number of applications [H9]. However, a comparison with other high-gradient separators is difficult in view of small magnetic induction and the small depth of the matrix (100 mm).

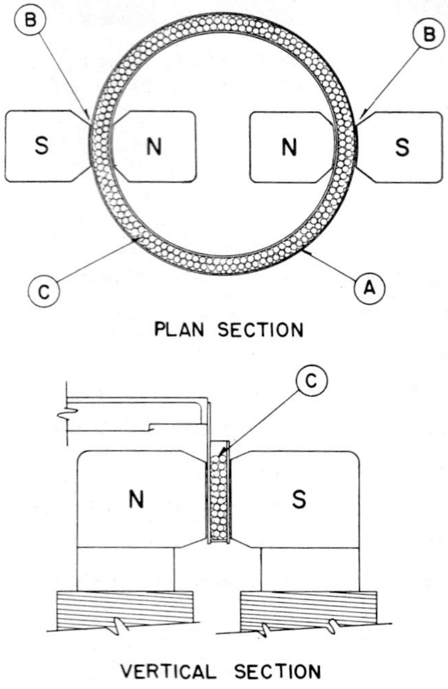

Fig. 2.35 Carpco-Amax carousel magnetic separator (courtesy of Carpco, Inc.).

II.4.2.3 Eriez high-gradient magnetic separator

The magnetic circuit which is practically identical to the one used in the Carpco magnetic separator is used in Eriez high-gradient magnetic separators; shown schematically in Fig. 2.36.

The background magnetic induction of up to 1.2 T in the air gap is generated by oil-cooled coils which are completely closed and thus protected from moisture. Matrix types commonly used include expanded metal and steel balls.

Commercial models range in capacity from 1 to 120 t/h and parameters of selected models are given in Table 2.7; a photograph of an industrial unit is shown in Fig. 2.37.

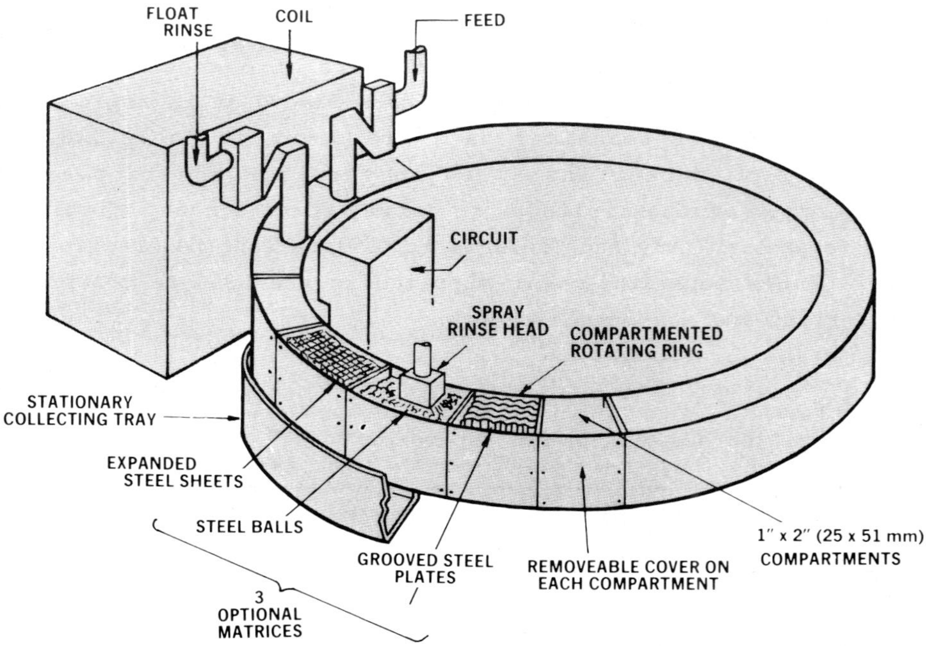

Fig. 2.36 Single-head Eriez high-gradient magnetic separator (courtesy of Eriez Magnetics, Inc).

TABLE 2.7

Eriez carousel magnetic separators

Model	Capacity (t/h)	Ring diameter (m)	Input power (kW)	Mass (t)
CF-10	1	0.76	3	5.7
CF-50	5	0.90	12	11
CF-100	10	1.24	32	13.6
CF-200	20	1.69	70	20
CF-400	40	2.70	140	30
CF-600	60	3.10	180	46
CF-1200	120	3.10	360	90 *)

*) two-deck unit

II.4.2.4 Boxmag-Rapid magnetic separator

Boxmag-Rapid manufactures a wet high-gradient magnetic separator HIW based on a design similar to the Carpco machine. An illustration of a two-rotor 8-head unit is shown in Fig. 2.38.

The matrix in this machine consists of magnetic stainless steel sections of trapezoidal wire bars placed at an angle in the carousel. Typical technical data are given in Table 2.8. HIW separators have been installed in Peru for the recovery of wolframite from the tailings, in Norway for the beneficiation of ilmenite and in Italy and Tanzania for silica sand beneficiation. A number of HIW separators were supplied to the USSR to beneficiate both ferrous and non-metallic ores.

TABLE 2.8

Boxmag-Rapid carousel magnetic separators

Model	Capacity (t/h)	Mass (t)	Input power (kW)
HIW1	5	6	8
HIW2	10	10	16
HIW4	20	18	18
HIW8	40	32	65

Fig. 2.37 Two-head Eriez magnetic separator (courtesy of Eriez Magnetics, Inc.).

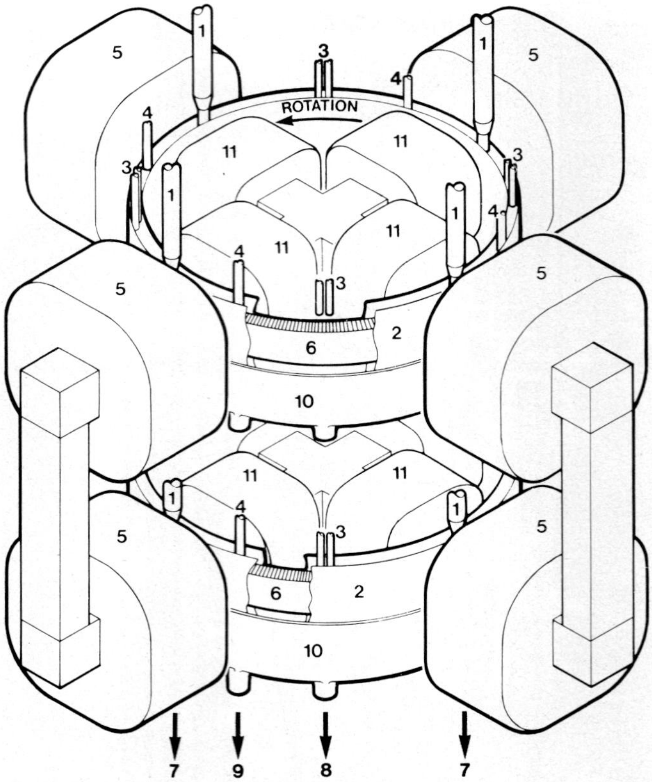

Fig. 2.38 HIW8 Separator (pipework may be arranged for two stage 4 zone separation or single stage 8 zone separation) (courtesy of Boxmag-Rapid).

Key
1. Feed pipe
2. Rotor
3. High pressure water jet
4. Low pressure water jet
5. Outer coil
6. Matrix
7. Non-magnetics discharge
8. Magnetics dicharge
9. Middlings discharge
10. Trough
11. Inner coil

II.4.2.5 MRVK magnetic separator

An unusual variation of the conventional design of a separator has been developed by the Ore Research Institute in Prague, Czechoslovakia [H10,H11]. The separator, designated MRVK, and illustrated in Fig. 2.39 consists of a carousel that rotates between vertical pole-pieces. The carousel is filled with a matrix, usually with steel rods, and the background magnetic induction is equal to approximately 0.9 T. The height of the matrix is 70 mm and the average throughput is 5 t h^{-1} per head, so that a six-zone unit has a capacity of 30 t h^{-1}. The principle of operation is identical to that of other conventional high-gradient magnetic separators.

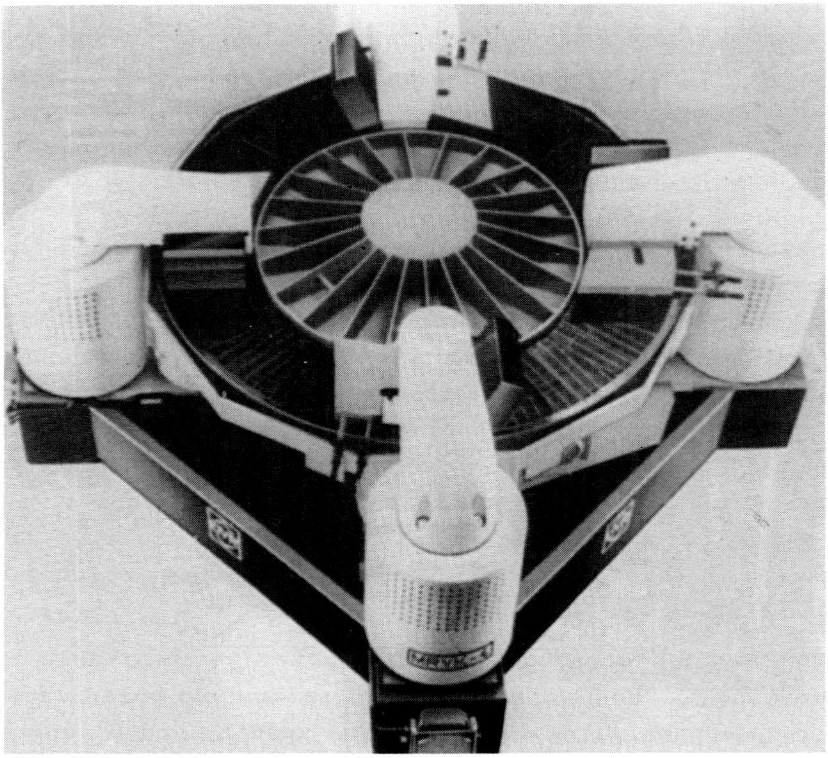

Fig. 2.39 4-zone high-gradient magnetic separator MRVK-4.

The main advantage of the magnetic circuit is that the gap between the pole pieces is constant for any size of the machine and is determined solely by the height of the matrix. On the other hand, owing to the unfavourable position of the coils on the yoke, the contribution from the coils to the magnetic induction in the working space is very small and a stray magnetic field is large. The selected specifications are summarized in Table 2.9.

The MRVK separators are used for upgrading siderite ore and glass sands in Czechoslovakia [Hl].

TABLE 2.9
The specifications of MRVK magnetic separators

	MRVK - 2	MRVK - 4	MRVK - 6
Background magnetic induction (T)	0.9	0.9	0.9
Throughput (t/h)	10 - 14	20 - 30	30 - 45
Water consumption (m^3/h)	12	24	36
Mass (t)	18	45	55

II.4.3 Solenoid magnetic separators

All high-gradient wet magnetic separators discussed so far produced the magnetic field by means of electromagnet with iron yokes. For small working volumes, and for moderately large magnetic induction available in such volumes this constitutes an efficient and relatively cheap design, although machines tend to be massive. The input power needed to saturate the iron poles is modest and reasonably high magnetic induction (~ 1 T) can be generated in the air gaps which guarantee sufficiently high throughputs (tens of tonnes per hour) of fine weakly magnetic material. From a discussion in section I.4.3, it appears that this design suffers from two fundamental drawbacks:

a) The maximum magnetic induction that is obtainable in the working volume is limited by the saturation polarization of iron (approximately 2 T). Since such an induction can be generated only in a very narrow gap, much lower inductions are usually generated in units with a practical meaning.

b) Scaling-up is very difficult. To increase the throughput, the gap between the poles must be increased, which results either in a dramatic drop of the magnetic induction in the working volume, or a substantial increase in the number of ampère-turns in order to maintain the same magnetic induction. The machine becomes extremely massive in relation to its capacity, magnetic induction becomes more and more limited by the width of the gap and the cooling of coils also becomes a problem.

To generate sufficiently large magnetic induction at large volumes, one has to use an iron-clad solenoid, as analysed in Section I.4.3.

II.4.3.1 Frantz Ferrofilter

The first magnetic separator that used the iron-clad solenoid was the so-called Frantz Ferrofilter, patented in 1937 by Frantz [F2].

This separator became the conceptual basis of cyclic high gradient magnetic separators in the kaolin industry. It consists of the iron-bound solenoid magnet shown in Fig. 2.40. The matrix consists essentially of screens fashioned from thin sharp ribbons of magnetic stainless steel. The filling factor of the matrix is 10 to 20 per cent and the background magnetic induction 0.15 T.

The device operates intermittently; the flow is interrupted, the magnetic field is reduced to zero and the magnetic particles are backwashed. Then the field is switched on again and the whole cycle is repeated.

Because of the low demagnetizing factor of the ribbon elements that are oriented parallel with the direction of magnetic field the magnitude of the background magnetic induction was sufficient for the recovery of fine ferromagnetic particles. However, in order to magnetize the randomly oriented filamentary matrix with a very high aspect ratio, much higher magnetic induction is needed. The modern versions of Frantz Ferrofilters are used for the removal of magnetic contaminants from industrial slurries.

A photograph of the pilot Frantz Ferrofilter with power supply, controls for automatic operation and alternative flow systems for gravity or pipeline feed, is shown in Fig. 2.41. The separator provides background magnetic induction up to 2.2 T.

As a means of achieving a high degree of retention of quasi-colloidal paramagnetic impurities from kaolin Frantz's concept was further extended, so that more powerful iron-clad solenoids were used in conjunction with randomly oriented steel wool, which is difficult to magnetize. The early developments of cyclic high-gradient magnetic separators are described in detail in [O4, K3, K9, K10, I1, I2, I3, W2].

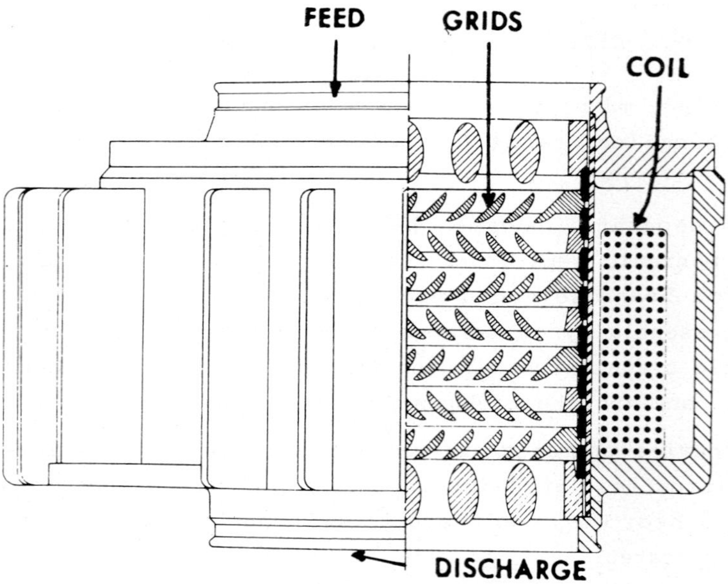

Fig. 2.40 Design elements of an air-cooled Frantz Ferrofilter for processing dry materials (courtesy of S.G. Frantz Co., Inc.)

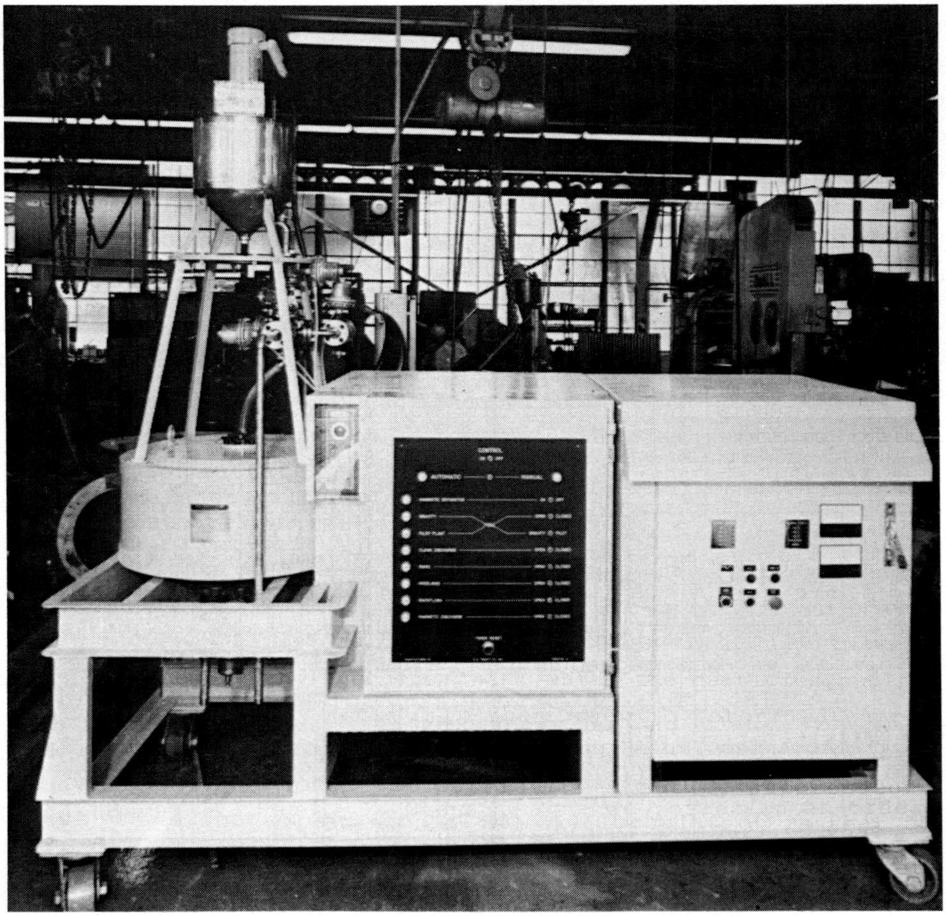

Fig. 2.41 High-intensity Frantz Ferrofilter (courtesy of S.G. Frantz Co., Inc.)

II.4.3.2 Cyclic high-gradient magnetic separators

The first industrial high-gradient magnetic separator which grew from Frantz's concept and from an investigation into the removal of weakly magnetic discolouring agents from clays [I3, I2, M12, K3]; was designed and built by Magnetic Engineering Associates (now Sala Magnetics) in 1969 [O6]. Fig. 2.42 shows schematically the essential elements of such a separator. The system comprises a canister filled with a matrix formed of compressed mats of magnetic stainless steel wool. The canister is placed in an iron-clad solenoid which generates a magnetic induction of up to 2 T. The thin fibres of the matrix create a high degree of the non-homogenity of the magnetic field and produce a large magnetic force acting on the magnetizable particles.

The slurry is pumped vertically upwards through the matrix with the magnetic field switched on. The magnetic particles from the slurry are trapped onto the surface of the magnetized fibres,

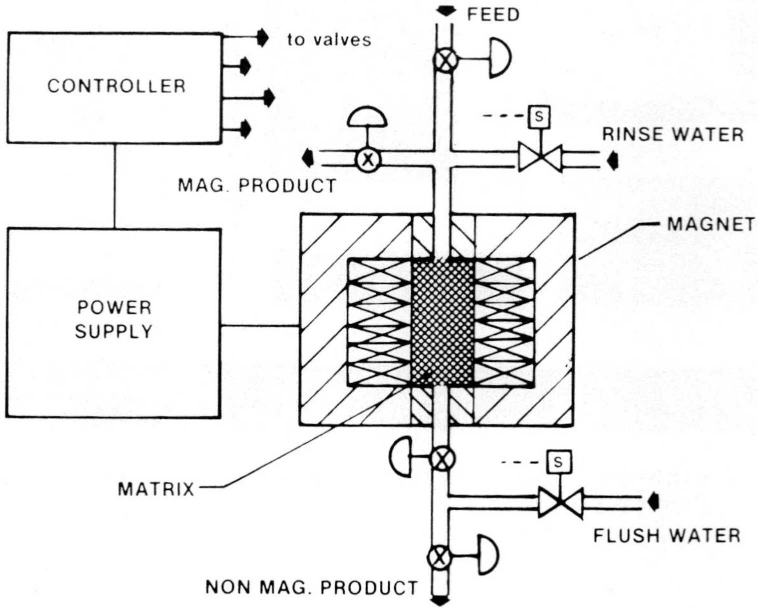

Fig. 2.42 Schematic diagram of a small cyclic SALA-HGMS magnetic separator (courtesy of Sala Magnetics).

while non-magnetic particles pass through the canister. When the matrix has become loaded with magnetic particles the flow is halted, the magnetic field is reduced to zero and the magnetic fraction flushed from the matrix.

Such cyclic devices are very useful in applications where a pulp to be processed contains a small fraction of magnetic particles, preferably very fine so that their mechanical capture by straining is avoided, and very weakly magnetic so that the blockage of the matrix by magnetic flocs is avoided and low flowrates can be used. In the opposite case the duty cycle of such a cyclic machine would be very low and the filamentary matrix would be susceptible to clogging as a consequence of the straining mechanism. These machines found a wide-spread application in the kaolin industry and in waste water treatment.

A wide range of cyclic high-gradient magnetic separators (Fig. 2.43) is manufactured by several companies.

A list of assorted cyclic high-gradient magnetic separators is given in Tables 2.10 and 2.11.

TABLE 2.10

Technical characteristics of various cyclic high-gradient magnetic separators, for kaolin beneficiation (matrix height 150 to 500 mm, max. magnetic induction 2 T)

Model	Mass (t)	Matrix diam. (m)	Matrix area (m^2)	Feed throughput	Power input (kVA)	Manufacturer
EO-20-84	340	2.1	3.6	115 m^3h^{-1}	400	Eriez
107-30-20		1.07	0.9			Sala
152-30-20		1.5	1.8			Sala
214-50-20		2.1	3.6			Sala
5 PEM		0.13	0.01		200	PEM
84 PEM		2.1	3.6	100 th^{-1}	400	PEM
120 PEM		3.0	7.1	200 th^{-1}		PEM

Fig. 2.43 2.1 m high-gradient cyclic magnetic separator (courtesy of Eriez Magnetics Inc.).

TABLE 2.1
Technical characteristics of cyclic high-gradient magnetic separators for steelmaking water purification (matrix height: 150 mm, maximum magnetic induction: 0.5 T, manufacturer: Sala Magnetics)

Model	Mass (t)	Matrix diam. (m)	Matrix area (m^2)	Feed throughput range (m^3 h^{-1})	Power input (kVA)	Canister diameter (m)
305-15-5	77	3	6.9	1300 - 4500	75	2.96
214-15-5	32	2.1	3.4	650 - 2200	56	2.1
152-15-5	20	1.5	1.7	350 - 1200	43	1.5
107-15-5	8	1.0	0.84	150 - 500	36	1.0
76-15-5	4	0.7	0.42	80 - 270	28	0.70
56-15-5	2.6	0.5	0.18	40 - 150	20	0.48
38-15-5	1.3	0.3	0.075	40 - 70	18	0.30
10-15-5	0.4	0.08	0.005	1 - 4	10	0.08

II.4.3.3 Continuous solenoid magnetic separators

The economic expansion of high-gradient magnetic separation to many mineral processing applications requires the utilization of fully continuous magnetic separation. Cyclic devices are limited in their use to separations in which the magnetic fraction is a small portion of the feed and to feeds which may be passed through the device at slow flowrates.

Solenoid continuous magnetic separators have eliminated the limitations of cyclic machines and several advanced devices have been built.

II.4.3.4 Sala-HGMS carousel separator

The Sala - HGMS carousel separator was developed from the Sala cyclic separator to meet the need for a continuous industrial machine capable of processing ores containing a high percentage of magnetics. [O1,M13,A7]. The first continuous device, Mark I, shown in Fig. 2.44, was constructed for the Francis Bitter National Magnet Laboratory at the Massachusetts Institute of Technology, to be used as a low capacity machine.

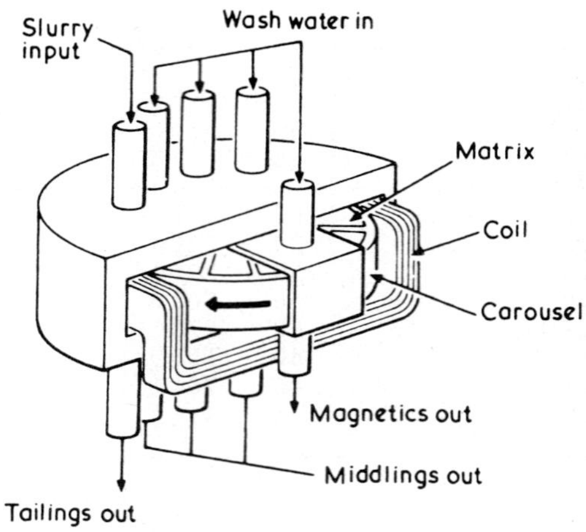

Fig. 2.44 Schematic view of Sala - HGMS Mark I carousel magnetic separator (after Oberteuffer, IEEE Trans. Mag. [Ol]).

The Mark II which was the first scalable prototype constructed had the principal characteristics of a large production machine: a double-saddle wedge type iron-clad coil surrounding the carousel ring and a sealing system isolating the matrix-containing chamber, as shown in Fig. 2.45.

The operational sequence is similar to that used in conventional continuous high-gradient separators and is illustrated in Fig. 2.46. The matrix is flooded during the entire feed, rinse and flush cycle.

The sealed system allows a close control of flow velocity through the matrix. In order to maintain this control feature for the separation of fine particles air-water interface force must be avoided. The seal system also allows the options of back flushing and back rinsing.

Another advantage of such a magnetic separator is that it is scalable to high process capacities. Magnetic field is parallel to the slurry flow. When increased capacity is desired, the width of the magnet is increased, while its height remains constant. Thus the surface area of the matrix can be enlarged without changing the characteristics of the magnetic circuit. The material cost for the

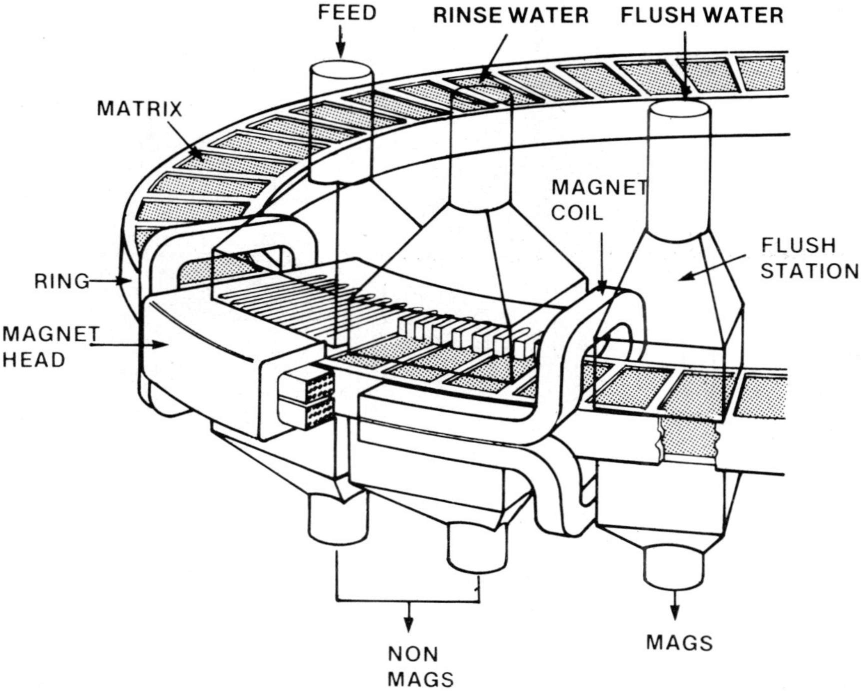

Fig. 2.45 Continuous Sala high-gradient magnetic separator (courtesy of Sala Magnetics).

magnet and the energy cost per tonne of processed ore will drop for large-size machines because capacity will increase more rapidly than costs.

The iron return frame reduces the magnetic flux leakage to places outside the magnetic head. In high-gradient magnetic separators with iron yokes the stray magnetic field is so high that despite a reversal of polarity of two adjacent magnetic heads, the magnetic induction in the vicinity of a flush station usually amounts to about 20 mT, which is a value sufficiently high to hamper the efficient flushing of the matrix. With iron-clad solenoids the magnetic induction outside the magnet is as low as 2 mT, depending on the matrix used.

Sala high-gradient separators are manufactured in various models. The number of magnetic heads ranges from one to four and magnetic induction varies from 0.5 to 2 T. Table 2.12 summarizes the basic technical data.

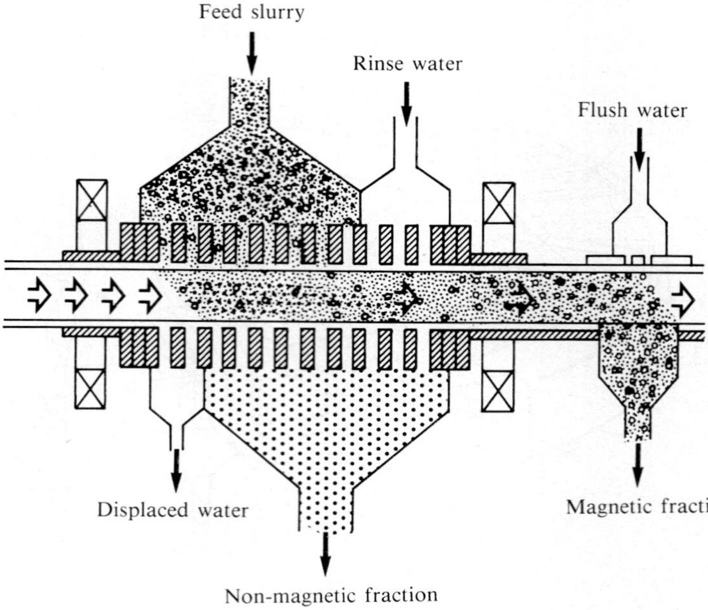

Fig. 2.46 Typical operational sequence in a continuous magnetic separator (after Arvidson et al. [A7]).

TABLE 2.12

Sala HGMS continuous magnetic separators (after [S17])

Model	Magnetic induction (T)	Carousel diameter (m)	Capacity per head (t/h)	Mass per head (t)	Input power per head (kW)	Basic mass (t)
120 A-B	0.5	2	3	1.5	25	4.5
	1.0			2.3	75	
	1.5			3.5	160	
	2.0			5.5	250	
185 A-B	0.5	3	25	4.5	25	5
	1.0			10	75	
	1.5			16	170	
	2.0			35	250	
350 ABCD	0.5	6	100	18	40	12
	1.0			45	125	
	1.5			95	175	
	2.0			200	250	
480 ABCD	0.5	8	200	36	100	18
	1.0			95	175	
	1.5			190	250	
	2.0			300	300	

Fig. 2.47 Sala HGMS carousel magnetic separator. Model 480 (Courtesy of Sala Magnetics).

Fig. 2.48 Sala HGMS carousel magnetic separator model 185 (courtesy of Sala Magnetics, Inc.).

Models 120 and 185 are available with one or two heads while models 350 and 480 are available with one to four heads. Machine heads in the same unit can work in parallel or in series and can be built for the same or different magnetic inductions [S17]; photographs of the separator are shown in Figs. 2.47 and 2.48. The separators can be used for processing iron ores, industrial minerals, coal and chemical wastes.

II.4.3.5 SOL magnetic separator

In the SOL separator built by Krupp GmbH [S18,S19], the saddle-shaped solenoid with the magnetic field perpendicular to the direction of the motion of the carousel, as used in Sala separators, was replaced by a simple axial solenoid that generates a magnetic field in the direction of motion of the carousel. The coil is wound with solid aluminium tape and is immersed in an oil cooling bath; no iron cladding is used. The two-head machine is illustrated in Fig. 2.49.

In order to provide a reasonably long retention time of the slurry in the magnetic field at a reasonably high throughput, the magnet has to be sufficiently long (approximately 1 m). This means that for a medium-sized separator, the coil is much longer compared to its diameter. Since the stray magnetic field in a long thin solenoid is small the influence of the external iron cladding is also small. The magnetic field must therefore be obtained directly from the current in the electrical conductor. The effect of the iron-cladding upon the characteristics of such an axial magnet is shown in Table 2.13. It can be seen that iron-cladding reduces the required number of ampère-turns by little and a small reduction in energy consumption does not justify an increase in the mass of the machine if cladding is used.

The main advantage of the SOL separator is its low mass, less severe tracking problems during the movement of the carousel through the magnetic field and an elimination of the need for robust bearings. The solenoid is split into a number of coils which generate distribution of magnetic induction in the magnet and thus partially reduce very high energy consumption.

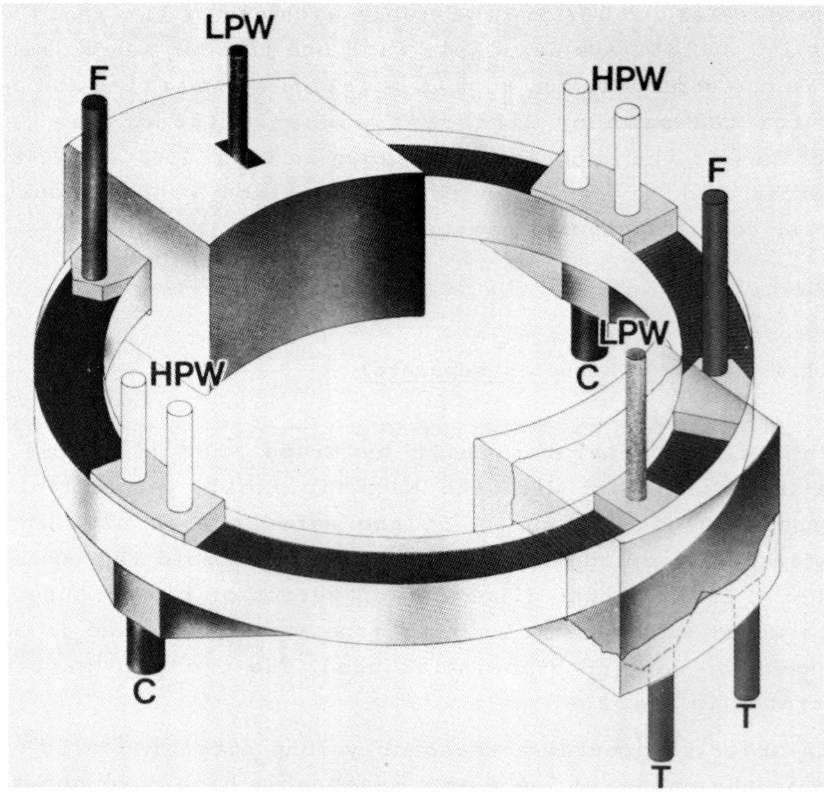

Fig. 2.49 SOL-magnetic separator (courtesy of Krupp-Polysius, GmbH).
F - feed, LPW - low pressure water (rinse), HPW - high pressure water (flush), C - mags, T - non-mags.

TABLE 2.13

The effect of steel cladding on technical parameters of an axial magnet (magnet length 0.9 m, gap width 0.56 m, gap height 0.26 m, background magnetic induction 1.2 T, capacity of the separator 30 t/h)

B = 1.2 T	With iron cladding	Without iron cladding
Number of ampère-turns (A)	8.85×10^5	1.0×10^6
Power input (kW) at 20°C	367	469
60°C	407	520
80°C	436	557
Energy consumption per tonne of ore (at 20°C)((kWh/t)	12.2	15.6

The SOL-separator is manufactured by Krupp GmbH in assorted sizes. A four-head unit is illustrated in Fig. 2.50; technical data are summarized in Table 2.14. Fig. 2.51 shows the detail of a carousel with matrix.

TABLE 2.14
SOL magnetic separators (Krupp-Polysius, GmbH)

Model	Carousel diameter (m)	Number of magnetic heads	Total capacity (t/h)	Mass (t)
12/12-9	1.2	2	5	1.5
16/12-24	1.6	2	20	
24/14-18	2.4	4	40	10
28/14-36	2.8	4	80	
36/16-36	3.6	6	120	
48/18-54	4.8	8	240	50

Notes: 1) matrix height : 100 mm for models 12 and 16
170 mm for other models
2) background magnetic induction not specified but probably between 0.6 and 0.8 T.

Fig. 2.50 SOL-magnetic separator (courtesy of Krupp-Polysius, GmbH).

Fig. 2.51 Detail of a carousel of SOL magnetic separator (courtesy of Krupp-Polysius, GmbH).

II.4.3.6 VMS magnetic separator

The VMS magnetic separator shown diagrammatically in Fig. 2.52 was developed in the Ore Research Institute in Prague, Czechoslovakia [H13, H14, C8].

Magnetic field is generated by the horizontally oriented solenoid (1) wound with a water-cooled hollow copper conductor. The coil is enclosed in steel-cladding (2) the upper and lower pole-pieces of which are perforated to allow passage of the slurry. The slurry is fed into the separator via the feed box (3) and then enters a rotating drum (4) which carries separation chambers (5) filled with matrix. The non-magnetic fraction is collected in a launder (6) while the middlings washed from the matrix at the rinse station (7) are collected in the launder (8). Magnetizable particles are attracted to the matrix, which subsequently leaves the magnetic field by rotation of the drum. The magnetic product is washed off from the matrix at the flush station (9) in a direction opposite to that of the feed and is collected in the trough (10).

Maximum background magnetic induction generated by the magnet is 1.7 T, stray magnetic induction outside the magnet is less than 5 mT. Rods of 3 and 5 mm diameter with spacing of between 1.5 and 2.5 mm, are used as a matrix, although woven wire-mesh has also been used.

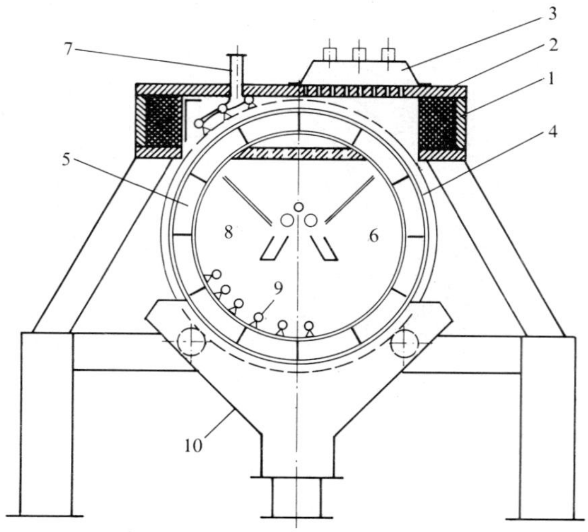

Fig. 2.52 VMS magnetic separator.

To increase the capacity of the machine, the width of the drum has to be increased. This means that the basic characteristics of the magnetic circuits, plus the length of the air gap (which determines the number of ampère-turns needed to achieve the required magnetic induction) remain unchanged. Furthermore, because of an increase in the length of the conductor, energy consumption increases only linearly with the drum length.

The VMS magnetic separator, schematically shown in Fig. 2.53 is manufactured by Železnorudné bane n.p. Spišská Nová Ves (Czechoslovakia); several models are shown in Table 2.15. Table 2.16 gives the specifications of a machine that was extensively tested in the USSR with the use of oxidized quartzites from the Krivoi Rog District and Kursk Magnetic Anomaly [C8].

The main advantages of the VMS magnetic separator are excellent ability of scale-up, high background magnetic induction at modest input power and reportedly good resistance of the matrix to clogging.

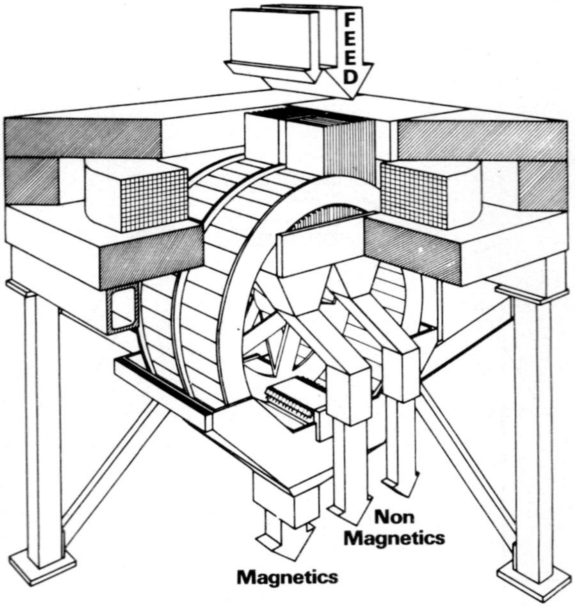

Fig. 2.53 Diagrammatic layout of VMS.

TABLE 2.15
VMS magnetic separators

Model	Magnetic induction (T)	Dimensions L x W x H (m)	Mass (t)	Input power (kW)	Capacity (t/h)
VMS 25	0.5	2.5 x 3.3	25	13	25 - 35
	1.0	x 4.5	37	48	
	1.5		48	190	
VMS 50	0.5	3.5 x 3.3	45	17	50 - 75
	1.0	x 4.5	56	66	
	1.5		101	225	
VMS 100	0.5	6.1 x 3.3	80	33	100 - 150
	1.0	x 4.5	110	132	
	1.5		250	550	

TABLE 2.16

Specifications of VMS 100-1T

Parameter	Value
Background magnetic induction	0.5 T
Power input:	
solenoid	22.4 kW
drum drive	3.6 kW
cooling circuit	4.2 kW
total	30.2 kW
Throughput - ore	100 $t\ h^{-1}$
- slurry	150 $m^3\ h^{-1}$
Specific power input	0.3 $kW\ t^{-1}$
Water consumption:	
rinse water	75 $m^3\ h^{-1}$
flush water	80 $m^3\ h^{-1}$
total	155 $m^3\ h^{-1}$
Specific water consumption (per tonne of feed)	1.6 $m^3\ t^{-1}$

II.5 SUPERCONDUCTING MAGNETIC SEPARATORS

Although high-gradient magnetic separators that use the resistive magnets meet the technological requirements of the mining industry, their cost-effectiveness is impaired by high energy consumption and by considerable mass owing to massive iron yokes or cladding. These disadvantages however can be overcome by using superconducting magnets which offer a number of advantages.

a) Superconducting magnets can produce high magnetic induction in large volume at low energy consumption. If operated in persistent mode, they consume no energy at all, while in non-persistent mode they consume only a small amount of energy as a consequence of energy dissipation in the resistive leads that supply the electric current. A factor of five to ten in power reduction is easily possible for a superconducting machine, compared with a conventional machine of the same processing capacity [W3]. The power is required mainly for liquifaction of helium rather than for ohmic losses in the coil of the magnet.

b) Since iron cladding is not needed the machine can be made physically smaller.

c) The superconducting coils can be arranged in such a way that a sufficiently large field gradient can be generated without using a matrix, while still maintaining high magnetic induction outside the magnet.

The advantages of resistive and superconducting magnets are summarized in Table 2.17, their comparison is shown in Table 2.18.

TABLE 2.17

Advantages of resistive and superconducting magnets

Resistive magnets	Superconducting magnets
Robust construction	Energy-saving
Reliablity	Size and mass reduction
Easy scaling-up	Elimination of return frame
	Higher magnetic field

Although the advantages of superconducting magnets seem to be convincing, a number of problems still hamper a genuine large-scale application of superconductivity in mineral processing, viz:

1) The actual saving of energy is not as high as was assumed, due to the energy needed for cooling.

2) The method of cooling has not yet been proved: refrigeration is still unreliable, even under laboratory conditions, and direct cooling by liquid helium, though reliable, is expensive and complicated, since the helium vapours must be recovered if the process is to be economically viable.

3) Since a magnetic separator is usually a small part of a complex flowsheet, any saving in energy can be easily offset by increased unreliability and unexpected expenses caused by forced shut-downs.

4) The need for magnetic induction greater than 2 T has never been convincingly demonstrated in matrix separators, and it seems improbable that, except for some isolated applications induction greater than 2 T will be required. It is claimed that by enhancing magnetic induction to well over 2 T it would be possible to increase the flow velocity $\mathbf{v}_o$ at the same rate (and thereby the production rate) by keeping $\mathbf{B}/\mathbf{v}_o$ constant.

Although this outcome of a simple theoretical model has been supported by isolated experiments, it is doubtful whether this rule would be generally valid, even in those applications in which non-magnetic fraction is the valuable product, and it is certainly not true for those applications in which a magnetic fraction is the useful product.

TABLE 2.18

Comparison of resistive and superconducting magnets for magnetic separators

	Resistive solenoid magnet	Superconducting magnet
Energy consumption	High	Low
Magnetic induction	Limited by the available cooling, usually up to 2 T	Up to 15 T
Mass	High	Low
Conductor	Hollow copper conductor	Superconductor (Nb-Ti)
Cooling	Direct water cooling	To 4 K (liquid helium)
Iron return frame	Important for short coils	Not required
Reliability of operation	High	Low to modest
<u>Capital costs</u>		
Magnet	High	High
Power supply	High	Low
Cooling	Low to medium	High
<u>Running costs</u>	High	Low to medium
Cyclic matrix separator	Well proven	In operation
Continuous matrix separator	Well proven	Not designed, yet
Open-gradient separator	Not suitable	Two types a) drum separator b) split-pair separator

Although superconducting magnetic separators have been built in various laboratories, and large-scale cyclic HGMS and reciprocating canister HGMS machines are in operation superconductivity has made no considerable impact on mineral processing as yet.

In the following paragraphs a concise description of some of existing or proposed superconducting magnetic separators is given.

II.5.1 Superconducting matrix separators

II.5.1.1 Cyclic magnetic separators

The first superconducting magnet that was used as a magnetic separator was a simple solenoid with a warm bore, which was filled with the matrix. The separator, schematically shown in Fig. 2.54 was designed and built by Magnetic Corporation of America and described in [S12,S20]. The separator had a working space 127 mm in diameter and 600 mm long, in which the canister, 500 mm long, was situated. The canister contained ferromagnetic stainless steel wool as a matrix. The magnetic separator was used by Watson et al. [W4,W5] for brightening kaolin by removing the iron and titanium oxide particles.

A similar system was used by Dörr et al. [D9] for coal desulphurization. The magnet consisted of 10 single coils, each

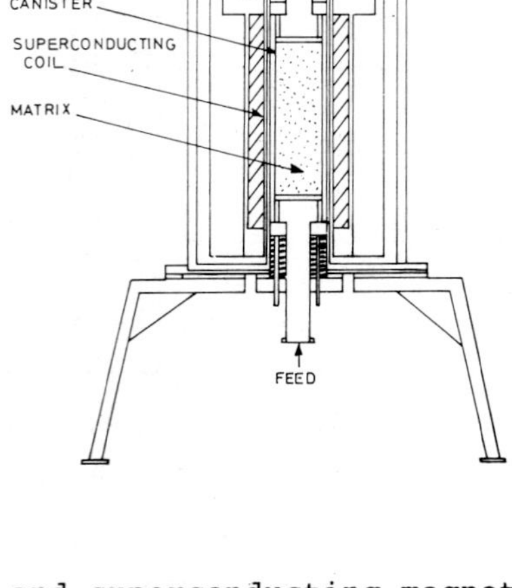

Fig. 2.54 Cryostat and superconducting magnet (after Stekly [S12]).

with an internal diameter of 108 mm and an external diameter of 210 mm, immersed in liquid helium. The magnet generated a magnetic induction of 6 T.

The CEN Saclay (France) in cooperation with the Ecole Supèrieure de Gèologie, Nancy, designed and built a self-contained, superconducting magnetic separator, having a high degree of reliability, minimum maintenance and low running costs; a schematic view of the separator is shown in Fig. 2.55 [H15].

This magnetic system differs from the previously described units in that it requires no external liquid helium supply. It has its own integral helium liquifier and a fully closed helium circuit. Service intervals for the cryogenerator and cryostat are about 2000 hours. The cooling system which employs a two-stage process (to 60 and 15 K) followed by the Joule-Thompson cycle, takes 24 hours to cool the magnet from room temperature to the operating temperature. The coil is wound with NbTi wire, its length is 800 mm, and its inside and outside diameters are 168 and 257 mm, respectively. The magnet operates at 150 A and generates the magnetic induction of 5 T.

The common feature of all superconducting magnetic separators discussed above is that they are cyclic units that require the

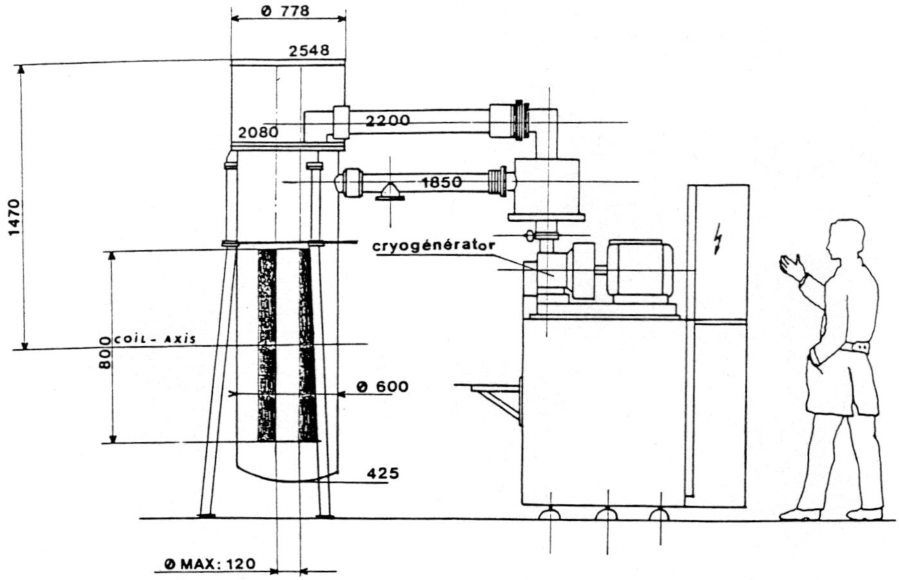

Fig. 2.55 Schematic view of the separator: cryostat and cryogenerator (after Huout et al. [H15]).

intermittent switching on and off of the magnetic field. Although superconducting coils that use available superconductors have been developed for charge times as low as 2 seconds [S12], the energy consumption during charge or discharge and increased cooling requirements indicate that it is optimum to keep the coil charged continuously. Although most developments are towards continuous machines, Eriez Magnetics designed a laboratory cyclic machine (Figs. 2.56 and 2.57) and later a large 2 m-diameter production-scale, cyclic superconducting separator [A8]. The superconducting magnet cooled by liquid helium to 4 K generates magnetic induction of 2 T at a energy consumption of 10 to 20 per cent of that for conventional magnets and represents an annual saving of $ 150 000 [A15]. The charge and discharge time constants are of the order of 60 seconds so that a high duty cycle in envisaged. The new separator measures 4 x 4 x 2 m and weighs 200 tonnes, occupies 34 per cent less space and weighs 47 per cent less than conventional separators with the same parameters [A15]. The unit is used for kaolin beneficiation at J.M. Huber Corporation's plant in Wrens, Georgia.

Fig. 2.56 Laboratory-type cyclic superconducting separator (courtesy of Eriez Magnetics, Inc.).

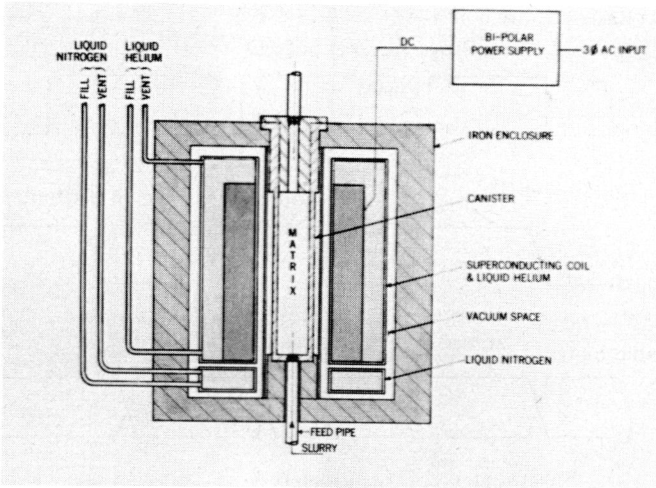

Fig. 2.57 Schematic diagram of a cyclic superconducting magnetic separator (courtesy of Eriez Magnetics, Inc.).

II.5.1.2 Reciprocating canister magnetic separators

Principle of operation

When conventional HGMS operating procedures are used, large superconducting magnets require long charge and discharge times, leading to low duty cycles. This major restriction was overcome when several investigators [S12,W6] proposed that two active canisters should be alternately moved into and out of the magnetic field. In this operation, one canister processes the slurry while the second canister is cleaned in a region of zero magnetic field. The cycles in single matrix and double matrix magnetic separators are shown in Fig. 2.58.

In the reciprocating version of the separator, it is possible to reduce the dead time to the time required to change the canisters, viz. as short as two seconds [W7]. For conventional systems the dead time is usually of the order of 200 sec.

Systems of this type, with the use of two canisters, have been described in the literature [S12, W6, R1]. In these systems, the

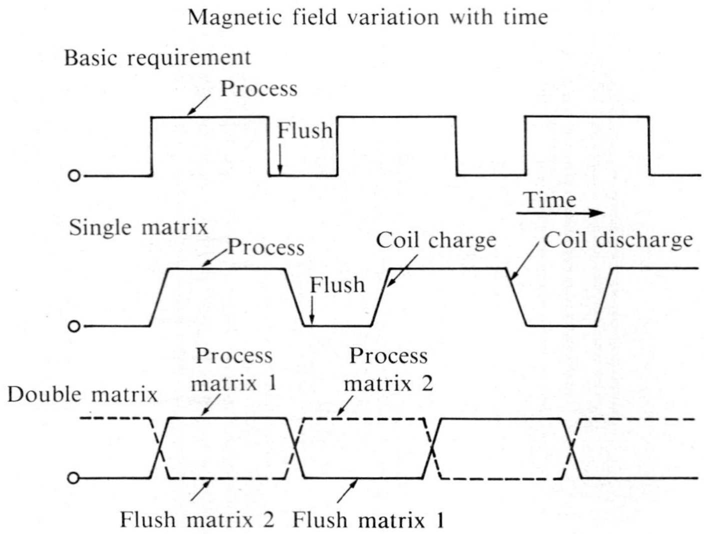

Fig. 2.58 Cycles in single-canister and double-canister magnetic separators (after Stekly [S12]).

two matrix-filled canisters are placed in a canister train surrounded by non-active matrix-filled dummy canisters whose purpose is to reduce the force required to move the active canisters from the field. For example, to remove a canister 2 m long and 1 m in diameter filled with 5 per cent of ferromagnetic steel wool from a solenoid generating a magnetic induction of 8 T a force of 3.5 to 5 x 10^6 N is required [W7]. By using a suitable arrangement of dummy canisters (Fig. 2.59), this force can be reduced to about 10^4 N [R1].

As the magnetic properties of the active filter vary according to the filling factor of the matrix, it is not advantageous to fill the dummy canisters in the same manner as the active canisters. Furthermore, in order to facilitate the bleeding and draining of the active canisters and to maintain access to the bottom of the active canisters (for the purpose of cleaning them) Chrz at al. [C9] proposed the use of a set of removable steel plates in the dummy canisters. By varying the size and composition of these plates it is possible to achieve a homogeneity of the canister train and to compensate for possible changes in the matrix filling factor and for the presence of a slurry in the active canister.

Further improvements to the double canister reciprocating system were suggested by Watson et al. [W7, W8]. The improvement is achieved by dynamically reducing the force on the magnet using an additional externally applied magnetic force which is varied as the canister train is moved. This is achieved by sensing the position of it and adjusting the externally applied force to restrain the magnet within narrow limits, thus allowing the superconducting magnet to be left in the persistent mode during the operation. The electrical leads can thus be designed for zero current, which will reduce the requirement for cryogenic liquids. Even greater saving is achieved in refrigerator capacity by a reduction in the support strength for the magnet, with an estimated helium saving of the order of 3 to 4 l/h.

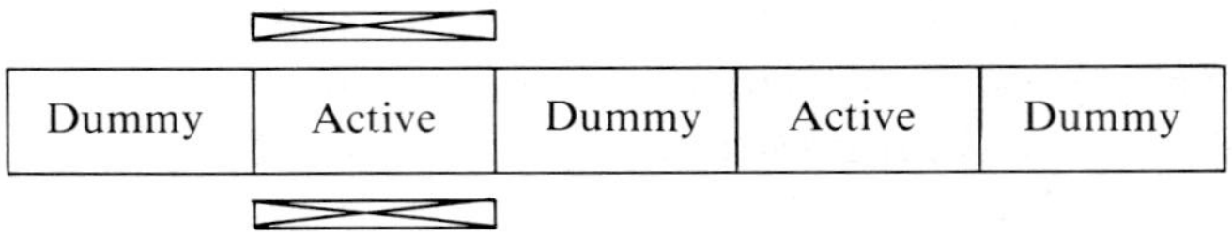

Fig. 2.59 Reciprocating canister train.

As described above, in a magnetic separator assembly consisting of two active canisters and three dummy canisters (Fig. 2.59), separation is interrupted only for the time needed to replace one canister with the other. The configuration of the active and dummy canisters should represent a body of constant magnetic susceptibility and infinite length and therefore zero forces should act upon the system during the canister train movement. Although in the practical situation the forces will be non-zero, they can be further minimized by altering the passive magnetic compensation, i.e. by changing the length and filling factors of the dummy filters. However, there are residual forces of the order of 5 kN with rates of change of 30 kN s^{-1} [W7]. A possible means of further reducing the residual forces is a contra-reciprocating canister system, as shown in Fig. 2.60.

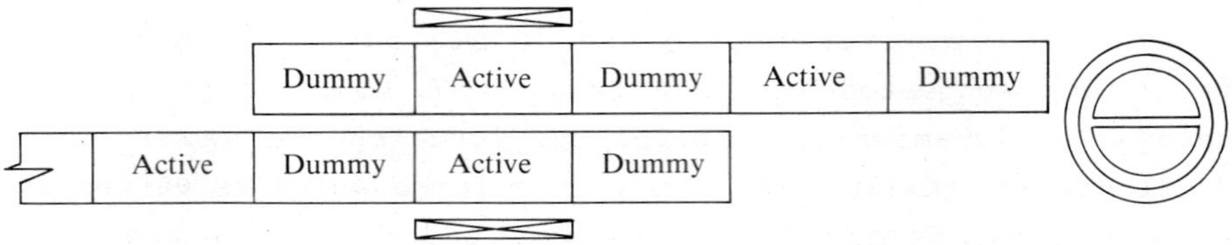

Fig. 2.60 Contra-reciprocating canister train (after Watson et al. [W7]).

However, the system requires synchronization and is mechanically complex. A more responsive system is the one that employs a compensator electromagnet capable of creating a further magnetic field interacting with the magnetic field of the superconducting magnet. The tendency of superconducting magnets to change position in reacting to the residual forces is sensed and the direction and amplitude of current flow in an auxiliary coil, positioned at one end of the solenoid, is changed proportionally. The system is schematically shown in Fig. 2.61. The field produced by the auxiliary coil will interact with and attract or repel the solenoid field to counter-balance the residual forces.

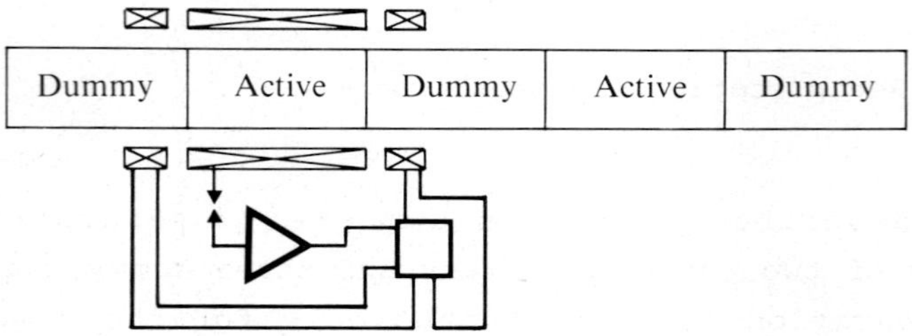

Fig. 2.61 Reciprocating canister train and load compensating system (after Watson et al. [W7]).

These dynamically balanced superconducting magnetic separators of commercial size appear to have considerably reduced cooling requirements compared with statically balanced systems. It has been estimated that a 5 Tesla reciprocating canister system, with a production rate equivalent to a conventional magnetic separator generating 2 T in a working volume of 2.1 m diameter, will have a refrigeration input power of about 20 kW [P7].

However, economic and technical reasons put an upper limit on the magnetic field in the reciprocating canister separator to about 5 T. As shown by Watson, in large-bore superconducting magnets (diameter greater than 0.3 m) the coil stresses become significant and fabrication costs limit the maximum practical magnetic induction to about 5 T. This is also the approximate upper limit at which displacement of fine matrices ceases to be reversible [P7] in field gradients at the extremities of solenoids during canister withdrawal.

The development of reciprocating magnetic separators

A small-scale reciprocating canister superconducting separator was used to reduce the iron and potassium from a slurry of kaolin [R1]. The machine was operated on a semi-continuous basis for a total of 844 cycles without any problems and demonstrated that reciprocation into and out of a magnetic field generated by a superconducting solenoid was a viable technique for kaolin beneficiation. No problems occurred with matrix blockage during the test program, showing that an active canister can be adequately flushed when moved into a region of low stray magnetic field. Reciprocation had no apparent detrimental effect on the matrix ability to recover its original dimensions once all compressive stresses had been released.

The first superconducting magnetic separator of production size to be installed on the plant was the reciprocating machine built in Czechoslovakia by the Research Institute of Food and Refrigeration Engineering [F7, K35]. The horizontal assembly of the separator comprises two active canisters having a volume of 0.18 m^3, filled with stainless-steel wool, and three dummy canisters. The reaction forces in the system were so reduced that the reciprocating stroke of the canister train was achieved merely by the hydraulic force of wash water, and no mechanical drive was employed.

The superconducting winding consists of eight parallel sections wound with a multifilamentary Nb-Ti conductor. The winding is cooled by liquid helium in a cryostat of conventional design. Magnetic induction at the centre of the warm bore (diameter 560 mm, length 1320 mm) is 5 T, while induction at the shielded flush station is 0.02 T. The separator for treating 5 tonnes of solids per hour, at a power input of 30 kW, was installed in 1985 at the West Bohemian Ceramic Works in Kazněјov, Czechoslovakia, for the beneficiation of kaolin.

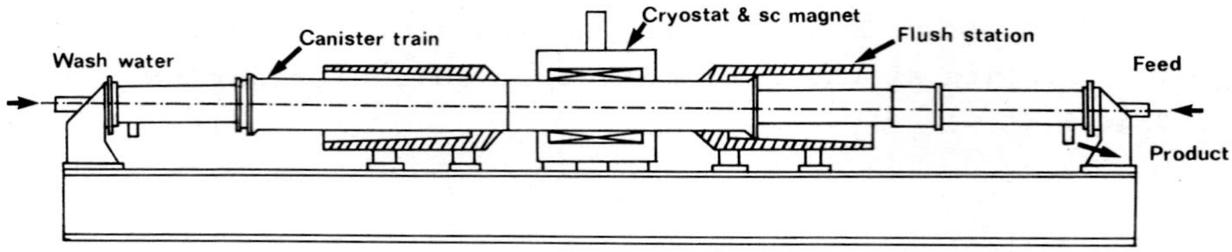

Fig. 2.62 Czechoslovak reciprocating superconducting magnetic separator.

A pilot-plant-size mobile reciprocating superconducting separator also was designed by Cryogenic Consultants Ltd.[G9]. A 0.3 m-bore superconducting magnet operating at 5 T is projected to treat up to 6 tonnes of kaolin slurry per hour. The principal features of the design are the use of closed-cycle refrigeration to eliminate the dependence on cryogens, the use of iron to shield the magnet and an electromagnetic drive for the reciprocator. The machine is intended to be used for pilot plant tests on mining sites.

II.5.1.3 Superconducting separator with a current carrying matrix

In the previous paragraphs, we have seen that the cyclical operation of a magnetic separator requires switching off of the magnet or removal of the matrix. Switching off the magnet is a lengthy process resulting in a short duty cycle that impairs the high efficiency of a separator. The removal of the matrix is mechanically complicated and could, amongst other problems, induce flux jumping, resulting in the quenching of the magnet. Some of these difficulties can be removed in a superconducting magnet separator by using the matrix carrying the electric current.

The notion of capture of magnetizable particles by current-carrying conductors was introduced by Parker and others [P23, E4, Y3]. Watson et al. [W31, W32] proposed a magnetic separator in which the current-carrying matrix is exposed to the external magnetic field. Particles captured by the matrix can be removed by switching off the current in the wires and the rinsing of the matrix is performed while the magnet remains energized. The dead times of such a machine can thus be reduced to a minimum. The important advantage of the separator is that it does not have any moving parts, while high current densities in the matrix wires and large input power lessen the effectiveness of the design.

II.5.2 Superconducting open-gradient magnetic separators

Superconducting separators discussed so far fall into a category of "settling" separators, in which magnetic particles settle on ferromagnetic matrix elements. These are subsequently rinsed to remove mechanically entrained non-magnetic particles and then flushed to collect the magnetic product. The presence of a ferromagnetic matrix, however, brings about a problem associated with keeping the matrix clean and passable long enough under industrial conditions. One possible approach to eliminate the problems inherent in matrix machines is to employ flow-through separators in which magnetic particles are continuously deflected in a non-uniform magnetic field to separate collecting points, according to their magnetic susceptibility and size. These deflection-type magnetic separators are often referred to as open gradient magnetic separators (OGMS) and include a drum separator, quadrupole separator, linear multi-pole separator and magneto-centrifugal separator.

Although these new open-gradient separators reflect a remarkable improvement in the understanding of the basic physical principles of magnetic separation, their use in the minerals industry will probably be limited. However, matrix separators will play a dominant role in the manipulation of weakly magnetic, finely-ground particles on the industrial scale.

The basic principles of open-gradient magnetic separation are outlined in Chapter III; the following sections contain a brief description of the developed open-gradient magnetic separators.

II.5.2.1 Quadrupole magnetic separator

The first quadrupole magnetic separator to be used in mineral industry was designed and built by Aubrey and Funk [A13]. The magnetic system generated 0.4 T on the pole faces and was used as a replacement of a drum magnetic separator for the recovery of magnetite, although weakly magnetic minerals were tested as well. The design consisted of two concentric tubes, the axis of which coincided with the quadrupole axis. The feed stream was introduced through the inner pipe and the magnetic particles were deflected radially through the gap in the inner tube into the outer tube. Non-magnetic material was collected and discharged through the lower section of the centre tube. As the results had been so encouraging manufacture of a large-scale unit was planned and replacement of the conventional coils by the superconducting ones was envisaged. However, this proposal did not materialize and five years later the superconducting quadrupole magnetic separator was developed by Cohen and Good [C10,C11]. The gradient of the magnetic field was generated directly by a superconducting winding of quadrupole configuration, as shown in Fig. 2.63.

The cross-section (Figs. 2.63 and 2.64) indicates that the annular configuration provides a magnetic field force acting

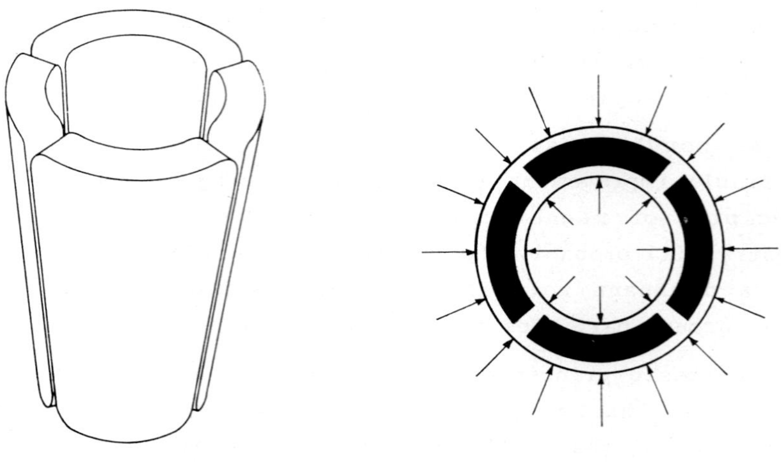

Fig. 2.63 Outline geometry of the quadrupole coils. The arrows indicate the direction of magnetic attraction. (After Cohen and Good [C10]).

radially outwards from the inside and radially inwards from the outside of the cryostat.

The magnet itself was wound from NbTi wire, and was epoxy cast with a glass fibre former. The highest magnetic induction measured on a surface of cryostat was 1.1 T [C12] and the gradient amounted to about 30 T m^{-1}.

The slurry could be fed tangentially into arched channels, both inside and outside the solenoid. The form of the external channel in shown in Fig. 2.65.

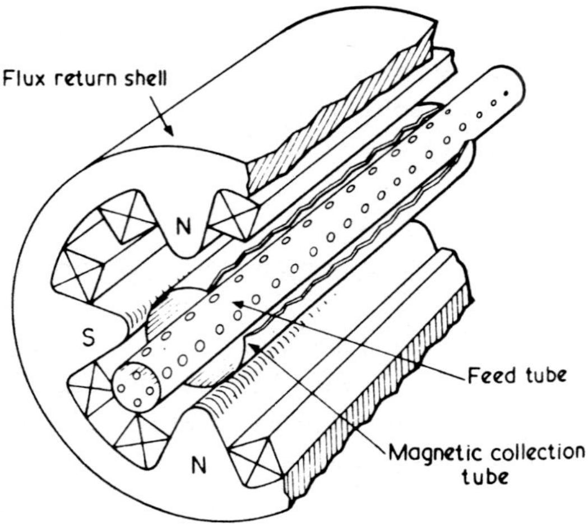

Fig. 2.64 Schematic cross-section of quadrupole magnetic separator (Reproduced from E. Maxwell, Cryogenics, April 1975, p.179 [M43], by permission of the publishers, Butterworth & Co., Ltd.)

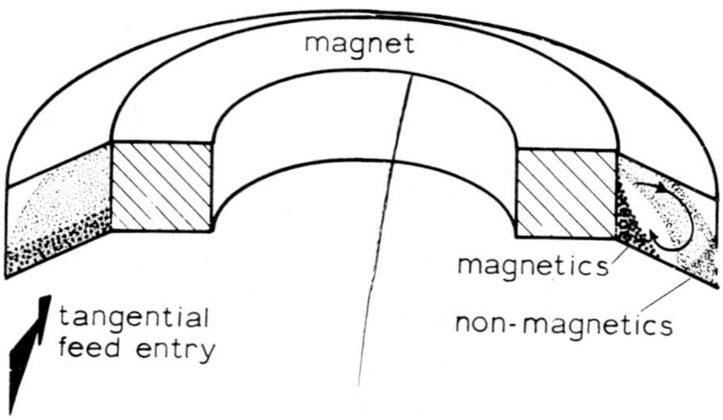

Fig. 2.65 Slurry channel with magnet on inner side (after Cohen and Good [C10]).

The bottom of the channel is not horizontal but sloping downwards away from the magnet. The magnetizable particles move in the proximity of the magnets while non-magnetic particles travel near the outer wall of the channel. At the exit, magnetic and non-magnetic particles leave as separate streams.

Although the system worked well on some artifically prepared suspensions, an application to slurries of natural ores at practically acceptable densities was impaired by a non-selective separation and by a low value of magnetic force generated by the magnet. Although theoretical studies have shown [H16, R2] that a gradient of the magnetic field, of an order of magnitude greater than that achieved by Cohen and Good, can be obtained, and that by a proper choice of hydrodynamic regime a selective separation of diluted slurries can be achieved, the quadrupole separator was not a commercial success and was replaced by linear multipole systems.

II.5.2.2 Linear multipole magnetic separator

The linear multipole superconducting magnetic separator, shown schematically in Fig. 2.66 was proposed by Schönert et al. [S21]. The system consists of an array of vertically positioned coaxial superconducting coils that are energized in opposite directions. The contours of constant magnetic induction and the flux lines for a simple two-solenoid system are shown in Fig. 2.67.

It can be seen that the magnet develops a strong force radially inwards towards the cryostat, which consists of a vertical cylinder. The horizontal radial magnetic force is substantially greater than the forces in the vertical or axial directions.

In the simplest form of the dry separator the ore is permitted to fall vertically as a curtain around the entire circumference of the magnet. The magnetic force diverts the ore inwards and cylindrical splitters divert the fractions into separate containers (Fig. 2.68)

The arrangement for wet separation is shown in Fig. 2.69.

Several small-scale linear multipole magnetic separators were built and tested [S21, C13, H17]. Magnet modelling has shown that an optimum magnet can be designed that will generate magnetic forces several times larger than those of the existing solenoid system and with a larger processing capacity.

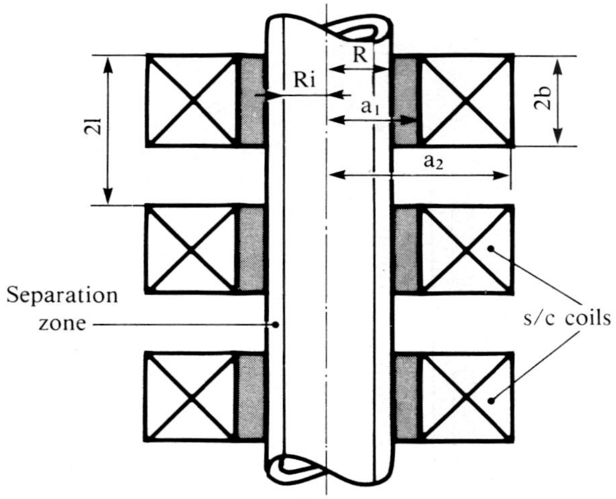

Fig. 2.66 Schematic view of linear multipole OGMS system.

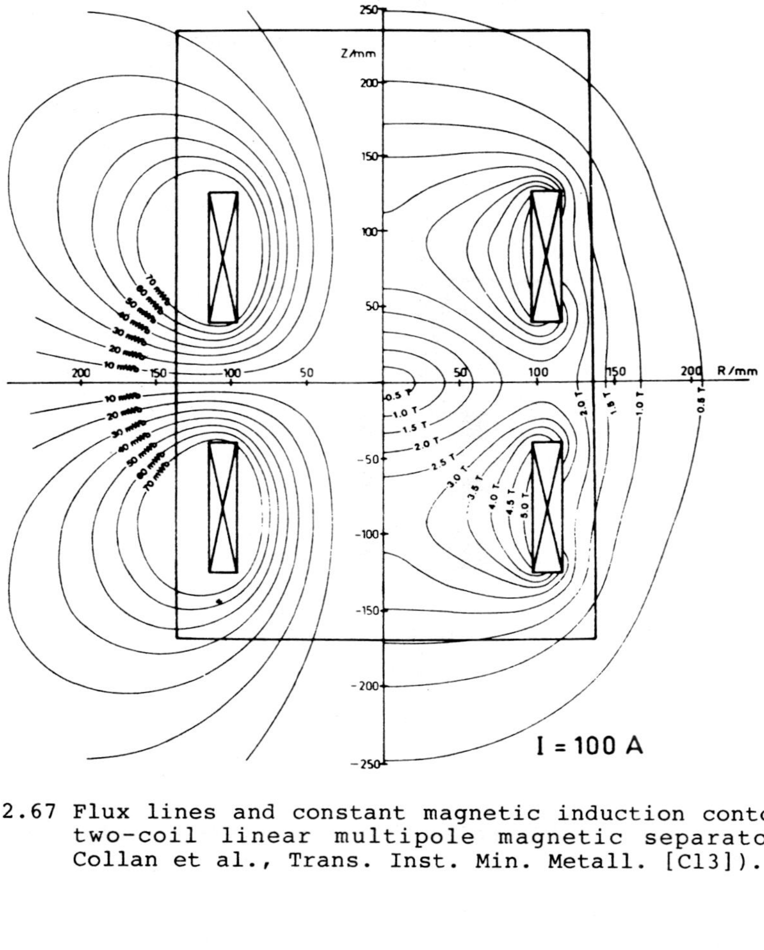

Fig. 2.67 Flux lines and constant magnetic induction contours of a two-coil linear multipole magnetic separator (after Collan et al., Trans. Inst. Min. Metall. [C13]).

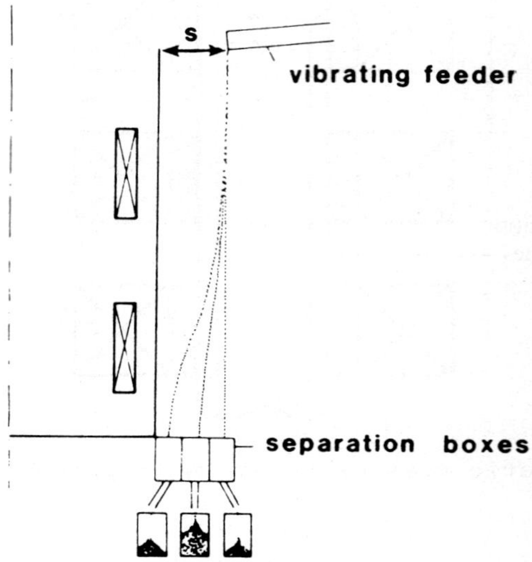

Fig. 2.68 Schematic picture of dry-magnetic separation experiments (after Collan et al., Trans. Inst. Min. Metall. [C13]).

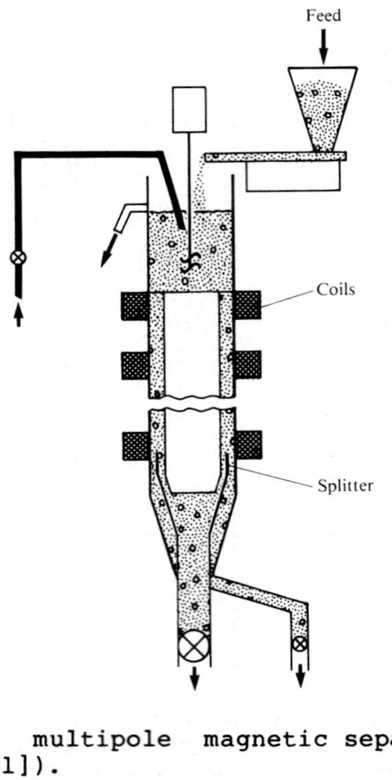

Fig. 2.69 Wet linear multipole magnetic separator (after Schönert et al. [S21]).

Design of a linear multipole magnetic separator

A detailed analysis of a split-pair magnetic separator has been published by Cheremnykh et al. [C14] Gerber and Watmough [G26] and by Kopp [K36, K37]. It has been shown by Gerber and Watmough [G26] that the geometry of a split-coil systems is completely determined by the parameters $\alpha = a_2/a_1$, $\beta_1 = b_1/a_1$ and $\beta_2 = b_2/a_1$, where a_i and b_i $(i = 1,2)$ are defined in Fig. 2.70.

Computations have shown that in the optimum configuration coils have approximately square cross-sections and the overall separator geometry is characterized by a set of optimum parameters: $\alpha \cong 1.37$, $\beta_1 \cong 0.25$ and $\beta_2 \cong 0.625$.

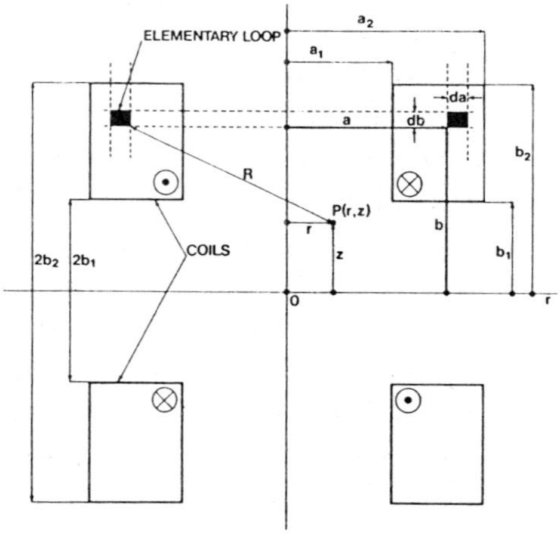

Fig. 2.70 Schematic view of the split-coil magnet (after Gerber and Watmough [G26]).

The attractive features of the linear multipole separator led not only to the development of laboratory units and of the theoretical models of the process, but also to the pilot-plant-size separator designed and built by Cryogenic Consultants Ltd. (CCL). These machines were tested by several mining companies in South Africa and despite their unreliable operation [C12, C15, S4] they have shown that linear superconducting multipole magnetic separators can successfully treat a selected class of minerals, possibly under industrial conditions, provided improvements in design and manufacture are introduced and a reliable cooling system is developed.

The CCL separators

The MK III separator developed by the CCL (Fig. 2.71) consists of a reverse pair winding that provides a strong attractive field around the outside of a cylindrical cryostat. The cryostat diameter is 365 mm and magnetic induction on the surface of the cryostat is 3 T, with gradients in the region of 80 T m^{-1}. The refrigerator is built into the top of the cryostat so that the whole system can be operated entirely without cryogens. After switching on a minimum of 40 hours is required for the magnet to reach the operating temperature. The spatial profiles of the product of the magnetic induction and field gradient are shown in Fig. 2.72.

One of these machines was installed at Foskor, South Africa, for the development of dry magnetic separation of phosphate minerals. Research work indicated [R3] that successful separation could be obtained but to ensure an economic performance considerable improvements in the cost benefit ratio were required. Consequently, it was decided to build a production-scale machine. The magnet was designed as a linear dipole in a rectangular cross-section cryostat. The advantages are that with linear geometry, there are two separating zones, both of which are against flat surfaces. An adjustable splitter must be placed between the magnetic and non-magnetic fractions as the stream of ore drops past the magnet. In linear geometry this can be a straight edge that can be moved closer to or further away from the magnet. With circular geometry, radial movement of the splitter is difficult; this was a disadvantage of the prototype MK III separator. The new magnet has a total separation length of 6 metres, both sides of the 3 m-long dipole being used for mineral processing [G10, G11]; technical details are given in Table 2.19 [G10].

The cross-section of the linear cryomagnet is shown in Fig. 2.73; the anticipated spatial profiles of the force density are given in Fig. 2.72.

Fig. 2.71 The linear multipole magnetic separator MK III.

TABLE 2.19

Technical data of the proposed magnetic separator CRYOFOS [G11]

Magnet mass	110 kg
Magnet length	3000 mm
Magnet width	< 67 mm
Magnet height	122 mm
Cryostat length	3850 mm
Cryostat width	85 mm
Cryostat height	446 mm
Designed magnetic induction of the surface of cryostat	3.2 T

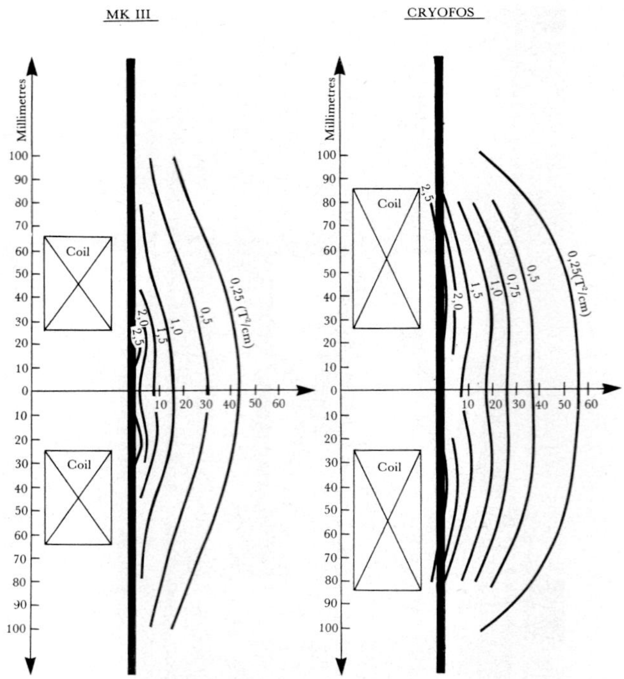

Fig. 2.72 Spatial profiles of the product of magnetic induction and field gradient for MK III and Cryofos magnets (after Roux et al. [R3]).

The magnet is cooled by liquid helium passing through heat-exchangers at the top and bottom of the magnet. The radiation shields are cooled by helium gas. Refrigeration for the magnet is provided by an improved version of the CCL refrigerator built on to one end of the cryomagnet.

Although the magnet has not reached operational status, the scientific and engineering challenge of the project well demonstrates the degree of sophistication in magnetic separation.

The Oxford Instruments Magnet

The cryogenic systems in the CCL magnetic separators use very small quantities of liquid helium to keep the magnets at temperatures of about 4 K. These systems rely on two stages of continuous recompression of helium gas, compared with refrigeration

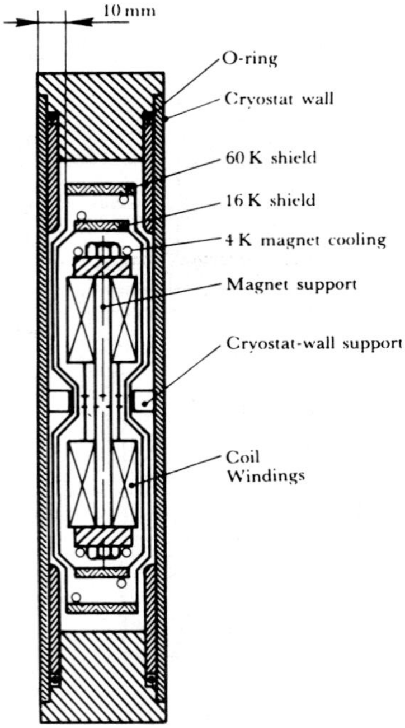

Fig. 2.73 The cross-section of the linear cryomagnet CRYOFOS (after Good and Kopp [G11]).

by expansion and heat exchange, and have an important attribute of being independent of external supplies of liquid helium. However, in the event of a breakdown of the compressor or a failure in power supply to it, the reserve of liquid helium can cause the magnet to quench within a few seconds, and it then can take a minimum of 40 hours to cool the system sufficiently to resume operation.

Oxford Instruments Ltd. built a cryogenic system MK IV that permits continued operation of the magnet for several hours without an external power supply. The magnet is maintained at 4.2 K by boiling liquid helium but a closed cycle refrigerator is employed to minimize the liquid helium boil-off and to eliminate consumption of liquid nitrogen. By this method, the hold-time of liquid helium is more than six months with the magnet in persistent mode [C16].

The magnet is immersed in a large reservoir of liquid helium, surrounded by two radiation shields and contained in a stainless steel vacuum vessel. The radial distance between the magnet and the

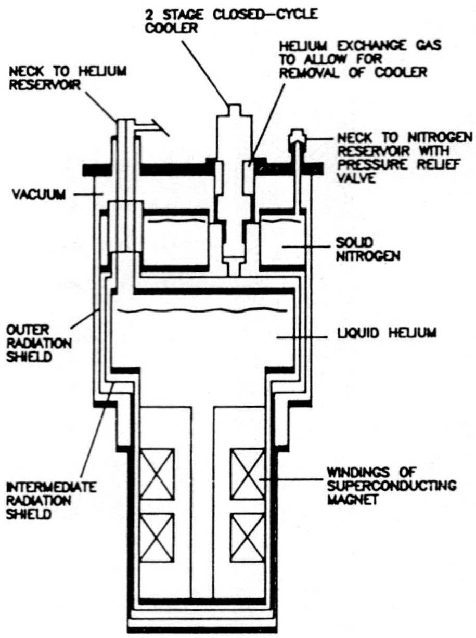

Fig. 2.74 Vertical section through the cryostat of the MK IV magnetic separator (after Cohen [C12]).

outside of the vessel is 17 mm, the outer diameter of the magnet winding is 386 mm and the outer diameter of the cryostat is 420 mm. The cross-section of the magnet system is shown in Fig. 2.74 [C12]. The magnet is built with two alternative fixed gaps. This is achieved by switching two supplementary coil sections in or out of the system and by dividing two main coils into multiple sections. A variety of field profiles can thus be tested in wet and dry applications. For a wet separator the minimum coil gap is 30 mm which gives the highest field and localized field gradients. For dry separation the gap can be increased to 50 mm. The fixed gradient can thus be reduced while the range of the magnetic force increases.

II.5.2.3 Superconducting drum magnetic separator

The magnetic separator developed by KHD Humboldt Wedag AG employs a multipole magnet and combines the advantages of a superconducting magnet with those of a conventional drum separator. As seen from Figs. 2.75 and 2.76, eight poles of the magnet system were arranged along one third of the circumference of the drum so that these coils may be regarded as one-third of a 24 pole magnet [J2]. Four race-track coils, wound from NbTi wire, are placed in a liquid helium tank and generate 3.2 T on the drum surface. The magnet is installed in the cylindrical part of the asymmetric helium tank, which is surrounded by the cylindrical vacuum tank and the rotating drum. This drum is supported in friction bearings by means of solid pipes.

The system is cooled by a refrigerator. The compressor unit compresses gaseous helium from 160 kPa to 1.5 MPa, which is then

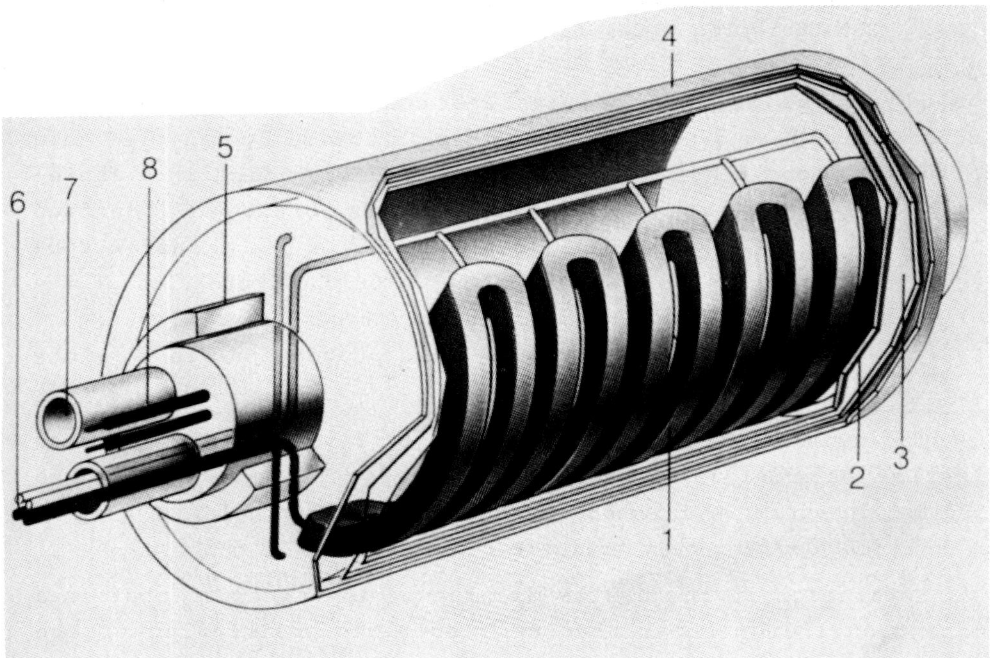

Fig. 2.75 Superconducting drum magnetic separator DESCOS (courtesy of Humboldt Wedag AG).
1 Magnetic coils
2 Radiation shield
3 Vacuum tank
4 Drum
5 Plain bearing
6 Helium supply
7 Vacuum line
8 Current supply

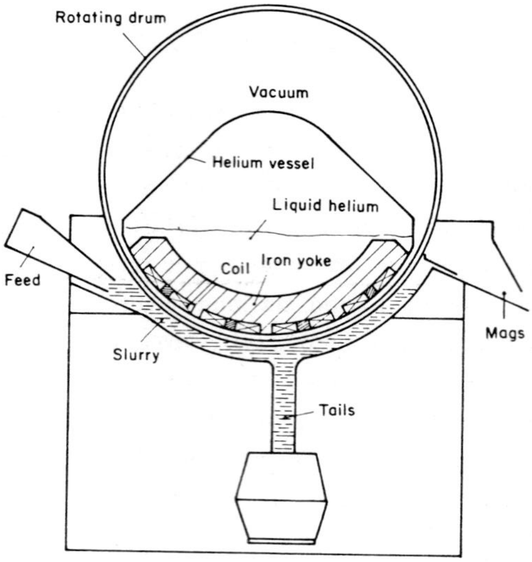

Fig. 2.76 Cross-section of the superconducting drum separator (Reproduced from K.P. Jüngst et al., Cryogenics 24 (1984), 648, [J2], by permission of the publishers, Butterworth & Co, Ltd.)

cooled to the ambient temperature and directed to the heat exchangers, where a portion of the gas is cooled by recycled cold helium to about 8 K. It is then partially liquified in a Joule-Thompson valve by expansion to 160 kPa. This portion of liquified helium is directed from the storage tank via the transfer tube system to the helium tank of the magnet. For the protection of the helium vessel against excessive pressure rises caused by magnet quench or vacuum breakdown three-stage protection via a safety line was provided.

The drum of the separator is made of fibre-carbon reinforced plastic. The outside diameter of the drum is 1216 mm and the length 1500 mm. The drum is driven over a serrated rubber belt, at a speed of rotation continuously variable between 2 and 30 rpm.

Separation in the DESCOS machine is identical to that in a conventional drum separator. By a suitable positioning of the swivel-type magnet system inside the drum, the machine can be used either for wet or dry processes. The magnetic field on the drum covers one third of the circumference and magnetic induction up to 3.2 T is generated over an axial length of 1 m.

The DESCOS separator, the overall view of which is shown in Fig. 2.77 has been tested on a number of ores and industrial minerals [A14]. By the removal of strongly magnetic impurities, high-quality concentrates of bauxite, fluorspar, apatite etc. were obtained. The machine can also be used to concentrate weakly magnetic ore like hematite and wolframite.

Fig. 2.77 Superconducting drum magnetic separator DESCOS (courtesy of Humboldt Wedag AG).

II.5.2.4 Superconducting helical flow magnetic separator

The concept of a centrifugal superconducting magnetic separator was introduced by Abdelsalam and Boom [A25, A26]. The centrifugal separator consists of a channel and magnet. The channel, schematically shown in Fig. 2.78, consists of two concentric cylinders with one inlet and two outlets. The centrifugal force plays the role of a gravitational force and is created by rotation of particles with the fluid around the axis of the magnet. With the magnetic forces acting inwardly, the magnetizable particles are attracted inwards while non-magnetic particles are deflected outwards under the effect of the centrifugal force. The magnetic and non-magnetic products are separated at the end of the channel into two separate outlets.

The separator generating magnetic induction up to 1.6 T is suitable for beneficiation of strongly magnetic ores of coarse particle size (greater than 100 μm), at high flowrates (up to 4 m s^{-1}). However by increasing the flux density to 5 T, the range of recoverable minerals could be extended to moderately weak magnetic ores.

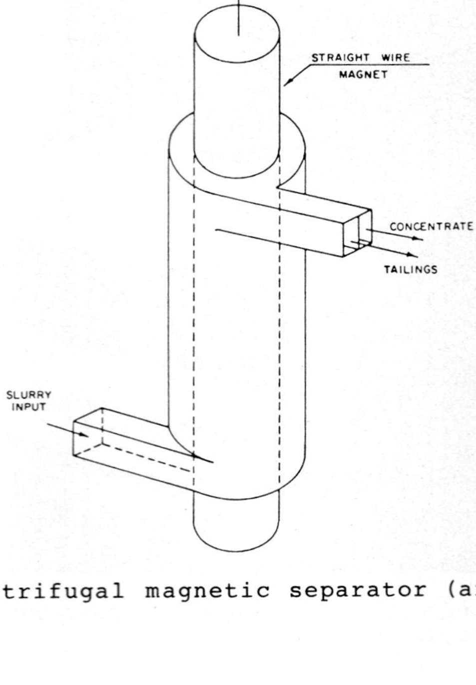

Fig. 2.78 The centrifugal magnetic separator (after Abdelsalam [A25]).

II.6 LABORATORY MAGNETIC SEPARATORS

A rule that applies more to magnetic separation than to any other beneficiation technique is that the quickest and surest way of finding out how an ore will behave in a magnetic separator is to run it through the separator. Theoretical models of the process, which would help us predict the efficiency of low-intensity magnetic separation, are scarce, while a wealth of theoretical descriptions of high-gradient magnetic separation applies mainly to oversimplified situations, with little significance for practical problems. The information obtained from these models can sometimes be even misleading and this, combined with the fact that magnetic and some physical and chemical properties of the ores are difficult to ascertain, indicates that laboratory tests are mandatory before any conclusions can be drawn as to the potential of the ore for upgrading by magnetic means.

A wide spectrum of laboratory devices is used to analyse the magnetic separability of small samples while an even larger number of bench-made magnetic separators have been built to simulate a particular process in a particular production-scale machine. Their general significance is limited and we shall confine ourselves in this section to those machines that are of general applicability and which are commercially available.

II.6.1 Davis tube

The Davis tube is a laboratory machine designed to separate small samples of strongly magnetic ores into magnetic and non-magnetic fractions. It was developed in 1921 and no significant changes have been made in its design since that time. Although it is neither well-designed as a separator nor as an analytical instrument, it is widely used for estimating the amenability of an ore to magnetic separation. The Davis tube separator, shown in Fig. 2.79, has been described in detail by Schultz [S22]. It consists of an inclined tube 25 mm in diameter through which a few grams of feed material are passed. To conduct a tube test, a sample of suitably prepared ore is poured into the water-filled tube. As wash-water flows through the tube, it is rotated to and fro until the non-magnetic ore particles are gradually rejected and until the proper degree of separation has been obtained. Magnetic force holds strongly magnetic particles in a magnetic field while less magnetic material is washed down the tube; the magnetic concentrate and the tailings may be recovered for examination.

The usual practice in the examination of strongly magnetic materials has been to perform a single Davis tube test, continuing the test for a prescribed time or until no more non-magnetic material can be washed out of the tube, in the judgement of the operator. The effect of various variables in the Davis tube on the separation efficiency, as obtained from various laboratories, is summarized below [S22].

Fig. 2.79 Davis tube separator.

Magnet poles. All machines investigated had coaxial 90° circular conical magnet poles. Although the pole gap varied slightly from 11 mm to 19 mm this variation had no significant effect on separation efficiency.

Magnetic field. The magnetic induction between the magnet poles must be sufficient to minimize the loss of strongly magnetic particles. Magnetic induction higher than 0.4 T should be used and induction up to 0.6 T can be obtained with commercial units.

Tube clearance. Tube wall thickness plus clearance between tube and magnet poles constitute the distance separating the pulp column from the poles. This distance usually ranges from 1 mm to 8 mm and the increased clearance tends to reduce the efficiency of separation.

Tube diameter. The internal diameter of a tube usually ranges from 24 to 27 mm and the separation efficiency of ferromagnetic ores is affected by variation in the tube's i.d. For less magnetic minerals, recovery increases with the decreasing i.d. of the tube.

Sample quantity. Although samples of 3 to 50 g can be used, those of 10 to 20 g are commonly employed. The sample quantity is directly proportional to the cross-sectional area of the tube and inversely proportional to the concentration of a ferromagnetic component in the feed.

Tube agitation. A stroke of 25 to 80 mm is usually used while the angle of rotation ranges between 45° to 120°, the most common being 90°. Rotation does not influence separation efficiency, but it facilitates movement of the rejected tailing particles down the tube to the outlet, thereby simplifying concentrate collection. Some machines have adjustable oscillation rates, usually between 60 and 120 cycles per minute. However, no effect of oscillation on separation efficiency was found for ferromagnetic ores, although for less magnetic minerals slower oscillaton rates gave better results.

Water. Water flowrate usually ranges between 100 and 1000 cm^{3} min^{-1} and may have an effect on separation characteristics.

Time. The duration of a test is usually from 3 to 20 minutes.

Usual procedure: [S22]

Pole pieces:	conical, 90° included angle
Pole gap:	9 to 19 mm
Magnetic induction:	min. 0.4 T
Tube:	cylindrical, glass, 25 mm i.d., 1.6 mm wall thickness
Tube position:	axis 40 to 50° from horizontal
Tube agitation:	stroke 25 to 80 mm, rotation 60° to 120°
Procedure:	three or more tube tests on identical ore samples (different oscillation rates, water rates or times are needed to obtain reliable magnetic separation data)
Test sample:	10 g of ore
Operation:	oscillation rate: 60 to 150 cycles per min water rate: 50 to 500 cm^3 min^{-1} time: 3 to 20 min

II.6.2 Frantz isodynamic separator

The Frantz isodynamic separator is a magnetic analytical device that allows fractionation of even weakly magnetic materials. This separator, shown in Fig. 2.80, utilizes specially shaped pole-pieces to produce a force that is almost independent of distance in a region between pole-pieces.

A narrow chute is placed between the pole-pieces and the material to be separated passes down the inclined chute under the influence of gravity. The chute is inclined both parallel and perpendicular to its length so that a component of gravitational force may compete with the magnetic force. Under the influence of gravitational and magnetic forces the grains migrate to one side or the other of the chute, which at one part is divided into two passages by a splitting ridge, which directs the products into two separate containers.

The separator can be operated in vertical and inclined modes. In the vertical mode the separator is suitable for separating free flowing ores in the range 150 to 500 μm. Although rapid separations are possible using this mode of operation, repeated passes are usually necessary to obtain optimum results.

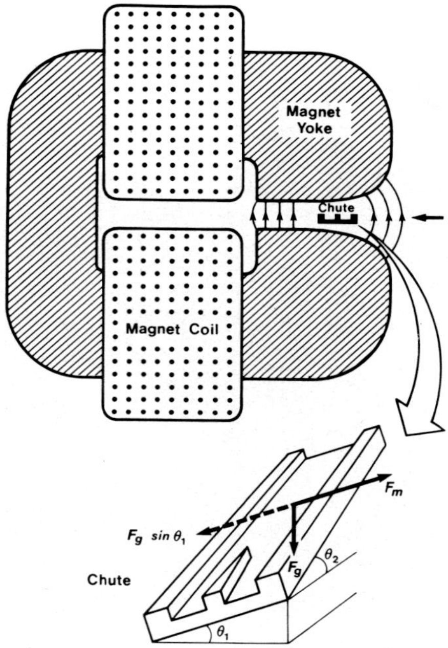

Fig. 2.80 Frantz isodynamic separator

In the inclined position (Fig. 2.81) the separator can separate pulverized ore as fine as 38 μm. The feed rate is much slower than that obtained with a vertical feed. The forward slope of the chute and the intensity of longitudinal vibration should be adjusted according to the particle size.

Magnetic field strength can be varied by adjusting the current from 0 to 2 A. The side slope can be varied continuously from 0 to 90° and can also be given a reverse slope by as much as 90° for the separation of diamagnetic minerals. Side slopes varying by less than 5° usually produce no significant differentials and angles greater than 15° C are usually employed [F8]. The forward slope and feed rate control the flow rate of the sample. A forward slope of about 15° is suitable for most separations.

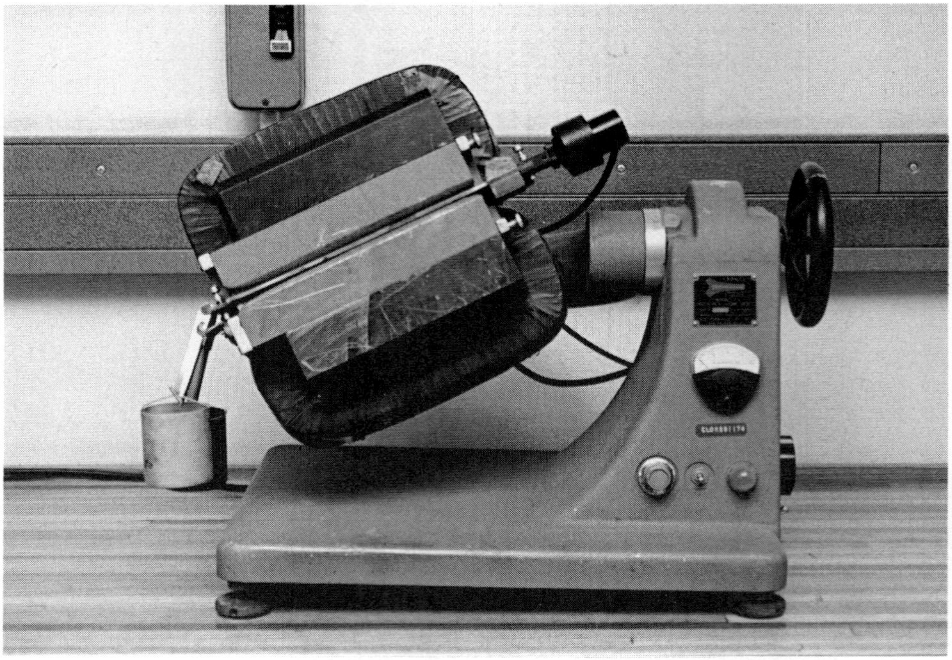

Fig. 2.81 Frantz isodynamic separator.

II.6.2.1 Theory of separation

Paramagnetic minerals

Inclination of the chute is measured by the transverse slope θ and the longitudinal slope ϕ . The pole design is such that changes in the magnetic field exactly balance changes in the field gradient to yield a constant product **HdH/dx.** The forces acting on a particle are depicted in the vector diagram shown in Fig. 2.82 [M9, N3]. Particles enter at point **A**, the centre of the tray being in line with the splitter. If the gravity ($\mathbf{F}_g$) and magnetic ($\mathbf{F}_m$) forces are balanced to cause the paramagnetic particle to move along **AB**, this particle has a 50 per cent chance of reporting to mags or non-mags.

The force balance is thus

$$F_g = F_m \tag{2.31}$$

or

$$mg\sin\Theta = \frac{1}{2}\mu_0 \kappa V \nabla (H)^2 \tag{2.32}$$

where **m** is the particle mass, θ is the transverse slope, **g** the gravitational constant, κ the volume magnetic susceptibility and **H** the magnetic field strength. By re-arrangement we obtain

$$\sin\Theta = \mu_0 \chi \frac{H}{g} \cdot \frac{\partial H}{\partial x} \tag{2.33}$$

where χ is the specific magnetic susceptibility. Since the pole design gives **HdH/dx** constant, the force on a paramagnetic particle is constant throughout the gap and the action of the separator on a single particle is determined solely by the mass susceptibility and is independent of the size of a particle and its mass.

The magnitude of the term **HdH/dx** is determined by the electric current **I** in the electromagnet; eq. (2.33) can thus be rewritten as

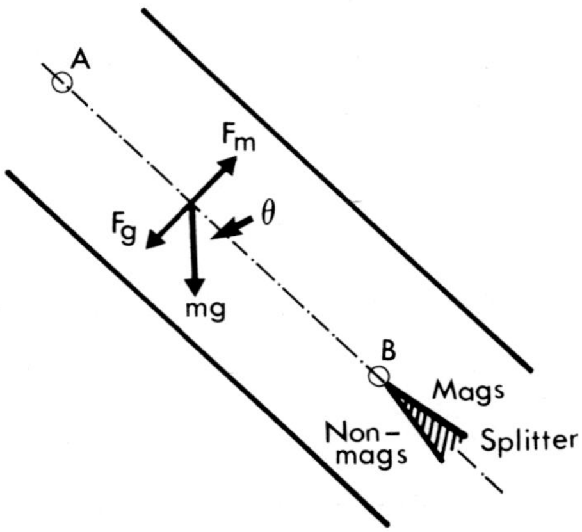

Fig. 2.82 Forces acting on a particle in Frantz isodynamic separator.

$$\frac{\sin \Theta}{\chi} = kI^n \tag{2.34}$$

where **n** is the constant close to 2 and **k** is the proportionality constant.

In order to determine the exact values of these constants the Frantz separator must be calibrated with compounds having well defined values of magnetic susceptibilities. Chemical compounds usually used for calibration are listed, together with their magnetic susceptibilities, in Table 2.20.

TABLE 2.20
Chemical compounds used for a calibration of a Frantz isodynamic separator [M9, T3]

Substance	Chemical formula	Specific magnetic susceptibility	
		$(cm^3g^{-1}) \times 10^{-6}$	$(m^3kg^{-1}) \times 10^{-7}$
Mohr's salt	$FeSO_4(NH_4)_2 \cdot SO_4 \cdot 6H_2O$	32.4	4.1
Nickel sulphate	$NiSO_4 \cdot 7H_2O$	16.0	2.0
Copper sulphate	$CuSO_4 \cdot 5H_2O$	5.9	0.74
Cobalt sulphate	$CoSO_4 \cdot 7H_2O$	37.0	4.6
Ferrous sulphate	$FeSO_4 \cdot 7H_2O$	41.4	5.2
Manganese chloride	$MnCl_2$	122.0	15.4
Manganese pyrophosphate	$Mn_2P_2O_7$	102	12.9

Mc Andrews [M9] found that for $I < 0.9$ A $n = 2.0$ and $k = 52.0 \times 10^3$; eq. (2.34) can thus be rewritten

$$\frac{\sin \Theta}{\chi} = 52.0 \times 10^3 I^2 \tag{2.35}$$

if χ is expressed in cgs units ($cm^3\ g^{-1}$), or

$$\frac{\sin \Theta}{\chi} = 4.14 \times 10^6 I^2 \tag{2.36}$$

if χ is expressed in SI units ($m^3\ kg^{-1}$). However, the value of constant **k** differs for different separators and each unit must be calibrated before the use of eq. (2.34) is attempted.

Magnetic susceptibility χ can thus be obtained from eqs. (2.35) or (2.36) by plotting the cumulative per cent recovery to mags versus current. A typical "magnetic profile" [D10] is shown in Fig. 2.83. The mean magnetic susceptibility of a sample can be defined as the value of χ for which 50 per cent of the sample reports to the magnetic product, as indicated in Fig. 2.83. For samples containing several magnetic components separate values of magnetic susceptibilities can be ascribed to each component. Although the method used is arbitrary, the accuracy is usually sufficient for the purpose of analysis of the magnetic separation process.

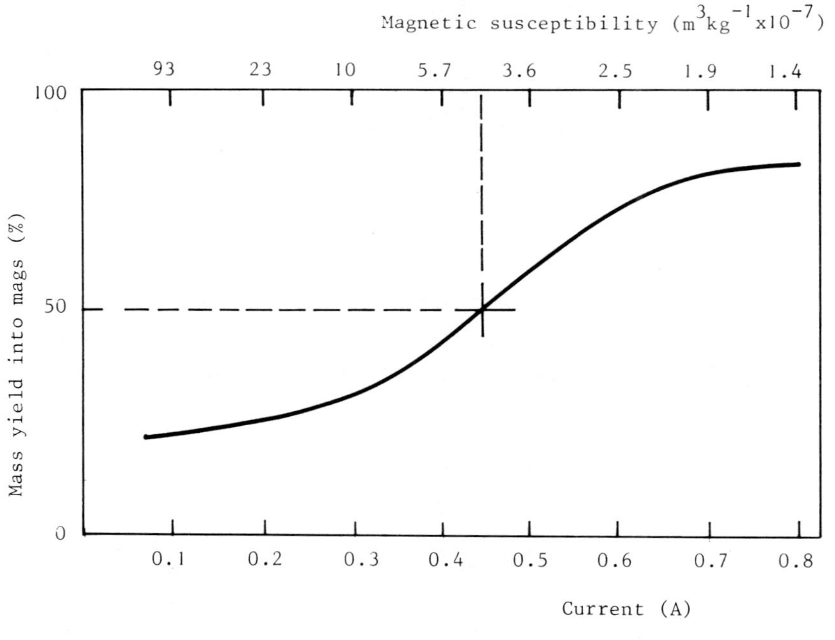

Fig. 2.83 Magnetic profile of the magnetic concentrate from WHIMS of uranium-gold cyanidation residues.

"Wet" Frantz separator

The recent modifications to the Frantz separator for wet processing [P6, S50] have proved capable of treating fine particles (5 to 10 μm) and have exploited fluid properties to effect separation.

Todd and Finch [T7] have shown that in a wet system, for field-independent susceptibility eq. (2.36) can be written (in SI)

$$\sin\Theta = 3.83\times 10^{6}\frac{\kappa_p-\kappa_f}{\rho_p-\rho_f}(1-\kappa_f)I_{50}^2 \qquad (2.37)$$

where I_{50} is the value of current at which 50 per cent of the sample reports to the mags.

Field-dependent susceptibility

The application of the Frantz separator to the determination of magnetic-field-dependent magnetic susceptibility was developed by Nesset and Finch [N3].

As follows from eq. (1.54) the field dependence of magnetic susceptibility can be expressed by the Owen-Honda equation, which can be rewritten as

$$M = M_0 + \kappa_\infty H \qquad (2.38)$$

where M_0 is the spontaneous magnetization (in A m^{-1}) and κ_∞ is the volume magnetic susceptibility extrapolated to the infinite field strength. Inserting eq. (2.38) in eq. (2.32) we obtain

$$\sin\Theta = \frac{\mu_0 M_0}{2g\rho}\frac{\partial H}{\partial x} + \frac{\mu_0 \chi_\infty}{g}H\frac{\partial H}{\partial x} \qquad (2.39)$$

It is seen that the magnetic force on the particle is not constant throughout the gap. Eq. (2.39) reduces to the paramagnetic case (eq. (2.33)) when $M_0 = 0$ and $\chi = \chi_\infty$.

To enable eq. (2.39) for the determination of M_0 and χ_∞, the magnetic field strength **H** and the field gradient **dH/dx** must be known as a function of current. Nesset and Finch have shown [N3] that eq. (2.39) can be rewritten, for the particular separator used in [N3], as

$$\sin\Theta = 2.33\frac{\mu_0}{\rho_p}M_0 I_{50} + 48.7\frac{\mu_0}{\rho_p}\chi_\infty M_0 I_{50}^2\times 10^3 \qquad (2.40)$$

Magnetization $\mathbf{M}_0$ and magnetic susceptibility χ_∞ can thus be obtained from measurements of $\mathbf{I}_{50}$ at various side slopes θ, fitted to eq. (2.40).

The values of magnetization and susceptibility determined by this method compare well with those given by the vibrating sample magnetometer (VSM) [N3]. Although the Frantz technique is more tedious than the VSM technique, it allows us to determine magnetic susceptibilities of each component in a mixture and the effect of ferromagnetic impurities is eliminated.

II.6.3 Laboratory high-gradient magnetic separators

A wide range of low and high-intensity laboratory magnetic separators is commercially available. Their application to mineral processing enables optimum performance to be determined and subsequently results can be scaled-up to the production size machines with confidence.

On the other hand, remarkable progress in the development of high-gradient magnetic separation in recent years has led to the situation whereby a number of commercial testing laboratories, universities and industrial installations have built their own units. These are often difficult to scale-up and results obtained from laboratory tests sometimes do not materialize on a large scale. The matrix height, the mutual orientation of the direction of magnetic induction, pulp velocity, background magnetic induction, volume of rinse and flush water and pulp density are parameters that must be translated with confidence from laboratory tests into production size machines. Laboratory units that allow a thorough investigation of the effect of these parameters on the separation efficiency are briefly described.

Fig. 2.84 shows a laboratory electromagnetic separator that can generate continuously variable background magnetic induction up to 1.2 T. A separation canister with a volume of 275 cm^3 can accommodate a matrix up to 200 mm in height. An improved unit generates up to 2 T, operates on 70 V dc at a current of 26 A and the input power is 1.8 kW.

In these separators the directions of the magnetic field and slurry velocity are mutually perpendicular and tests performed on these units simulate the operation of a large-scale continuous carousel separator equipped with an electromagnet and iron yoke.

In modern high-gradient magnetic separators that use an iron-clad solenoid to generate the magnetic field, the vectors of magnetic induction and pulp velocity are parallel.

The operation of production-scale cyclic and continuous solenoid separators was simulated by testwork on laboratory separators that use the same orientation of the field and flow.

Sala Magnetics Inc. manufactures a laboratory cyclic separator that generates 2 T in a 100 mm-diametre bore. It has a matrix 150 mm long, a mass of 1860 kg and a power input of 100 kW.

Fig. 2.84 Laboratory magnetic separator Eriez L-4.

A similar unit is marketed by Eriez Magnetics, Inc. The specifications are summarized in Table 2.21.

TABLE 2.21
Specifications of the ELS-4 magnetic separator

Background magnetic induction	max. 2 T, continuously variable
Power input	70 kW
Current	950 A
Voltage	75 V d.c.
Conductor	square hollow copper conductor
Cooling water	28 litres per minute
Temperature rise	30° C
Matrix diameter	127 mm
Matrix height	102 mm
Canister volume	1286 cm^3
Mass	726 kg

CHAPTER III

THEORY OF HIGH-GRADIENT MAGNETIC SEPARATION

In view of the diversity of magnetic separation techniques the theoretical descriptions of the magnetic separation process are of limited general applicability, Moreover, many branches of magnetic separation still constitute highly empirical technology.

Wet high-gradient magnetic separation has received more theoretical attention in recent years than any other branch of magnetic separation. A notable number of papers dealing with the problem of particle capture in high-gradient magnetic separation led to an understanding of the interaction between a particle and a matrix element. However, completely general treatment of the magnetostatic and hydrodynamic behaviour of an assembly of the ore particle in a system of matrix elements, in the presence of a strong external magnetic field is a theoretical problem of considerable complexity.

The problem can be looked upon from both the macroscopic and microscopic points of view. The macroscopic studies are aimed at the phenomenological description of the separation process, the prediction of its dynamic behaviour and the development of methodology and techniques for design, calculation and optimization.

The microscopic theories in magnetic separation are intended to provide information and insight about mechanisms of particle capture and conditions under which capture may be facilitated. The complex problems facing the microscopic theories can be simplified considerably by considering, in the first instance, the interaction between a single particle and a single matrix element. The theory of multi-element matrix magnetic separators can then be developed as an extension of the single element model or a phenomenological description can be used. A detailed description of the theoretical models of wet high-gradient magnetic separation is given in a book by Gerber and Birss [G1] and in a review paper by Birss and Parker [B13]. A brief survey of the theoretical description is given in the following sections.

III.1 THEORY OF SINGLE-WIRE PARTICLE CAPTURE

Historically, the limiting trajectory approach to high gradient magnetic separation is based on the fundamental picture of the particle transfer to solid surfaces, as used, for instance, in deep bed filtration. The method is based on the Lagrangian description of particle motion, i.e. attention is focused on a single particle trajectory that can be described by the second law of mechanics. In most cases the thermal random force on particles is neglected so their path is deterministic and the analysis applies to relatively large particles and strong flows only. There is no conceptual difficulty in incorporating the diffusion effect into the theory and the equation of motion then becomes a Langevin equation.

In the absence of diffusion, the solution of the equation of motion is sufficient for predicting particle trajectories around a collector. The exact path depends on the initial conditions, while the particle trajectory that divides the approaching particle paths terminating in capture on the collector surface from those which pass by, is called the limiting trajectory (see Fig. 3.1). In principle, it is possible to extend the trajectory equation approach to allow for particle diffusion. When a flowing particle is also subjected to Brownian motion, its trajectory is no longer deterministic and the limiting trajectory approach cannot be

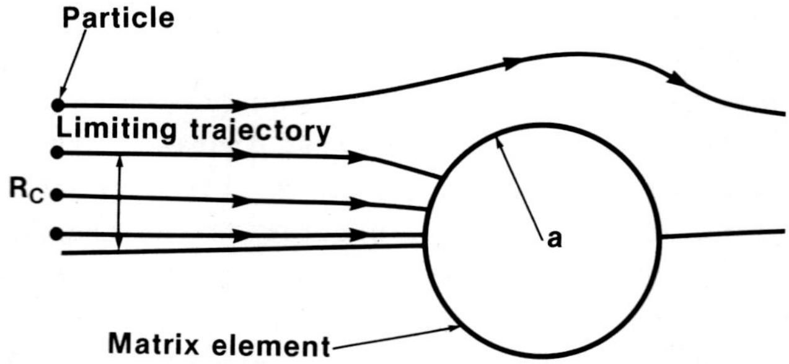

Fig. 3.1 Capture radius $\mathbf{R_c}$ for small paramagnetic particles captured by a ferromagnetic wire.

applied because limiting trajectories no longer exist. Instead, one can assign, for a particle initially at a given position, a probability of particle capture.

III.1.1 Geometrical configuration

The early model of high-gradient magnetic separation was developed by Watson [W9], who followed the procedure proposed by Zabel [Z2] to describe particle behaviour in electrostatic filters.

Watson considered a ferromagnetic wire of radius **a** placed along the z - axis in diamagnetic fluid, as shown in Fig. 3.2. A uniform magnetic field $\mathbf{H}_o$, strong enough to saturate the wire, is applied in the **x** direction. Spherical paramagnetic particles of radius **b**, volume **V** (= $4/3\,\pi\,b^3$), density ρ_p and volume magnetic susceptibility κ_p are carried by a fluid of viscosity η that flows with uniform velocity $\mathbf{v}_o$ in the negative **x**-direction.

This orientation is one of three possible arrangements between the directions of magnetic field and particle flow velocity with respect to the ferromagnetic wire. Since it is assumed that the fibre is always orthogonal to the external magnetic field, the fluid flow can be either parallel (longitudinal (L) orientation) or perpendicular (transverse (T) orientation) to the direction of magnetic field. The third alternative configuration, when the initial fluid flow is parallel to the wire axis is known as the axial (A) configuration. These three geometrical configurations are illustrated in Fig. 3.3.

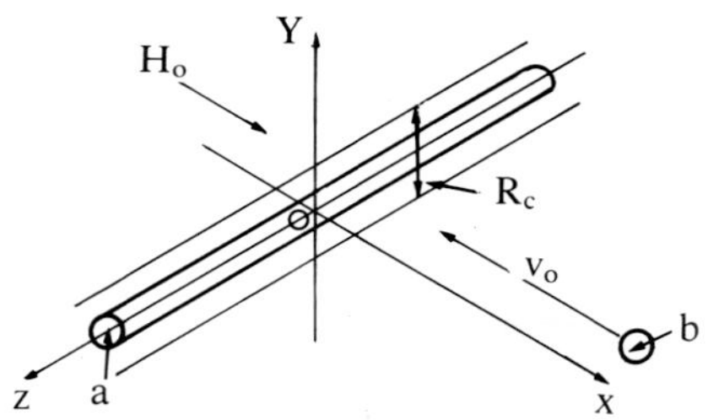

Fig. 3.2 A ferromagnetic wire of radius **a**, placed axially along the **z** axis, in a uniform magnetic field $\mathbf{H}_o$ applied in the **x** direction, interacts with a paramagnetic particle of radius **b** in a moving fluid. The fluid flows past the wire with a velocity $\mathbf{v}_o$ in the negative **x** direction.

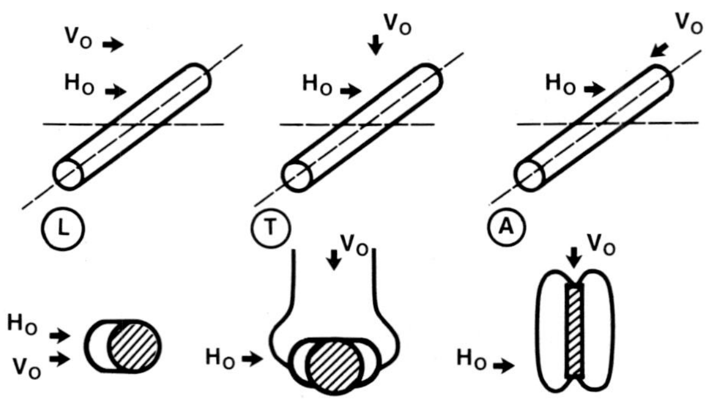

Fig. 3.3 Representation of three geometrical configurations.

In order to calculate the motion of a paramagnetic particle travelling in a fluid stream in the presence of a magnetized fibre one assumes that the particle experiences a magnetic force $\vec{F}_m$, a hydrodynamic drag force $\vec{F}_d$ and a gravitational force $\vec{F}_g$ in the direction of gravitational acceleration $\vec{g}$. The general configuration of the particle motion problem in single-wire approximation is shown in Fig. 3.4

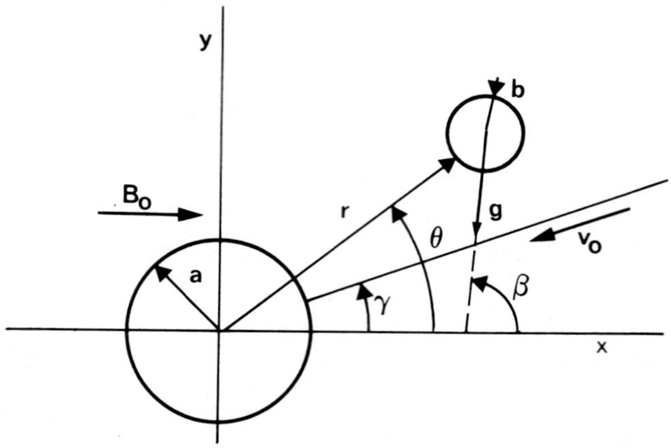

Fig. 3.4 Coordinate system for particle-matrix collector interaction.

III.1.2 Magnetic force

In order to obtain the magnetic force acting on a small paramagnetic particle in the vicinity of a magnetized wire, one expresses the magnetic field, external to and generated by the magnetized wire, as a function of wire radius.

If the current density is zero outside the wire, then curl $\vec{B} = 0$ and the vector of magnetic induction $\vec{\mathbf{B}}$ can be expressed as the gradient of a magnetic scalar potential ϕ, namely $\vec{B} = -\text{grad}\,\phi$ [J5]. Since div $\vec{B} = 0$ then also div grad $\phi = 0$, and the magnetostatic problem is reduced to the solution of Laplace's equation for ϕ :

$$\nabla^2\phi = 0 \tag{3.1}$$

Thus, the potential ϕ must satisfy Laplace's equation as well as the boundary conditions, viz. that the tangential components of $\vec{H}$ and the normal components of $\vec{B}$ be continuous at the surface of the wire.

Having solved the Laplace's equation with appropriate boundary conditions, one then arrives at the components of the magnetic field outside the wire, in polar coordinates [S1], as defined in Fig. 3.4:

$$H_r = H_0 \cos\theta + H_0 A \frac{a^2}{r^2} \cos\theta \qquad (3.2)$$

$$H_\Theta = -H_0 \sin\theta + H_0 A \frac{a^2}{r^2} \sin\theta \qquad (3.3)$$

where

$$A = \frac{\mu_w - \mu_f}{\mu_w + \mu_f}$$

μ_w and μ_f are the magnetic permeabilities of the wire material and fluid medium, respectively. For a ferromagnetic wire in an aqueous fluid $\mu_w >> \mu_f$ so that $A \cong 1$ for most practical purposes.

The first terms in eqs. (3.2) and (3.3) are the radial and azimuthal components of the external magnetic field, while the second terms are the components produced by the magnetization of the wire. The field due to magnetization causes the distortion of the background field $\mathbf{H}_o$ in the vicinity of a matrix element. This matrix material is in general, magnetized to magnetization $\mathbf{M}$, which is proportional to the external field $\mathbf{H}_o$ until it saturates. Above the saturation value ($\mathbf{M}_s$), $\mathbf{M}$ is constant and does not increase with $\mathbf{H}_o$.

If the external field $\mathbf{H}_o$ is larger than the field required to magnetically saturate the wire ($\mathbf{H}_s$), then $\mathbf{H}_o$ in the second term of eqs. (3.2) and (3.3) must be changed to $\mathbf{H}_s$.

If the parameter $\mathbf{K}$ is introduced so that

$$K = \frac{M}{2H_0} \qquad (3.4)$$

and assuming that magnetization of a wire is given as

$$M = \frac{2\mu_f(\mu_w - \mu_o)}{\mu_o(\mu_f + \mu_w)} H_o \qquad (3.5)$$

equations (3.2) and (3.3) can be rewritten in a general form

$$H_r = H_0 \left(1 + K\frac{a^2}{r^2}\right) \cos\theta \tag{3.6}$$

$$H_\theta = -H_0\left(1 - K\frac{a^2}{r^2}\right) \sin\theta \tag{3.7}$$

For $H_o < H_s$ $K = 1$ and for $H_o > H_s$ $K = M_s/2H_o$.

As follows from the evaluation of the magnetic energy of a system comprising a fluid and a particle placed in a non-homogeneous magnetic field (Aharoni [A1], Eisenstein [E1,E2], Rowlands [R4, R5, R6] and Gill and Malone [G12]) the magnetic force on a small paramagnetic particle in the vicinity of a magnetized wire can be written as (see eq. (1.7))

$$\vec{F}_m = \frac{1}{2}\mu_0 V\kappa\nabla(H^2) \tag{3.8}$$

where $\kappa = \kappa_p - \kappa_f$ is the difference between volume susceptibilities of the particle and fluid.

By inserting eqs. (3.6) and (3.7) into eq. (3.8) and assuming a spherical particle, the components of the magnetic traction force are given by [G1]

$$F_{mr} = -\frac{4\pi\mu_0\kappa Ma^2b^3}{3r^3}\left(\frac{Ma^2}{2r^2} + H_0\cos 2\theta\right) \tag{3.9}$$

$$F_{m\theta} = -\frac{4\pi\mu_0\kappa Ma^2b^3}{3r^3} H_o \sin 2\theta \tag{3.10}$$

If the net volume magnetic susceptibility $\kappa = \kappa_p - \kappa_f$ is positive, then the radial component of the magnetic traction force between the particle and the wire is attractive, provided that

$$\frac{Ma^2}{2r^2} + H_0 \cos 2\theta > 0 \tag{3.11}$$

The component $\mathbf{F}_{mr}$ becomes repulsive, for $\kappa_p > \kappa_f$, in two symmetrical spatial regions, the boundaries of which are determined by a critical angle θ_c [B13]

$$\Theta_c = \arctan\left(\frac{1+K\frac{a^2}{r^2}}{1-K\frac{a^2}{r^2}}\right) \tag{3.12}$$

As shown in Fig. 3.5 angle θ_c determines the sections of the wire surface that are attractive or repulsive to paramagnetic or diamagnetic particles. It is apparent from eq. (3.12) that paramagnetic particles, with positive values of magnetic susceptibility will be collected at angles smaller than θ_c and repelled at angles greater than θ_c. On the other hand, diamagnetic particles with negative susceptibility will be attracted to the matrix element at angles $\theta > \theta_c$ and repelled in a region where $\theta < \theta_c$. This situation is summarized in Fig. 3.6, which shows the contours of a magnetic force acting on a small paramagnetic particle in the vicinity of an infinite ferromagnetic wire. The wire surface can be divided into four regions, two of which, perpendicular to the applied magnetic field are repulsive for paramagnetic particles (and attractive for diamagnetic particles), while two, parallel to the applied field, are attractive for paramagnetic particles (and repulsive for diamagnetic particles). The hyperbolae in Fig. 3.6 indicate the lines of zero radial force.

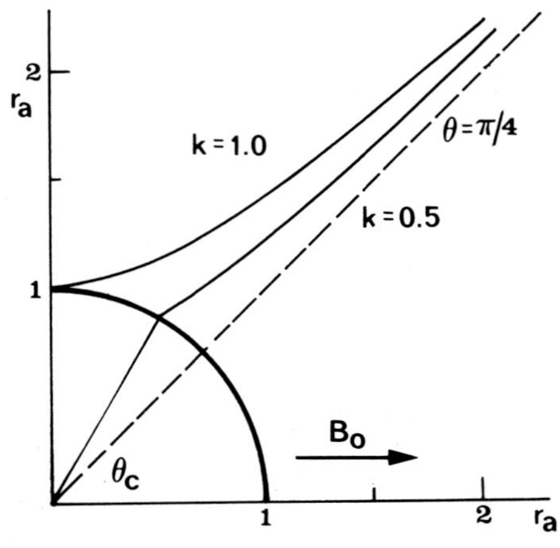

Fig. 3.5 Attractive and repulsive regions of space adjacent to a wire matrix element.

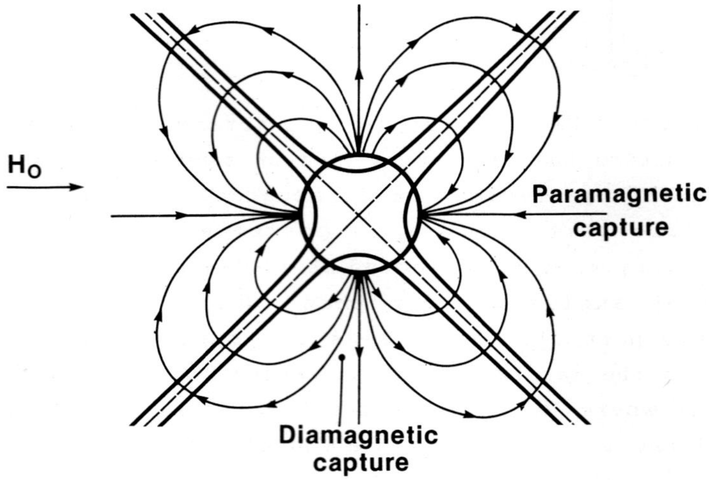

Fig. 3.6 Magnetic force field about a cylindrical wire.

In the case of the elliptical wire magnetized along the minor axis the repulsive region for paramagnetic particles is compressed into narrow zones near the sharp ends, whilst the attractive regions are spread out along the flat sides, as shown in Fig. 3.7. However, when the elliptical wire is magnetized along the major axis, the zones of attraction of paramagnetic particles are considerably reduced.

The expression (eq. 3.8) for the magnetic force is a product of two factors: the first factor, $\kappa \mathbf{V}$ reflects the physical properties of mineral particles; the second factor, magnetic force density

$$f_m = \mu_o \nabla(H^2) \tag{3.13}$$

describes the properties of a magnetic separator.

Gerber [Gl] argues that magnetic force density characterizes a magnetic separator (as far as its extraction capability is concerned) and claims that it is a single parameter that can be attributed to a magnetic separator and utilized for its classification. This approach treats magnetic field strength and field gradient in a symmetrical way, although their effect on the physical properties of the mineral particles and on the selectivity

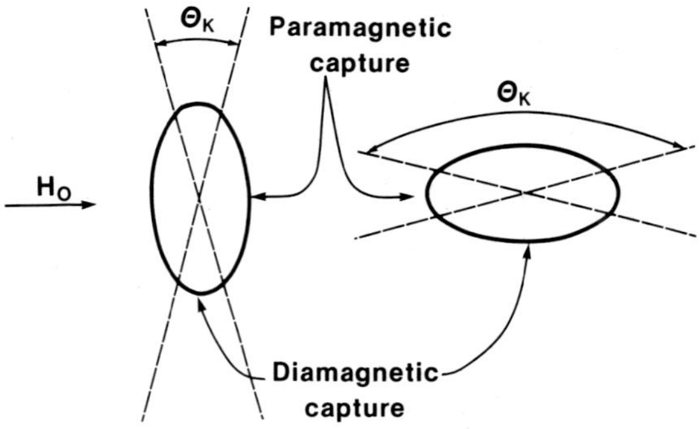

Fig. 3.7 Regions of attraction and repulsion of a paramagnetic particle in the vicinity of an elliptical wire magnetized along its minor and major axes.

of separation is substantially different. Most minerals having a parasitic ferromagnetic component exhibit field-dependent magnetic susceptibilities. Magnetic properties of the mineral particles therefore depend on the operating value of magnetic induction and not on the gradient of the field. Likewise, the applied magnetic field induces an agglomeration of particles which markedly affects the recovery and grade of the magnetic fraction. It is clear that mineral particles behave differently at different values of the applied magnetic field, though at the same value of the magnetic force density. The latter thus constitutes an inappropriate classification parameter.

III.1.3 Competing forces

The hydrodynamic drag acting on a particle travelling in a fluid stream is evaluated by calculating the flow field around a circular cylinder. To simplify analysis, the effects of the fluid boundary layer are usually neglected and two limiting cases are considered:

(a) High Reynolds numbers, where the inertial forces are much greater than the frictional forces and the potential flow equation is valid

(b) Reynolds number much less than unity where the inertial forces are small compared with the frictional forces so that viscosity is active in the entire flow field and laminar flow conditions prevail.

The Reynolds number is a ratio of the inertial force, associated with mass and velocity, and of the viscous or drag force in a fluid system. For a fluid of density ρ_f, dynamic viscosity η, or kinematic viscosity ν, flowing over a wire of radius **a**, with a linear velocity $\mathbf{v}_o$ the Reynolds number (Re) is given by

$$Re = \frac{2 v_0 \rho_f a}{\eta} = \frac{2 v_0 a}{\nu} \tag{3.14}$$

For high Reynolds numbers (Re >>1) the inertial forces predominate and the effect of the fluid viscosity will be negligible. The fluid motion can be approximated by the irrotational flow, the velocity can be written as a gradient of a scalar potential and the flow is called a potential flow [L7,B14]. The expression for the velocity of the potential flow is , however, a realistic approximation only at reasonably large distances from the front of the wire; in the vicinity of the wire and at the rear approximation is poor. Fig. 3.8 shows the course of the streamlines and the velocity profile for the potential flow.

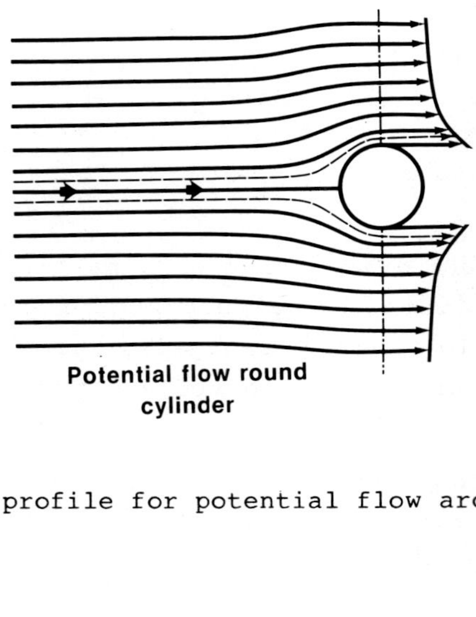

Fig. 3.8 Velocity profile for potential flow around a cylinder.

The equations for the velocity components $\mathbf{v}_r$, $\mathbf{v}_\theta$ of the potential flow past a cylindrical wire are

$$v_r = v_0\left(1-\frac{a^2}{r^2}\right)\cos(\theta-\gamma) \tag{3.15}$$

$$v_\theta = -v_0\left(1+\frac{a^2}{r^2}\right)\sin(\theta-\gamma) \tag{3.16}$$

When the inertial forces are small compared to the viscous forces, Re << 1 and the flow has a viscous (or creeping or laminar) character. The velocity profile for such a laminar flow is shown in Fig. 3.9.

Lamb [B14, Z2, G1] obtained a solution of the Navier-Stokes equation that satisfied the boundary conditions at the wire surface (disappearance of the components of velocity: $\mathbf{v}_r = \mathbf{v}_\theta = 0$ for $\mathbf{r} = a$). The fluid velocity components are then

$$v_r = v_0\,\frac{\ln\left(\frac{r}{a}\right)-0.5\left[1-\left(\frac{a}{r}\right)^2\right]}{2.002-\ln Re}\cos(\theta-\gamma) \tag{3.17}$$

$$v_\theta = -v_0\,\frac{\ln\left(\frac{r}{a}\right)+0.5\left[1-\left(\frac{a}{r}\right)^2\right]}{2.002-\ln Re}\sin(\theta-\gamma) \tag{3.18}$$

As mentioned above, flow velocity at the wire surface becomes zero in the viscous flow model, while in the potential flow model

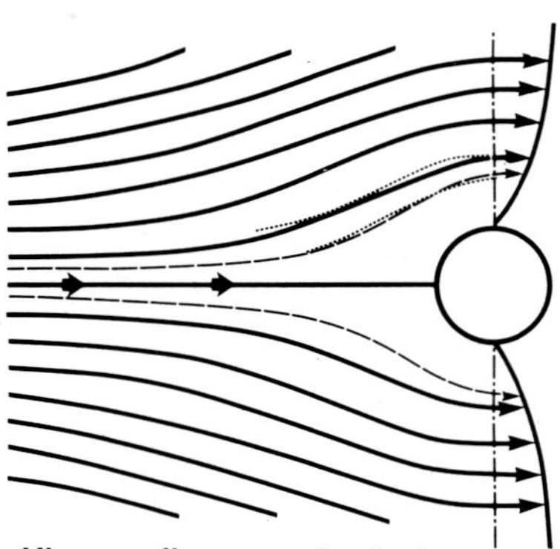

Fig. 3.9 Velocity profile for viscous flow around a cylinder.

it can reach values of up to $2\mathbf{v}_o$. These considerable differences between the two models are especially noticeable in the collection process in wet high-gradient magnetic separation. It has been shown [C17] that particles in viscous (laminar) flow begin to move away from the wire much further upstream than those in potential flow. Because the viscous flow field "sees" the wire much further upstream, particle transport to the region of the highest magnetic force is greatly diminished. A more realistic description of the capture process with viscous flow has been demonstrated experimentally by Schewe et al. [S23].

Viscous drag force is obtained from Stokes' equation (eq. (1.15)), although when a particle is accelerating in the fluid, it experiences drag forces that are additional to that given by Stokes law [B41]. The components of this drag force can be written

$$F_{dr}=6\pi\eta b\left[v_r-\frac{dr}{dt}\right] \tag{3.19}$$

$$F_{d\theta}=6\pi\eta b\left[v_\theta-r\,\frac{d\theta}{dt}\right] \tag{3.20}$$

The components of fluid velocity $\mathbf{v}_r$ and $\mathbf{v}_\theta$ are given by eqs. (3.15) and (3.16) or eqs. (3.17) and (3.18), depending upon the value of the Reynolds number.

Gravitational force is given by eq. (1.14), its components, in polar coordinates, can be written as [G1]

$$F_{gr}=\frac{4\pi b^3}{3}\,(\rho_p-\rho_f)\;g\cos\,(\theta-\beta) \tag{3.21}$$

$$F_{g\theta}=-\frac{4\pi b^3}{3}\,(\rho_p-\rho_f)\;g\sin\,(\theta-\beta) \tag{3.22}$$

The components of the inertial force are [G1]

$$ma_r=\frac{4}{3}\pi b^3\rho_p\left[\frac{d^2r}{dt^2}-r\left(\frac{d\theta}{dt}\right)^2\right] \tag{3.23}$$

$$ma_\theta=\frac{4}{3}\pi b^3\rho_p\left[r\frac{d^2\theta}{dt^2}+2\frac{dr}{dt}\cdot\frac{d\theta}{dt}\right] \tag{3.24}$$

III.1.4 Equations of particle motion

The dynamic behaviour of the particle in a magnetic separator can be described by an equation of the form

$$m\vec{a}_p = \vec{F}_d + \vec{F}_m + \vec{F}_g \tag{3.25}$$

where $\mathbf{ma}_p$ is the inertial force, $\mathbf{F}_d$ the viscous drag force, $\mathbf{F}_m$ the magnetic force and $\mathbf{F}_g$ the gravitational force. The complete equations of motion were obtained by a number of authors [W9, G1, C17, L8, L9]. If the analysis is limited to the motion of small particles ($b < 75\ \mu m$) for which the gravitational force is much smaller than the drag force, i.e. condition

$$\frac{2\rho_p b^2}{9\eta} \ll 1 \tag{3.26}$$

is satisfied [W9], then the inertial and gravitational terms in eq. (3.25) can be neglected and the radial and azimuthal components of the equation of motion (3.25) can be written, according to Birss and Parker [B13], as

$$\frac{dr_a}{dt} = \frac{v_o}{a}\left(1 - \frac{1}{r_a^2}\right)\cos(\theta - \gamma) - \frac{v_m}{a}\left(\frac{K}{r_a^5} + \frac{\cos 2\theta}{r_a^3}\right) \tag{3.27}$$

$$r_a \frac{d\theta}{dt} = -\frac{v_o}{a}\left(1 + \frac{1}{r_a^2}\right)\sin(\theta - \gamma) - \frac{v_m}{a}\cdot\frac{\sin 2\theta}{r_a^3} \tag{3.28}$$

where $\mathbf{r}_a = \mathbf{r/a}$ and

$$v_m = \frac{2\mu_o(\kappa_p - \kappa_f)MH_o b^2}{9\eta a} \tag{3.29}$$

is the parameter introduced by Watson [W9], having the dimension of $\mathbf{ms}^{-1}$ and termed magnetic velocity.

When $\gamma = 0$, equations (3.27) and (3.28) describe the motion of a particle in the longitudinal configuration (Fig. 3.3 a), while when $\gamma = \pi/2$ they describe the motion in transverse configuration

(Fig. 3.3 b). By combining eqs. (3.27) and (3.28) we obtain the time-independent equation [W9]

$$\frac{1}{r_a}\cdot\frac{dr_a}{d\theta}=-\frac{\left(1-\frac{1}{r_a^2}\right)\cos(\theta-\gamma)-\frac{v_m}{v_o}\left(\frac{K}{r_a^5}+\frac{\cos2\theta}{r_a^3}\right)}{\left(1+\frac{1}{r_a^2}\right)\sin(\theta-\gamma)+\frac{v_m}{v_o}\cdot\frac{\sin2\theta}{r_a^3}} \tag{3.30}$$

The particle trajectories can thus be described by an equation of a general form

$$\frac{1}{r_a}\cdot\frac{dr_a}{d\theta}=f\left(\frac{v_m}{v_o},\frac{M}{2H_o},r_a,\theta\right) \tag{3.31}$$

The path taken by the particle depends only on $\mathbf{v}_m/\mathbf{v}_o$, $\mathbf{M}/2\mathbf{H}_o$, the initial position and the initial velocity of a particle. Eq. (3.27) shows that $\mathbf{M}/2\mathbf{H}_o$ is the coefficient of a term which varies as $\mathbf{r}_a{}^{-5}$. This short-range term is important only near the wire, far away from the wire the path is a function of only $\mathbf{v}_m/\mathbf{v}_o$. This means that for large values of $\mathbf{v}_m/\mathbf{v}_o$ the efficiency of the particle capture by the wire is practically independent of $\mathbf{M}/2\mathbf{H}_o$. Capture efficiency thus depends only on $\mathbf{v}_m/\mathbf{v}_o$ and not on $\mathbf{v}_m$ or $\mathbf{v}_o$ separately [W10].

The complete equations of motion (eqs. (3.27) and (3.28)) have to be solved numerically, by employing a standard procedure such as a fourth-order Runge-Kutta technique [C18]. Trajectories can be divided into two kinds, depending on whether the particle is captured or not. The borderline case of a trajectory for which the particle is just captured is called the limiting or critical trajectory and the capture radius $\mathbf{R}_c$ is defined in Fig. 3.1. A normalized capture radius $\mathbf{R}_{ca}$, defined as $\mathbf{R}_{ca} = \mathbf{R}_c/\mathbf{a}$, is usually used.

Fig. 3.10 shows the particle trajectories with the magnetic field set to zero. The capture mechanism is inertial impaction, gravity or direct interception. Fig. 3.11 shows the typical trajectories of a paramagnetic particle in L-configuration at non-zero magnetic field. Fig. 3.12 depicts the limiting trajectories in L - configuration of paramagnetic and diamagnetic particles, as a function of $\mathbf{v}_m/\mathbf{v}_o$.

The dynamic behaviour of a magnetizable particle near an isolated ferromagnetic wire, as described by theoretical models [L8, L9, W10, C18, C17, G1, B13, C19, S24], and based on the solution and analysis of the equations of motion (eqs. (3.27),(3.28)) have been confirmed by experiments [S23, L9, G13, G14, G15, R7, T4, L10, F9].

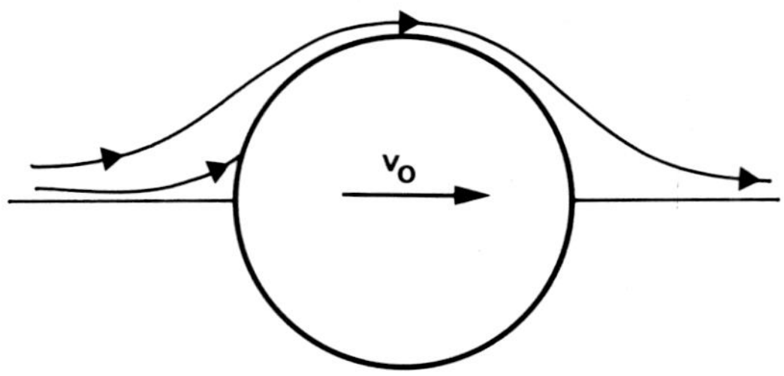

Fig. 3.10 Particle trajectories at zero magnetic field.

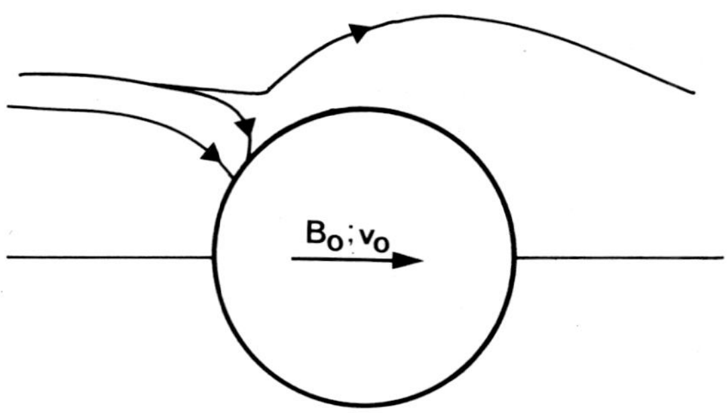

Fig. 3.11 Trajectories of a paramagnetic particle in L-configuration, in non-zero magnetic field.

III.1.5 Solutions of equations of motion

As mentioned earlier, the problem of particle capture in single-particle single-wire system, has been treated in several papers [W9, W10, L9, B15, C20, U1]. The theoretical models used in the description of the capture process usually use the so-called "inviscid" approximation of the fluid flow, in which the fluid is considered to be ideal when interacting with the wires, but viscous when interacting with particles.

Other authors [C18, C17, B16, G16] assumed the flow to be viscous in relation to both particles and wires. Most of these

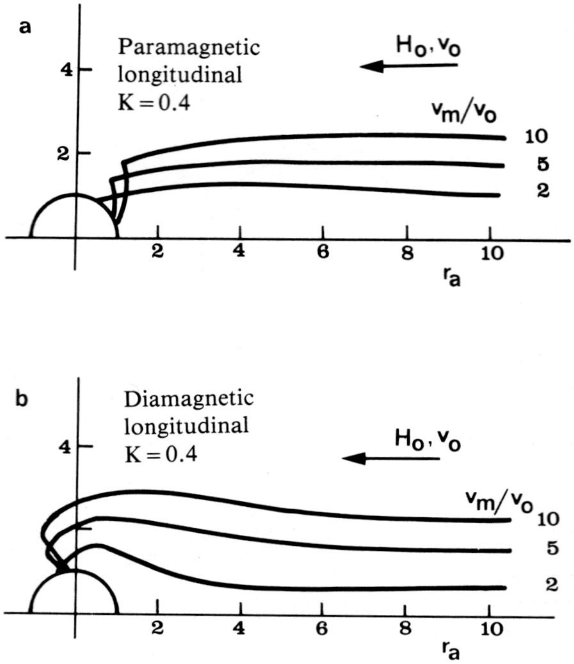

Fig. 3.12 Limiting trajectories for paramagnetic (a) and diamagnetic (b) particles in L-configuration, $\mathbf{K}$ = 0.4, as a function of $\mathbf{v}_m/\mathbf{v}_o$ (after Birss and Parker [B13]).

models are concerned with the L- and T- configurations of fluid flow and magnetic field. The analytical solutions are, however, difficult to obtain and these models are predominantly of numerical character. Under certain simplifying assumptions the approximate solution of the equation can be found, as demonstrated by Uchiyama and Hayashi [U2]. Although their solutions are only approximate they show a good correlation with the predictions of the numerical computations. The analytical solutions for the capture radius in L- and T- configurations, for the case of potential flow are summarized in Table 3.1 [B13, U2].

From an inspection of the equations in Table 3.1 it is clear that the dependence of the capture radius $\mathbf{R}_{ca}$ upon $\mathbf{v}_m/\mathbf{v}_o$ can be divided into two categories: at low values of $\mathbf{v}_m/\mathbf{v}_o$

$$R_{ca} \sim \frac{v_m}{v_o}$$

TABLE 3.1

Approximate analytical expressions for normalized capture radius $\mathbf{R}_{ca}$ for paramagnetic particles in L- and T-configurations, in the potential flow approximation [U2, B13, G1]

Configuration	$\frac{v_m}{v_o}$	R_{ca}	
Longitudinal	$> \frac{\sqrt{2}}{1-K}$ >3	$1.3\left(\frac{v_m}{v_o}\right)^{\frac{1}{3}} - (0.87+0.052K)\left(\frac{v_m}{v_o}\right)^{-\frac{1}{3}}$	(3.32)
	>0.1 <2	$\frac{v_m}{2v_o}\,\frac{(1-K^2)^{\frac{1}{2}}+K(\pi-\arccos K)}{1+\frac{4K}{5}\cdot\frac{v_m}{v_o}}$	(3.33)
Transverse	>3	$1.3\left(\frac{v_m}{v_o}\right)^{\frac{1}{3}} - (0.87-0.36K)\left(\frac{v_m}{v_o}\right)^{-\frac{1}{3}}$	(3.34)
	>0.1 <2	$\frac{v_m}{2v_o}\,\frac{(1-K^2)^{\frac{1}{2}}+K(2\pi-\arccos K)}{1+\frac{4}{5}K\frac{v_m}{v_o}}$	(3.35)

while at large values of $\mathbf{v}_m/\mathbf{v}_o$

$$R_{ca} \sim \left(\frac{v_m}{v_o}\right)^{\frac{1}{3}}$$

On the other hand, the empirical studies [B13, C20] indicate that for large values of $\mathbf{v}_m/\mathbf{v}_o$ $\mathbf{R}_{ca}$ varies as a 3/8 power of $\mathbf{v}_m/\mathbf{v}_o$

$$R_{ca} = 1.2\left(\frac{v_m}{v_o}\right)^{\frac{3}{8}} - \frac{(1-K)}{5}\left(\frac{v_m}{v_o}\right)^{\frac{1}{2}} \tag{3.36}$$

in T-configuration, while in L-configuration the following relation was derived [B13]

$$R_{ca} = \left(\frac{v_m}{v_o}\right)^{\frac{3}{8}} + \frac{K}{5}\left(\frac{v_m}{v_o}\right)^{\frac{2}{5}} \tag{3.37}$$

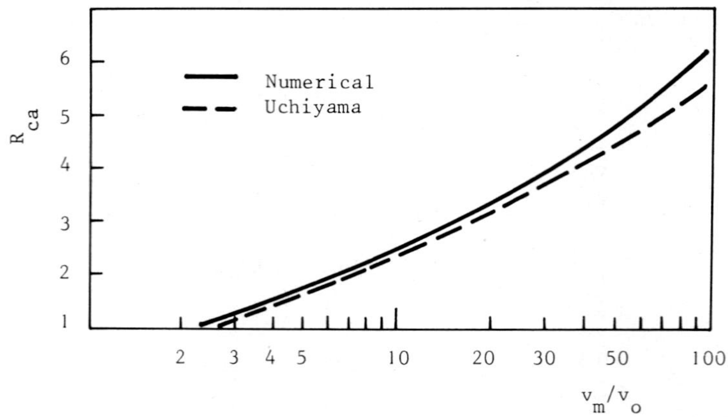

Fig. 3.13 Normalized capture radius $\mathbf{R}_{ca}$ as a function of $\mathbf{v}_m/\mathbf{v}_o$, for paramagnetic particles, in T-configuration, for $\mathbf{v}_m/\mathbf{v}_o > 2$ and $\mathbf{K} = 0.2$. (After Birss and Parker [B13]).

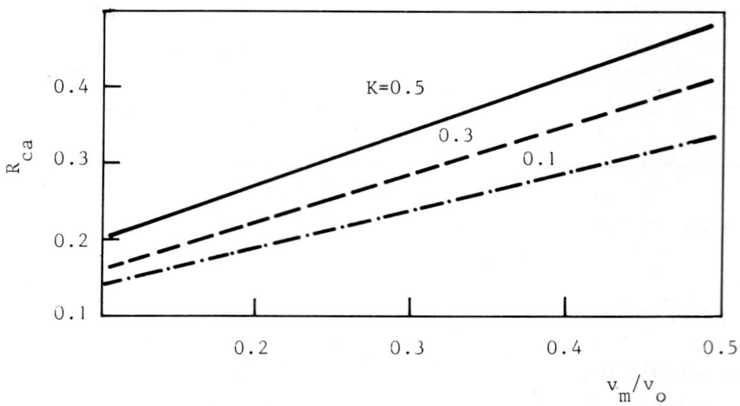

Fig. 3.14 Normalized capture radius $\mathbf{R}_{ca}$ as a function of $\mathbf{v}_m/\mathbf{v}_o$, for paramagnetic particles, in L-configuration, for $\mathbf{v}_m/\mathbf{v}_o <$ 0.5. (After Birss and Parker [B13]).

The comparison of the normalized capture radius, in T-configuration as a function of $\mathbf{v}_m/\mathbf{v}_o$ (for $\mathbf{v}_m/\mathbf{v}_o > 2$), as obtained from analytical and numerical analyses is shown in Fig. 3.13.

For low values of $\mathbf{v}_m/\mathbf{v}_o$, of the order of 0.1 in L-configuration both analytical and empirical approaches give a linear relationship between $\mathbf{R}_{ca}$ and $\mathbf{v}_m/\mathbf{v}_o$ [B13], as shown in Fig. 3.14.

It can thus be anticipated that an approximate condition of strong capture would be

$$\frac{v_m}{v_o} \gg 1 \qquad (3.38)$$

and this criterion will be derived in Section III.1.9 [W9, W10, W11]. However, the experimental results show that a much less stringent condition can still lead to a reasonably high degree of capture.

III.1.6 Comparison of capture radius for potential and viscous flow approximations

As mentioned in Section III.1.3 potential flow is a realistic approximation at a large distance from the wire, while viscous flow gives a better description of the capture process in the vicinity of the wire. As shown in Figs. 3.15 and 3.16, the particles in viscous flow have begun moving away from the wire much further upstream than those in potential flow. The viscous flow field "sees" the wire much further upstream and thus the particle transport to the region of the highest magnetic force is greatly diminished [C18]. Another feature of the viscous flow is that the captured particles respond much more sharply to the magnetic force.

Cummings et al. [C18] found that in L-configuration capture radius that had been calculated with the viscous flow model was appoximately 1/3 of that calculated for the potential flow model.

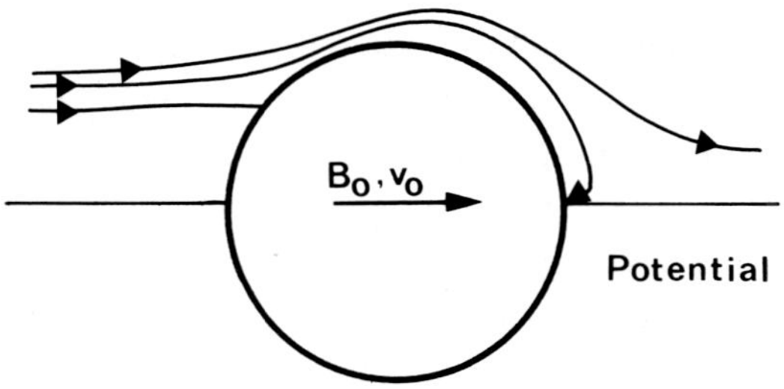

Fig. 3.15 Particle trajectories for potential flow in L-configuration.

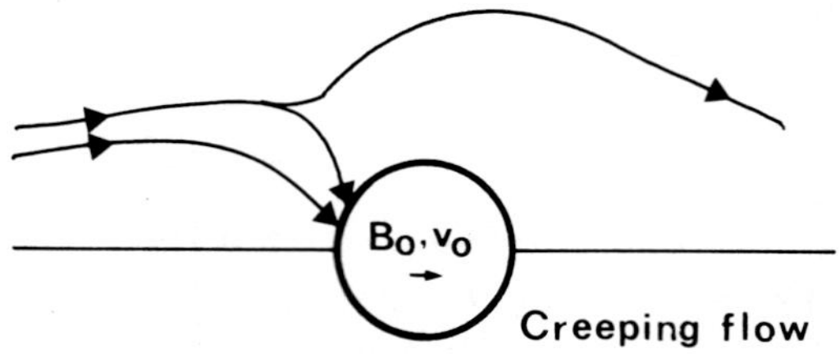

Fig. 3.16 Particle trajectories for viscous flow in L-configuration.

On the other hand, Paul [P8] showed that there was a little difference between potential and viscous flow predictions. Although capture radius calculated from potential or viscous flow approximations is very similar, particle trajectories can differ substantially.

The comparison of theoretical models with the experimental results of magnetic capture of red blood cells by a single wire [P9] has shown that the particle motion can well be described by the potential flow model. Only a slight improvement to the fit between theory and experiment is observed using a viscous flow correction.

Similarly, Birss et al. [B16] inferred from their theoretical analysis of the axial magnetic filter that the predictions for capture efficiency, using viscous and potential flow fields, are not greatly different. Schewe et al. [S23] observed that in T - configuration the experimental results can be best described by the viscous flow model, while the potential flow model gives a poor fit of observed trajectories.

Friedlaender et al. [F10] investigated the motion of a particle in the vicinity of a magnetized single sphere and found that the particle trajectories and capture efficiency could be well described in terms of a viscous flow model, while agreement between theory and experiment was excellent. The viscous flow model was also found to give a better description of the particle motion in an infinite array of ferromagnetic wires [R8].

The significant difference in particle trajectories and frequently in the capture radius for potential and viscous flow fields indicates that a correct modelling of the fluid flow is essential for the understanding of the process of particle capture. This is particularly important for real matrices of magnetic separators, where the fluid flow through the matrix, rather than magnetic properties of the matrix, determine the performance of the separator [W12].

III.1.7 Axial configuration

The analysis of a particle motion in axial (A) configuration, where the wire axis and the fluid flow are mutually parallel, and the magnetic field is perpendicular to them (see Fig. 3.3), does not require, in contrast to other cases, numerical computations, and results can be derived in an analytical form.

It has been shown by Birss et al., Gerber and Uchiyama et al. [B15, B17, B16, G16, U1, U3, U4, G1, B13] that the motion of a particle in the axial configuration can be described in terms of normalized cylindrical polar coordinates $\mathbf{r}_a$, θ, $\mathbf{z}_a$, where $\mathbf{r}_a = \mathbf{r}/\mathbf{a}$ and $\mathbf{z}_a = \mathbf{z}/\mathbf{a}$. The wire axis and the fluid flow are parallel to each other and to the z-axis, as shown in Fig. 3.17.

The equations of the particle motion (eqs. (3.27) and (3.28)) reduce in this configuration to the simplified forms

$$\frac{dr_a}{dt} = -\frac{v_m}{a}\left(\frac{K}{r_a^5} + \frac{\cos 2\theta}{r_a^3}\right) \tag{3.39}$$

$$\frac{d\theta}{dt} = -\frac{v_m}{a} \cdot \frac{\sin 2\theta}{r_a^4} \tag{3.40}$$

$$\frac{dz_a}{dt} = \frac{v_o}{a} \tag{3.41}$$

where the last equation (3.41) describes the motion of the particles along the length of the wire.

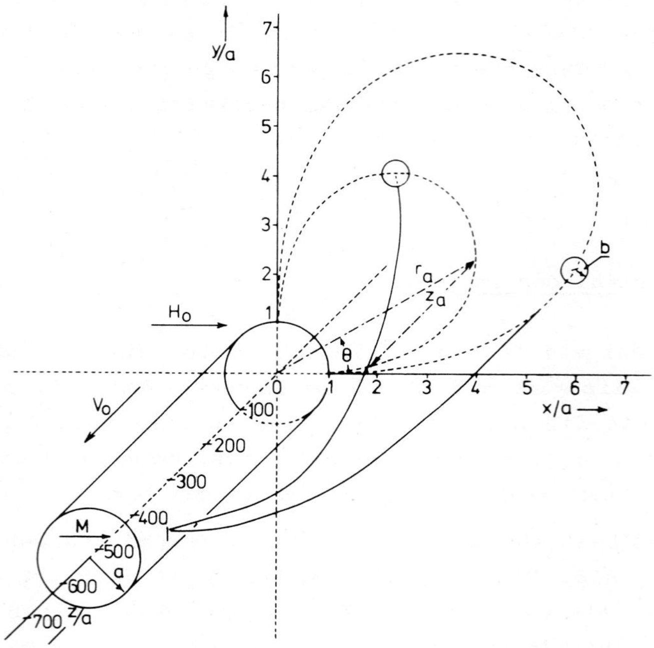

Fig. 3.17 The axial configuration (after Birss et al. [B15]).

The elimination of time from eqs. (3.39), (3.40) and (3.41) produces a set of differential equations

$$\frac{dr_a}{d\theta} = \frac{K}{r_a \sin 2\theta} + r_a \cot 2\theta \tag{3.42}$$

$$\frac{dz_a}{d\theta} = \frac{v_o}{v_m} \cdot \frac{r_a^4}{\sin 2\theta} \tag{3.43}$$

From these equations the particle trajectories can be determined analytically. The solutions of the non-linear Bernoulli equation (3.42) and a linear equation (3.43) can be written as

$$r_a = (-K\cos 2\theta + C|\sin 2\theta|)^{\frac{1}{2}} \tag{3.44}$$

$$z_a = \frac{v_o}{v_m}\left[\frac{K^2}{2}\, ln\, \frac{\tan\theta}{\tan\theta_i} - K(r_{ai}^2 - r_a)^2 - \frac{(K^2 + C^2)}{2}\,(\cos 2\theta_i - \cos 2\theta)\right] \tag{3.45}$$

where $\{\mathbf{r}_{ai},\ \theta_i,\ \mathbf{z}_a = 0\}$ and $\{\mathbf{r}_a,\ \theta\ ,\ \mathbf{z}_a\}$ are the initial and current particle positions, respectively. The integration constant **C** may be determined by inserting the initial position coordinates $\{\mathbf{r}_{ai},\ \theta_i\}$ in eq. (3.44).

The solution (eq. (3.44)) describes a family of trajectories projected onto the (x, y) plane; two of these curves are shown in Fig. 3.17. The solution (eq. (3.45)) determines the progress of the particle along the z-axis and two trajectories starting at two different initial positions are shown by solid lines in Fig. 3.17.

III.1.8 Magnetic separation of ultrafine particles

As described above the capture of relatively large particles in a high-gradient magnetic separator can be explained by the standard theory. When the particle size decreases, the Brownian thermal motion gains in prominence and at a certain particle size the conventional theory becomes deficient and the generalized theory has to be applied.

The generalized theory of HGMS was developed by Gerber et al. [G17, T5, G18] who incorporated diffusion into the capture mechanism and established the particle size criterion thus allowing the generalized and conventional theories to become coincidental.

The trajectories of individual ultrafine particles suspended in a fluid are of little significance for the description of the capture process in HGMS since, owing to diffusion, it is not known whether a given particle will be captured by the matrix, or will escape. The statistical approach must therefore be adopted and the theory has to be formulated in terms of the distribution of particle concentration [G17].

The system of particles has to satisfy the equation of continuity

$$\frac{\partial c}{\partial t} + \nabla \vec{J}_p = 0 \tag{3.46}$$

where **c** is the particle concentration and J_p is the total flux of particles. In the absence of any fluid flow, this flux consists of two contributions: $\vec{J}_p = \vec{J}_D + \vec{J}_F$. The term J_D is the flux of particles as a result of thermal diffusion and is given by: $\vec{J}_D = -D_p \nabla c$. The other term, J_F, is a result of the action of external forces that impose a drift velocity $\vec{v}_d$ on the system of particles; thus $\vec{J}_F = c\vec{v}_d$. By inserting the expressions for J_D and J_F in eq. (3.46), and assuming that the Einstein diffusion coefficient D_p does not depend on the concentration of particles we obtain

$$\frac{\partial c}{\partial t} = D_p \nabla^2 c - \nabla(\vec{v}_d c) \tag{3.47}$$

For the ideal suspension of particles D_p is given by the Nernst-Einstein relation $D_p = ukT$, where **k** is the Boltzmann constant, **T** the absolute temperature and **u** the particle mobility given in Stokes' approximation as

$$u = (6\pi\eta b)^{-1} \tag{3.48}$$

Drift velocity is obtained as

$$\vec{v}_d(\vec{r}) = u\vec{F}(\vec{r}) \tag{3.49}$$

where **F** is the total force on a particle and is given as a sum of all forces F_i acting on a particle in a suspension

$$\vec{F} = \sum_i \vec{F}_i \tag{3.50}$$

Equations (3.47) and (3.49) are general equations describing the capture process in HGMS. Since form of these equations is independent of particle size they are valid over the whole range from macroscopic to molecular dimensions. For ultrafine particles, eq. (3.49) can be inserted in eq. (3.47) and the resulting HGMS diffusion equation may be solved. For larger particles the first term on the right-hand side of eq. (3.47) can be neglected and drift velocity becomes the actual particle velocity $d\vec{r}/dt$. Eq. (3.49) then becomes an equation of motion in the sense of the conventional HGMS theory while the mass conservation equation (3.47) is automatically satisfied.

Gerber et al. [G17] solved the diffusion equation (3.47) and have shown that the build-up of particles on the magnetized wire can have a dynamic or static characters, or both, depending upon the relationship between the particle radius **b** and the critical radius $\mathbf{b}_{crit}$, given for various minerals in Table 3.2

Particles with $\mathbf{b} < \mathbf{b}_{crit}$ are captured dynamically as a cloud around the wire while particles with $\mathbf{b} > \mathbf{b}_{crit}$ form a static build-up surrounded by a dynamic build-up. This situation is shown in Fig. 3.18. If the capture process is extended to include the retention of particles in a statistical sense, there is no lower limit in particle size on their capture.

TABLE 3.2

Critical particle radius $\mathbf{b}_{crit}$ for water-suspended minerals (after Takayasu et al. [T5] and Gerber et al.[G17])

Mineral	(SI)	$\mathbf{b}_{crit}$ (Å)	$\mathbf{b}_{conv}$ (μm)
Fe		14	0.03
Fe_3O_4		22	0.04
$Mn_2P_2O_7$	4.73×10^{-3}	100	0.15
$\alpha - Fe_2O_3$	1.47×10^{-3}	147	0.2
CuO	2.42×10^{-4}	266	0.3
Al	2.08×10^{-5}	541	0.5
Al_2O_3	-1.81×10^{-5}	1090	0.75
Au	-3.45×10^{-5}	771	0.6

As the particle size decreases, the conventional HGMS theory becomes inadequate for the capture process and the generalized HGMS theory must be used. Gerber et al. [G17] have shown that the conventional theory of particle capture in HGMS can be applied for the particle size satisfying the inequality

$$2|F|b \geq kT \qquad (3.51)$$

where $|F|$ is the magnitude of the total force acting on the particle (eq. (3.50)). The illustrative values of the threshold radius $\mathbf{b}_{conv}$ of a particle to be captured on a ferromagnetic wire ($\mathbf{a}$ = 50 μm, $\mathbf{B}$ = 1 T, $\mathbf{J}$ = 1.7 T) are shown in Table 3.2.

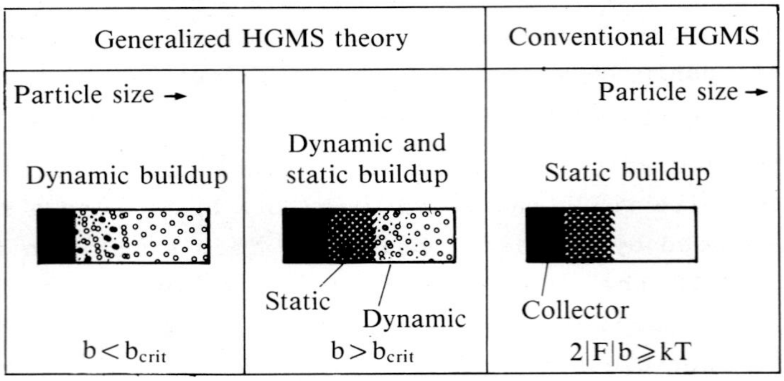

Fig. 3.18 Relation between particle size and type of particle retention on a magnetized wire (after Gerber [G18]).

III.1.9 Particle build-up

Previous analyses of particle trajectories in the vicinity of a magnetized wire considered the capture process in the initial stage, i.e. with no loading of the wire. Although the trends and the functional relations predicted from these calculations appear to be valid generally, the full picture of the capture process in a matrix can be obtained only by including the effect of particle stability on the surface of the wire and the way particles are deposited on the surface.

The determination of the exact manner of the particle capture and build-up is a difficult problem. As the suspension flows through the matrix, some of the particles, under the influence of a number of forces come into contact with the matrix elements and are deposited onto these elements or onto already deposited particles. Consequently the geometry, the structure and the surface characteristics of the matrix may be modified significantly. These changes, in turn, affect the flow of the suspension through the matrix and the subsequent capture. Further complications may also arise, due to the dislodgement of the captured particles and their possible re-capture. As a result, the dynamic behaviour of the separation process varies with time as well as with other variables.

The force that is principally responsible for the particle

capture is the magnetic force. Particles are attracted towards the wire if the radial component of the magnetic force $\mathbf{F}_{mr} < 0$. Thus, the relation $\mathbf{F}_{mr} = 0$ defines the critical angle θ_c at which the radial component of the magnetic forces becomes repulsive, the angle being given by eq. (3.12). The stability of a particle on the wire surface is determined by the balance of the tangential components of the magnetic, drag and other forces. As follows from Fig. 3.19, the paramagnetic particle will be retained if the radial force is attractive (i.e. $|\theta| < |\theta_c|$) and if the tangential force is such that the particle is attracted towards the stagnation point, i.e. $d\theta/dt < 0$. Under these conditions the angle θ_p for which a paramagnetic particle will be retained, is given as [Gl]

$$|\theta_p| \leq \arcsin\left[1 - \left(\frac{v_o}{v_m}\right)^2\right]^{\frac{1}{2}} \qquad (3.52)$$

It follows from eq. (3.52) that retention will occur only if $\mathbf{v}_m/\mathbf{v}_o > 1$, which is equivalent to eq. (3.38), given in Section III.1.5. If the whole monolayer of particles from $-\theta_c$ to $+\theta_c$ is to be stable, then for $|\theta_p| < |\theta_c|$ we obtain, from eqs. (3.12) and (3.52):

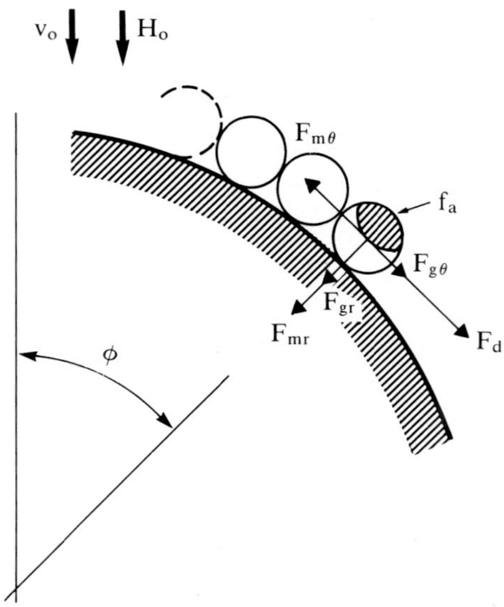

Fig. 3.19 Force components acting on a spherical particle on the wire surface in L-configuration.

$$\frac{v_m}{v_o} \geqslant \left(\frac{2}{1-K}\right)^{\frac{1}{2}} \qquad (3.53)$$

As we shall see later, these retention criteria overestimate a value of $\mathbf{v}_m/\mathbf{v}_o$ and a high capture efficiency can be observed for $\mathbf{v}_m/\mathbf{v}_o$ much less than unity.

The first theoretical model of the particle build-up on the magnetized wires was developed by Luborsky and Drummond [L11]. The build-up of particles in L-configuration on ribbon - shaped matrix elements was investigated. The authors assumed that for $\mathbf{v}_m/\mathbf{v}_o >> 1$ the build-up profile fans out symmetrically in a 90° angular sector over the front face of the wire so that when **n** monolayers of particle are deposited, the wire radius increases from the value **a** to $\mathbf{a}_n = \mathbf{a} + 2\mathbf{bn}$.

The inability of the model to describe the limiting build-up volume was later rectified by the same authors [L12]. The saturation volume of the up-stream side build-up was determined by the locus of a zero net azimuthal force acting on the particles in the build-up profile. The drag force (eqs. (3.19) and (3.20)) was modified to include the effect of the boundary layer on a surface of wire.

An improved model of particle build-up was developed by Watson [W13] who introduced a friction-like azimuthal force acting on the particles. This force is a function of the wire surface roughness which enhances the particle stability on the surface. The predictions of Watson's model are comparable to those obtained from simpler models [L11, L12].

III.1.10 Model of Nesset and Finch

Nesset and Finch [N2, N4] developed a static model of particle capture in HGMS, in which the fluid drag force includes the changing boundary layer thickness over the front portion of a wire. The magnetic field dependence of magnetic susceptibility, according to the Owen-Honda relation (eq. (1.54)) was also incorporated in the model.

III.1.10.1 Particle build-up

Fig. 3.19 illustrates the forces acting on a particle on the surface of the wire in L-configuration. It is assumed that if the net resultant force is towards a region of greater attractive force then the particle will remain attached to the wire; if not, the particle will move toward a region of lesser attractive force and ultimately will be swept away from the wire. The position where the forces are in equilibrium determines the equilibrium profile of the build-up.

The magnetic force acting on a particle on the surface of a wire, as shown in Fig. 3.19, is given by eqs. (3.9) and (3.10). The equations are also valid for field-dependent susceptibility and the explicit equations for the case when magnetic susceptibility can be expressed by the Honda-Owen equation (1.54) were derived by Hollingworth et al. [H30]. The fluid drag on a spherical particle can be written as

$$F_d = 4\pi b^2 f_a \tau_o \tag{3.54}$$

where τ_o is the shear stress at the wall of a wire and $\mathbf{f}_a$ describes the integral effect of τ_o acting over the upper half of the spherical particle, only. If τ_o is assumed constant then $\mathbf{f}_a = \pi/8$ [B18].

The velocity profile is a function of ϕ , an angle measured from the front stagnation point, as shown in Fig. 3.20. Shear stress reaches a maximum at about 60°. Velocity $\mathbf{v(x)}$ is given by

$$v(x) = 2v_o \sin\phi \tag{3.55}$$

The shear stress on the wall ($\mathbf{y}$ = 0) is equal to

$$\tau_o = \eta\left(\frac{\partial v}{\partial y}\right)_o \tag{3.56}$$

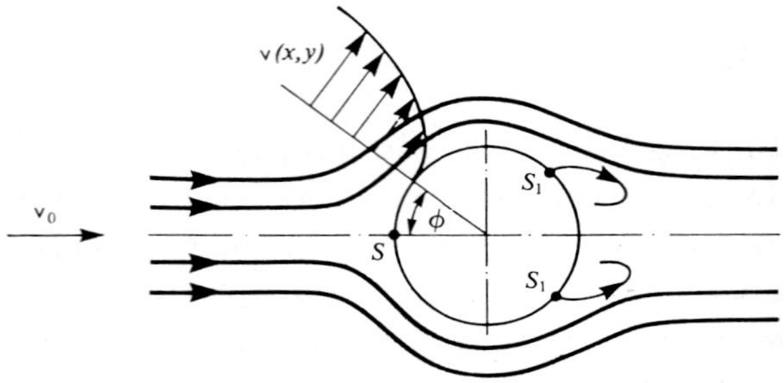

Fig. 3.20 Boundary-layer separation and stagnation points on a matrix collector.

while $(\partial v/\partial y)_o \simeq v/\delta$ where δ is the laminar boundary layer thickness

$$\delta = 5\left(\frac{\eta a}{\rho v}\right)^{\frac{1}{2}} \tag{3.57}$$

Assuming that the potential flow is given in the form of Blasius series

$$v(x) = u_1 x + u_3 x^3 + u_5 x^5 + \ldots.. \tag{3.58}$$

where **u** are the coefficients, and expanding $\sin\phi$ in eq. (3.55) into series, we obtain, through eq. (3.56)

$$\tau_o = \left(\frac{\eta v_o^3 \rho_f}{8a}\right)^{\frac{1}{2}} (9.861\phi - 3.863\phi^3 + 0.413\phi^5 - 0.0261\phi^7 \ldots..) \tag{3.59}$$

where ϕ is given in radians.

The last equation shows that the shear force is proportional to $\mathbf{v}_o^{3/2}$ and $\eta^{1/2}$ in contrast with Stokes law (eq. (1.15)) in which the dependence on $\mathbf{v}_o$ and η is linear.

Equations (3.54) and (3.59) describe the fluid drag force on a spherical particle on the surface of cylindrical wire and take into account the changes in boundary layer behaviour on the upstream side of the wire.

If the particle is to remain attached to the wire, it must experience a net attraction, both radially towards the wire and tangentially towards the front stagnation point. As follows from Fig. 3.19, the net radial and tangential forces can be written as

$$F_r = F_{mr} + F_{gr} \tag{3.60}$$

$$F_\phi = F_{m\phi} + F_d \tag{3.61}$$

By solving for ϕ in eqs. (3.60) and (3.61) at various **r** for the conditions $\mathbf{F}_\phi = 0$ and $\mathbf{F}_r = 0$, the loci of critical angles ϕ_c can be determined [N4] and these define the region of particle attraction; loci of critical angles are shown in Fig. 3.21. The two retention criteria are satisfied only in the cross-hatched region.

The calculated build-up profiles, as shown in Fig. 3.22 are in good agreement with experimentally observed build-up shapes [F11]. Both experiment and theory agree on a fan shape; both show build-up of particles beyond a 90° sector as a consequence of the presence of short-range term $\mathbf{Ma}^2/\mathbf{r}^2$ in eq. (3.9). Experimental evidence suggests accummulation to angles greater than ϕ_c some distance from the wire than predicted.

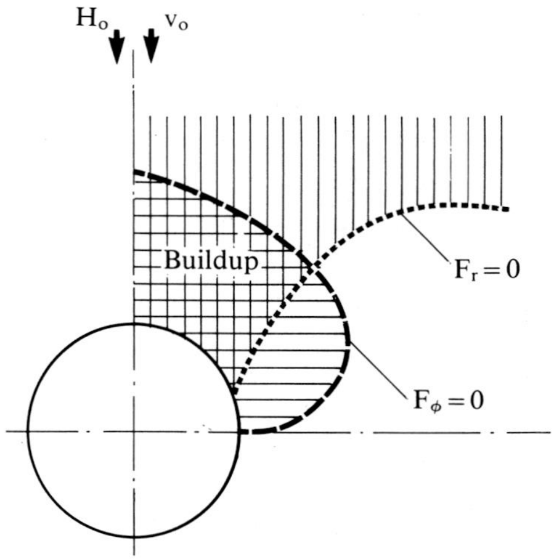

Fig. 3.21 Loci of zero total radial and azimuthal forces around the wire (after Nesset and Finch [N2]).

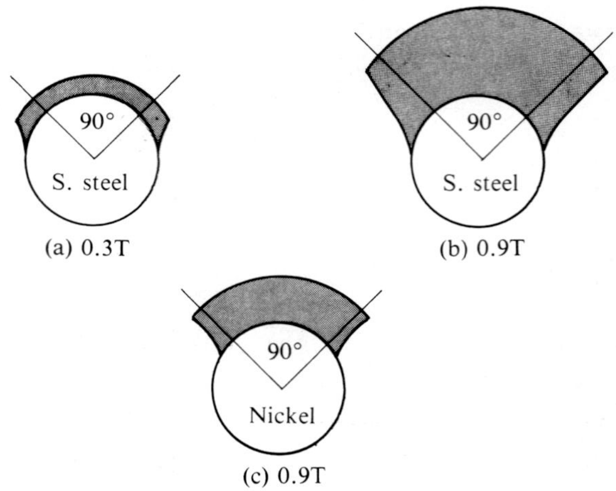

Fig. 3.22 Equilibrium build-up profiles for various magnetic inductions and matrix materials (after Nesset and Finch [N4]).

A detailed comparison of theory with experiment [N5] indicates that the boundary-layer model of the fluid drag is more realistic than the Stokesian approach, which gives about 50 per cent higher drag force than the Blasius solution. Furthermore, the model of Nesset and Finch is found to adequately predict the upstream build-up profiles in single-wire HGMS in L-configuration.

III.1.10.2 Matrix loading

As follows from Fig. 3.21 the fully loaded build-up is approximately a 90° segment of a constant radius **r**. The magnitude of **r** can be determined from the condition $\mathbf{F}_{\phi} = 0$. After rearranging eq. (3.61) Nesset and Finch [N2] obtained (at $\phi = 45^{o}$)

$$r_a^2 = \left(\frac{N_L}{C}\right)^{\frac{4}{5}} \tag{3.62}$$

where **C** is a constant, equal to 2.5 and

$$N_L = \frac{b\kappa B J_s}{\mu_o \rho_f^{\frac{1}{2}} V_o^{\frac{3}{2}} \eta^{\frac{1}{2}} a^{\frac{1}{2}}} \tag{3.63}$$

$\mathbf{N}_L$ is termed the loading number and is proportional to the ratio of magnetic to fluid shear forces in equilibrium at relative build-up radius r_a. The maximum build-up of particle on a wire is thus a function of N_L.

The volume of the build-up per unit length of wire normalized to the wire volume is given by

$$\gamma_v = \frac{1-\epsilon}{4}\left[r_a^2 - 1\right] \tag{3.64}$$

where ϵ is the porosity of the build-up. Replacing r_a we obtain

$$\gamma_v = \frac{1-\epsilon}{4}\left[\left(\frac{N_L}{C}\right)^{\frac{4}{5}} - 1\right] \tag{3.65}$$

and

$$\gamma_m = \frac{1-\epsilon}{4}\left[\left(\frac{N_L}{C}\right)^{\frac{4}{5}} - 1\right]\frac{\rho_p}{\rho_w} \tag{3.66}$$

where γ_v and γ_m are the so-called volume and mass loading, respectively, ρ_w is the density of the matrix material.

Parameter N_L, can be expressed in terms of magnetic velocity (eq. (3.29) as

$$N_L = \frac{9a}{2b}\,Re^{-0.5}\,\frac{v_m}{v_o} \tag{3.67}$$

where **Re** is the Reynolds number for the wire given by eq. (3.14). It can be seen that for $a/b \rightarrow 1$ and $Re \rightarrow 1$ $N_L \rightarrow v_m/v_o$.

The loading number N_L can be used to determine the threshold minimum particle size that can be recovered by a magnetized wire. The criterion of the lower technological limit can be obtained from the condition that the volumetric loading γ_v of the wire is equal to zero, which is satisfied for N_L = 2.5, as follows from eq. (3.65). The practical limit can be loosely determined from the requirement that the recovery is acceptable. Since it is difficult to express recovery in terms of this model, certain simplifications have to be made. Nesset and Finch [N2] derived a condition N_L = 25 as a practical lower limit of the capture, under the assumption that recovery can be expressed in the form

$$R = 0.75\,\frac{\gamma_m \rho_w}{L_m} \tag{3.68}$$

where $\mathbf{L}_m$ is the matrix loading (mass of the magnetic fraction per unit volume of the matrix). Since the value of $\mathbf{N}_L$, and thus of the minimum particle size to be captured is sensitive to the correct choice of the expression for recovery, the criterion $\mathbf{N}_L = 25$ often does not reflect the real situation in a magnetic separator and can give erroneous results, because of the inadequacy of eq. (3.68).

Let us illustrate the sensitivity of the $\mathbf{N}_L$ - criterion to the definition of recovery by the following example. Recovery into the magnetic fraction can be written as

$$R = \frac{\text{mass of mags} \times \text{grade of mags}}{\text{mass of feed} \times \text{grade of feed}}$$

Since γ_m = mass of mags/mass of matrix and the grade of mags/grade of feed is the concentration ratio ($\mathbf{E}_c$), we can write

$$R = \frac{kE_c\gamma_m\rho_w}{L_F} \qquad (3.69)$$

where $\mathbf{L}_F$ is the mass of feed per unit volume of the matrix and $\mathbf{k}$ is a constant (= 0.75 in eq. (3.68)).

In the recovery of uranium from the leaching residues, $\mathbf{L}_F \cong 1600\ \text{kg m}^{-3}$ and $\mathbf{E}_c \cong 10$. Thus for $\mathbf{R} = 100$ per cent we obtaîn from eq. (3.69) $\gamma_m = 0.027$, which in turn gives $\mathbf{N}_L \cong 4$. Consequently, this criterion does not differ very much from that for the technological limit.

III.1.11 Capture of diamagnetic particles

It has been shown in Section III.1.2 that the dipolar magnetic force $\mathbf{F}_m$, between a small particle of volume magnetic susceptibility κ_p and a cylindrical wire is given, in polar coordinates, by eqs. (3.9) and (3.10). If the net magnetic susceptibility $\kappa = \kappa_p - \kappa_f$ is positive then the force between the particle and the fibre is attractive provided that the radial component $\mathbf{F}_r$ is negative. This is the case of capture of paramagnetic particles, under the condition that eq. (3.11) is satisfied.

On the other hand, in two symmetrical curved volumes, the boundaries of which are determined by a critical angle θ_c given by eq. (3.12), the radial force acting on paramagnetic particles ($\kappa_p > \kappa_f$) becomes repulsive. If the particle is diamagnetic the radial force will be attractive for $\theta > \theta_c$ in the first quadrant, as shown in Fig. 3.5.

The value of θ_c depends upon both $\mathbf{J}_s$ and $\mathbf{B}_o$ as shown in Fig. 3.23. In the limiting case of $\mathbf{K} = 1$ ($\mathbf{B}_o = 0.85$ T for $\mathbf{J}_s = 1.7$ T), the angular sector of the wire surface available for capture of diamagnetic particles becomes very small. At high values of background magnetic field the volume of the space available for a capture of diamagnetic particles can be however significant, as follows from Fig. 3.23. Since the effective area of the matrix wire for a capture of paramagnetic particles will be correspondingly reduced this can negatively affect the loading capacity and performance of a magnetic separator. This argument again underlines the fact that an effort to increase the efficiency of magnetic separation simply by increasing the magnetic field (or magnetic velocity) can often result in a reduction not only of the grade of the magnetic fraction, but also of the recovery of a valuable mineral into the magnetic product.

As the value of magnetic susceptibility of diamagnetic materials is several orders of magnitude lower than that of most paramagnetic materials, the probability of the capture of diamagnetic particles is substantially lower than that of paramagnetic particles of the same size. If, however, $\mathbf{B}_o > \mathbf{J}_s$ and $(|\kappa|b^3)_{dia} = (\kappa b^3)_{para}$ then the probability of the capture of

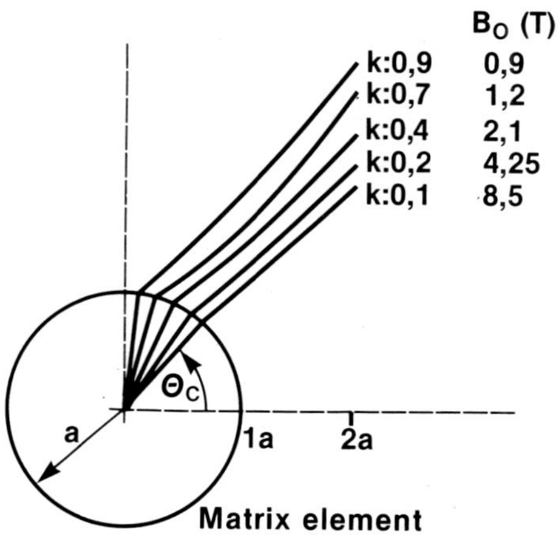

Fig. 3.23 Variation of the critical angle θ_c with $\mathbf{K}$ and $\mathbf{B}_o$ (for $\mathbf{J}_s = 1.7$ T).

diamagnetic and paramagnetic particles will be similar. Thus, the quartz particles, $\kappa = 10^{-5}$, $\mathbf{b} = 50\ \mu m$ have the same capture probability as the hematite particles, $\kappa = 10^{-3}$, $\mathbf{b} = 10\ \mu m$. Therefore, the capture of gangue particles can appreciably impair the grade of the magnetic fraction at high magnetic fields.

It has been shown by Birss and Parker [B19] that there are certain distinct features of the capture of paramagnetic and diamagnetic particles. The maximum radial magnetic force for paramagnetic particles occurs for the ratio $\mathbf{a}/\mathbf{b} = 2.69$ [O1], while for diamagnetic particles the optimum value of this ratio is a function of parameter **K**. For example, for $\mathbf{K} = 0.8$ $\mathbf{a}/\mathbf{b}$ is found to have the optimum value 1.31 [B19]. Furthermore, in the axial configuration the capture time for the diamagnetic particles is always greater than that for paramagnetic particles because the repulsive nature of the short range term in eq. (3.9) for diamagnetic particles produces a much wider excursion than that for paramagnetic particles. However, computed particle trajectories, in transverse and longitudinal configurations indicate that the trajectories for diamagnetic particles are essentially identical, for high $\mathbf{v}_m/\mathbf{v}_o$ ratios, to those of the paramagnetic particles of the equivalent $|\kappa_p - \kappa_f|$ and $|\mathbf{v}_m/\mathbf{v}_o|$ values in L- and T-configurations, respectively, while for lower values of $|\mathbf{v}_m/\mathbf{v}_o|$ the capture efficiencies differ considerably [B19].

The first experimental confirmation of the diamagnetic capture was reported by Friedlaender et al. [F12]. They showed that by dissolving a paramagnetic salt in water, κ_f can be made positive and sufficiently large, so that $\kappa = |\kappa_p - \kappa_f|$ is large and the attractive magnetic force is sufficient to recover the diamagnetic particles suspended in a solution. The video system was used to monitor a capture of diamagnetic aluminium oxide particles ($\chi_p = -1.44 \times 10^{-6}\ cm^3 g^{-1}$, or $\kappa_p = -7.2 \times 10^{-5}$ (SI), $\mathbf{b} = 7\ \mu m$). The recovery of latex particles suspended in a paramagnetic liquid was studied by Glew et al. [G19] and good agreement of the experimental results with the theories of the dynamic build-up of particles on single wires was obtained.

Although the recovery of diamagnetic particles by magnetic separation is feasible in principle, the simultaneous capture of paramagnetic particles prevents a magnetic separator from being a useful device for separating diamagnetic particles from a mixture of minerals. In order to separate diamagnetic particles from the paramagnetic ones, a selective matrix must be designed. The design of such matrices and their application to the recovery of diamagnetic minerals was investigated by Kelland and co-workers [T6, K12]. They showed that by filling the paramagnetic collection regions in a sheet matrix with epoxy, selective diamagnetic capture is enhanced and paramagnetic capture suppressed. The selectivity of separation can also be enhanced by using the elliptical wire array in which the magnetic field is applied parallel to the major axis of the cross-section (see Fig. 3.7).

III.1.12 Magnetic separation at moderate Reynolds numbers

Experimental investigation into particle capture in L-configuration has indicated [F11] that for moderate values of Reynolds number for wire (**Re** 6 to 40), the material can be captured at the rear of the wire. Theoretical treatment by Watson [W14, W15] has shown that at higher velocities of the slurry a pair of stable standing vortices which appear on the downstream side of a wire can induce a capture of the particles on the trailing side of a matrix element.

An analysis of equations of the particle motion [W14, W15] shows that downstream capture cannot occur in the vortex system unless $\mathbf{K} \cong 1$. Under these conditions, the capture cross-section is small because either $\mathbf{v}_o$ or $\mathbf{a}$ is large, so that $\mathbf{v}_m/\mathbf{v}_o \ll 1$. On the other hand, for higher values of $\mathbf{v}_m/\mathbf{v}_o$, capture at the rear of the wire must be preceded by the capture on the front of the wire. The particles are then swept around to the wire. Watson also found that to achieve downstream capture, particle inertia must be included in the equations of motion, so that the particles cross the streamlines. Also, the condition $\mathbf{v}_m/\mathbf{v}_o \ll 1$ must be satisfied so that the particles stay close to the wire as the repulsive region for paramagnetic particles is traversed. The fact that the particle must first be captured on the upstream side of the wire means that the capture radius for the rear side must be smaller that that for the upstream side

$$R_c^{(rear)} \leq R_c^{(front)}$$

Fluid behaviour around the downstream side of the cylindrical wire, as a function of the Reynolds number, is schematically shown in Fig. 3.24. It can be seen that for $5 < \mathbf{Re} < 45$ the boundary layer begins to separate and a stationary wake is formed. The wake consists of two parts, twin vortices and trail and is fed by the boundary layer which separates from the wire surface.

When Reynolds number Re = 30 is reached, the wake begins to oscillate. For Re > 45 it starts to shed alternately from either of the vortices and passes downstream, where it assumes a configuration of the Kármán vortex street [T11].

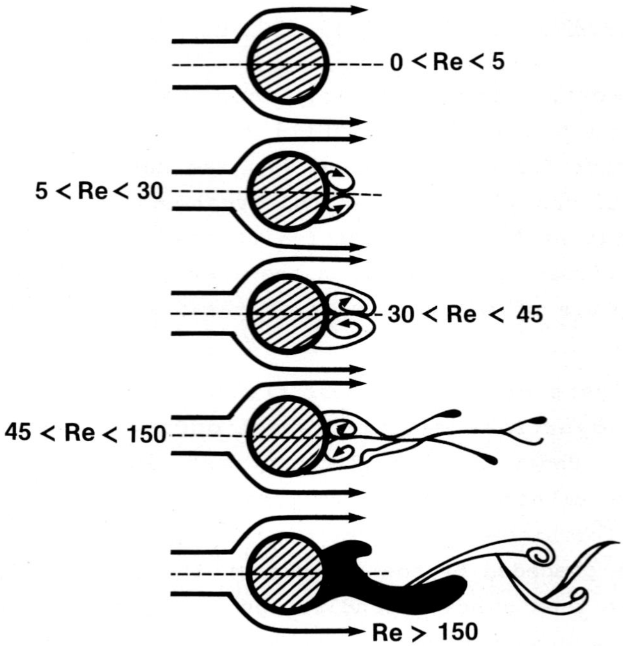

Fig. 3.24 Vortex formation on a circular cylinder as a function of Reynolds number.

The predictions of Watson's model were tested experimentally and it was observed by Friedlaender et al. [F11] that at a low slurry velocity particles collect only on the upstream side of the wire. As the slurry velocity increases, particles also collect on the downstream side, and at an even higher velocity they do not collect on the upstream side at all. These experimental observations are illustrated in Fig. 3.25.

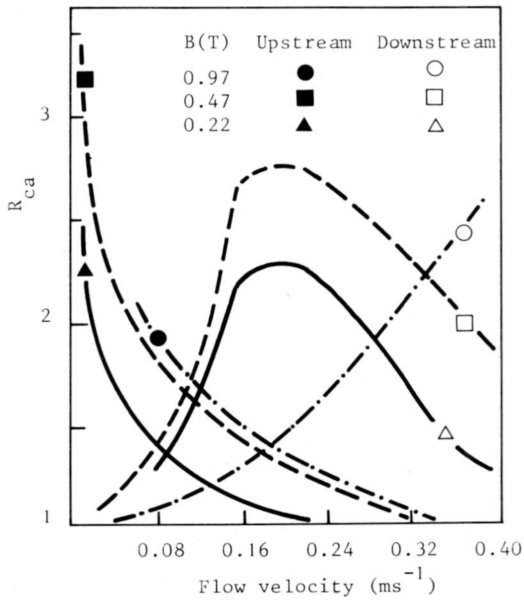

Fig. 3.25 Saturation value of normalized build-up radius $\mathbf{R}_{ca}$ of upstream and downstream collection of $Mn_2P_2O_7$ particles as a function of slurry velocity (after Friedlaender et al. [F11]).

Hollingworth and Finch [H19] observed that the upstream build-up follows the loading trend,as given by eq. (3.65), in contrast to the downstream build-up. Furthermore, they found that build-up increased to **Re** $\simeq$ 40 but diminished, when the wake started to shed.

III.2 THEORY OF CAPTURE IN MULTI-WIRE MATRICES

The matrix of a real magnetic separator consists of a large number of individual collectors. In general, the study of magnetic separation has been confined solely to the investigation of magnetic filtration by the use of two different, yet often complementary approaches: macroscopic (phenomenological) and microscopic. The phenomenological approach describes the dynamic behaviour of a magnetic filter with a set of partial differential equations and characterizes the filtration mechanism by means of several model parameters the value of which, for a given application, can be obtained from appropriate small-scale experiments. The solution of these equations provides the basis of design, scale-up and optimization.

Although this approach affords only a limited insight into the physical process of magnetic filtration, it represents a practical and rational methodology for the design of industrial-scale magnetic filters, provided the laboratory experiments, from which the relevant parameters are obtained, can be conducted with sufficient accuracy.

The ultimate objective of the microscopic physical approach is to derive the mathematical models from the first principles which quantitatively describe the complete dynamics of magnetic separation. The single-wire model can thus be extended to a more complicated model, describing an infinite array of parallel wires. The third stage of the model describes the behaviour of random wire matrices. The microscopic theories of magnetic filtration are intended to provide information and insight about the mechanism of particle capture and the effect of particle build-up on the dynamics of magnetic filtration. It is thus clear that the generality and mathematical complexity of the description of the magnetic filtration process increase and the modelling at any of these levels of complexity is far from being completed at present. Much work remains to be done before the results of these theoretical analyses can be translated directly for the practical design calculation.

So far, in this Section, only magnetic filtration has been considered. Magnetic filtration is a process for the removal of magnetizable fine and colloidal particulates from dilute liquid suspensions by passing them through a magnetized matrix. It is clear that such a process is much simpler than magnetic separation, as the term is understood in mineral processing. The suspensions treated in mineral processing are invariably slurries with a high solids content that often comprise a wide spectrum of minerals with widely differing particle-size distribution, particle shape, and magnetic and surface properties. Although in magnetic filtration the magnetic fraction is usually the waste, in magnetic separation, as applied to minerals processing, the magnetic fraction is often the valuable product, and besides a high recovery of the valuable mineral an acceptable grade of the concentrate must be guaranteed. Therefore, although high capture efficiency is required, the selectivity of the separation process must be maintained by recovering the valuable mineral and by rejecting the unwanted gangue minerals.

Since the phenomenological model of magnetic filtration does not, from the first principles, contain a mechanism of selective filtration, its applicability to most mineral dressing processes is therefore limited.

Although the microscopic models of magnetic filtration have borrowed heavily from the deep bed filtration models, the extension of the single-wire trajectory calculations to the real matrix situation, the approach common in deep-bed filtration microscopic models, is not strictly applicable to magnetic filtration. Owing to the presence of the external (magnetic) force, which is absent in deep bed filtration, the capture mechanisms in single-wire and multi-wire situations differ substantially. Thus, the trajectory models of magnetic filtration, although they do not lack a mathematical elegance, are of limited use, particularly in the magnetic separation of complex minerals. Consequently, we shall limit our description of these models to general remarks only, and the interested reader can find more detailed information in the references cited.

III.2.1 The phenomenological model

In contrast to the Lagrangian methods in which the trajectories of individual particles are followed, Eulerian methods are concerned with the distribution of particles in space, and for non-spherical particles, orientation distribution as well. In its most general form, such distribution is described by a set of coupled differential equations known as Fokker-Planck equations. Provided the distribution of particles throughout the suspension is known, the flux of particles towards the surface of the matrix element can be readily calculated. In most cases of interest, especially for colloidal particles dispersed in fluids, the Fokker-Planck equation reduces to the usual continuity equation [A9].

The phenomenological approach to the magnetic filtration

process was initially used by Kaiser et al. [K13]. Mathematical analyses by Watson [W16, W17], Akoto [A10] and Uchiyama and Kurinobu [U5] are also based on the same approach. These early models have since been considerably refined by Collan et al. [C21, C22] and by Gerber [G20].

The fundamental equations describing the particle retention in a filter are the macroscopic conservation equation and the rate equation. Let us consider particles travelling through an absorption filter, as illustrated in Fig. 3.26. By equating the difference between the numbers of particles entering and leaving the volume Δx per unit time, to the time rate of increase of the number of particles, the conservation equation is obtained as

$$\frac{\partial N(x,\tau)}{\partial \tau} = v_o \frac{\partial c(x,\tau)}{\partial x} \tag{3.70}$$

where $\mathbf{N}(\mathbf{x}, \tau)$ is the number of captured particles in the separator between $\mathbf{x}$ and $\mathbf{x} + \Delta\mathbf{x}$, at time τ, $\mathbf{c}(\mathbf{x}, \tau)$ the number of suspended particles in the volume of fluid $\epsilon \Delta x$, ϵ the porosity of the matrix, $\mathbf{v}_o$ the superficial velocity of a suspension and τ the displacement time defined as $\tau = t - \epsilon x/v_o$.

The rate equation is usually expressed in the form of the Iwasaki relation [R9]

$$-\left(\frac{\partial c}{\partial x}\right)_t = \lambda c \tag{3.71}$$

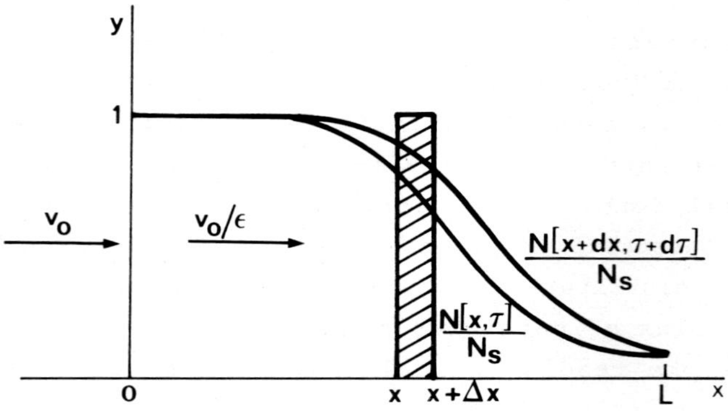

Fig. 3.26 Normalized build-up $\mathbf{N}/\mathbf{N}_S$ as a function of filter length **L.**

namely, that the profile of the particulate concentration in the liquid phase throughout a filter is logarithmic. λ is commonly referred to as the filter coefficient and has dimensions of reciprocal length: $\lambda = 1/\ell$, where ℓ is the characteristic absorption length of the matrix material.

The solution of the rate equation can thus be written as

$$\frac{c(x)}{c_{in}} = \exp(-\lambda x)$$

$$\frac{c_{out}}{c_{in}} = \exp(-\lambda L) \tag{3.72}$$

where $\mathbf{c}_{in}$ and $\mathbf{c}_{out}$ are the influent and effluent concentrations.

During the filtration process, particles are continuously captured from the fluid flowing through the filter onto the matrix surface. Thus, the instantaneous concentration of trapped particles increases according to the equation

$$\frac{dN(x,\tau)}{d\tau} = v_o \lambda c(x,\tau) \tag{3.73}$$

If we assume that at time $\mathbf{t} = 0$ the matrix was clean, then

$$N(x,\tau) = N(0,\tau)\exp(-\lambda x) \tag{3.74}$$

As follows from eqs. (3.72), the condition $\exp(-\lambda L) \ll 1$ must be met, if filtration is to be effective. This is equivalent to $L \gg \ell$ and this limiting situation is called "strong coupling limit".

The assumption that the filter coefficient λ is constant and greater than zero ceases to be true at time τ_s

$$\tau_s = \frac{N_s}{\lambda v_o c_{in}} \tag{3.75}$$

where $\mathbf{N}_s$ is the saturation concentration. At the front end of the filter, λ drops to zero and for $\tau = \tau_s$ both $\mathbf{c}$ and $\mathbf{N}$ decrease exponentially with $\mathbf{x}$ from the front to the rear of the filter

$$c(x,\tau_s) = c_{in}\exp(-\lambda x)$$

$$N(x,\tau_s) = N_s\exp(-\lambda x) \tag{3.76}$$

After the front of the filter becomes saturated, the portion of the filter, of length $\mathbf{x}_s$, which is saturated to $\mathbf{N}_s$ increases, so that the active part of the filter $(\mathbf{L} - \mathbf{x}_s)$ diminishes. The concentration profiles $\mathbf{c}(\mathbf{x}, \tau)$ and $\mathbf{N}(\mathbf{x}, \tau)$ remain constant, for $\tau > \tau_s$ [C22]

$$c(x,\tau) = c_{in}\exp[(x_s - x)\lambda]$$

$$N(x,\tau) = N_s\exp[(x_s - x)\lambda] \qquad (3.77)$$

The concentration profiles of eqs. (3.77) are shown in Fig. 3.27.

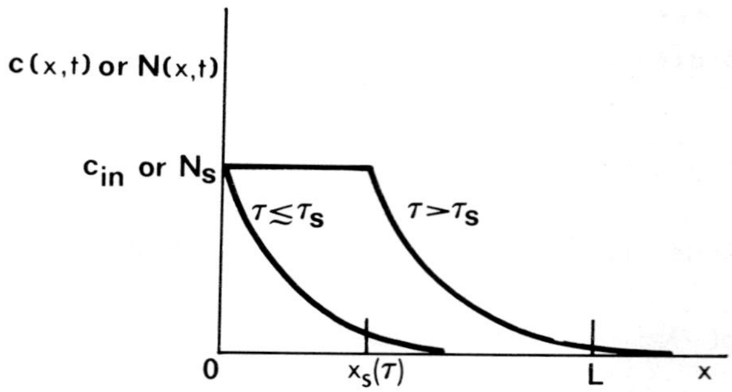

Fig. 3.27 Build-up profiles in a random filter.

Since nothing is known about the physical processes taking place inside the filter in the phenomenological approach, all information about the filter must be derived from its breakthrough curve, i.e. from the output concentration as a function of time; for the present situation, the breakthrough curve is shown in Fig. 3.28.

As indicated, from $\tau = 0$ to $\tau = \tau_s$ the output concentration stays constant. Once the inlet has become saturated, $\mathbf{c}_{out}(\tau)$ starts to increase exponentially. Finally, at time T $\tau = \tau_s + \mathbf{L}/\mathbf{v}_s$, where $\mathbf{v}_s$ is the velocity of the saturation front, the saturation point $\mathbf{x}_s$ reaches the outlet and the filter is fully saturated.

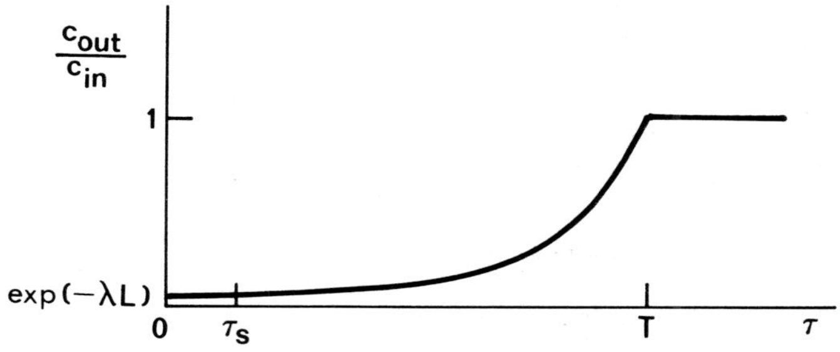

Fig. 3.28 Breakthrough curve of a filter with a constant absorption coefficient λ .

Filter performance

The formula (3.72) relating the number of particles per unit volume, $\mathbf{c}_{out}$, that leave the filter, to the number of particles per unit volume, $\mathbf{c}_{in}$, that enter the filter, can be rewritten, for longitudinal and transverse filters, as [W9]

$$\frac{c_{out}}{c_{in}} = \exp\left(\frac{-\alpha f R_{ca} L_a}{\pi}\right) \tag{3.78}$$

where $\alpha = 2$ for an ordered filter and $\alpha = 4/3$ for a random filter, under the assumption that only two-thirds of all the wires are oriented perpendicularly to the background magnetic field. $\mathbf{L}_a = \mathbf{L}/\mathbf{a}$, $\mathbf{R}_{ca}$ is the normalized capture radius for the appropriate configuration given in Table 3.1 and f the filling factor of the matrix.

The assumption that two-thirds of the wires are aligned orthogonally to the magnetic field leads to a capture efficiency one order of magnitude higher than that obtained by experiment. Birss et al. [B20, S27] have shown, however, that if the isotropic fibre distribution is considered, the mean capture radius associated with a large number of wires is greatly reduced in comparison with the case when the magnetic field is orthogonal to the wire. Eq. (3.78) assumes a modified form, viz.

$$\frac{c_{out}}{c_{in}} = \exp(-0.25fR_{ca}L_a) \tag{3.79}$$

for the longitudinal configuration and

$$\frac{c_{out}}{c_{in}} = \exp(-0.12fR_{ca}L_a) \tag{3.80}$$

for the transverse configuration.

The important feature of these equations is that, in contrast to the earlier theories [W9], but in agreement with some experimental observations, it predicts a less efficient performance in T-configuration as compared to L-configuration [B20].

III.2.2 Microscopic models

The single-wire theory has been extended in a number of publications [U2, E2, S28, G21, R8] to include the combined effect of the field and flow effects in a regular array of wires on the particle capture.

Hayashi and Uchiyama [H20] calculated the particle trajectories and capture efficiencies in the case of two capturing layers of regularly arrayed wires. Their results show that the particle trajectory for a double layer is greatly altered from that for a monolayer. However, the intrinsic capture radius for the back wire is not changed from that for the front one. In the case of weak capture ($\mathbf{v}_m/\mathbf{v}_o < 1$) the capture efficiency of the double layer of the wires is essentially twice that of an individual layer. In the case of strong capture, however, the performance is less than twice the value of an individual layer by an amount dependent on the extent to which the geometrical masking of the second layer is present. It was inferred that in the case of more than two layers, the capture radius for the wires in the third, the fourth, etc. layer will not be changed intrinsically from that for the front one. The only factor to be taken into account is the masking of the capture region of the back wires by the front ones.

Greiner and Hofmann [G22, G23, G24, G25] developed a multiwire model of ordered magnetic filters that takes into account the effect of the geometry of the matrix upon the capture efficiency. The magnetic and hydrodynamic fields in the matrix were computed in the periodic approximation from the basic microscopic equations for the matrix. The results show that for an ordered wire matrix approximately the same initial filter efficiencies are obtained in axial and transverse configurations, in contrast to single-wire models, which predict a better performance for the transverse configuration [B15].

Reger et al. [R8] developed an analytical theory of the particle capture by an infinite wire array in transverse configuration, at low Reynolds numbers. The results, based on a laminar flow model, give lower capture efficiency than that following from potential flow calculations. The model makes it possible to determine the capture efficency as a function of $\mathbf{v}_m/\mathbf{v}_o$, wire spacing and build-up.

III.3 PARTICLE CAPTURE BY FERROMAGNETIC SPHERES

Investigations into particle trajectories and capture efficiency in high-gradient magnetic separation have been mostly limited to single-wire systems or to multiple-wire matrices. The results obtained from these studies can provide information for the practical application of magnetic separation to mineral beneficiation. Many types of filamentary matrices that are used in large-scale magnetic separators (steel wool, expanded metal, wire mesh, rods) can be treated in the framework of multiple-wire models and some conclusions drawn from single and multiple wire models can be used in the application of the above-mentioned matrices.

At the same time, a ball matrix consisting of a collection of randomly packed equisized ferromagnetic spheres has been used in a number of industrial applications, in cyclic and continuous magnetic separators. Steel ball matrices have been employed in the purification of paper mill condensates [A11], in cooling the water at nuclear power plants [H21] and in the reprocessing of nuclear fuels [W18]. The ball matrix was also found to be efficient for mineral treatment, as demonstrated by the recovery of uranium and gold from leaching residues [C7, C23, L13, S29] and of copper and lead from flotation tailings [S30].

It has been shown in a number of experimental studies [F10, W18, S29, C21] that the capture efficiency of the ball matrix is comparable to that of the steel wool matrix, despite the fact that the gradient of the magnetic field generated by the ball matrix is one or two orders of magnitude smaller than that produced by the steel wool. Although it is accepted that the ball matrix often gives similar results to those obtained with filamentary matrices at moderate magnetic fields, it is claimed [W12] that the performance of the ball matrix at high magnetic fields is relatively poor. There are no published experimental results to support this assertion and a comparison of the performance of different matrices published by different authors cannot always be done with sufficient confidence. The conditions under which experiments are performed are often not identical: parameter like particle size distribution, matrix loading, matrix depth and flow velocity are not usually kept constant by different investigators and their variation influences the interpretation of results.

III.3.1 Particle capture on a single sphere

The first attempts to investigate the particle capture and build-up on single spheres were published by Friedlaender et al. [F10, F15, F16]. If the inertial and gravitational forces can be neglected, the equation of motion of magnetizable particle in the vicinity of a ferromagnetic sphere can be written as

$$\vec{F}_m + \vec{F}_d = 0 \qquad (3.81)$$

Equation (3.81) was solved numerically [F10]; typical particle trajectories in transverse configuration are shown in Fig. 3.29. It was found that the laminar flow fits the experimental data much better than the potential flow, particularly in the proximity of the collecting sphere, where the two flow models deviate most. The capture cross-section was found to depend linearly on $(\mathbf{v}_m/\mathbf{v}_o)^{1/4}$, in good agreement with the experimental results.

The particles are attracted towards the sphere if the radial component of the magnetic force $\mathbf{F}_{mr}$ is negative. Therefore $\mathbf{F}_{mr} = 0$ defines the curve $\theta_c(r_a)$, delimiting the attraction area in the paramagnetic mode, in the main symmetry plane. The critical angle

θ_c is then given as [F15]

$$|\theta_c| = \arcsin\left[\frac{2\left(1+2K\dfrac{a^3}{r^3}\right)}{3\left(1+K\dfrac{a^3}{r^3}\right)}\right] \tag{3.82}$$

which is the equivalent of the relation (3.12) derived for a cylindrical wire.

Paramagnetic particles will thus be attracted towards the sphere within two conically-shaped surfaces obtained by the rotation of the delimiting curve, defined by eq. (3.82), around the direction of **H**. Diamagnetic particles will be attracted in the remaining space. This situation, in the longitudinal configuration is shown schematically in Fig. 3.30.

A typical experimental build-up of paramagnetic particles in a transverse configuration is shown in Fig.3.31.The angle γ depends on the fluid velocity but not on the magnetic field, and again indicates the dominant role of the fluid shear stress in particle retention.

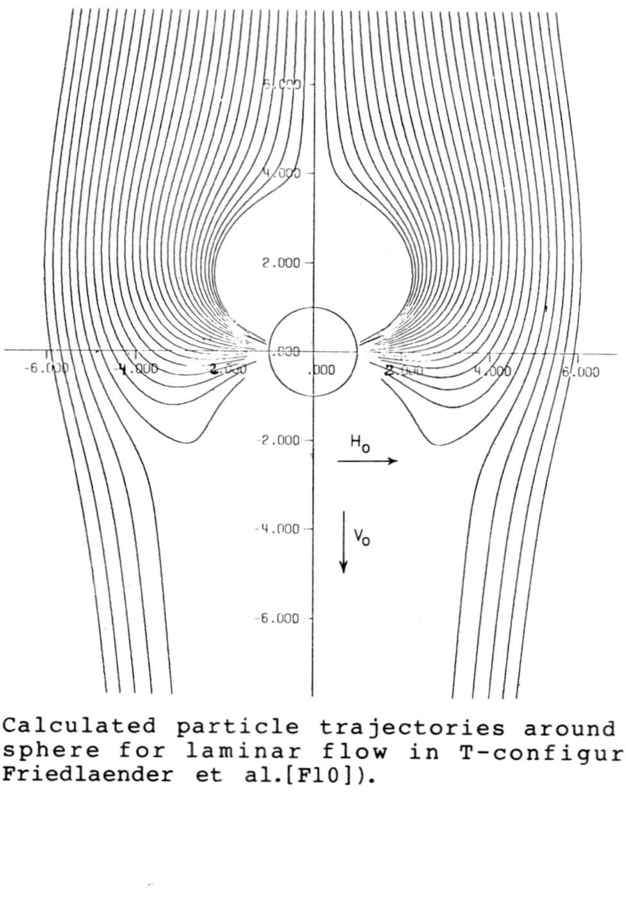

Fig. 3.29 Calculated particle trajectories around a magnetized sphere for laminar flow in T-configuration (after Friedlaender et al.[F10]).

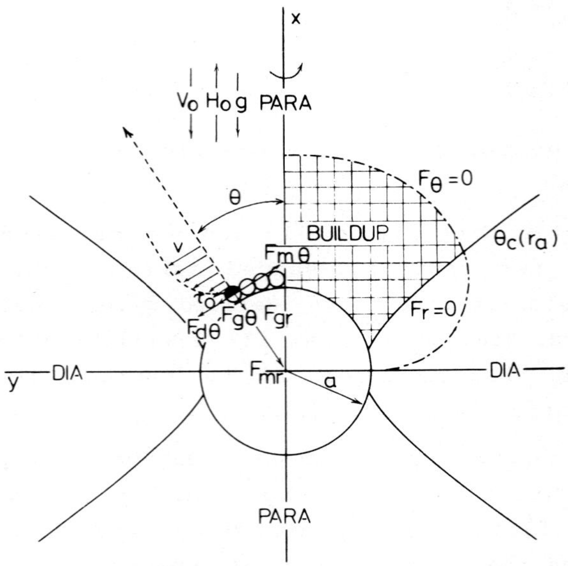

Fig. 3.30 Attraction and repulsion areas for paramagnetic particles in the main symmetry plane of a spherical collector (after Friedlaender et al. [F15]).

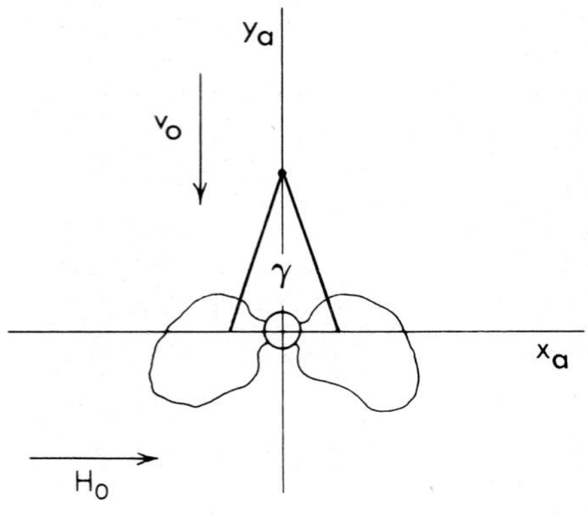

Fig. 3.31 Experimental build-up of paramagnetic particles in T configuration.

III.3.2 Particle capture in a ball matrix

Mention was made in the preceding paragraph that Friedlaender's model of a particle capture on a single ferromagnetic sphere predicted that the capture cross-section in such a system continued to increase linearly with the magnetic field, even after the sphere has reached its saturation magnetization. In most applications of the ball matrix to the recovery of the magnetizable particles from the slurry, however, very little increase in the capture efficiency is observed for magnetic fields greater than a certain threshold value, close to the field required to saturate the matrix. Watson and Watson [W12] developed a model of particle capture in a ball matrix which implies that the capture cross-section does not increase indefinitely with increasing background magnetic field but remains constant near the value attained when magnetic saturation occurs.

In Watsons' model, the single ball used by Friedlaender et al. [F10] is surrounded by the non-magnetic spheres of equal size. These balls serve to create the fluid flow pattern near the ball similar to that occurring within the ball matrix. In the framework of this Happel model [H22] the Kuwabara method [K14] was used to express fluid velocity. Having solved the equation of motion (eq. (3.81)) Watson and Watson calculated the particle trajectories in the vicinity of the ferromagnetic sphere and subsequently, the probability of particle capture. It was found that the probability of capture of paramagnetic particles suspended in a fluid by a ball matrix increases with the background magnetic field at a constant flowrate, until the background field was high enough to magnetically saturate the ferromagnetic sphere. For a magnetic field greater than this threshold value, the capture probability decreases with increasing background field.

Watson and Watson suggest two basic reasons for such behaviour. First, the particles in this model are never at a distance greater than 0.163 a (as follows from the filling factor of a system of randomly packed spheres, $f = 0.636$) from the surface of the sphere. In this region the attractive force on the particle is always dominated by the short-range term in eqs. (3.9) and (3.27). Once the sphere is saturated this term becomes constant.

Secondly, for magnetic fields greater than the saturation field the attractive region near the ball available for the retention of paramagnetic particles, decreases with the increasing magnetic field.

A study similar to that of Watson and Watson [W12] was made by Moyer et al. [M14, M15]. They also assumed that the system of fluid and matrix elements is composed of spherical cells, each containing one sphere, adjacent to which is a fluid shell. The viscous drag force and magnetic force were determined by establishing the fluid velocity profile within the assembly of spheres and the magnetic field of the ferromagnetic spheres in the uniform background magnetic field. Although the equation of motion, eq. (3.81) is similar to that of a particle in the vicinity of a single ferromagnetic sphere, flow ($\mathbf{v}_o$) and magnetic ($\mathbf{v}_m$) velocities were renormalized to allow for the influence of neighbouring spheres on the flow pattern and on the magnetic field of the assembly.

Numerical integration of the equation of motion has revealed that the capture cross-section $\mathbf{R}_c$ for the filling factor $\mathbf{f}$ greater than 0.03 decreases with increasing $\mathbf{f}$. The effect of the neighbouring spheres is thus to diminish $\mathbf{R}_c$ since $\mathbf{R}_c$ is confined to the fluid shell which itself shrinks with increasing $\mathbf{f}$. The results also suggest that the longitudinal configuration is most effective for the capture of diamagnetic particles while the transverse configuration is more advantageous for the recovery of paramagnetic particles. These findings are summarized in Figs. 3.32 and 3.33.

The capture radius also depends on the background magnetic field, as shown in Figs. 3.34 and 3.35. In the limit of a low filling factor $\mathbf{R}_{ca}$ grows nearly linearly with $(\mathbf{v}_m/\mathbf{v}_o)^{1/4}$, in agreement with the work of Friedlaender et al. [F10] for a single sphere. For higher filling factors, $\mathbf{R}_{ca}$ reaches a constant value, in accordance with the conclusions of Watson and Watson [W12].

Moyer et al. [M15] argue that the observed saturation of capture efficiency occurs at the magnetic field strengths far below those required to magnetically saturate the balls comprising the matrix. This conclusion indicates that the levelling off of the performance of magnetic separators at higher magnetic fields is unrelated to the magnetic saturation of the matrix elements, in contrast to the conclusions of Watson and Watson [W12].

On the other hand, the conclusions of Moyer et al. well

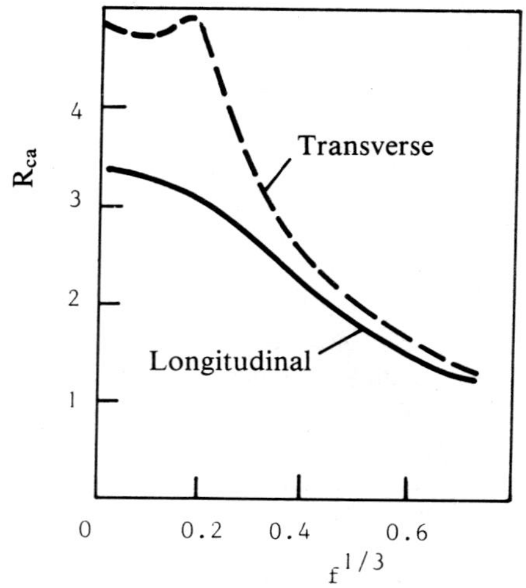

Fig. 3.32 Normalized capture radius for paramagnetic particles as a function of the filling factor f of the ball matrix (after Moyer et al. [M15]).

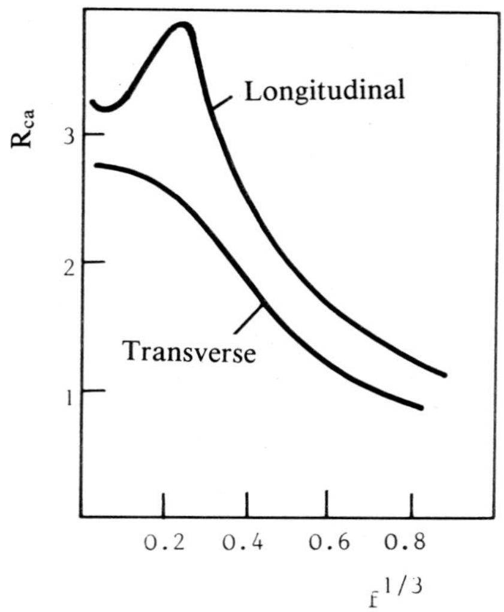

Fig. 3.33 Normalized capture radius for diamagnetic particles as a function of the filling factor f of the ball matrix (after Moyer et al. [M15]).

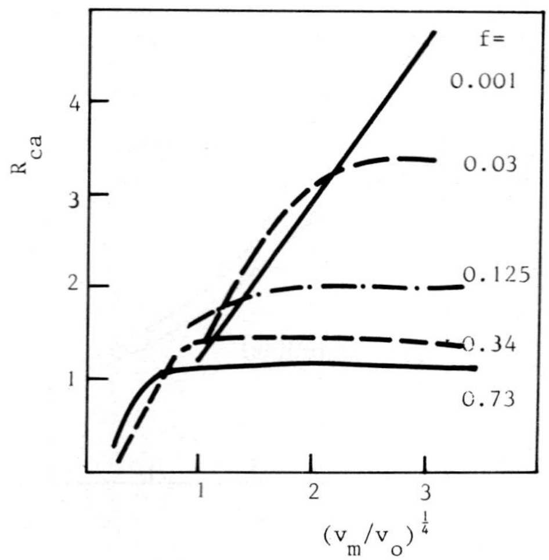

Fig. 3.34 Normalized capture radius in L-configuration for paramagnetic particles as a function of $\mathbf{v}_m/\mathbf{v}_o$, for assorted values of the filling factor of the ball matrix (after Moyer et al. [M15]).

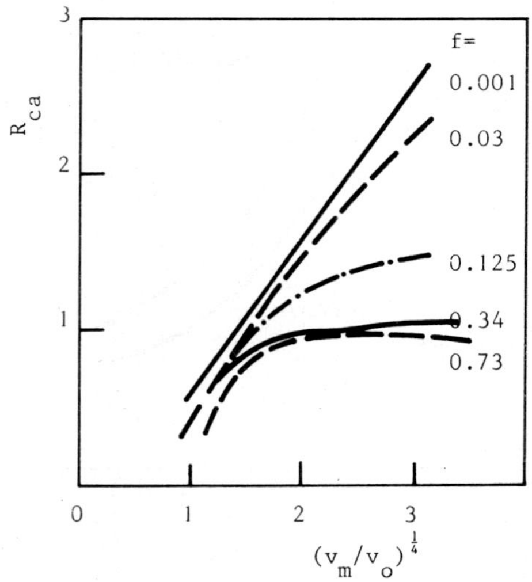

Fig. 3.35 Normalized capture radius in L-configuration for diamagnetic particles, as a function of $\mathbf{v}_m/\mathbf{v}_o$, for assorted names of the filling factor of the ball matrix (after Moyer et al. [M15]).

confirm the insensitivity of the efficiency of magnetic separation to the value of the saturation polarization of the matrix material, observed experimentally [S29, S31, S32, Y1]

The models discussed above show from the first principles that, at a real matrix of sufficiently low porosity, the increase of the background magnetic field does not lead to an increase of particle recovery from a suspension.

Although the adverse effect of the increase of the magnetic field can be expected for the matrices, the filling factor of which is sufficiently large, a similar phenomenon has been observed in a number of experimental studies where a steel wool matrix was used. According to the single and multiple wire models in such a highly porous matrix (filling factor up to 0.05) the capture probability for paramagnetic particles increases with magnetic field, even after the collectors have been magnetically saturated. In such a matrix, the freely-flowing particles are not restricted to flow near the collector and consequently particle trajectories are mainly determined by the long-range term in eqs. (3.9) and (3.27). This term continues to increase with the magnetic field after the collectors have become magnetically saturated [W12].

Since numerous experimental results as well as results from the theoretical study by Moyer et al., contradict this outcome of the generally accepted theoretical models, modifications to the overall physical picture of particle capture in a real matrix are needed.

III.4 PARTICLE CAPTURE IN A REAL MATRIX

The efficiency of particle capture in a real matrix of a high-gradient magnetic separator has been described by several models, all of which borrowed heavily from the theoretical description of deep-bed filtration.

The microscopic models of particle retention in a real matrix usually incorporate the single-collector model into a multiple collector situation and calculate the critical trajectories of particles or critical cross-section radii for either ordered or random matrices. The development of trajectory models of multicollector magnetic separation follows closely the physical models developed in deep-bed filtration, although owing to the presence of an external magnetic force in a magnetic separator, such an approach is not strictly permissible.

The particle capture in any filter bed is a two-step process, consisting of transport and attachment. The transport process is in general caused by a combination of numerous interactions: convection, gravitational and buoyant and by Brownian motion. In magnetic separation the magnetic traction force is considered to be the only dominant mechanism of particle capture, and it is usually assumed that the critical capture radius is equal to zero for $\mathbf{B} = 0$ (or $\mathbf{v}_m = 0$) (Table 3.1).

In a single-collector approach this is an acceptable approximation which, however, fails in a multicollector situation. It can be argued that the fundamental 'non-magnetic' mechanisms of particle transport, namely interception, gravity, inertia, hydrodynamic interaction and Brownian motion will bring most particles in a suspension into contact with the matrix in a magnetic separator, provided that certain conditions, usual in deep-bed filtration, are met.

It can be further argued that the magnetic traction force which plays a critical role in the particle capture in single-collector separation is not needed for the transport of particles in a real matrix.

Once a particle reaches a surface of the matrix element, the attachment step becomes important. At this stage, a combination of surface and static dipolar magnetic interaction between the matrix element and the particles and between the particles themselves will determine the efficiency of particle attachment to the matrix.

In the magnetic separation of mineral mixtures the situation is more complex than in magnetic filtration, because, in addition to a high capture efficiency of magnetizable particles, a high grade of the magnetic product must often be guaranteed. Therefore, gangue particles must be removed from the matrix surface by the erosion force of the fluid flow and by the repulsive character of surface interactions between gangue particles and the matrix and the particles of the valuable mineral. The selectivity of magnetic separation is thus determined by completely different considerations, namely, that unwanted, less strongly magnetic particles must be swept away from the matrix surface by the erosion fluid force, in contrast to the trajectory models, in which the non-magnetic particles never reach the matrix surface.

The role of basic parameters like particle size, matrix collector size, matrix magnetization, magnetic field strength and flow velocity in trajectory and in filter-bed models can differ substantially and it is not surprising that the agreement between trajectory models and experimental results from a real matrix is usually very poor.

Hencl et al. [H47] recognized that a modified physical picture was needed and although some fundamental arguments are invalid, the paper can serve as a first point for a more realistic description of a process of particle capture in a real matrix. Later Svoboda and Ross [S51, S52] developed a model of particle transport and attachment in a matrix of HGMS.

III.4.1 Particle deposition

A suspended particle following a streamline of the flow may come in contact with a matrix element by virtue of its finite size, as shown in Fig. 3.36, case A. This transport process is called interception and straining. If a particle has sufficient inertia it will maintain its trajectory and will come in contact with a

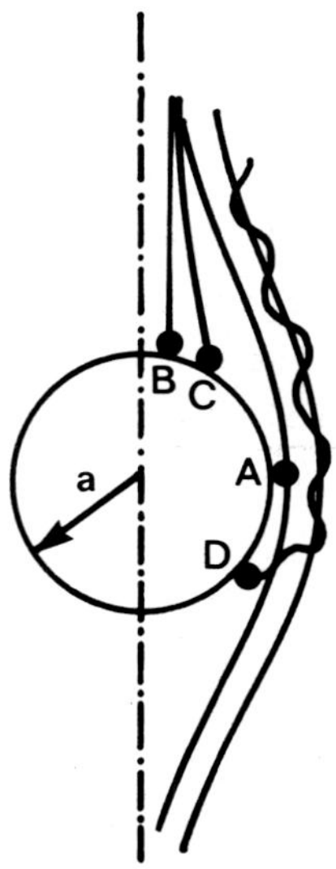

Fig. 3.36 Collision mechanisms in deep-bed filtration.

collector (case B). If the density of the suspended particles is greater than that of water, the particles will follow a different trajectory, due to the influence of the gravitational field (case C in Fig. 3.36). This gravitational transport process is often called sedimentation.

Particles in a suspension are subject to the Brownian motion of the suspending medium and the diffusion can be, for small particles, an important transport mechanism (case D). Particles can also experience a greater liquid velocity on one side than on the other, thus causing a particle to rotate. A pressure difference develops laterally to the flow direction and the resultant force causes the particle to move across the flow field. This hydrodynamic effect appears as a random drifting motion across the streamlines and contributes to the retention of the particles on the matrix [R9, I8].

The fundamental equations that describe the particle retention in a filter bed are the macroscopic conservation equation (eq. (3.70)) and the rate equation (eq. (3.71)). The solution of the rate equation is given by eq. (3.72).

In order to show that a particle passing through a matrix has a probability approaching unity for it to come into contact with the matrix, eq. (3.72) must be rewritten as:

$$\frac{c_{esc}}{c_{in}} = \exp(-\lambda L) \tag{3.83}$$

where c_{esc} is the concentration of particles that escaped collision with the matrix. The filter coefficient λ must be now expressed as a function of the collision efficiency, and not of the collection efficiency. The overall collision filter coefficient will by equal to the sum of filter coefficients, as a result of all collision mechanisms:

$$\lambda_{tot} = \lambda_g + \lambda_i + \lambda_n + \lambda_h + \lambda_d + \lambda_c \tag{3.84}$$

where λ_g is the filter coefficient caused by gravitational collision, λ_i caused by interception, λ_n caused by inertial collision, λ_h caused by hydrodynamic effect, λ_d caused by diffusion and λ_c caused by wedging.

The collision probability of a particle in a matrix was

calculated from eqs. (3.83) and (3.84), where the expressions for the individual terms in λ_{TOT} were obtained from [W37, I8, H48]. Fig. 3.37 shows the dependence of the percentage of particles that collided with the matrix, as a function of particle diameter, for a steel wool matrix. It can be seen that with fine steel wool even submicron particles collide at a high rate of probability with the matrix.

The effect of matrix collector size on collision probablity is shown in Fig. 3.38. These figures imply that collision probability approaches 100 per cent for particle sizes ranging from a fraction of a micrometre to several tens of micrometres, depending on the matrix element diameter and on the superficial velocity of a slurry.

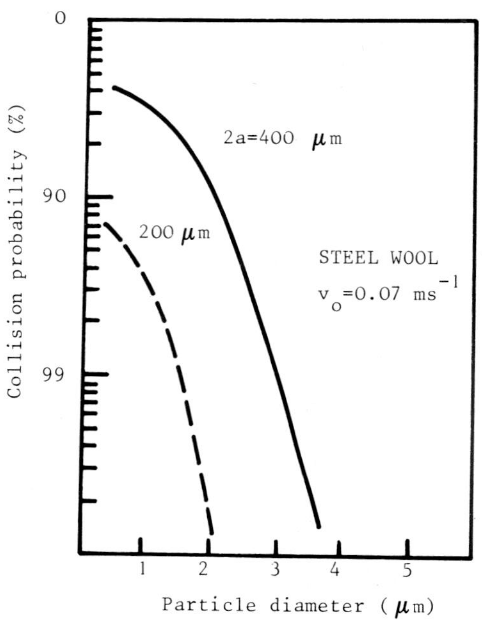

Fig. 3.37 Collision probability of particles in a steel wool matrix as a function of particle diameter.

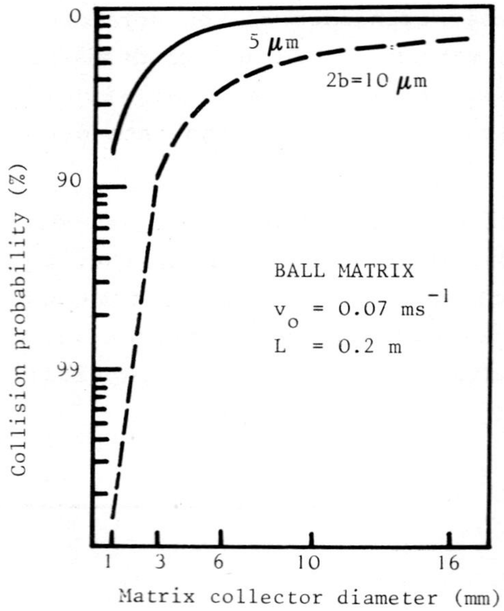

Fig. 3.38 Collision probability of particles in a ball matrix as a function of ball diameter.

III.4.2 The retention of particles

Total interaction between the particles, matrix and fluid medium must be attractive if the result of the collision is the retention of a particle on the matrix. The probability of particle retention is thus determined by the van der Waals-London interaction $\mathbf{V}_A$ (eq.(3.107)), electrical double-layer interaction $\mathbf{V}_R$ (eq.(3.104)), magnetic interaction

$$V_M = -\frac{4\pi\chi B^2 b^3}{3\mu_0} - \frac{32\pi^2\chi b^3 a^3 B J_s}{9\mu_0 r^3} \qquad (3.85)$$

and hydrodynamic shear stress $\mathbf{V}_D$ can be calculated from eqs. (3.54) and (3.59), taking into account that $\mathbf{V}_D = \int_a^\infty \mathbf{F}_d dr$.

The necessary condition for a particle, once it collides with the collector, and remains attached to the matrix, is that the total interaction energy $\mathbf{V}_T = \mathbf{V}_A + \mathbf{V}_R + \mathbf{V}_M + \mathbf{V}_D$ is negative. If we neglect the surface interactions $\mathbf{V}_A$ and $\mathbf{V}_R$, the condition for a particle of radius **b** to remain attached to the matrix is

$$b > \frac{0.49\mu_0 a^{\frac{1}{2}} v_o^{\frac{3}{2}} \rho_f^{\frac{1}{2}} \eta^{\frac{1}{2}}}{\chi B J_s} D \tag{3.86}$$

where

$$D(\theta) = \left(\theta - \frac{\theta^3}{3!} + \frac{\theta^5}{5!} - \frac{\theta^7}{7!} \cdots\right)$$

In a typical case of magnetic separation of uraninite particles, at **B** = 1 T, J_s = 2 T, in a ball matrix and $\mathbf{v}_o$ = 0.1 m s^{-1}, we obtain a limiting particle radius **b** = 0.3 μm. Therefore, the condition of particle retention by the matrix, namely $\mathbf{V}_T \leq 0$, is satisfied even for particles much smaller than those usually observed experimentally. On the other hand, it follows from Figs. 3.37 and 3.38 that the limiting size of particles that can be captured by the matrix, and which is compatible with experimental observations, can be obtained from the condition that collision probability approaches unity, namely

$$\frac{c_{esc}}{c_{in}} \ll 1 \tag{3.87}$$

Consequently, this means that the limiting particle size is independent of the applied magnetic field (provided $\mathbf{V}_T \leq 0$), field gradient and matrix magnetization. However, this limiting particle size is a function of a collector size, matrix length and flow velocity, in the same way as collision probability in deep-bed filtration.

The inclusion of surface interactions in total interaction energy has shown that the value of limiting particle size is independent of surface interactions. However, these surface interaction are dominant for very small particles (**b** < 10^{-8} m) and can induce capture of diamagnetic particles [S52].

III.5 MAGNETIC FLOCCULATION

The demand for mineral products is subject to two factors affecting the efficiency of beneficiation: a decrease in the quality of economically accessible ores and an increase in the degree of fineness of their valuable components. Consequently, the development of technology for the extraction of valuable minerals from finely disseminated ores or tailings is imperative. It is understood that, in the development of such technology, certain limitations, such as energy sources, environmental factors, and economic considerations, will need to be taken into account.

Since the recovery of fine particles is still very inefficient in many systems, they must often be discarded as slimes. Vast quantities of valuable minerals are often lost in the form of slimes: one-fifth of tungsten, copper, uranium and other metals, and one-third of phosphates [S33]. Unless new technology is developed, the increasing amount of finely-dispersed minerals being mined will lead to increased volumes of waste thus representing a substantial loss of valuable components and causing environmental problems.

Methods for the treatment of fine particles are still in the initial phase of development, and conventional methods usually fail for particles smaller than 20 μm. It has been shown by Mills [M16] that out of 25 methods, only four can be applied to the treatment of particles smaller than 10 μm, and that particles finer than 5 μm can be manipulated only by high-gradient magnetic separation. Even then, particles with diametres of a few micrometres can be concentrated only under limiting conditions (high magnetic fields and very low flowrates), thereby considerably reducing the throughputs. It is clear that this approach is more suitable for filtration and for the removal of undesirable admixtures, as demonstrated by the application of HGMS to the purification of kaolin, rather than for the concentration of fractions containing valuable minerals.

If the results obtained by HGMS or flotation are unacceptable as a direct consequence of small particle size, the solution to the problem may lie in the formation of aggregates comprising fine particles. In the past, "flocs" were successfully formed by the addition of inorganic electrolytes or polymers, e.g., in the removal of colloidal particles from effluents. In the beneficiation of minerals, this technique can be used successfully only when the process of flocculation is selective, as proved in the treatment of low-grade iron ores [C24].

Flocculation can also be induced by the application of external forces to a system of particles, e.g., by stirring, or by the application of magnetic or electrical fields. Magnetic flocculation has been used for several years to create flocs of ferromagnetic particles, the magnetic induction required to induce this process being of the order of 0.01 T. Experimental observations on the magnetic flocculation of ferromagnetic minerals and its application to the mining industry have been described in a number of papers [L14, B21, A12, B22, B23, B24, B25, L15] and in the monograph by Karmazin et al. [K15]. The theoretical description of mutual coagulation of ferromagnetic particles is also well established [D11, J6, L16, L15].

Although investigations into the magnetic flocculation of paramagnetic and diamagnetic particles (which constitute the major portion of current mineral resources) started only recently [H23, W19, S34], new systematic studies of the theoretical and experimental aspects [S35, P11, W20, S36, P12, V2, S37] show that feebly magnetic fine particles can be expected to flocculate in currently available magnetic systems.

In the following sections we shall briefly outline the present status of the theory of magnetic flocculation of strongly and weakly magnetic minerals, while its role in mineral processing will be reviewed in Section IV. 15.

III.5.1 Magnetic flocculation of strongly magnetic minerals

The tendency of an assembly of ferromagnetic particles in the magnetic field of a magnetic separator to reduce its magnetostatic energy often results in the formation of aggregates of the particles. The effect of magnetic flocculation constitutes the inherent part of any process of beneficiation of finely ground strongly magnetic materials and directly affects the efficiency of material separation.

In magnetic separation, the process of magnetic flocculation improves the performance of a magnetic separator and increases the recovery of the magnetic concentrate. In magnetic desliming, thickening, flotation, filtration and other processes performed in a magnetic field, magnetic flocculation increases the throughput and selectivity of the devices. In the process of classification magnetic flocculation exhibits a negative influence, prevents the particles from being separated according to their particle size and reduces the efficiency of classification [K15].

III.5.1.1 Theory of magnetic flocculation

We shall consider a suspension of identical spherical ferromagnetic grains in a non-magnetic fluid. Under the influence of sufficiently large external magnetic fields such grains tend to associate in long chains. Each particle carries a magnetic moment $\vec{\mu}_m = \mathbf{V}\vec{\mathbf{M}}$ where $\mathbf{M}$ is the magnetization and $\mathbf{V}$ the volume of a particle. The energy of magnetic interaction between grains (1) and (2) separated by the distance $\mathbf{r}$ is

$$V_M = \frac{1}{\mu_o}\left[\frac{\vec{\mu}_m(1)\vec{\mu}_m(2)}{r^3} - \frac{3(\vec{\mu}_m(1)\vec{r})\ (\vec{\mu}_m(2)\vec{r})}{r^5}\right] \qquad (3.88)$$

If magnetization of the grains is sufficiently large, the electrical double layer interaction and the van der Waals interaction can be neglected. Total interaction energy thus consists of the magnetic dipolar term (eq. (3.88)). Effective magnetic interaction in the suspension is obtained by averaging $\mathbf{V}_M$ over all orientations of the colloidal particles. Thus [C25, C26]

$$\overline{V}_M = -kT\ ln < \exp\left(-\frac{V_M}{kT}\right) > \qquad (3.89)$$

where $\mathbf{k}$ is the Boltzmann constant and $\mathbf{T}$ is the absolute temperature and

$$< \exp\left(-\frac{V_M}{kT}\right) > = \frac{1}{64\pi^3}\int \exp\left(-\frac{V_M}{kT}\right) d\Omega_1 d\Omega_2 d\Omega_H \qquad (3.90)$$

where $\mathbf{d}\Omega_i$ (i = 1,2,H) represents an element of solid angle describing the orientation of each of the dipoles or of the external field. It was shown by Chan et al. [C25, C26] and by Scholten et al. [S38] that by expanding the exponential in eq. (3.89) into the modified spherical Bessel functions and the Legendre polynomials, and by integrating, we obtain the average interaction energy, in the weak interaction limit, i.e. $\mu_m^2/\mu_o \mathbf{r}^3 < \mathbf{kT}$ as

$$\overline{V}_M = -\frac{\mu_m^4}{3\mu_o^2 r^6 kT} \qquad (3.91)$$

while in the strong interaction limit ($\mu_m^2/\mu_o \mathbf{r}^3 > \mathbf{kT}$)

$$\overline{V}_M = -\frac{2\mu_m^2}{\mu_o r^3} \qquad (3.92)$$

We can thus see that although the magnetic dipole-dipole interaction (eq. (3.88)) can be either attractive or repulsive, depending on the relative orientation of the dipoles, the orientational averaging process gives the energetically favourable (attractive) orientations heavy weighing.

If the applied magnetic field is sufficiently large the attractive magnetic interaction energy will be greater than the sum of repulsive interactions (e.g. electric double layer interaction) and the particles will tend to form aggregates. Another condition to be satisfied is that the time of flocculation must be smaller than the residence time of the particles in the magnetic field. As discussed in the following paragraph the flocculation time can be made sufficiently short by exposing the suspension to a reasonably large magnetic field. Furthermore, the strength of the aggregates must be adequately large to prevent a breakage by the fluid shear stress. The rate of flocculation is thus determined by the interplay between dipolar magnetic interaction induced by the magnetic-field and the hydrodynamic interaction which tends to disintegrate the flocs.

The process of magnetic flocculation proceeds through several stages, as shown in Fig. 3.39. At small values of the applied magnetic induction **B** i.e when the hydrodynamic forces are greater than the magnetic force, the formation and disintegration of the aggregates proceeds simultaneously. The degree of magnetic flocculation of the solids (defined as a ratio of the number of

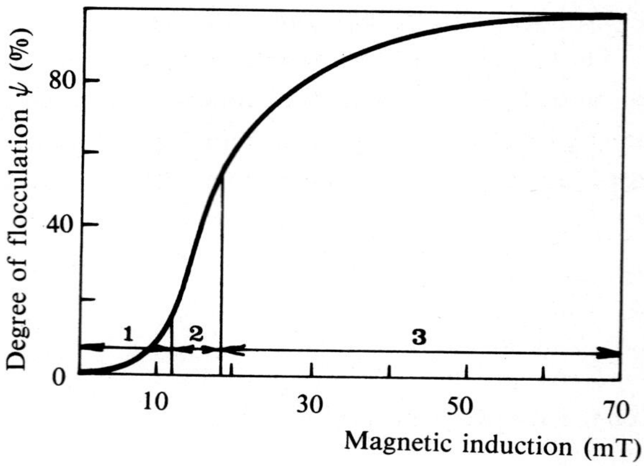

Fig. 3.39 Effect of the applied magnetic field on the degree of flocculation of ferromagnetic particles.

magnetic particles in a floc to the initial number of magnetic particles in the suspension) is proportional to $\mathbf{B}^2$ in this phase [K15] and corresponds to interval 1 in Fig. 3.39. At this stage, the grade of the flocs is high and flocculation is selective.

With a further increase of **B**, the magnetic interaction becomes dominant, the equilibrium between magnetic and mechanical forces is perturbed, flocculation proceeds as an avalanche, and the grade of the flocs is low. This phase is denoted by 2 in Fig. 3.39. When the concentration of solids in a suspension drops below a certain threshold value, the strength of magnetic interaction is reduced, the characteristic time of flocculation increases and ultimately the equilibrium of magnetic and mechanical forces is established and the rate of flocculation converges to zero, as indicated by phase 3 in Fig. 3.39.

The rate of flocculation is proportional to the concentration of solids **c** at a given moment, so that [K15]

$$\frac{dc}{dH} = -kc \tag{3.93}$$

where **k** is a constant, depending upon the magnetic and surface properties of the particles, and their size. By integrating and selecting appropriate initial conditions we obtain for the degree of flocculation ψ_f

$$\psi_f = [1 - \exp(-kH)] \qquad (3.94)$$

Eq. (3.94) describes the behaviour of a suspension in the interval 3 and its validity was confirmed by the experimental studies of magnetic flocculation of magnetite [K15]. By a proper choice of experimental conditions it is thus possible to generate high-grade magnetite aggregates or to induce fast flocculation, resulting in high recovery of the magnetic phase.

Strength of the aggregates

Aristaksesyan and Karmazin [A20] have shown that in a magnetic field a suspension of magnetic particles is exposed to the pressure proportional to $\mu_s - \rho_s \cdot d\mu_s/d\rho_s$ in the direction perpendicular to the direction of the applied magnetic field, and to the drag, proportional to $\mu_s + \rho_s \cdot d\mu_s/d\rho_s$ in the direction of the field. Here μ_s and ρ_s are the magnetic permeability and density of a suspension, respectively. This is confirmed by the elongated shape of the flocs. The above expressions also imply that the condition of the formation of aggregates is the discontinuity of the magnetic permeability of a suspension ($d\mu_s/d\rho_s \neq 0$).

It was shown by Lantto [L15] that apparent remanent polarization of a cluster of ferromagnetic spherical particles can be written as

$$J = \frac{\mu_o H_c}{D} \qquad (3.95)$$

being thus proportional to the coercive force $\mathbf{H_c}$ of the bulk material and inversely proportional to the demagnetizing factor.

In a more general case of particles with an arbitrary shape

$$J = \frac{\mu_o H_c}{D + n/c - 1/3} \qquad (3.96)$$

where $\mathbf{n}$ is a factor by which the demagnetizing factor of a particle is smaller than 1/3. With a suitable combination of values of $\mathbf{D}$, $\mathbf{n}$ and $\mathbf{c}$, the value of $\mathbf{J}$ can be made very large.

The tensile strength of a floc can be written as

$$\sigma = \frac{J^2}{2\mu_o} \tag{3.97}$$

When the cluster has the form of an ovoid with the aspect ratio 10, then **D** = 0.02. Tensile strength is then [L15]

σ = 6.25 kPa for H_c = 2000 A m^{-1}
σ = 25 kPa for H_c = 4000 A m^{-1}

III.5.2 Magnetic flocculation of weakly magnetic minerals

In contrast to strongly magnetic minerals, the magnetic flocculation of paramagnetic and diamagnetic minerals has not been studied extensively. The theoretical models that predict the behaviour of weakly magnetic particles of quasi-colloidal dimensions, in the presence of an external uniform magnetic field have been published only recently [S34, S35, S36, P11, P12, V2, V3, M18].

III.5.2.1 Stability of colloidal suspensions

The criteria of the stability of colloidal suspensions are largely governed by the interplay of repulsive and attractive forces between particles. A detailed description of this interplay is the basis of the theory of stability of hydrophobic colloids as given by Derjaguin and Landau [D12] and Verwey and Overbeeck [V4] (DLVO theory). The DLVO theory views the stability of colloids as arising from electrical charges on identical particles and a consequent electrical repulsion between them. Opposing this repulsion are the attractive forces of the universal van der Waals type. If the repulsion force can be sufficiently reduced so that the attractive interaction becomes dominant then the particles colliding with each other will stick together to form quasi-stable aggregates.

An interfacial potential is accompanied by a characteristic distribution of charge between the phases. Since the system has an overall electrical neutrality the net charge must be zero. The charge on a particle, together with the equal and opposite charge in a suspension, comprise the electrical double layer (EDL). It was shown by Hogg et al. [H25] that the interaction of EDL between two dissimilar particles, for the condition of low surface potential and low thickness of EDL compared to the particle size, is given, under the assumption of a constant surface potential, by

$$V_R = \frac{\epsilon b_1 b_2(\psi_1^2+\psi_2^2)}{4(b_1+b_2)}\left[\frac{2\psi_1\psi_2}{(\psi_1^2+\psi_2^2)} ln\frac{1+e^{-Xh}}{1-e^{-Xh}} + ln(1-e^{-2Xh})\right] \qquad (3.98)$$

where $\mathbf{b_1}$, $\mathbf{b_2}$ and ψ_1, ψ_2 are the radii and surface potentials of particles, respectively, ϵ the dielectric constant of the medium, $\mathbf{X}$ the Debye - Hückel factor for reciprocal length, and $\mathbf{h}$ the distance between the surfaces of the interacting particles.

In the case of identical spherical particles eq. (3.98) reduces to

$$V_R = \frac{\epsilon b \psi^2}{2}\, ln\,(1+e^{-Xh}) \qquad (3.99)$$

When the thickness of EDL is not small (i.e. when bX ≮ 1) V_R can be expressed in a simple approximate form

$$V_R = \pi\epsilon b^2\psi^2 \frac{\exp(-Xh)}{(2b+h)} \qquad (3.100)$$

Similarly, the London-van der Waals interaction for dissimilar spheres can be expressed as

$$V_A = -\frac{A}{6}\left[\frac{2a_1a_2}{h^2+2(a_1+a_2)h} + \frac{2a_1a_2}{h^2+2(a_1+a_2)h+4a_1a_2} + ln\frac{h^2+2(a_1+a_2)h}{h^2+2(a_1+a_2)h+4a_1a_2}\right] \qquad (3.101)$$

where $\mathbf{A}$ is the Hamaker constant. For identical particles, eq. (3.101) simplifies to

$$V_A = -\frac{A}{6}\left[\frac{2a^2}{h^2+4ah} + \frac{2a^2}{h^2+4ah+4a^2} + ln\frac{h^2+4ah}{h^2+4ah+4a^2}\right] \qquad (3.102)$$

However, eqs. (3.101) and (3.102) are not valid when the London forces operate over distances comparable to or large than 0.1 λ (λ being the wavelength of the intrinsic electronic oscillations of the atoms), for then the interaction is considerably reduced by the retardation effect [S39, R10].

When a suspension of colloidal particles is placed in an external magnetic field there is, in addition to the interactions considered so far, a magnetic dipolar interaction, described by eq. (3.88).

In the strong interaction limit [C25, S34] the average magnetic interaction energy can be written as

$$V_M = -\frac{8\pi\mu_o\kappa_1\kappa_2 b_1^3 b_2^3 H^2}{9(h + b_1 + b_2)^3} \qquad (3.103)$$

μ_o is the magnetic permeability of vacuum, κ the volume magnetic susceptibility of a particle and **H** the magnetic field strength.

As follows from the extension of DLVO theory to magnetic flocculation [S34], the total energy of interaction of colloidal particles in the external magnetic field is given by the sum of electrical double-layer energy $\mathbf{V_R}$, van der Waals interaction energy $\mathbf{V_a}$ and magnetic dipolar interaction energy $\mathbf{V_M}$

$$V_T = V_R + V_A + V_M \qquad (3.104)$$

Although electrical double layer energy is repulsive for the interaction of particles of the same mineral, it can be either repulsive or attractive for the interaction between the particles of a valuable mineral and the gangue. The van der Waals interaction is usually attractive, but it has been shown [V5] that in a three-component system where two minerals 1 and 2 are suspended in a liquid medium 3, and an inequality $A_{11} < A_{33} < A_{22}$ between the Hamaker constants is satisfied, the overall Hamaker constant is negative, and the van der Waals interaction becomes repulsive. Magnetic interaction can be either attractive or repulsive, depending on the signs of the magnetic susceptibilities of the interacting particles. However, eq. (3.103) implies that the energy of magnetic interaction of identical particles does not depend on the sign of the magnetic susceptibility so that eq. (3.104) describes the behaviour of paramagnetic as well as diamagnetic suspensions in an external magnetic field.

A curve of the total interaction energy $\mathbf{V}_T$ as a function of the normalized interparticle distance $\mathbf{s}$ ($\mathbf{s} = \mathbf{r}/\mathbf{a}$) is shown in Fig. 3.40. A potential barrier $\mathbf{V}_{max}$ separates two minima: a closer and deeper primary minimum (PM) and a more distant and shallow secondary minimum (SM) of depth $\mathbf{V}_{min}$.

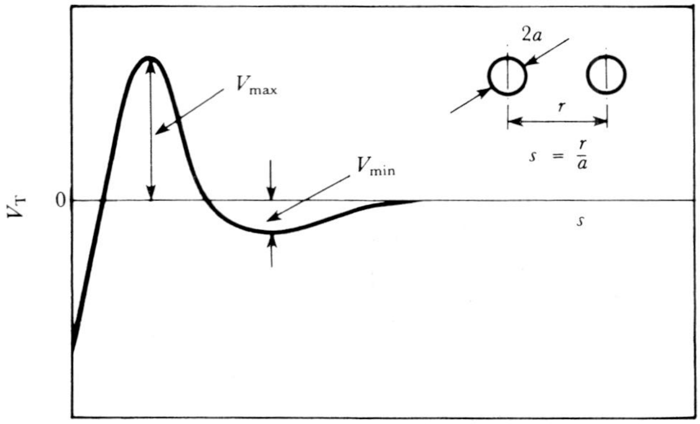

Fig. 3.40 The general form of the total potential energy for a pair of interacting colloidal particles.

The height of the barrier determines the stability of a suspension. If the barrier is very low, there is nothing to prevent contact between the particles, and flocculation will take place at a rate governed by the frequency with which the particles collide. If there is a secondary minimum in the curve of a potential energy, particles can flocculate into this more distant potential well; in this instance, the potential barrier need not be suppressed. A necessary and sufficient condition for flocculation is that the secondary minimum be sufficiently deep.

The existence and position of the primary and secondary minima can be affected by changes in the magnetic field, the magnetic susceptibility of the particles and the pH value of the suspension. The total interaction energy $\mathbf{V}_T$ is shown in Fig. 3.41 as a function of magnetic induction and particle size. The most important part of the diagram is region IV in which, owing to high magnetic induction, the potential barrier in supressed, the total interaction becomes attractive, and particles can flocculate into the primary minimum.

It follows from Fig. 3.41 that, for any particle size, a value for the magnetic induction can be found at which the potential barrier disappears. A typical plot of total interaction energy of two hematite particles of 5 μm diameter is shown in Fig. 3.42. Such a threshold magnetic field, at which the flocculation into a primary minimum starts, can be determined provided that the total potential energy of interaction $\mathbf{V}_T$ is negative [S34], or that the stability ratio $\mathbf{W}_p$, which is given by eq. (3.105) is smaller than, or equal to, unity [S35].

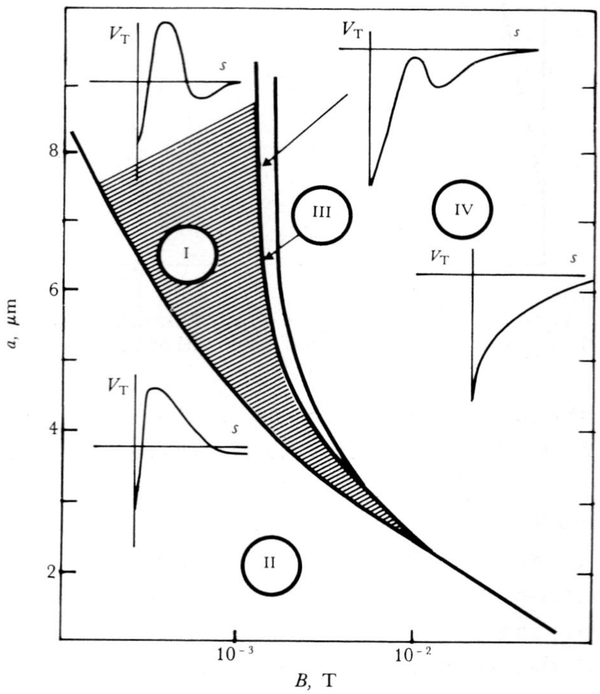

Fig. 3.41 Regions in which primary and secondary minima are found in the a, B plane for hematite (ψ = 30 mV, $\mathbf{X}$ = 10^5 m^{-1}):
(I) potential barrier present and secondary minimum sufficiently close
(II) secondary minimum too distant
(III) and (IV) attractive interaction, flocculation into primary minimum would be dominant
(after Svoboda and Žofka [S36, S37]).

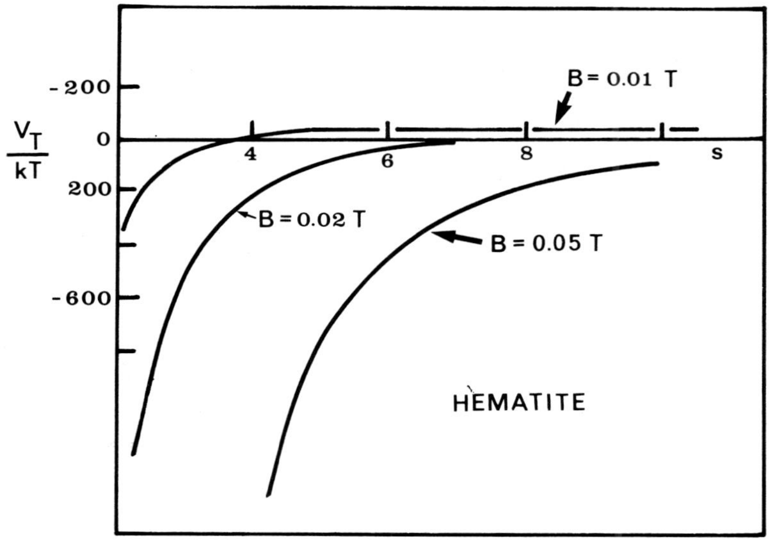

Fig. 3.42 Curves of total potential energy of two hematite particles in an external uniform magnetic field. The curves were computed for a system described by the following parameters: **b** = 2.5 x 10^{-6} m; ψ = 25 mV; χ = 2.7 x 10^{3} m^{-1}; κ = 2 x 10^{-2} (SI); **A** = 5 x 10^{-20} J [S34].

$$W_P = 2\int_2^{\infty} \exp\left(\frac{V_T}{kT}\right)\frac{ds}{s^2} \qquad (3.105)$$

The stability ratio is defined as the ratio of the rate of rapid coagulation to the rate of slow coagulation.

The values for the threshold magnetic field $\mathbf{B}_F^p$, at which magnetic flocculation into the primary minimum begins are shown in Fig. 3.43 as a function of particle size for various minerals. It can be seen that as particle size and magnetic susceptibility increase, the threshold magnetic field decreases.

For fine particles with a volume magnetic susceptibility smaller than 10^{-3} (SI), the threshold magnetic field is too high for magnetic flocculation to be practicable with the magnetic circuits available at present. This is particularly true for the magnetic flocculation of diamagnetic materials, which require a magnetic induction of the order of tens or hundreds of tesla.

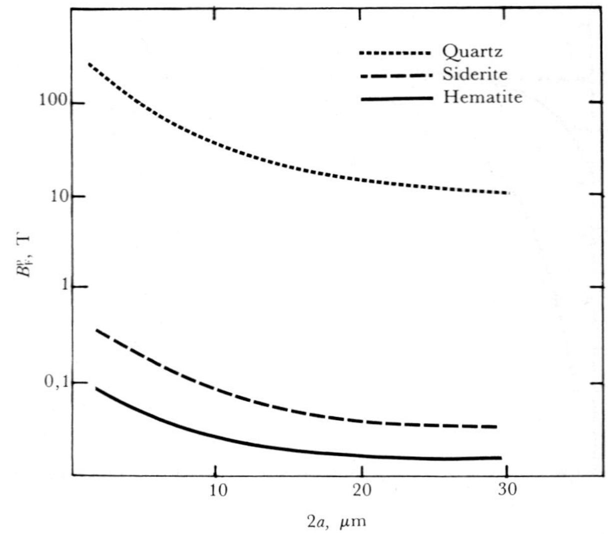

Fig. 3.43 Threshold magnetic field at which flocculation into a primary minimum begins versus particle size [S37].

III.5.2.2 Flocculation in the secondary minimum

As shown in Fig. 3.41, there is a secondary minimum in addition to the primary minimum in regions I, II, and III. In an intermediate region III, the potential barrier is suppressed completely, and secondary minima can occur. Although in regions I and II, the potential barrier prevents particles from flocculating into the primary minimum, particles can flocculate into the secondary minimum. The secondary minimum in region II is too distant from the origin, and, consequently, the aggregates will be packed too loosely. Furthermore, the secondary minimum is too shallow, and the stability of the flocs will be low.

In region I, the secondary minimum can be deep enough and close enough to the origin to render it suitable for the practical application of flocculation. The region of this secondary minimum is fairly sensitive to the particle size, magnetic induction, magnetic susceptibility, and pH value of a suspension, and, for a given mineral, can be determined exactly only by the location of zeros for the first derivate of eq. (3.104).

The mere existence of a secondary minimum does not imply an observable destabilization of a suspension. The other condition to be satisfied is sufficient depth of the secondary minimum; the analysis indicates [S36] that, for V_{min} greater than 1 kT (where kT is the unit for thermal energy), the value of the stability ratio (W_s) for flocculation in a secondary minimum as given by eq. (3.106) will be close to unity, and the flocculation into the secondary minimum will be rapid.

$$W_S = \frac{s_m \int_{s_m}^{\infty} \exp\left(\frac{V_T}{kT}\right)\frac{ds}{s^2}}{1 - \exp\left(\frac{V_{min}}{kT}\right)} \qquad (3.106)$$

where s_m is the position of the secondary minimum.

The threshold magnetic field B_F^s required to start the flocculation into the secondary minimum can be determined from the condition $W_s = 1$. The dependence on a particle radius of threshold magnetic field B_F^s of flocculation in the secondary minimum, which was determined from a condition of rapid flocculation ($W_s = 1$), is shown in Fig. 3.44. A comparison of these values with those of the threshold magnetic field B_F^p for flocculation in the primary minimum (Fig. 3.43) indicates that the magnetic field needed to induce rapid flocculation in the secondary minimum is substantially lower.

Fig. 3.45 shows the ratio B_F^s/B_F^p as a function of the particle size. If this ratio is smaller than unity, flocculation into the secondary minimum will be dominant, and in some instances it will be possible for a threshold magnetic field B_F^s that is an order of magnitude lower than B_F^p to be achieved.

It can be seen that, the higher the surface potential ψ and the lower the thickness of a diffuse double layer represented by $\mathbf{x}^{-1}$, the more advantageous it is for particles to flocculate into the secondary minimum. It is also of interest that the curves in Fig. 3.45 are not dependent on magnetic susceptibility, and are therefore universal for various minerals and sensitive solely to surface interactions.

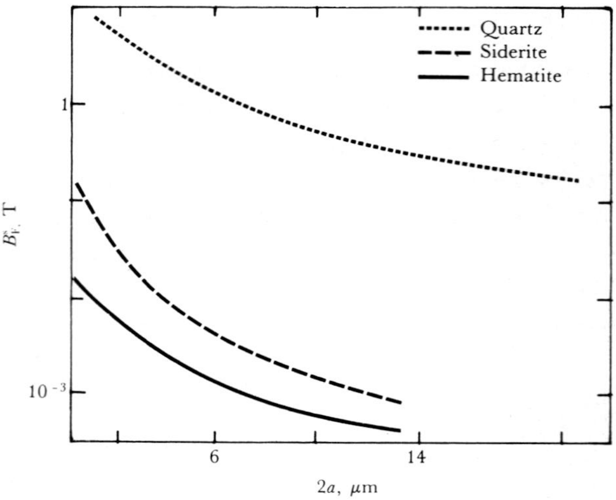

Fig. 3.44 Threshold magnetic field for flocculation into a secondary minimum versus particle diameter [S37].

We have seen that under suitably chosen conditions, the interaction potential between a pair of colloidal particles can develop a significant ($\gtrsim$ 1 kT) secondary minimum, while maintaining a large (>> 1 kT) primary maximum. Although the primary maximum will prevent irreversible coagulation into the primary minimum, the particles can undergo reversible flocculation into the secondary minimum. The stability of the aggregates flocculated into the secondary minimum is determined by the lifetime of a floc which is a useful quantity not only for the information it furnishes about the rate of redispersion under various conditions, but also because it provides us with a sensitive probe of the interaction (mechanical, magnetic and hydrodynamic) between colloidal particles.

III.5.2.3 The kinetics of flocculation

It has been shown that the stability of suspensions depends on the magnitude of the potential barrier and on the existence and depth of the secondary minimum. In addition, however, the determination of how quickly flocculation will take place is necessary.

Fuchs [F18] has shown that the time taken for the number of particles to be halved is given by the relation

$$t_{\frac{1}{2}}=\frac{3\eta W}{4kTN_o} \tag{3.107}$$

where η is the viscosity of the medium, and $\mathbf{N}_0$ is the initial concentration of the particles.

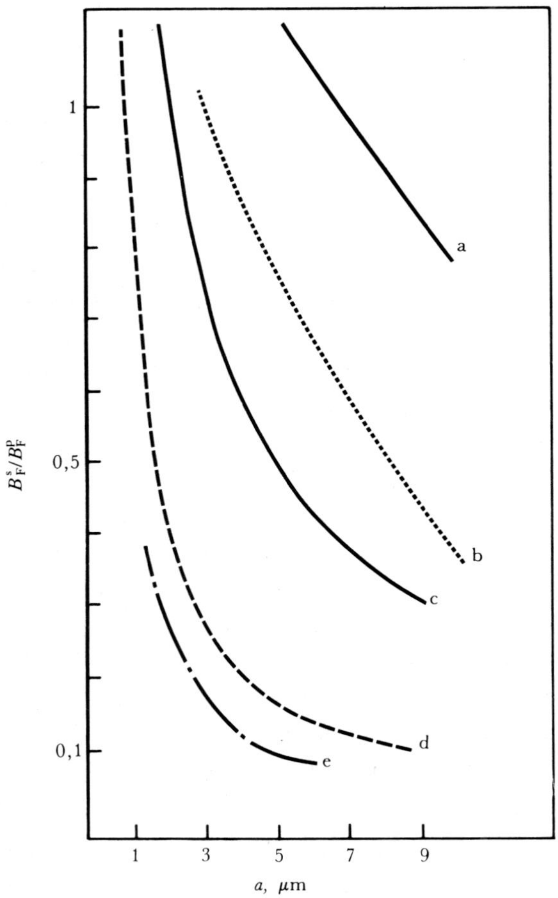

Fig. 3.45 The ratio $\mathbf{B}_F^s/\mathbf{B}_F^p$ as a function of particle size for various combinations of ψ and $\mathbf{X}$.
(a) ψ = 10 mV, X = 10^5 m^{-1}
(b) ψ = 30 mV, X = 10^5 m^{-1}
(c) ψ = 10 mV, X = 10^6 m^{-1}
(d) ψ = 30 mV, X = 10^6 m^{-1}
(e) ψ = 50 mV, X = 10^6 m^{-1}
[S37].

It can be noted that the half-time of flocculation is directly proportional to the stability ratio and indirectly proportional to the initial concentration of the particles. To achieve a reasonably short flocculation time low values of **W** and high values of $\mathbf{N}_0$ must be used. This can be fairly difficult, particularly for suspensions containing particles with a diameter of a few tens of micrometres.

The stability factor can attain values smaller that unity, which means that stability is diminished by the attractive forces. Such a situation arises at the limit of high ionic strengths or of small surface potentials, or at sufficiently high magnetic fields.

In chemical flocculation, values of **W** smaller that 10^{-1} are reached with difficulty [V4]. This significantly limits the possibility that flocculation can be accelerated, and in many instances it excludes the use of flocculation, mainly because flocculation time becomes impractically long. It has been shown [S35] that, in magnetic flocculation, the values of the threshold magnetic field determined from the condition that $\mathbf{W} = 1$ for rapid flocculation, allow reasonably short flocculation times to be achieved only for particles of a diameter equal to or smaller than 1 μm (for a concentration of solids of 10 per cent by volume). The time factor becomes decisive for larger particles, and much higher magnetic fields are needed for a flocculation time that would allow a practical application of flocculation, for which the stability ratio reaches low values [S35]. This can be seen in Fig. 3.46, where the stability factor for flocculation into a primary minimum is shown as a function of magnetic field for various sizes of hematite and siderite particles. Values for **W**, of the order of 10^{-4}, can be achieved at reasonable magnetic fields, so that the flocculation time can be reduced in the same ratio.

The stability ratio for flocculation in the secondary minimum is shown in Fig. 3.47, from which it can be seen that values of $\mathbf{W}_s$ lower that 10^{-1} cannot be obtained and that, consequently, the flocculation time cannot be substantially reduced. Moreover, the curves of $\mathbf{W}_s$ shown in Fig. 3.47 have upper and lower limits, since the secondary minimum exists only in limited regions of the magnetic field. Hence, an increase in the magnetic field cannot increase $\mathbf{W}_s$; on the contrary, the secondary minimum disappears.

As a means of accounting for the dynamics of particle aggregation in a magnetic field, Parker et al. [P11, P12] developed a kinetic model of magnetic flocculation by constructing a force balance equation similar to the equations of motion of a particle in the vicinity of a magnetized wire (see eqs. (3.27) and (3.28)). The force balance equation involves the spatial derivatives of the potentials $\mathbf{V}_R$, $\mathbf{V}_A$ and $\mathbf{V}_M$, as well as Stokes' expression for hydrodynamic drag.

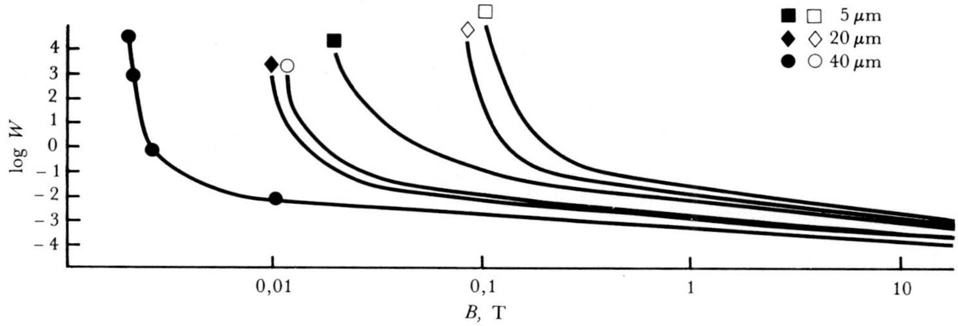

Fig. 3.46 Dependence of log $\mathbf{W}_p$ on a magnetic induction for hematite (solid symbols) and siderite (outlined symbols) [S37].

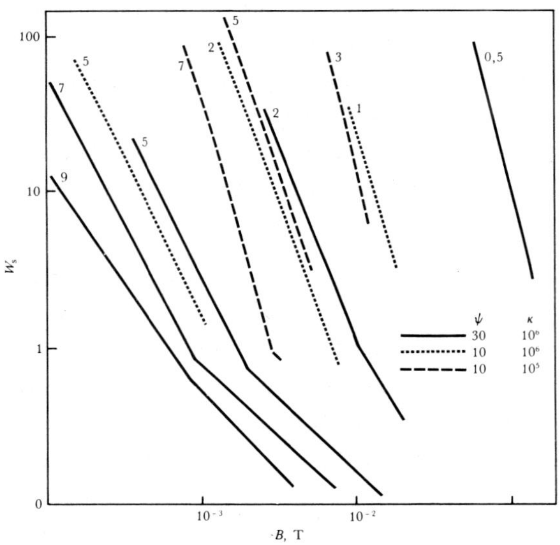

Fig. 3.47 Dependence of $\mathbf{W}_s$ on a magnetic induction for siderite particles of the radius (μm) shown beside each curve [S37].

Particle trajectories were calculated and it was shown that the concepts of flocculation frequency and of half-time of flocculation have characteristics similar to parameters for describing rapid flocculation in the absence of a magnetic field. It was argued by Parker et al. [P12] that the condition $\mathbf{V}_T = 0$ [S34] is not a precise definition of the magnetic flocculation threshold field since the magnetic interaction is angular dependent. However, as has been shown by Chan et al. [C25, C26] the orientational averaging process gives the energetically favourable orientation by the magnetic dipoles heavy weighing. The magnetic interaction energy $\mathbf{V}_M$ is thus predominantly negative and the angular effects as proposed by Parker et al. [P12] are negligible.

III.6 OPEN-GRADIENT MAGNETIC SEPARATION (OGMS)

Although several open-gradient superconducting magnetic separators have been built and tested on laboratory and pilot-plant scales, a description of the separation process has been limited to a simplified physical picture of dry magnetic separation in a split-pair OGMS system. Considerable attention has been paid to the optimization of the coil configuration [H16, C14, G26, P15] and the results of assorted experimental tests of dry separation process were reported in the literature [S21, H17, H18, C13, R3, C28]. Wet separation has received much less attention, and the only systematic experimental investigation was limited to quadrupole magnets. Split-pair OGMS has been used in wet separation only in a few isolated experiments [S21, C13, S4].

The performance of the open-gradient separator depends on the forces acting on the particles - the magnetic force $\mathbf{F}_m$, the force of gravity $\mathbf{F}_g$ and the fluid drag $\mathbf{F}_d$. In the wet separation of small particles gravity can be neglected, but in dry separation all these forces must be included in the mathematical model. The particle trajectories can be calculated on the assumption that the particles are completely independent of one another. Such calculations can give some idea of the potential uses of the separator, although a practical design of the equipment must be based predominantly on experimental tests.

III.6.1 Wet magnetic separation

In wet open-gradient magnetic separation a slurry is passed at constant axial velocity $\mathbf{v}_o$ along the annular space outside the walls of the cryostat and parallel to its z axis. The particle stream is subjected to both radial and axial magnetic traction forces within this annular space as a result of non-uniform fields generated by the split-pair. Boehm et al. [B31] showed that radial and axial equations of motion can be written

$$m\frac{d^2r}{dt^2}=\frac{1}{2}\mu_o\chi V\nabla_r(H^2)-6\pi\eta b\frac{dr}{dt} \qquad (3.108)$$

$$m\frac{d^2z}{dt^2}=\frac{1}{2}\mu_o\chi V\nabla_z(H^2)-6\pi\eta b\left(\frac{dz}{dt}-v_o\right) \qquad (3.109)$$

If particle inertia can be disregarded then a time-independent equation follows from eqs. (3.108) and (3.109).

$$\frac{dr}{dz}=\frac{\nabla_{ra}(H'^2)}{\frac{v_o}{v_m}+\nabla_{za}(H'^2)} \qquad (3.110)$$

Where $\mathbf{H'} = 2\mathbf{r}_i\mathbf{H}/\mathbf{I}$, $\mathbf{r}_i$ is the mean radius of the coils and $\mathbf{I}$ is the applied electric current in each loop.

In analogy with HGMS (Section III.1.4), magnetic velocity is given by the expression

$$v_m=\frac{\mu_o\chi b^2(NI)^2}{36\eta a^3} \qquad (3.111)$$

where **N** is the number of turns in each coil.

The solution of eq. (3.110), for a given set of coils (which determine the values of $\nabla_r(H^2)$ and $\nabla_z(H^2)$ depends only on the ratio $\mathbf{v}_m/\mathbf{v}$, and on the initial position of a particle. Although in analogy with HGMS it is again possible to define a limiting capture radius and to determine the trajectories of particles to be collected by a magnet, the usefulness of such a model for practical application is limited. The interparticle collisions, the character of the flow field and particle build-up affect the collection process to such a degree that simplified single-particle dynamics is of minor importance.

III.6.2 Dry magnetic separation

A detailed analysis of particle trajectories in dry OGMS was made by Collan et al. [C13] while Kopp [K16, K17] developed a rudimentary kinetic model that can be used to calculate the separation distance and the throughput of the magnetic separator.

The set-up used in dry magnetic separation experiments is shown in Figs. 2.68 and 3.48. A curtain of dry particles falls under gravity, a distance **s** from the outer wall of a cryostat. The stream of particles is exposed to the above-mentioned three forces and is divided into two or more components, which are then collected in the separation boxes.

During the deflection process the particles collide with each other and establish an average stream velocity, which depends on

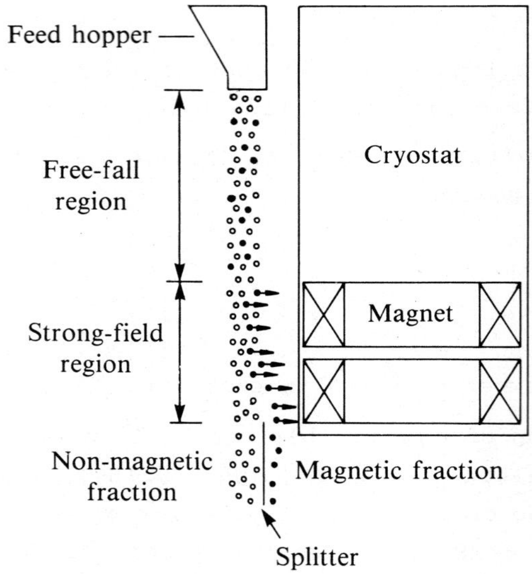

Fig. 3.48 Schematic diagram of dry OGMS with a split-pair magnet.

the drop height and the square of the particle radius. These conditions tend to randomize the process, reducing the transverse velocity of the magnetizable particles and sweeping some of the "non-magnetic" particles along with the mags.

Under the joint action of the viscous drag and the force of gravity given by eqs. (1.15) and (1.14) respectively, the particles will accelerate with the effective acceleration $\mathbf{g}(1 - \mathbf{v}/\mathbf{v}_t)$ until they reach a terminal velocity

$$v_t = \frac{2b^2}{9\eta}(\rho_p - \rho_f)g \tag{3.112}$$

For a small drop height **h** we can write

$$v \cong v_o\left(1 - \frac{v_o}{3v_t}\right) \tag{3.113}$$

where $\mathbf{v}_o = (2\mathbf{gh})^{1/2}$.

The magnetic force of the form given by eq. (1.8) acts on a magnetizable particle in the direction of the field gradient towards the magnet. The average deflecting force decreases rapidly as the distance from the cryostat wall increases, which implies that the trajectories of the magnetic particles crucially depend on their initial position. As shown in [C13, K17], the range of the magnetic force is about 5 cm and the representative value of the magnetic force density $\mathbf{B}(\nabla\mathbf{B})_r$ can be as taken equal to about 100 T^2 m^{-1}.

The "magnetic acceleration" is defined as

$$a_m = \frac{F_m}{m} = \frac{\chi}{\mu_o}|B\nabla B| \tag{3.114}$$

For specific magnetic susceptibility in the range between 10^{-8} and 10^{-6} m^3kg^{-1}, magnetic acceleration varies between 1 and 100 ms^{-2}, or

$$0.1 \leq a_m/g \leq 10$$

For $\mathbf{a}_m < 0.1$ **g**, the magnetic force is too weak to significantly influence the motion of particles, while for $\mathbf{a}_m > 10$ the force is too strong and the separation will be non-selective.

If the density of the feed stream is so low that collisions can be neglected, the particles will acquire a magnetic velocity $\mathbf{v}_m = \mathbf{a}_m\mathbf{t}$ at right angles to the direction of fall. A "magnetic displacement" is thus written [K17]

$$d_m = \frac{1}{2}a_m t_R^2 = \frac{a_m L^2}{2v^2} \tag{3.115}$$

where $\mathbf{t}_R$ is the retention time of a particle in a magnetic field, approximately equal to $\mathbf{t}_R \cong \mathbf{L}/\mathbf{v}$, where $\mathbf{L}$ is the length of the magnetic field region.

If, however, the interactions among the particles cannot be neglected, then the interparticle collisions will tend to equalize the velocities of the particles. If the stream thickness is $\mathbf{t}_s$, the number of collisions that an average particle will make in traversing the stream will be $\mathbf{n}_c \cong \mathbf{t}_s/\mathbf{l}_f$, where $\mathbf{l}_f$ is the mean free path

$$l_f = \frac{b}{3N_oV} \tag{3.116}$$

$\mathbf{N}_o$ is the number of particles per unit volume and $\mathbf{V}$ the mean particle volume.

The magnetic displacement is then given by

$$d_m = \frac{2L}{v}\left(\frac{a_m b}{6N_oV}\right)^{\frac{1}{2}} \tag{3.117}$$

As can be seen this magnetic displacement has a different functional dependence than the displacement for the collisionless case (eq. (3.115)). It also follows from eq. (3.117) that the acceptable displacement, under the common conditions ($\mathbf{L}$ = 0.2 m, $\mathbf{v}_o$ = 2.8 ms^{-1}, $\mathbf{a}_m = \mathbf{g}$ = 10 ms^{-2}, χ_1/χ_2 = 10) can be obtained only for particles greater than 100 μm in diameter.

The throughput of a falling curtain separator per unit length of a magnet is given [K17]

$$Q = t_s \rho_p N_o V v \tag{3.118}$$

Since the thickness of the stream $\mathbf{t}_s$ must be smaller than a relative magnetic displacement $\Delta\mathbf{d}_m$ between particles that are

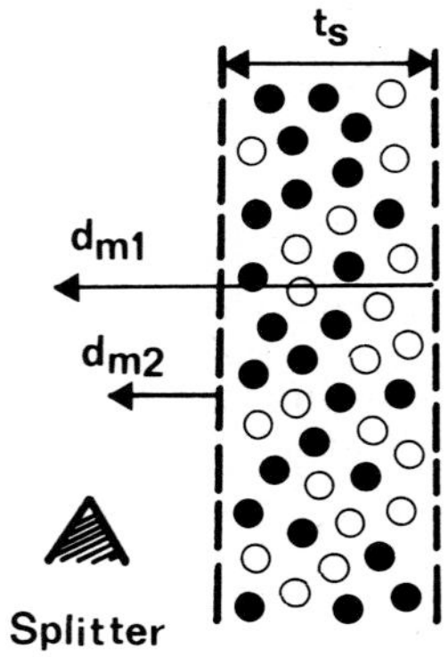

Fig. 3.49 Displacement and stream thickness.

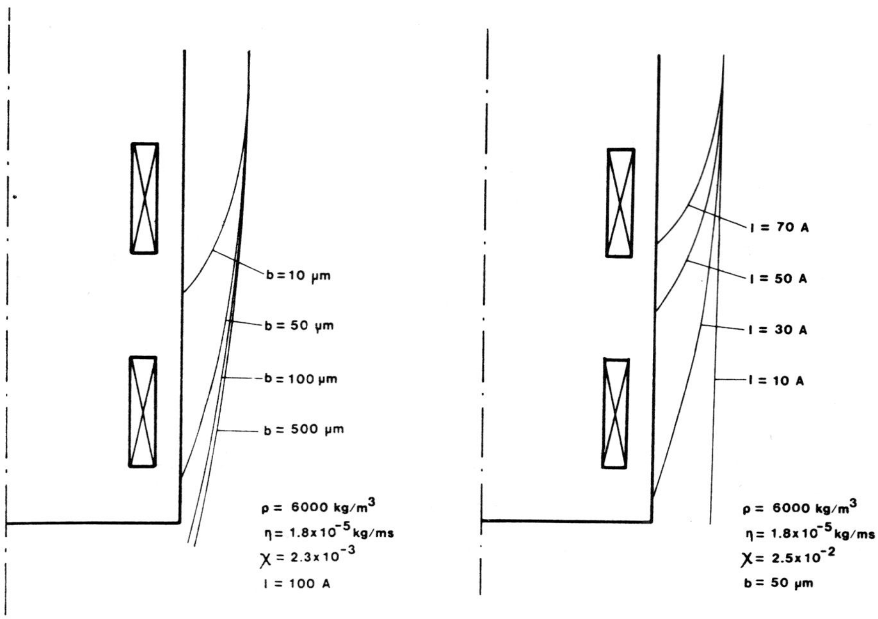

Fig. 3.50 Calculated trajectories of particles in split-pair dry OGMS for different values of particle radius (left) and magnet current (right) (after Collan et al. [Cl3]).

more strongly magnetic and less strongly magnetic (Fig. 3.49), then the optimized throughput becomes

$$Q = \Delta d_m \rho_p N_o V v \tag{3.119}$$

A numerical evaluation of eq. (3.119) shows that the capacity of a machine can vary from 5 to 30 t/h per metre length of a magnet.

The examples of the computed trajectories of particles in a split-pair OGMS for various magnet currents and particle sizes, are shown in Fig. 3.50 [C13].

CHAPTER IV

PRACTICAL ASPECTS OF MAGNETIC SEPARATION

IV.1 SELECTION OF MAGNETIC SEPARATION TECHNIQUE

IV.1.1 Introduction

The selection of the most appropriate magnetic technique to be used in a given application is a formidable task. There are numerous types of magnetic separators, each one being suitable for only a limited range of applications. The choice is influenced by a wide variety of factors and an adequate application of magnetic separation may well be considered an art rather than a science. Since the successful operation of a magnetic separator depends, to a great degree, on practical experience, the results of separation cannot generally be predicted, except for certain applications under clearly specified conditions.

Magnetic separation cannot, as yet, be performed successfully simply on the basis of theoretical models and experience is still the foundation, particularly in mineral processing. At present, there is no branch of magnetic separation where the theoretical treatment of the subject could be considered entirely adequate.

Each embodiment of a magnetic separation technique has distinct merits as well as limitations; no single technique is applicable to all applications of magnetic separation. Very often there is just a single 'right choice' that is superior in all respects to other possible solutions to the separation problem, thereby implying that there is always a high probability of making a 'wrong choice' where a particular type of separator or the selected operating conditions are totally unsuitable for the proposed duty [M25]. The evaluation of magnetic separation as a means of beneficiation, or generally of recovering solids, liquids or gases, is often based solely on the data collected from a few brief tests, not always carried out in a suitable magnetic separator. The likelihood of rejection of a magnetic separation technique is thus very high although its adaptability and potential are greater than of any other filtration or beneficiation technique.

Magnetic separation is used in a great variety of industries whose requirements can differ considerably. The selection of a proper magnetic separation technique is further complicated by the fact that the magnetic separation operation is usually just a part of a complicated flowsheet so that further limitations are imposed upon the choice of suitable magnetic separation equipment.

For the selection of a magnetic separator to be used in a given application the objective of the process is to be well defined. In principle, magnetic separation is used in three distinct processes:

1. Magnetic filtration
   a) recovery of liquid (the magnetizable solids are discarded)
   b) recovery of magnetizable solids (liquid is discarded).
2. Removal of more strongly magnetizable particles from a mixture with less strongly magnetizable particles.
3. Concentration of valuable more strongly magnetizable particles from a mixture with less strongly magnetizable particles.

The choice of a magnetic separator and the selection of appropriate operating conditions, differ very widely between these three processes. Although in well-established applications of low-intensity magnetic separation, the difference between these processes is well understood, in newly-developed, high-intensity or high-gradient magnetic separation, a certain degree of confusion exists as to the choice of the optimum experimental conditions. Since the operating conditions common in magnetic filtration are sometimes directly used for the concentration of valuable magnetizable minerals from complex ores, this sometimes results in the inefficient performance of a magnetic separator and in consequent rejection of magnetic separation as a means of beneficiation.

IV.1.2 The selection of a magnetic separation technique

Magnetic separation, like any other type of separation, is greatly influenced by a number of variable factors. Provided that optimum conditions have been established, these factors can be used to greatly improve the separation process. Perhaps the most relevant starting point in the selection of the most suitable separation technique is to consider the particle size distribution of the solids tó be treated. The situation is shown schematically shown in Fig. 4.1.

If the size of particles to be treated by magnetic separation is greater than, say, 75 μm then we have an option of a dry or wet process. If, however, the particles are finer than 75 μm,

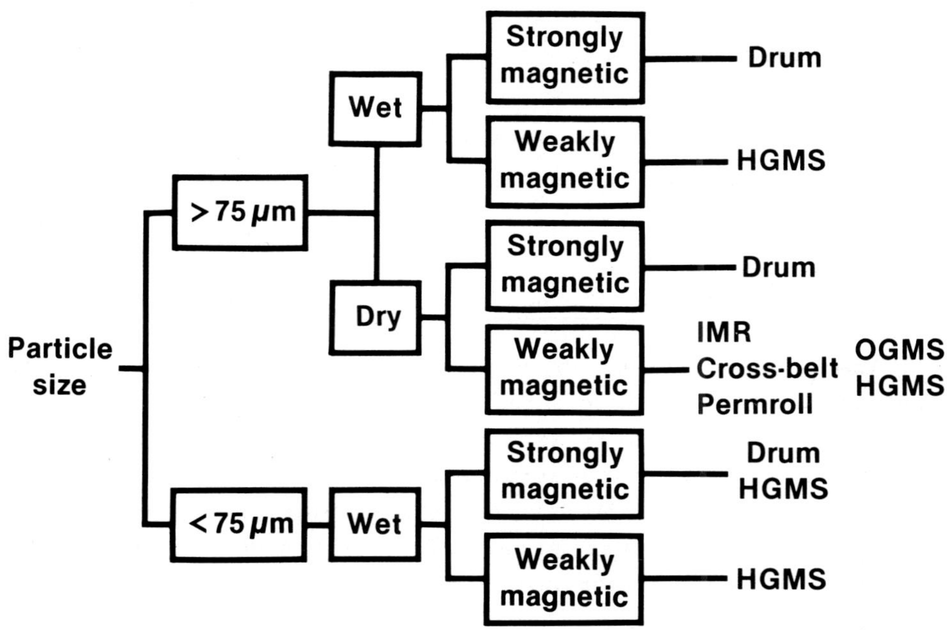

Fig. 4.1 Classification of magnetic separation processes.

the wet technique is more appropriate, although this threshold particle size can be smaller or greater than 75 μm, depending on the amounts of fines present and on the physical properties of the feed.

A further choice of magnetic separation techniques depends on the magnetic properties of the mineral to be treated. For strongly magnetic minerals, low-intensity drum magnetic separators can be used, in either dry or wet modes of operation.

Coarse, weakly magnetic minerals can be treated by wet, high gradient magnetic separation. However, there is an upper limit imposed on the particle size by the type of matrix and particles greater than 1 mm usually cause clogging, even of a coarse matrix.

In dry mode weakly magnetic particles can be recovered by a variety of high-intensity magnetic separators, including the induced magnetic roll, cross-belt separator, permanent magnetic roll separator, open-gradient magnetic separator and dry, high-gradient magnetic separator.

If the particles are finer than, say, 75 μm, the best approach is to use wet magnetic separation, although in some cases dry separation can be attempted. Fine, strongly magnetic particles are usually recovered by a wet low-intensity magnetic separator while for weakly magnetic minerals the wet high-gradient magnetic separation is best.

The schematics of the decision making process for selecting the magnetic separation technique is shown in Fig. 4.2.

IV.1.3 HGMS or WHIMS or HIMS ?

Thanks to the advent of new powerful, high-intensity and high-gradient magnetic separators the application of magnetic separation methods has been greatly extended, and several matrix magnetic separators have been constructed. Hand in hand with a rapid development in separation devices, a wide array of names have been used to classify these separators, although the operating principles of most of them remain the same.

To emphasize the difference between these machines and the classical, low-intensity magnetic separators, the early continuous matrix magnetic separators were termed high-intensity separators.

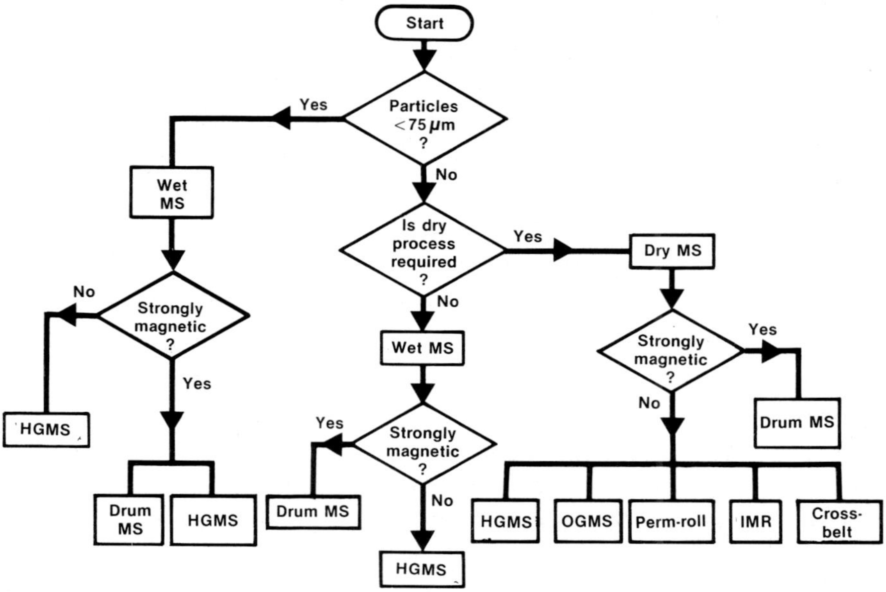

Fig. 4.2 Selection of the magnetic separation technique.

The term is deceiving, since many closed-gradient separators (e.g. induced magnetic roll) or novel open-gradient magnetic separators (e.g. split-coil separators) generate the same magnitude of magnetic induction as these matrix separators.

With the arrival of powerful solenoid matrix separators, which in their cyclic version are used for kaolin brightening, the main emphasis shifted to their capability for generating high gradients of magnetic field as a result of using a fine steel wool matrix. These separators were termed high-gradient magnetic separators (HGMS) and the same term was later assigned to the continuous version of solenoid matrix separators. Nevertheless, Kolm et al. [K11] in their classical paper, described all conventional matrix separators of the Jones type as high-gradient magnetic separators in order to highlight the major difference between these modern machines and classical low-intensity (drum) separators or high-intensity low-gradient separators (e.g. IMR). In spite of this, however, the term high-gradient magnetic separator has been reserved mainly for cyclic solenoid magnetic separators with steel wool matrices which received particular attention in theoretical modelling.

Concurrently, the extensively used matrix separators incorporating electromagnets with iron yokes were called high-intensity magnetic separators (HIMS or WHIMS). Confusion was increased when the continuous solenoid separators (e.g. Sala HGMS machines) were termed high-gradient magnetic separators, although the gradient generated often did not exceed that generated by the Jones-type units. To typify the confusion, Gerber and Birss [G1] state that "in mineral processing, the continuous HGMS machines are in competition with the wet high-intensity magnetic separators (WHIMS)".

It must be emphasized that there is no difference between HGMS and WHIMS magnetic separators. They generate the same magnitude of magnetic field and the value of the gradient is also the same, since this is determined by the type of matrix, which, in turn, is dictated by the application. As shown in previous sections, excessive emphasis on the field gradient usually leads to non-selective separation, to low grade of magnetic concentrate or to unacceptable losses of the non-magnetic product in the magnetic fraction. Therefore, the use of very fine steel wool (high gradient) matrices should be avoided in mineral processing and should give way to coarser but more selective matrices (meshes, screens, rods, balls), usually used in WHIMS machines (in terms of Gerber and Birss's classification).

It has been observed but never sufficiently admitted that, apart from very low selectivity, steel wool matrices have a capture efficiency seldom exceeding that of coarse matrices, even in such demanding applications as kaolin brightening. These observations support the model [S51, S52] in which it is shown that the gradient of the field is of secondary importance only. This is because the majority of particles in a slurry collide with the matrix elements due to the mechanical interactions present in classical deep bed filtration. Although the presence of the strong field gradient can increase the capture efficiency of a small fraction of particles which would otherwise escape without colliding with the matrix, its significance is only marginal. Of prime relevance is the sufficiently strong background magnetic field, which generates a strong static dipolar magnetic interaction between the matrix element and the colliding mineral particle.

To summarize, the terms high-intensity magnetic separation and high-gradient magnetic separation describe, in principle, the same process, in which a strong magnetic field is used to magnetize the matrix and to generate a strong local magnetic induction in the vicinity of the matrix collectors. The emphasis, either on high intensity or on high gradient, is deceiving as these parameters must be selected according to the application. Furthermore, the maximum values of these quantities do not necessarily mean a successful separation. More often than not the moderate values of the field and of the gradient are the best choice.

Nevertheless, in an effort to unify the nomenclature, we are of the opinion that the term "high gradient" when applied to all matrix separators clearly indicates the difference between matrix magnetic separators and other magnetic separation techniques. On the other hand, the term "high intensity" can cause confusion because high intensity of the magnetic field can be achieved in various magnetic separation machines (e.g. IMR, cross-belt, permanent magnetic roll), even though the scope of these machines differs broadly.

Likewise, emphasis on the steel wool matrix (by coining a separate name for this special technique) is not justified, as the physical principles of particle collection in all matrix separators are the same, and, in many applications, particularly in mineral processing, the use of a steel wool matrix is undesirable.

IV.2 DRY MAGNETIC SEPARATION

Dry magnetic separation is a technique dating back to the beginning of the century. It has been practised successfully in the beneficiation of strongly magnetic minerals, as well as feebly magnetic minerals such as wolframite, cassiterite, ilmenite etc. Although these operations have been very successful with regard to metallurgical performance and the ability to meet product specifications, the operation itself has several fundamental disadvantages that should not be overlooked.

For dry magnetic separation to be successful the ore must be completely (i.e. not just partially) dry, and this usually involves an expensive heating stage. For particles smaller than a certain threshold size (here loosely chosen as 75 μm) the attractive electrostatic and van der Waals forces among the particles will cause sticking and the formation of aggregates thus adversely affecting the selectivity of magnetic separation. Care must also be taken to ensure control of possible dust hazards, an expensive precaution both in capital and operating costs. The production of dry tailings is sometimes cited as an advantage by virtue of their easy disposal. However, experience sometimes contradicts this statement [D8], as dry tailings prove more expensive to dispose of than wet tailings.

The particle-size distribution, and particularly the amount of fines present, have a strong influence on flow characteristics of the feed and the capacity of dry magnetic separators. In practice, when excessive fines are present, the material will not flow through the air gap of the dry IMR [H9], and the greater the amount of fines present in the feed, the more severe are the dust problems associated with dry magnetic separation. Dry processes often require the ore to be sized into several size fractions each of which must be spread in a monolayer over the separating surface [B27, A17]. Actually, when magnetic separation is performed on samples sized into, say, three fractions, the efficiency of separation is usually much improved in comparison with unsized samples.

As a rule, a large proportion of fines necessitates higher rotating speeds of a roll or drum. Although increase in speed will improve the grade of the magnetic product and the throughput of the separator, it will also increase the losses of the magnetic minerals. These losses must be compensated by using the multiple passes of the non-magnetic product, during which a lower rotating speed is used compared to the first pass. Dry magnetic separators also offer a possibility of thermo-magnetic separation of ferromagnetic minerals with different Curie temperatures [W22], a factor so far not fully explored.

IV.2.1 Electrostatic phenomena in dry magnetic separation

During powder-handling operations the powder invariably becomes electrostatically charged. The electrostatic charge may in some circumstances alter the dynamics of the separation process. The powder particles come into contact with various surfaces and become charged by the process known as triboelectrification [B28]. When two dissimilar materials make and then break contact, charge transfer from one of the materials to the other occurs. The triboelectric charging behaviour of some materials is sensitive to an atmospheric environment, whilst other materials are insensitive. The influence of relative humidity upon powder charging is complex. However, at high values of relative humidity high levels of negative charging occur. At low values of relative humidity, although fine particles still charge negatively, the coarse particles show a change of charging polarity such that the net charge conveyed by the powder is positive [B28].

When a particle charges during handling, the charge transferred at the particle contact point will redistribute itself over the particle surface by electrostatic forces. The rate at which redistribution proceeds depends upon the electrical relaxation time of the particle material. Relaxation time σ_R is given by the product $\rho_R \epsilon_d$ where ρ_R is the particle resistivity and ϵ_d the dielectric constant of a particle. Insulating materials with ρ_R above 10^{13} Ω m have a relaxation time of minutes or even hours, so that appreciable charge redistribution often does not occur during the magnetic separation process. Semi-insulating and conducting particles attain surface charge equilibrium relatively quickly. For spherical particles the final equilibrium surface charge density is uniform, but for particles of non-spherical shape final redistribution is non-uniform.

The existence of surface charge on a particle give rise to an electrostatic field which has a maximum intensity at the particle surface and decays inversely with the square of distance away from the surface. The maximum surface charge density as a function of particle radius is

$$\sigma = \epsilon_o E = 8.18 \times 10^{-6} b^{-0.3} \tag{4.1}$$

where **E** is intensity of the electric field (in V m^{-1}), ϵ_o the dielectric constant of air, **b** the particle radius (in m) and σ the surface charge density (in C m^{-2}).

As a general rule, if electrostatic phenomena arise in a powder handling situation they are likely to become increasingly significant as particle size is decreased. Particles charged triboelectrically, either positively or negatively, may adhere strongly to surfaces. Upon contact with surfaces, charged particles may adhere strongly by a combination of electrostatic and van der Waals forces.

Derkach [D1] has shown that the force of molecular attraction between two spherical particles is given by

$$F_A = \frac{4A\pi\sigma_t b_1 b_2}{b_1 + b_2} \tag{4.2}$$

where **A** is the coefficient (taking into account the contact area, the degree of humidity etc.) and σ_t the surface tension of a particle.

The effect of moisture on the metallurgical results of dry magnetic separation is illustrated in Fig. 4.3. It can be seen that the grade of the magnetic concentrate is seriously impaired by the presence of moisture and the permissible moisture content decreases with decreasing particle size.

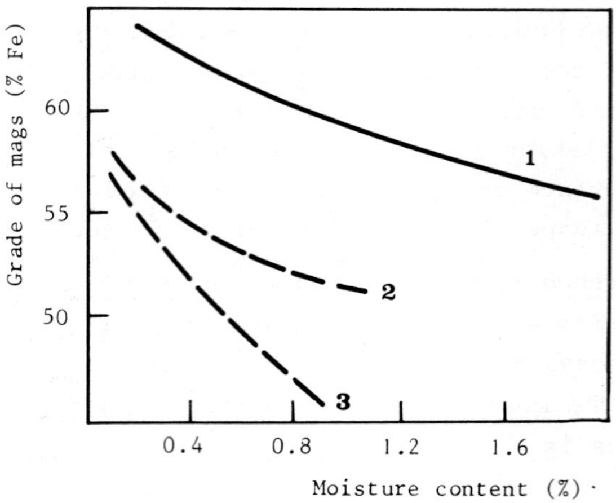

Fig. 4.3 The grade of the magnetic concentrate as a function of the moisture content in the ore
1) magnetite ore, dry cross-belt separator
2) and 3) martite ore, IMR
(After Derkach [D15])

IV.2.2 Comparison of dry magnetic separation techniques

The technique of dry high-intensity magnetic separation has undergone an extensive development in the last few years. Various types of magnetic separators, based on different principles of the generation of magnetic field have been developed and to make a correct choice of the most suitable technique is a difficult task. A comparison of the separation efficiencies of different techniques is usually absent in the literature. Furthermore, every comparison is only conditional since a relative merit of each technique is a function of a mineral to be treated, and of its physical properties.

McCabe [M10] tested the performance of cross-belt, induced roll and permanent roll (SmCo) magnetic separators, on samples of andalusite and alaskite. The aim of the tests was to reduce the Fe_2O_3 content to a minimum while minimizing the loss of Al_2O_3 or SiO_2.

The results show that a permanent roll magnetic separator equipped with an 0.7 mm belt produces the non-magnetic product with consistently higher Fe_2O_3 content than any of the other separators that were tested.

Although the grade of the product improves with the use of the 0.2 mm belt the presence of static electricity causes some non-magnetic particles to be lost to the magnetic fraction.

The induced roll magnetic separator compared favourably with permanent magnetic roll separator, it was not adversely affected by a static charge and it produced the non-magnetic fractions containing the lowest amount of Fe_2O_3.

Although the cross-belt separator is not as efficient as the other separators, it does have a few advantages:

a) only one pass is necessary as opposed to four passes on the permanent roll separator and IMR
b) since the four electromagnets are independently adjustable, selective removal of different magnetic minerals is possible
c) it has a low-intensity magnet for the scalping of ferromagnetic impurities.

Several dry magnetic separators were tested by Roux et al. [R3] in the beneficiation of pyroxenite ore to yield apatite concentrate. The average metallurgical efficiencies of different types of separators are compared in Fig. 4.4 while Fig. 4.5 shows the relation between phosphate recovery and the particle size for IMR, matrix high-gradient magnetic separator and open-gradient superconducting magnetic separator.

A thorough practical experience with these magnetic separators was gained at Foskor and their comparison is summarized in Table 4.1.

A gradual deterioration in the results was observed with IMR, caused by abrasive wear on the epoxy filler between the iron discs of the rolls. Furthermore, because the separation distance between the magnetics and non-magnetics is very small a precision setting of a splitter is required. As the mineralogy of the ore varies, constant attention to the optimization of roll speeds and splitter settings is required, which makes it difficult to maintain good metallurgy [R3].

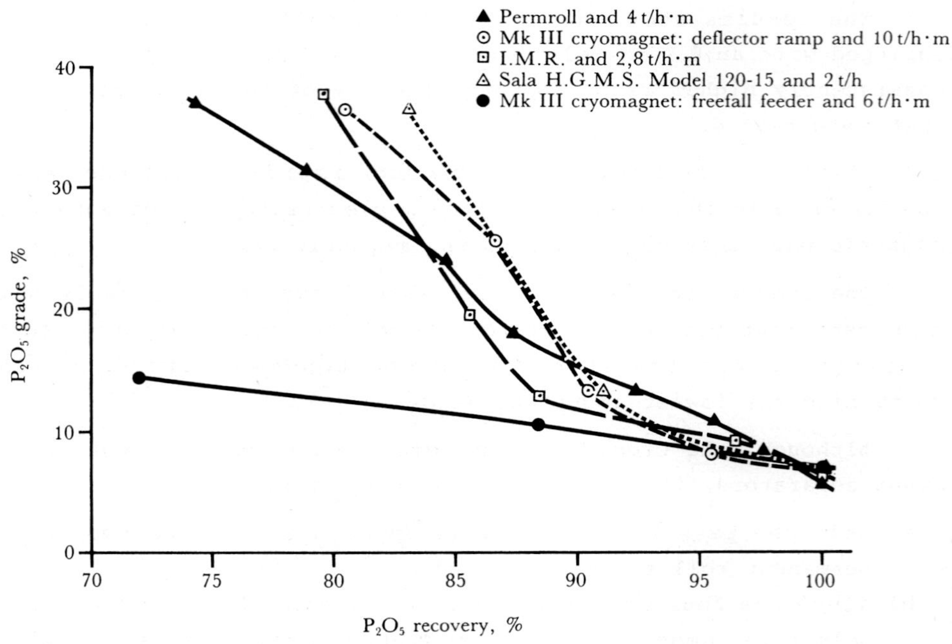

Fig. 4.4 Comparative metallurgical efficiencies of various dry magnetic separators as applied to the apatite recovery (after Roux et al. [R3]).

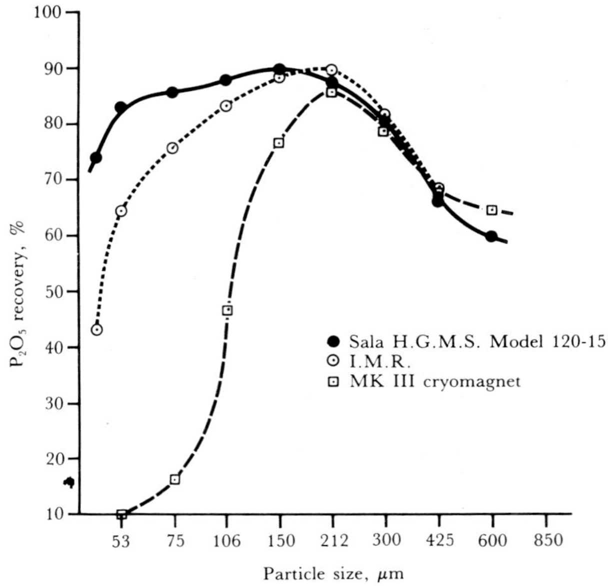

Fig. 4.5 Phosphate recovery as a function of particle size for various magnetic separators.

A Sala HGMS Model 120-15 carousel separator operated reliably and well for 1000 hours and it was found that single-stage separation could be performed to yield metallurgy equivalent to that of two stage IMR separation. The expanded metal matrix showed no measurable wear, while the metallurgy was shown to be adversely affected by oversize material, which gradually accumulated in the matrix. In order to keep coarse particles out of the feed and to avoid the onerous task of cleaning the matrix canisters it became necessary to fit a second screen that could be kept under continuous visual observation. This problem requires serious attention in a full scale plant.

An extensive series of experiments on Sm-Co permanent magnet roll separators has shown that the feed rate, roll speed and thickness of magnet and steel discs must be optimized for each separation stage.

The open-gradient, split-coil superconducting separator tested at Foskor is shown in Fig. 4.6. The ore feed rate can be controlled

TABLE 4.1

Comparison of different types of magnetic separators, as applied to phosphate recovery (after Roux et al. [R3])

Attribute	Type of separator			
	Induced-magnetic roll	Permanent-magnet roll	High-gradient matrix	Open-gradient superconducting
1. Stage of development	Very high	Early	High	Early
2. Track record in practice	Proved	A few prototypes in plant service	A few machines in plant service	None
3. Capacity per machine	Low, 4 t/h	Low, 8 t/h	Very high, 400 t/h	High, 60 t/h
4. Ore distribution system in plant	Very complex	Very complex	Very simple	Simple
5. Mass, kg/(t/h) of feed	2190	440	4116	41
6. Power consumption, kWh per ton of ore	2,0	0,6	5,6	0,5
7. Tolerance to ferromagnetics	Low, scalper required	Low, scalper required	High, no scalper required	Very low, scalper required
8. Tolerance to oversize particles	None, rolls severely damaged	Impervious to oversize	Low, matrix clogs	Very high
9. Method of cooling	By air	None required	By water	By liquid helium
10. Effect on environment	Plant requires air conditioning	None	Slight heat output	Slight heat output
11. Maintenance	High on rolls and bearings	Could be high on rolls, low on bearings	Low	Low
12. Metallurgical performance				
12.1. Sensitivity to ore variation	High	Low	Low	Very high
12.2. Recovery of fine apatite	Poor	Very poor	Good	Very poor

by sliding the outer cone up and down the collar to decrease or increase the feed opening. The skirt was fitted to prevent ore particles from reaching the surface of the magnet because of their collisions with other particles, rather than as a result of the pull of the magnetic field.

The splitter could be moved up or down to vary the fraction of the feed collected as non-magnetics. This system, like any other that involved the simple free fall of the ore stream gave reasonably good metallurgical results only at very low feed rates. A much more satisfactory ore-presentation system is that, in which the ore slides down a ramp. This imparts an outward velocity to the particles, thereby counteracting to a considerable extent, the sweeping across of non-magnetic particles by the magnetic ones; the situation is shown schematicaly in Fig. 4.6. This approach made it possible to obtain metallurgical results comparable with those of IMR separators, at feed rates of up to 10 tonnes per hour per metre, or 3.5 times the optimum feed rate of an IMR. As with IMR, a four-stage flowsheet was developed, the feed rate, shape of ramp and splitter position being optimized for each stage [R3].

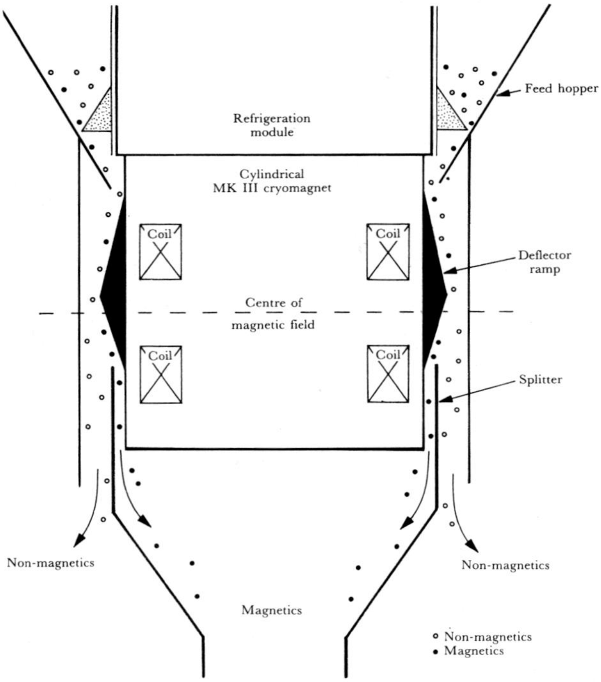

Fig. 4.6 Open-gradient superconducting separator with the ore-presentation system (after Roux et al.[R3]).

IV.3 WET MAGNETIC SEPARATION

As shown in Fig. 4.1 wet magnetic separation can be applied to coarse as well as fine particles, that are either strongly or weakly magnetic. Strongly magnetic minerals can be recovered by wet low-intensity machines which are by far the most common in use. The art of wet low-intensity magnetic separation is well established and will be described only briefly; the main emphasis will be put on the treatment of weakly magnetic materials by wet high-gradient magnetic separation (HGMS). As seen in Fig. 4.1 HGMS can be used to treat slurries containing coarse as well as fine particles, although most attention is paid to the application of HGMS to the difficult task of the treatment of fine particles down to the quasicolloidal sizes.

IV.3.1 Wet low-intensity magnetic drum separators

IV.3.1.1 Basic operations of drum separators

When applied to ore concentration, the use of wet drum separators is limited to the treatment of particles smaller than 10 mm. Drum separators are usually related to the grinding circuit as a means of liberating magnetic minerals and by their application are typically designed for three basic operations [A28].

1. Cobbers
2. Roughers
3. Finishers.

Cobbers

Drum separators used in cobbing operations are designed to obtain maximum rejection of a non-magnetic fraction and maximum recovery of the magnetic mineral. Typically, cobbers are applied to a rod-mill discharge product. Most iron-ore concentrating drum separators are applied in multi-drum configurations so that a high-grade magnetic concentrate is obtained. In a cobbing operation, two drums are typical, but as many as four drums have been used. A re-pulping box is incorporated and 900 and 1200 mm-diameter drums are usually used.

Roughers

Drum magnetic separators used in roughing or cleaning are applied after the size of the cobber concentrate has been further reduced. Because the cobber has significantly upgraded the ore, magnetic loading of the drum is considerably increased. The objective is again to obtain a high-grade magnetic concentrate while maintaining high recovery. Typically, a double-drum separator is used in countercurrent modes of operation, although a combination of concurrent and countercurrent tanks is often used.

Finishers

The finishing drum magnetic separators are designed to produce the highest possible iron content in the concentrate. The feed tank and feed arrangement are usually of semi-countercurrent design. Both 750 and 900 mm-diameter drums are used in finisher applications, while the feed size is usually reduced to less than 75 or 45 μm.

IV.3.1.2 Variables of low-intensity magnetic separation

Factors associated with the properties of the feed, to be considered in the selection of a wet-drum magnetic separator, are [A28]:

1. Feed rate (tonnes per hour)
2. Feed size (max. 10 mm)
3. Feed volume (m^3 per hour)
4. Concentration of magnetic particles
5. Degree of liberation
6. Size distribution
7. Desired recovery into the magnetic fraction.

As shown in Table 4.2 the maximum feed rate varies with the type of separator considered; maximum and recommended feed volumes are also given. Typical feed sizes for the three applications are:

Cobbers: -6 mm to -850 μm

Roughers: -500 μm to -300 μm

Finishers: -53 μm to -45 μm

TABLE 4.2
Feed rates and feed volumes in wet, low-intensity drum separators [A28]

Application	Feed rate ($t\ h^{-1}m^{-1}$)		Feed volume ($m^3h^{-1}m^{-1}$)	
	Max.	Recommended	Max.	Recommended
Cobber	75	45	100	70
Rougher	60	45	100	70
Finisher	15	15	35	20

IV.3.2 The importance of correct operating conditions in HGMS

Wet high-gradient magnetic separation is used in three different technological processes, as shown in Fig. 4.7. HGMS can thus be used either as a magnetic filter to remove single solid component from a suspension, or as a magnetic separator to remove an unwanted magnetic component from a mixture of minerals, or for a concentration of valuable weakly magnetic mineral from a mixture.

It is clear that the conditions for a successful separation in these three categories can differ widely and a distinct difference between these three approaches is often little understood.

Because it is usually assumed that recovery is proportional to the magnetic force $\mathbf{F}_m$, $\mathbf{F}_m$ is frequently made as large as possible, either by using the highest possible applied magnetic field, or by employing a very fine matrix to increase the gradient of the field, often by using both ways. The result of such an approach is usually unsatisfactory: the mass yield into the magnetic fraction becomes very high as a consequence of non-selective separation, due to an excessively high magnetic field, and of the mechanical straining due to a too fine matrix. The grade of the magnetic concentrate is often only marginally better than that of the feed and the recovery of a valuable metal into the mags is frequently only moderate. The importance of the optimization of the operating conditions is shown in Table 4.3 where two cases of uranium recovery from leaching residues by HGMS are analysed. It can be seen that by using a too fine matrix and a too high magnetic field, the mass yield into the magnetic product is too high, separation selectivity is low and even the recovery of uranium is not sufficient.

Although the unacceptability of such an approach is obvious, there is a number of papers reporting such curious results. For example, Kerkdijk et al. [K20] measured the efficiency of magnetic separation of metals from fly ash in terms of "the capture efficiency"; in other words, of mass yield into the magnetic fraction. It was found that at 1 T the capture efficiency approached 100 per cent (or all solids fed into the separator were captured) and it is not surprising that "the concentration of iron

TABLE 4.3
Comparison of mass balances of two cases of recovery of uranium from leach residues by HGMS

I. Optimum magnetic field (1 T)
Optimum matrix (6 mm balls)

	Mass yield (%)	Grade (ppm U_3O_8)	Recovery (% U_3O_8)
Mags	10	1000	52.6
Non mags	90	100	47.4
Feed	100	190	100

II. Excessively high magnetic field (5 T)
Excessively fine matrix (steel wool)

	Mass yield (%)	Grade (ppm U_3O_8)	Recovery (% U_3O_8)
Mags	40	210	45.2
Non mags	60	170	54.8
Feed	100	186	100

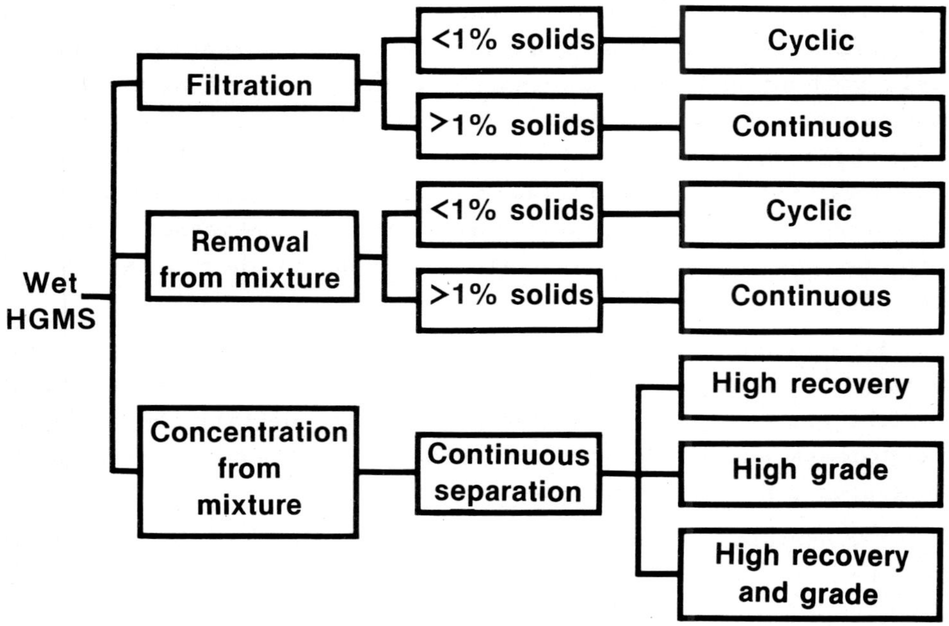

Fig. 4.7 The applications of wet high-gradient magnetic separation.

in the mags becomes identical to that of the starting material" [K20].

Such an approach is permissible in magnetic filtration where a single solid species is to be removed and the mass recovery (or capture efficiency) must be maximized. In the treatment of minerals, however, it is equally important not to retain the gangue particles, nor particles of the unacceptable grade because the efficiency of separation is determined by the interplay between the grade of the concentrate and the recovery of metal into the concentrate. The mass yield into the concentrate plays only a secondary role and the tendency is usually to make the mass yield as low as possible but compatible with acceptably high metal recovery. This would benefit the process in two ways: first, the selectivity of separation, and therefore the grade of the mags, will be high, and second, the mass of the concentrate, which usually undergoes a further treatment, is low.

The concept of the separation efficiency based on the maximization of the magnetic force usually fails even in a much simpler application of HGMS to the removal of magnetic impurities from mixtures.

Although it may be true that, in exceptional cases, the efficiency of extraction will rise with an increasing magnetic field without unduly enhancing the entrainment of valuable non-magnetic product in the mags , in most applications the operating parameters of a magnetic separator must be carefully chosen, in such a way that a high extraction efficiency of the impurities is guaranteed, without excessive losses of the non-magnetic product into the mags, caused by non-selective separation and mechanical straining.

It is evident from the above analysis that the first, and perhaps the most critical step in selecting a correct wet high-gradient magnetic separation technique is to identify the target of the process and to define the criteria that would distinguish successful and unsuccessful separations.

The decision-making process in such a selection is outlined in Fig. 4.8.

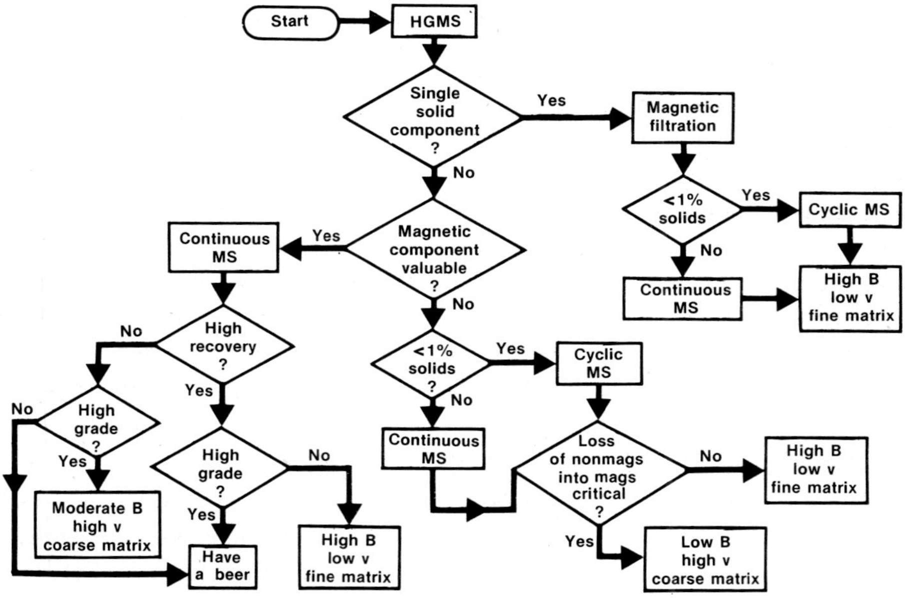

Fig. 4.8 Selection of the operating conditions in wet high-gradient magnetic separation.

IV.3.2.1 Magnetic filtration

Magnetic filtration is used primarily for the removal of fine and colloidal particulates from dilute fluid suspensions. Removal is carried out by passing suspensions through the magnetized matrix, which leads to the retention of particles throughout the matrix. The fluid stream may be either liquid or gas.

The process of magnetic filtration is analogous to that of deep bed filtration, the difference being that the external magnetic force is available in magnetic filtration. Although the magnetic force does not significantly increase the probability of a particle collision with the matrix [S52], the enhanced attachment efficiency caused by the presence of strong, static dipolar magnetic interaction between the matrix and a colliding particle increases significantly the efficiency of the particle capture. Furthermore, this allows us to use much higher flowrates of a suspension through the filter bed, as compared to deep bed filtration. The similarity between magnetic filtration and conventional deep bed filtration indicates that the basic rules common to deep bed filtration should be obeyed in magnetic filtration as well.

As the aim of magnetic filtration is the removal of a single solid component from a suspension, the brute force approach can be used so that all particles are collected on the matrix. Therefore, the largest possible magnetic force should be used which dictates that the maximum magnetic field, in combination with the finest possible matrix compatible with the particle size distribution, is to be employed. Simultaneously, to warrant the highest degree of extraction of solids the fluid shear stress on the deposited particles should be kept as low as possible. On the other hand, the flow velocity of a suspension through a matrix bed should be sufficiently large to ensure high throughput of a separator.

The performance of a magnetic filter depends on a large number of variables, including size distribution of suspended solids, their magnetic susceptibility, suspension concentration, filtration rate and the characterictics of the magnetic filter itself.

The selection of a matrix is dictated mainly by the fact that a straining mechanism of the particle retention is undesirable and should be avoided. Accordingly, the ratio of the matrix element radius to particle radius **a/b** should be larger than 10 but preferably between 100 and 500. This requirement contradicts the notion that the magnetic force should be maximised, i.e. that the **a/b** ratio should be approximately of the order of unity (see Section II.1.1). Although in magnetic filtration the ratio **a/b** in the range 10 to 50 is still acceptable, in applications of magnetic separation to mineral processing, the **a/b** ratio should be larger than 100 and preferably 300.

Mechanical straining usually leads to the formation of a surface mat of deposit on the matrix, which causes a high resistance to the flow. The tendency of suspended solids to clog the matrix can be reduced by using uniformly-sized matrix elements.

The basic rules for the selection of operating parameters of a magnetic filter can be summarized as follows.

Magnetic field

The magnitude of the applied magnetic field is dictated mainly by the magnetic susceptibility of the suspended solids. In filtering the ferromagnetic solids the magnetic induction from 0.3 to 0.5 T is fully sufficient, as demonstrated by the application of magnetic separation to the purification of steel mill waste waters and of the cooling water in power plants.

In more demanding applications, in which feebly magnetic solids are to be removed, much higher magnetic induction is usually required (up to 2 T), although little gain is usually obtained with fields exceeding 1.5 T. Although some theoretical models imply that the capture efficiency is directly proportional to the applied magnetic field, the validity of such argument is dubious. The magnetic properties of suspended solids usually change with the magnetic field, the magnetic susceptibilty decreases with increasing magnetic field, the matrix loading increases due to the reduction of the matrix surface area available for the paramagnetic capture and magnetic flocculation of the suspended solids becomes an important mechanism. As a result, the threshold magnetic field, at which the magnetic dipolar interaction between the deposited solids and the matrix collectors is larger than the erosion interaction due to fluid flow, can be very moderate and the increase of the field can either have no effect on the capture efficiency or the performance of a magnetic filter can even deteriorate.

Matrix

In order to avoid straining the ratio **a/b** should be of the order of 50 or larger. Therefore, if the mean particle size in a suspension is, say, 1 μm, a steel wool matrix with the filament diameter of 50 μm or greater can be used. However, even a much coarser matrix can often give results comparable to those obtained with very fine matrices.

Flow velocity

As stated, the highest possible ratio between the magnetic force and the erosion hydrodynamic force should be used so that the greatest extraction efficiency can be achieved. Although low flowrates are usually employed, the magnitude of the flowrate depends on the magnetic susceptibility of the solids to be removed. For the removal of ferromagnetic impurities from the steel-mill waste waters, interstitial velocities of up to 0.3 ms^{-1} are used, while in the applications where fine paramagnetic particles are to be collected the flowrate can be as low as 1×10^{-3} ms^{-1}.

On the other hand, since low flow velocity implies a low throughput of a magnetic filter the final choice of the flowrate will be determined by the interplay between the requirements on the purity of the effluent and on the capacity of the filter.

IV.3.2.2 Removal of magnetic impurities from mixtures

The removal of unwanted magnetic components from mixtures is a well-established application of magnetic separation, and a wide spectrum of minerals can be upgraded by reducing the content of magnetic impurities, thanks to the advent of efficient and cost-effective dry magnetic separators. Although dry separation is an answer in many applications, wet high-gradient magnetic separation has to be used for the beneficiation of finely ground minerals, and the brightening of kaolin or dolomite by HGMS is a well proven and established technology.

The common feature of all techniques for the removal of magnetic admixtures is that, as in magnetic filtration, the brute force approach can be used. This means that high magnetic field, fine matrix and low flow velocities (or low roll rotation velocities) should be used to achieve efficient removal of impurities.

Such a simple approach of maximizing the magnetic force and minimizing the competing (erosion or centrifugal) forces can, however, seldom be employed because there is generally a loss of the valuable non-magnetic fraction with the magnetic product.

Further, as discussed earlier, the combination of a high magnetic field and of a fine matrix may lead to the formation of the clusters containing both magnetic and non-magnetic particles. These clusters tend to form a surface mat that increases the resistance of the matrix to the flow and ultimately leads to the matrix blockage.

The loss of the non-magnetic product in the mags can be tolerated in exceptional cases, where the purity of the final, non-magnetic product is of prime importance. In most applications, the excessive loss of the product into the mags is not acceptable and careful selection of the operating conditions is required. The matrix should be sufficiently coarse to avoid mechanical straining; on the other hand, it should be reasonably fine so that the magnetic force is sufficiently large. Similarly, the value of the magnetic field must not be too large, and the slurry velocity must be sufficiently high to avoid the entrainment of the product in the magnetic fraction.

Although a simple theory of magnetic filtration teaches that efficiency of filtration increases as $\mathbf{v}_m/\mathbf{v}_o$ (or as $\mathbf{B}/\mathbf{v}_o$), its direct extension to the removal of impurities from mixtures is often not permissible, although in some cases the agreement of theory with experiment was achieved.

IV.3.2.3 The concentration of minerals

The concentration of valuable, weakly magnetic minerals from finely ground ores is the largest application of high-gradient magnetic separation. The application of modern magnetic separation techniques to the beneficiation of minerals has a great potential, since magnetic separation can produce high-quality concentrates from a large variety of minerals.

The objectives of the concentration of minerals by magnetic separation are threefold, as shown in Fig. 4.8.

a) High recovery of the valuable metal into the magnetic concentrate. In this case high magnetic field, in combination with low flow velocity, should be used. A fine matrix is usually recommended in this application but in many cases it has been observed that recovery can be lower than when a coarser matrix has been used.
b) High grade of the magnetic concentrate. In this case recovery is of secondary importance and the operating conditions should be chosen in such a way that selectivity of the separation process is maximum. In order to achieve high selectivity, only a moderate magnetic field should be used. An excessively high magnetic field causes the formation of flocs comprising both magnetic and non-magnetic particles and also enhances the possibility of the formation of the surface mat deposit.

This consideration also dictates the use of a reasonably coarse matrix and correct choice of the matrix is the most critical step in achieving a selective separation. High flow velocity of the slurry also enhances the grade of the magnetic product, although the increased flowrate can hardly rescue the situation when an unsuitable matrix is used.

c) High recovery and high grade. In this case, the engineer faces a problem of great complexity since contradicting requirements are often imposed on the operating conditions. The selection of the matrix, and of the values of magnetic field and flow velocity must be performed on the basis of the experimental programme, in which other variables that influence the separation process (e.g. pulp density, dispersion, etc.) are simultaneously investigated.

Although the problem does not lend itself to a general discussion, some fundamental points can be made. The matrix must not be too fine (e.g. steel wool) or too coarse (diameter of collector exceeding say 5 mm); the former is non-selective and often gives low recoveries while the latter usually exhibits insufficient recovery.

If the ore is not too fine, a high flow velocity usually improves the grade without reducing the recovery. For samples containing excessive amounts of fines (smaller than e.g. 10 μm) high flowrates will, however, reduce recovery to an unacceptable level. Although reasonably large magnetic fields should be used, experience shows that magnetic induction exceeding 1.0 T usually does not improve recovery even of fine, very weakly magnetic minerals, while the grade of the magnetic fraction, as a rule, drops.

IV.3.3 Cyclic or continuous separation ?

The choice between cyclic and continuous high-gradient magnetic separators is dictated mainly by the concentration of magnetically coverable particles in the feed. If the mass yield into the magnetic fraction is smaller than, say, 1 per cent, a cyclic separator can be a suitable device. If the feed rate is not too high, feed time can be longer than the time required for the valve setting and magnetic field decay and rise. Therefore the duty cycle of such a batch separator can be sufficiently large.

However, in the majority of applications of HGMS in mineral processing, concentration of the magnetic component is larger than, 1 per cent; furthermore, high flowrates are usually required to achieve a high grade of concentrate and high capacity. Under these conditions the matrix would be loaded within a matter of seconds and the duty cycle would be a low percentage. The situation can be illustrated by the following example. If we assume that a cyclic magnetic separator, with a canister diameter of 1 m, is used to recover uranium from the Witwatersrand residues, we can assume that the separation process is carried out under the following conditions: a ball matrix with a filling factor of $\mathbf{f} = 0.6$, matrix height 0.2 m, slurry density 40 per cent solids (by mass), solids density 2700 kg m^{-3}, interstitial flow velocity $\mathbf{v}_0 = 0.12$ m s^{-1}. The volume of the matrix is 0.09 m^3 and the permissible matrix loading $\mathbf{L}_M = 200$ kg m^{-3} (eq. 4.10). Therefore, 18 kg of the magnetic product can be captured before the allowed loading is exceeded. Since the mass yield into the magnetic fraction is usually 10 per cent, 180 kg of solids can be fed into the separator, an amount that corresponds to 0.34 m^3 of slurry. The volumetric flow $\mathbf{V}_0$ through the separator can be calculated from the relation

$$V_0 = v_0 S(1 - f)$$

where $\mathbf{S}$ is the cross-sectional area of separator, in our case $\mathbf{S} = 0.785$ m^2. For $\mathbf{v}_0 = 0.12$ m s^{-1} and $\mathbf{f} = 0.6$ we obtain $\mathbf{V}_0 = 0.04$ m^3 s^{-1}, so that it would take about 8 seconds for 0.34 m^3 of slurry to flow through the separator.

Since magnetic field decay required a minimum of 5 to 10 seconds, and since flushing takes 15 seconds (valve opening and closing also requires a few seconds), the resulting duty cycle of a cyclic magnetic separator when recovering uranium and gold from the leaching residues would be about 20 per cent. Duty cycle for the beneficiation of iron ore would be even lower.

Such a low utilization factor for an expensive cyclic device makes the matter costly and any additional expenditure for more expensive continuous separators and a more demanding operation becomes justified. A continuous magnetic separator that operates with 100 per cent duty cycle can treat a wide range of minerals with varying content of the magnetic particles. The comparison of cyclic and continuous high-gradient magnetic separators as applied in mineral processing is given in Table 4.4.

It is clear that the role of cyclic magnetic separators in the minerals industry is rather limited, as the duty cycle and the versatility of such devices is low, while the specific running and capital costs are high. Although the position of cyclic separators in the beneficiation of minerals is, in some exceptional cases (kaolin purification), well established, the most important field of application of cyclic separators is the filtration of cooling water in thermal power plants and of the steel mill waste water. The comparison of feed times for a cyclic high-gradient magnetic separator in different applications is summarized in Table 4.5. The analysis of the matrix loading and the feed time in a cyclic magnetic separator is given in Section IV.5.5.

TABLE 4.4
The comparison of cyclic and continuous magnetic separators

	Cyclic	Continuous
Duty cycle	Low to medium	100 %
Capacity	Low to medium	High
Processing cost per unit capacity	High	Moderate
Versatility	Low	High
Mechanical complexity	Low	High

TABLE 4.5
Comparison of some operating parameters for a cyclic magnetic separator

Application	Magnetic fraction in the feed (kgm^{-3})	Flow velocity (ms^{-1})	Feed time (hours)
Filtration of the thermal power plant water	0.0001	0.24	1000
Steel mill waste water	0.1	0.12	1.5
Kaolin purification	1 ÷ 3	<0.01	0.1
Hematite ore	250	0.16	0.0003
Uranium-gold leach residues	50	0.12	0.002

IV.3.4 The test programme in magnetic separation

Magnetic separation, and wet high-gradient magnetic separation in particular, is greatly influenced by a number of variable factors, some of which can have a decisive effect on the acceptance or the rejection of magnetic separation as a beneficiation technique. To establish optimum conditions, a test programme is usually necessary, which involves a considerable number of tests.

The factors that influence the efficiency of high-gradient magnetic separation to the greatest extent can be placed in four groups:

I. a) magnetic susceptibility of particles to be recovered
   b) magnetic susceptibility of gangue mineral
   c) particle size distribution
   d) degree of liberation

II. a) magnetic field
   b) matrix type
   c) matrix height
   d) matrix loading

III. a) pulp flow velocity
   b) pulp density
   c) pulp feed rate
   d) volume and flowrate of rinse water
   e) volume and flowrate of flush water

IV. Pulp pretreatment:
   a) desliming
   b) dispersion
   c) pH
   d) scalping

In low-intensity magnetic separation the length of the separation zone, depth of the slurry stream and the speed of drum rotation also affect the performance of a drum magnetic separator.

Because each of these variables can considerably affect the outcome of the magnetic separation tests, the following sections will discuss the role of variables in detail.

IV.4 MAGNETIC FIELD

IV.4.1 Magnetic field in a matrix magnetic separator

The magnitude of a magnetic field generated by the magnet of a magnetic separator is perhaps the most important parameter that signifies the potential of a separator for the treatment of various classes of magnetizable slurries. Some manufacturers of matrix magnetic separators claim that their machines generate a magnetic field that is, at first glance, higher than one would expect from the design of the magnet. The reason for this discrepancy is that the value of magnetic induction in the vicinity of a matrix element is sometimes stated, instead of the background magnetic induction.

The total external magnetic induction in the vicinity of a matrix element is given by the superposition of an essentially uniform (on the scale of the interaction distance of a particle and a matrix element) background magnetic induction $\vec{\mathbf{B}}_o$ and the induction due to the matrix element $\vec{\mathbf{B}}_m$ which is position-dependent:

$$\vec{B} = \vec{B}_o + \vec{B}_m \tag{4.3}$$

The induction $\mathbf{B}_m$ causes the distortion of the background induction $\mathbf{B}_o$ in the vicinity of each matrix element. This matrix material is magnetized to magnetic polarization $\mathbf{J}$ which is proportional to background magnetic induction $\mathbf{B}_o$ until it saturates. Above the saturation value $\mathbf{J}$ is constant, equal to $\mathbf{J}_S$ and does not increase further with $\mathbf{B}_o$.

The magnetic induction $\mathbf{B}_m$ which the matrix element generates is a direct result of its magnetization and has the form (in polar coordinates centered at the center of the matrix element, see Fig. 4.9) [S24]:

$$\vec{B}_m = \vec{J} f(\frac{a}{r}, \gamma) \tag{4.4}$$

where the dimensionless function $\mathbf{f}$ depends on the radius $\mathbf{a}$ of the matrix element and on the distance $\mathbf{r}$.

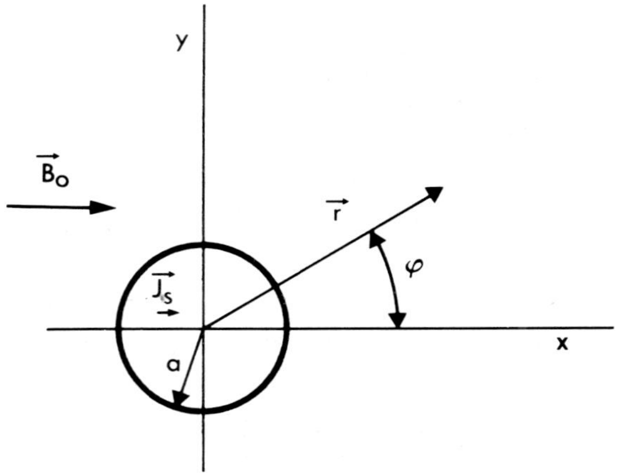

Fig. 4.9 The geometrical configuration for a calculation of magnetic induction in the vicinity of a matrix element.

Under the assumption that the matrix material is saturated, the total magnetic induction at point **r** outside the matrix can thus be calculated from the expression:

$$B = \left[B_o^2 + J_s^2 \left(\frac{a}{r}\right)^4 + 2B_oJ_s \left(\frac{a}{r}\right)^2 \cos 2\varphi\right]^{\frac{1}{2}} \qquad (4.5)$$

where **a** is the radius of a matrix element and $\mathbf{J_s}$ is the saturation polarization of the matrix material. Consequently the magnetic induction on the surface of the matrix element, for φ = 0, is equal to the algebraic sum of the saturation polarization of the matrix material (usually 2 T) and of the background magnetic induction. Thus, to generate 2.5 T on the matrix, a background magnetic induction of 0.5 T is needed. If the magnetic field is measured, for instance, by a Hall probe with a finite thickness, then 2.5 T measured 100 μm from the surface of the matrix element would require the background magnetic induction of 0.62 T.

In view of the fact that a very modest background magnetic field is needed to generate magnetic induction in the excess of 2 T in the intermatrix space, the information provided by some manufacturers on the magnetic field should be taken with a certain degree of caution.

IV.4.2 The effect of magnetic field strength on the performance of a magnetic separator

The performance of a magnetic separator is, to a great extent, determined by the magnetic field strength, although the role of the latter is often overstressed. As discussed in Chapter I, a necessary condition for a successful separation of magnetizable particles from a mixture is that the magnetic force is greater than the sum of the competing mechanical forces (eq. 1.1). Since the magnetic force acting on magnetizable particles is directly proportional to the magnetic field strength, one would expect that an increase in magnetic induction would also increase the efficiency of magnetic separation. Although such is the case in many fundamental studies, where a single-collector approach is used, there are numerous applications, mainly in mineral processing, where such a simple formula fails and an increase in the magnetic field can often have a negative effect on the performance of a magnetic separator.

In order to draw more general conclusions about the role of the magnetic field strength in magnetic separation we shall discuss separately three fundamental classes of the application of magnetic separation.

IV.4.2.1 Magnetic filtration

As discussed in Section IV.3.2.1 the brute force approach can be employed in magnetic filtration because the objective is to separate the magnetizable solids from the fluid stream at the highest possible efficiency. The magnitude of the magnetic field strength is determined by the magnetic susceptibility of the suspended solids, their size distribution, the required throughput of the filter and by the requirement imposed on the quality of the filter effluent.

The usual picture in magnetic filtration is that the efficiency of the removal of solids initially increases with the increasing field strength up to a certain point, after which filtration efficiency remains constant. The value of this threshold magnetic induction is a function of the flow velocity, particle size and magnetic susceptibility of the solids to be removed.

A typical example of such a straightforward relationship between the magnetic field and the separation efficiency is shown in Fig. 4.10. The figure illustrates the effect of the magnetic field on the efficiency of removal of suspended solids from the municipal waste waters seeded with magnetite. As seen, an almost total breakthrough occurred at 0.01 T, while the induction of about 0.05 T was necessary to hold the seeded flocs on the matrix. Such an "all-or-none" type response is typically obtained with magnetic seeding techniques [A19].

As shown in Fig. 4.11 a similar pattern occured in the removal of the suspended solids from steel mill waste waters. Magnetic induction of 0.3 to 0.4 T is usually sufficient and higher values of the magnetic field do not increase removal efficiency [H34, T9]. The threshold magnetic field is an order of magnitude greater than that needed in the seeding technique, as the particles in steel mill waste waters are usually much finer (~ 1 μm). Although the main constituents are iron oxides such as Fe_2O_3 and

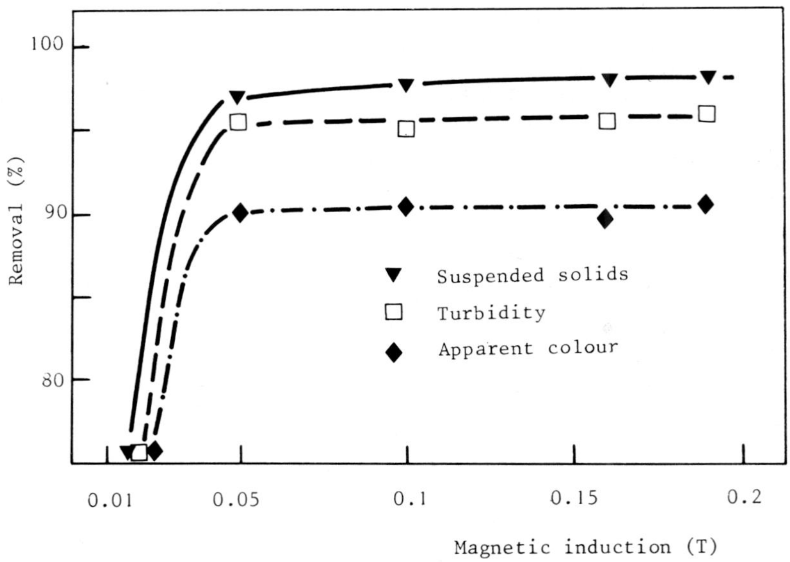

Fig. 4.10 The efficiency of the solids removal from municipal waste water seeded with magnetite, as a function of magnetic field. Flow velocity 0.06 ms^{-1}, magnetite concentration 300 mg/l, alum concentration 100 mg/l (after Allen et al. [A19]).

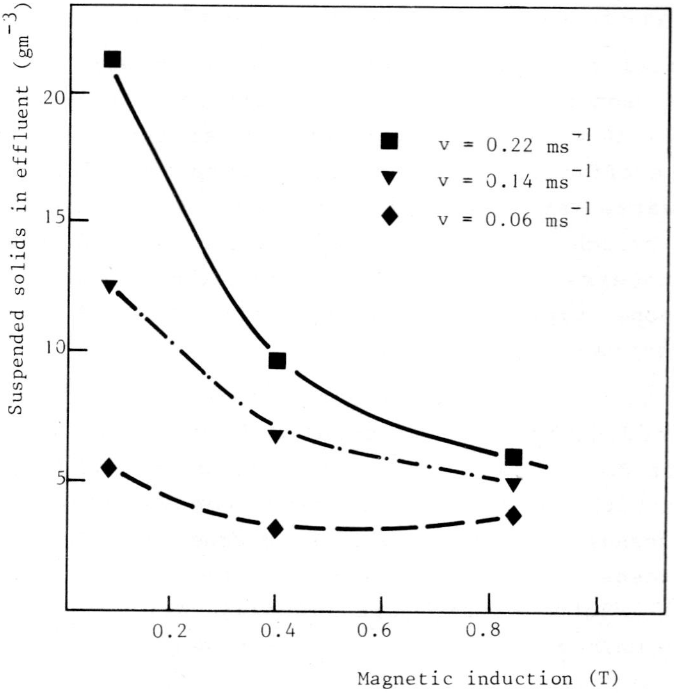

Fig. 4.11 The concentration of the suspended solids in the effluent from the magnetic filter, as a function of magnetic field. Feed: scale pit effluent in hot rolling mill. (After Harland et al. [H34]).

Fe_3O_4, many other components used as raw materials in the process also appear.

These two examples typify the situation usually encountered in magnetic filtration, namely, that the removal of solids is not generally improved by increasing the field strength over a certain value. In fact it has been observed in numerous cases that excessive magnetic field strengths can be detrimental to the contaminant removal. This phenomenon is illustrated in Fig. 4.12 where the turbidity of magnetite-seeded suspension is plotted as a function of the volume of sewage water being processed, at various field strengths. It can be seen that for a clean matrix the turbidity is almost independent of the applied magnetic field and that at magnetic inductions exceeding 1 T the turbidity rises as the volume of processed water increases, a trait not seen at lower fields [P19].

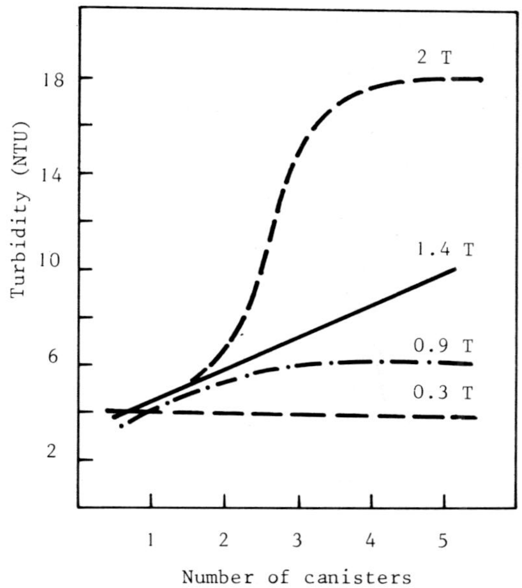

Fig. 4.12 Turbidity of the magnetite-seeded water effluent as a function of the volume of water processed (after Petrakis et al. [P19]).

The magnetic filtration of weakly magnetic solids requires much higher magnetic fields to obtain sufficiently high efficiency of the particle removal, as demonstrated by Fig. 4.13, where the retention of the asbestos fibres is plotted as a function of the magnetic field. Although high magnetic induction (1.6 T) is needed to achieve removal efficiency exceeding 90 per cent, the removal of particles greater that 70 per cent can be obtained at the fields as low as 0.2 T.

IV.4.2.2 Removal of magnetic impurities from mixtures

Although high magnetic field strengths are often desirable to bring about an efficient removal of magnetic impurities from mixtures, care must be exercised in the selection of the magnetic induction. In contrast to magnetic filtration where excessive magnetic field strength usually does not affect the performance of a magnetic filter, results of the removal of magnetic impurities from mineral mixtures can often be impaired by the undesirable

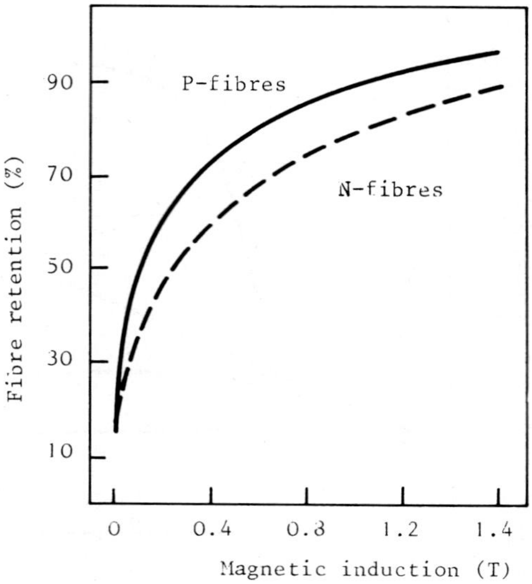

Fig. 4.13 The retention of asbestos fibres as a function of the magnetic field (after Timbrell [T10]).

losses of the valuable non-magnetic product due to entrainment in the magnetic fraction.

A typical case is illustrated in Figs. 4.14 and 4.15. It can be seen that the specification of the magnetic field strength for coal cleaning by magnetic separation requires a trade-off between the quality and the recovery of the product. Fig. 4.14 shows that at high magnetic inductions, substantial portions of the ash and sulphur were removed, at the expense, however, of coal recovery, as shown in Fig. 4.15.

Another example of mechanical entrainment of coal in a magnetic fraction as a function of the removal efficiency of sulphur and ash is depicted in Fig. 4.16.

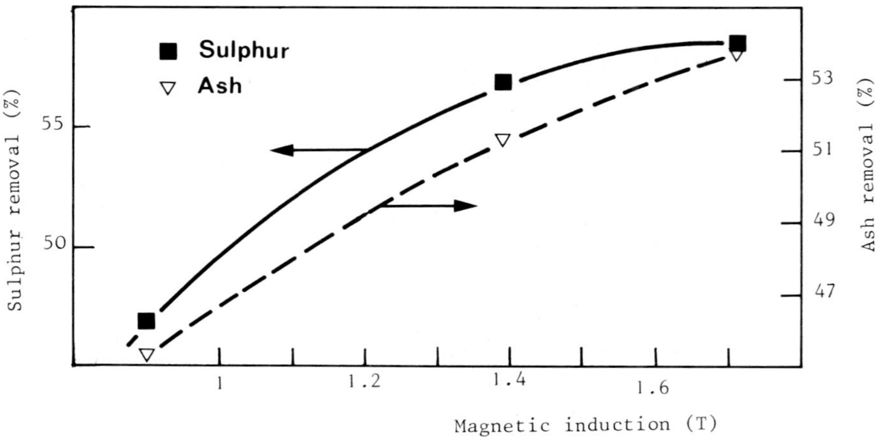

Fig. 4.14 The effect of the magnetic field strength on sulphur and ash removal from coal by a magnetic separator (after Wechsler et al. [W25]).

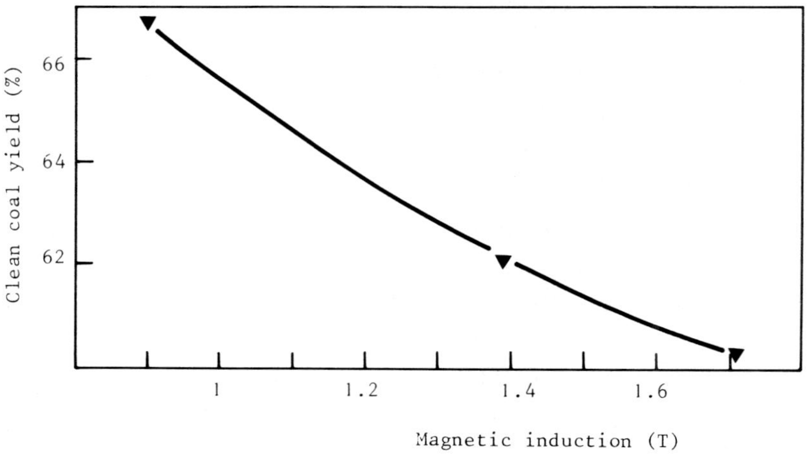

Fig. 4.15 Effect of magnetic field strength on clean coal yield from a magnetic separator (after Wechsler et al. [W25]).

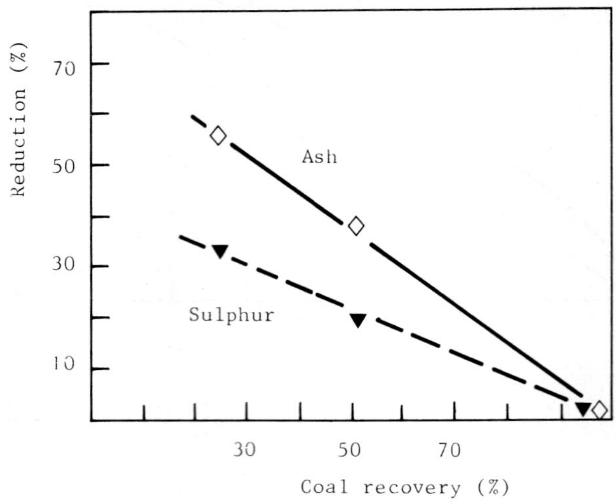

Fig. 4.16 The reduction in sulphur and ash versus recovery of clean coal (after Luborsky [L17]).

As in magnetic filtration, the removal efficiency of paramagnetic impurities does not increase indefinitely with increasing magnetic field strength, as predicted by simple theoretical models, but the levelling of the filtering efficiency at higher magnetic fields is usually observed; this is illustrated in Figs. 4.17 and 4.18.

The brightening of kaolin by the removal of weakly magnetic discolouring impurities is frequently cited as one of those applications in which the efficiency of the impurity removal depends only on the ratio $\mathbf{v}_m/\mathbf{v}_o$ (or $\mathbf{B}_o/\mathbf{v}_o$). Thus for high efficiency of extraction of the discolouring bodies, the ratio of the applied magnetic field to the slurry velocity must be kept high.

Although some experimental results [W4] support this conclusion of the theoretical model, in other investigations into kaolin brightening the advantages of using high magnetic field strengths were not so obvious. Fig. 4.19 shows the effect of the magnetic field on the iron oxide concentration in kaolin, for a fine fibre matrix [Y1]. The effect of the field on the removal of the impurities is very small, in the range between 0.8 T to 2 T, particularly in view of the fact that the head concentration was 2 per cent Fe_2O_3. For coarser fibre matrices the efficiency of extraction was independent of the field for values of magnetic induction greater than 0.8 T.

Similar behaviour of kaolin in a magnetic separator was observed by Svoboda [S4] and the effect of magnetic field on the iron oxide concentration is depicted in Fig. 4.20.

It can be seen that with a steel wool matrix magnetic induction greater than 0.8 T does not improve the grade of the kaolin product, while with a ball matrix this threshold magnetic field is higher, namely 1.2 T. Moreover, the extraction efficiency

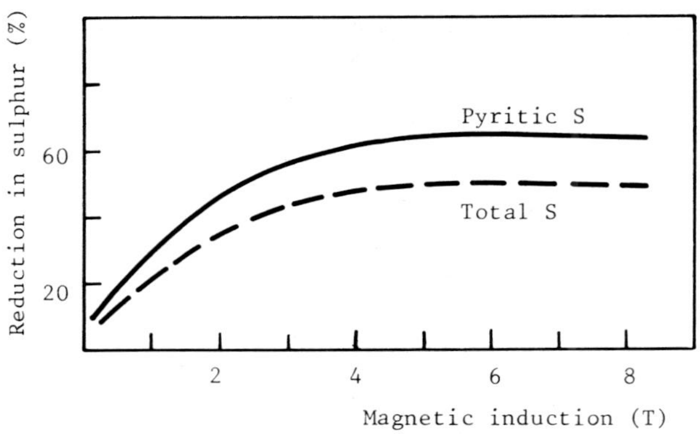

Fig. 4.17 The effect of the magnetic field strength on sulphur removal from coal by a magnetic separator (after Luborsky [L17]).

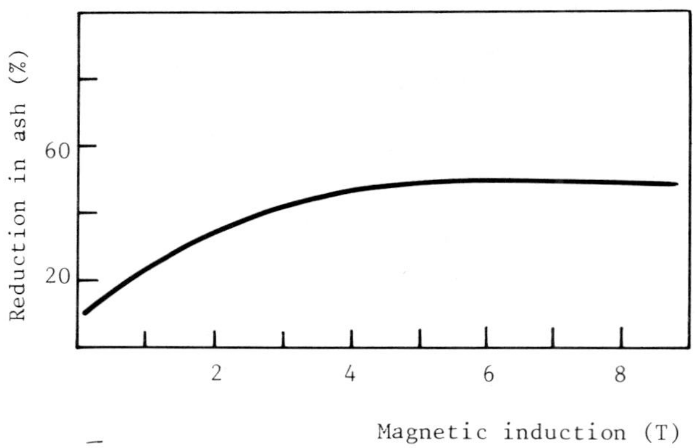

Fig. 4.18 The ash reduction as a function of magnetic field, from a coal by magnetic separator (after Luborsky [L17]).

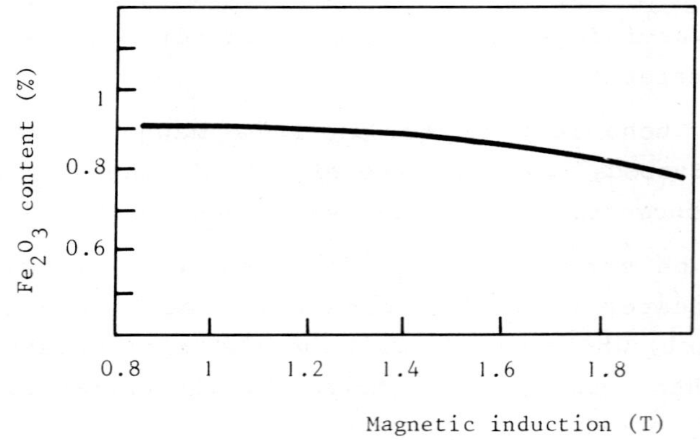

Fig. 4.19 The effect of magnetic field strength on the concentration of iron oxide in kaolin (China) (after Ying [Y1]).

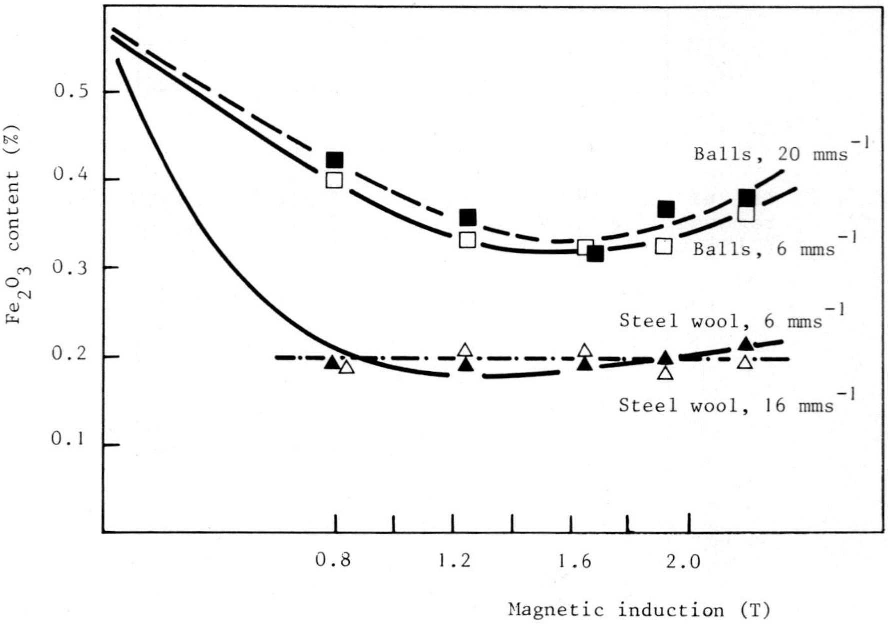

Fig. 4.20 The effect of magnetic field strength on the concentration of iron oxide in kaolin (South Africa) [S4].

tends to decrease at high fields (2.2 T). The adverse effect of high magnetic field strengths on the recovery of kaolin is illustrated in Fig. 4.21.

It will be seen that a trade-off between the quality and recovery of a non-magnetic product has to be made when selecting the magnitude of the applied magnetic field. In most applications there is always a threshold magnetic field strength which marks the maximum field for which the grade of the non-magnetic product still improves. Magnetic field strength smaller or equal to the threshold field should be used and the choice of the operating magnetic field will be further narrowed by the recovery-grade relationship.

In addition to the conflict between grade and recovery there are valid economic constraints on the selection of the magnetic field strength. In the design of magnetic separation systems the capital investment and processing costs increase rapidly as magnetic induction is increased. It is therefore desirable to use the lowest practical magnetic field strength consistent with acceptable product quality.

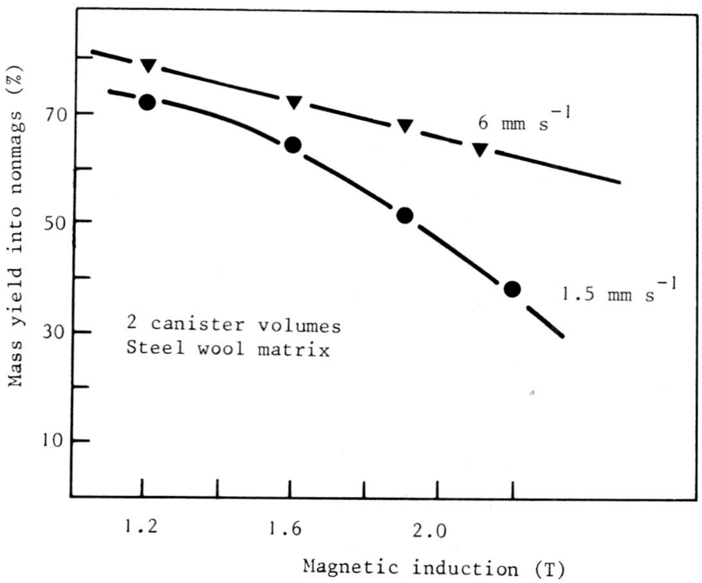

Fig. 4.21 The effect of magnetic field strength on kaolin yield from a magnetic separator [S4].

IV.4.2.3 The concentration of minerals

The selection of the operating magnetic field in the concentration of weakly magnetic minerals is a sensitive task and only a well defined objective of separation will determine the approach taken. The traditional conception is that an increase in the magnetic field strength will result in an increase of recovery of weakly magnetic minerals. There are, however, a number of factors that can cause the recovery into the magnetic product to decrease with increasing magnetic field, and there is usually an optimum magnetic field strength at which the recovery reaches the maximum, while with a further increase the recovery either remains constant or even decreases. This situation is illustrated in Fig. 4.22 where the recovery of wolframite is plotted as a function of the magnetic induction.

A similar effect of the magnetic field strength on the recovery of hematite is shown in Figs. 4.23 and 4.24. It can be seen that a modest magnetic induction of 0.5 T is sufficient to obtain very high recoveries.

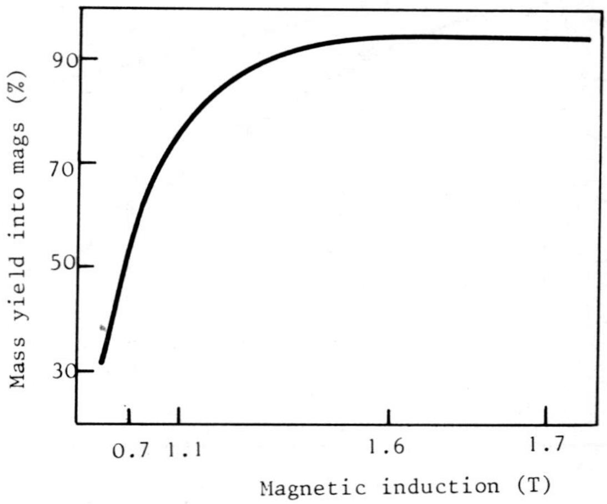

Fig. 4.22 Recovery of wolframite as a function of magnetic induction between grooved plates used as a matrix (after Masoner [M11]).

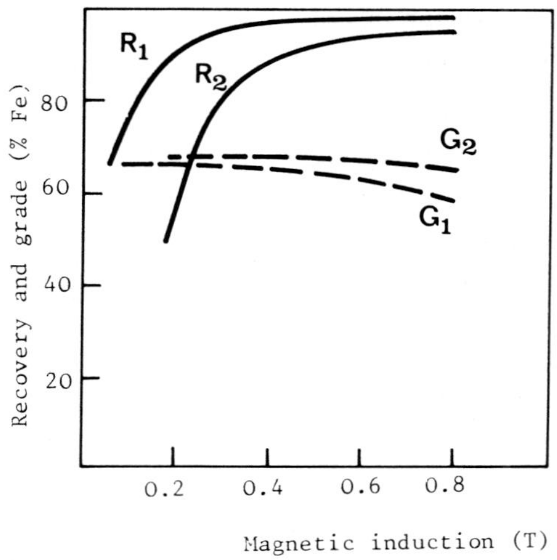

Fig. 4.23 Recovery of hematite (CRVD Brazil) and the grade of the magnetic concentrate as a function of magnetic field. 1 - longitudinal configuration, 2 - transverse configuration, matrix: 10 mm rods, matrix height: 80 mm. (After Hencl [H1]).

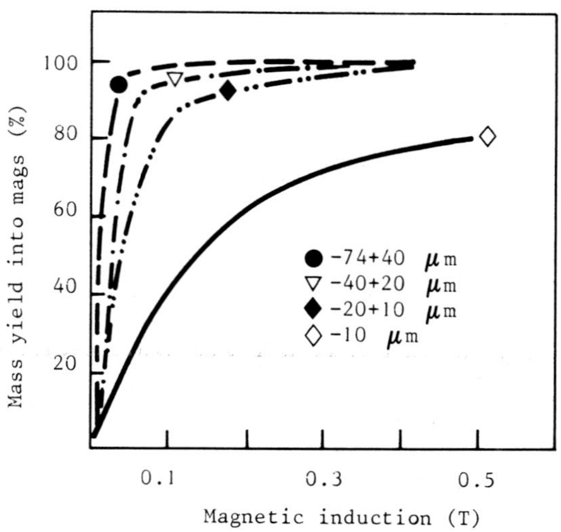

Fig. 4.24 Recovery of size fractions of hematite (Tou-Anzan, China) as a function of magnetic field. Slurry flowrate 0.08 ms^{-1}. (After Jurng et al. [J3]).

Similarly, magnetic field has a weak effect on the recovery of very weakly magnetic copper and lead minerals from the flotation tailings, as shown in Fig. 4.25. However, this observation applies mainly to the results obtained in L-configuration, while in T-configuration recovery tends to increase with the increasing magnetic field, at least at low values, as demonstrated by Figs. 4.23 and 4.25.

In numerous cases the recovery of a weakly magnetic mineral, after reaching a maximum at some intermediate value of the magnetic field strength, starts to drop with a further increase of the field. Fig. 4.26 depicts the recovery of cassiterite by HGMS as a function of the background magnetic field. It can be seen that after reaching a maximum at 1 T the recovery of cassiterite decreases with increasing magnetic field.

We have seen that an increase in the magnetic field strength initially leads to an increase in the recovery, but after reaching a maximum, the recovery tends to stay constant or even to decrease.

However, in contrast to magnetic filtration or to the removal of magnetic impurities from industrial minerals, recovery of a valuable mineral is not the only criterion of the performance of a

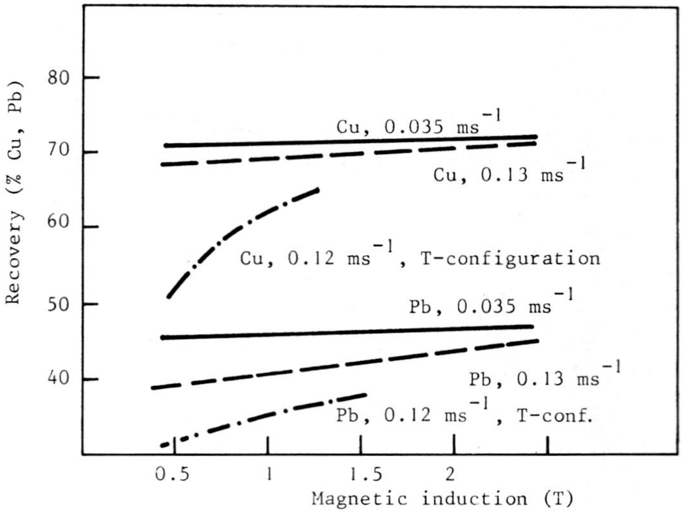

Fig. 4.25 Recovery of copper and lead from the Tsumeb flotation tailings by HGMS, as a function of magnetic induction. Matrix: 6 mm balls, matrix height 200 mm. [S30].

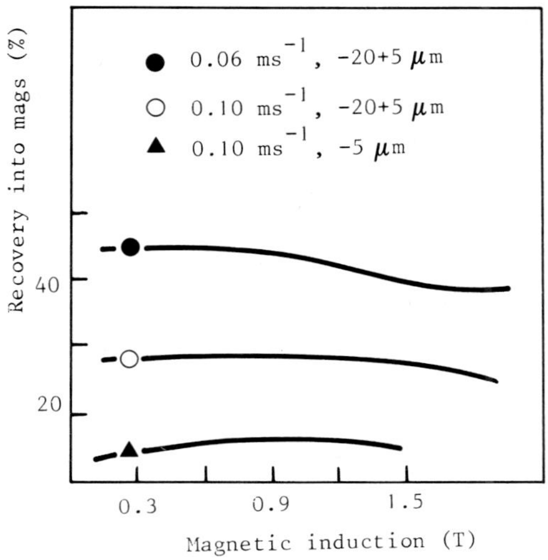

Fig. 4.26 Recovery of cassiterite (Bolivia) by HGMS as a function of magnetic field (after Arellano and Zambrana [A21]).

magnetic separator when applied to concentration. The grade of the magnetic product is usually equally important and therefore the magnetic field strength has to be selected in such a way that recovery and grade are optimised simultaneously.

The grade of the magnetic concentrate usually decreases with increasing magnetic field, as a consequence of non-selective magnetic capture caused by the entrainment of non-magnetic gangue particles in clusters of magnetizable particles, and by the capture of non-liberated grains containing magnetic inclusions. A decrease in the grade of the magnetic concentrate with increasing magnetic field is well established and represents a characteristic feature of a majority of magnetic separation applications to minerals concentration, as shown in Figs. 4.23 and 4.27. However, in some cases it has been observed that at low fields the grade of the mags initially increases and only after reaching the maximum does it begin to drop with an increasing magnetic field. This situation is illustrated by Figs. 4.28 and 4.29.

Obviously a selection of the applied magnetic field strength must be based on the interplay between the recovery and the grade of the concentrate and on the relative weight factors ascribed to these two metallurgical factors. The importance of the trade-off between recovery and grade is shown in Fig. 4.30.

We have shown that an increase of the applied magnetic field does not necessarily result in an increase of the extraction efficiency of a magnetic separator, although it could be expected from a simplified theoretical picture of magnetic separation. There are several factors that cause that too high magnetic field can be detrimental to the performance of a magnetic separator.

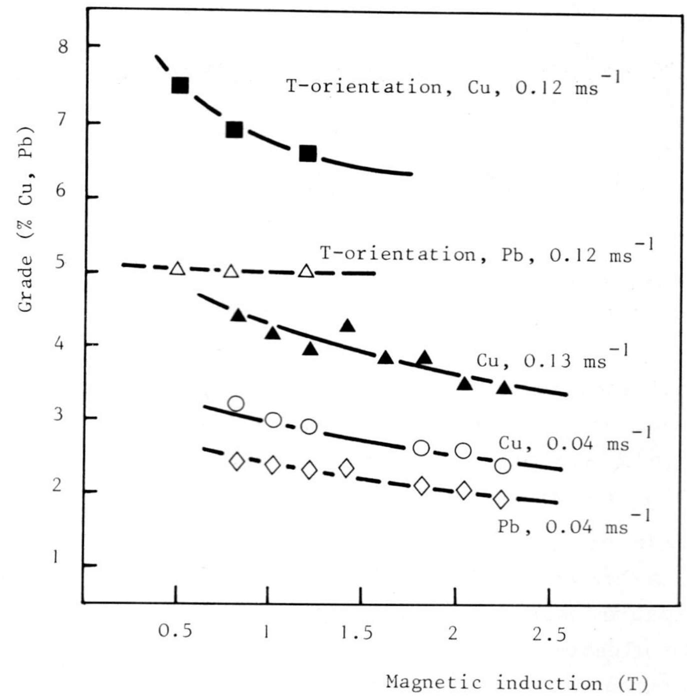

Fig. 4.27 The effect of the magnetic field on the grade of the magnetic concentrate from Tsumeb flotation tailings. Matrix: 6 mm balls. [S30].

We have seen in Section I.3.4 that most minerals treated by magnetic separation are not purely paramagnetic but possess a certain degree of magnetic ordering, which results in the field-dependence of their magnetic susceptibility. As follows from the Honda-Owen equation (eq. (1.54)), magnetic susceptibility usually decreases with increasing applied magnetic field, and the drop in the value of magnetic susceptibility, can often be quite appreciable, as Figs. 1.26 and 4.31 reveal. Therefore, the reduction in the magnetic susceptibility that accompanies the increase in the magnetic field can sometimes result in stagnation or even decrease in the recovery.

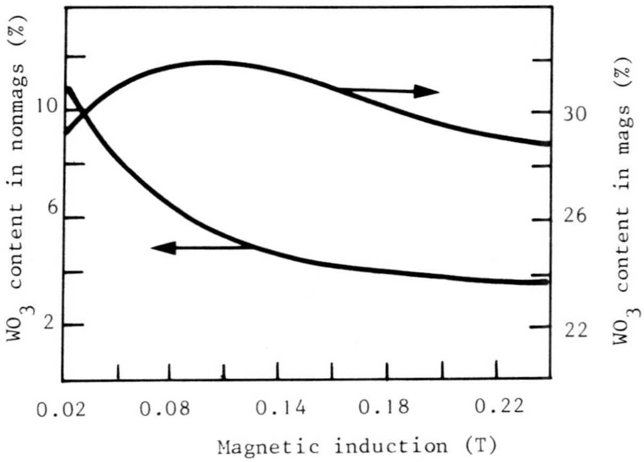

Fig. 4.28 The effect of the magnetic field on the grade of the magnetic concentrate from tungsten-tin slimes. Matrix: 8 mm balls. (After Karmazin et al. [K21]).

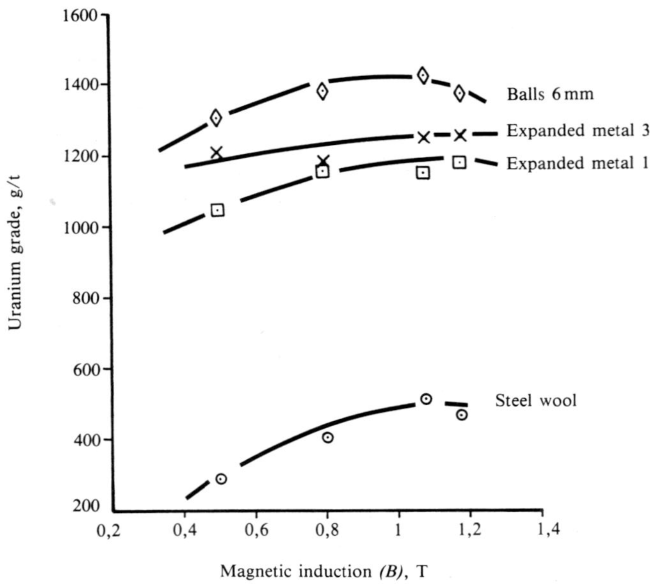

Fig. 4.29 The effect of magnetic field on grade of the magnetic concentrate from the uranium-gold leaching residues [S29].

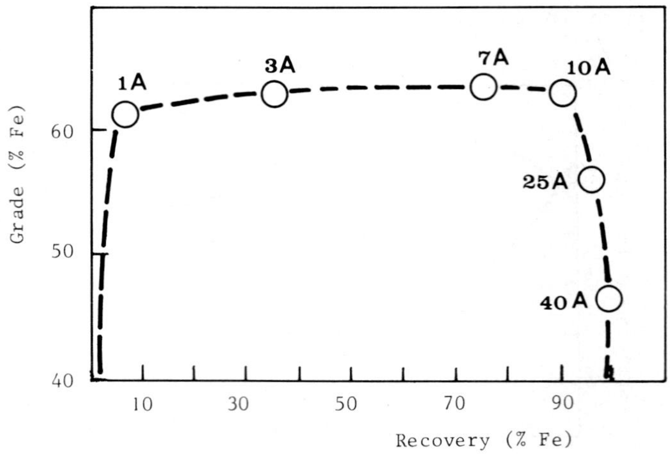

Fig. 4.30 The effect of magnetic field (expressed as the electric current in the coils of Jones magnetic separator) on recovery and grade of the magnetic concentrate from the hematite ore (after Stone [S16]).

Another factor that adversely affects recovery is the decrease of the surface of the matrix collectors available for the capture of paramagnetic particles, when the applied magnetic field is increased; the situation is depicted in Fig. 3.23. It is evident that, by increasing the field, the loading of the matrix increases as a consequence of a smaller available surface area of the matrix. The filling factor of the matrix increases as does the interstitial velocity of the slurry. Although it may lead to an improvement in the grade of the magnetic concentrate, it also results in a reduction of the recovery.

As discussed in Section I.4.5, an increase in the applied magnetic field beyond the saturation of the matrix material can cause a decrease in the gradient of the magnetic field. Again, a potentially beneficial effect of a higher magnetic field is offset by a possible reduction of the gradient.

High magnetic fields often induce the formation of clusters of magnetic particles through the process of magnetic flocculation. Since the formation of such flocs is usually non-selective this causes not only the deterioration of the grade of the magnetic concentrate but it may also cause the recovery to drop as a result of excessive matrix loading and of increased interstitial velocity of the slurry through the matrix.

We have thus shown that a higher magnetic field does not necessarily mean better results. The optimum operating magnetic field has to be determined in each particular case in such a manner that the recovery-grade relationship satisfies the objectives of the application.

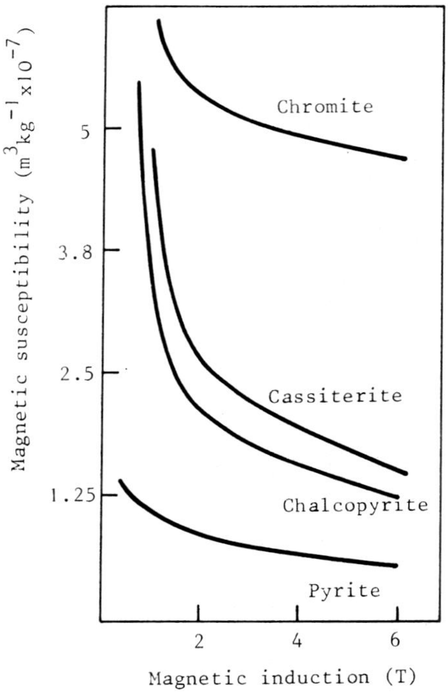

Fig. 4.31 Magnetic field dependence of magnetic susceptibility of assorted minerals (after Kukuck [K22]).

IV.5 MATRIX

IV.5.1 The optimum size of a matrix

The importance of a correct choice of matrix for the concentration of valuable mineral component cannot be overemphasized. The geometry and size of the matrix play a decisive role in the achievement of optimum recovery and concentrate grade and determine matrix loading, a factor most crucial in a large-scale magnetic separator.

In spite of the importance of a correct choice of matrix very little attention has been paid to the theoretical analysis of the optimum matrix size and geometry. The attention was mainly concentrated on maximizing the magnetic force in a single-collector approach, with the aim of applying the rules of the matrix selection to multi-collector matrices.

The magnetic force has been shown to reach a maximum [O1,A1] for a ratio of collector radius to particle radius, **a/b**, equal approximately to three. This maximum was found to be fairly flat, and it has been confirmed experimentally [C30] that, in the range **a/b** from 1 to 10, the change in the efficiency of capture is negligible. It has also been shown [C31] that the optimum ratio of collector size to particle size is not constant. Consequently the criterion that the particle size and collector size should match [O1, A1] is somewhat complicated. Similarly, the theoretical analysis [M26] of the spatial distribution of magnetic force implies that the optimum ratio of collector size to particle size lies between 5 and 10, this being a function of magnetic induction, particle size and type of matrix. The optimum value for **a/b** has also been shown to increase strongly with increasing amounts of deposited particles.

Although the results of the single-collector-single-particle studies are useful for a better understanding of the dynamics of a magnetizable particle in the vicinity of an isolated magnetized collector, their validity for a real matrix is very limited.

Attempts to maximize the ratio of traction magnetic force to drag force usually result in particle capture by a straining mechanism. If **a/b** < 10, the straining mechanism is dominant and leads to the formation of a surface mat of deposit on the matrix. This layer of deposit causes high resistance to the flow, increases the losses of the non-magnetic product in those applications where magnetic impurities are removed, or severely impairs the grade of the magnetic concentrate in those applications where the valuable minerals are concentrated into the magnetic fraction. Naturally, the straining mechanism is undesirable and should be avoided in the design of magnetic separators. For straining to be minimized, **a/b** should be larger than 10, but preferably should range from 100 to 500. The presence of a high level of mechanical straining in magnetic filtration of fine magnetite through a nickel sponge is shown in Fig. 4.32. It can be seen that straining is absent and particles are deposited uniformly throughout the whole filter for a coarse sponge with **a/b** = 1250, while with finer sponges, where **a/b** = 500 and 150, particles are predominantly distributed in the vicinity of the filter inlet [E3].

In the application of HGMS to the beneficiation of minerals, the matrix must be matched in such a way that recovery and grade are at their maximum simultaneously. The optimum collector size

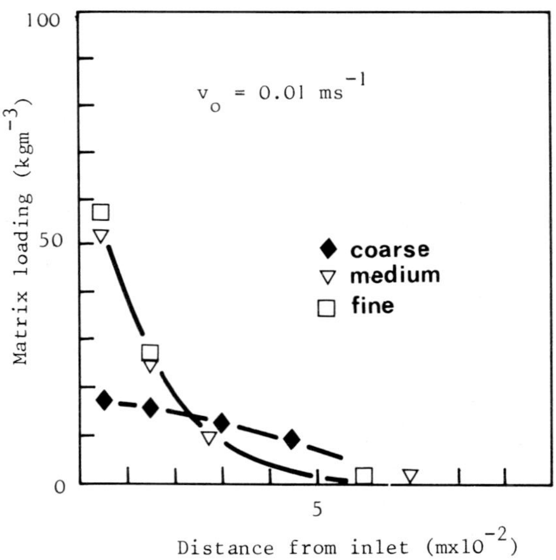

Fig. 4.32 Distribution of matrix loading $\mathbf{L_M}$ as a function of distance **L** from filter inlet (after Endoh and Yamaguchi [E3]).

becomes, therefore, a function of the size distribution and composition of the ore, and of the hydrodynamic conditions in the intermatrix space. The usual practice is a tedious trial-and-error search for the best shape and size for the matrix, during which several fundamental rules must be kept in mind. With a decreasing collector diameter, the recovery of the magnetic fraction usually increases whereas the grade of the concentrate decreases. With very fine matrices, the mass yield into the magnetic concentrate becomes very high and the grades too low, so that the magnetic concentrate may not differ very much from the feed. On the other hand, with fine matrices which possess a high level of mechanical capture, a simultaneous increase in the flow velocity and applied magnetic field allows one, to a certain degree, to improve the grade by washing out the mechanically entrained non-magnetic material while maintaining recovery of the magnetic material.

A number of experimental results showed that the most satisfactory metallurgical performance is obtained at the ratio of collector size to particle size ranging from 100 to 300. This is demonstrated in Fig. 4.33, in which the grade of the magnetic concentrate is plotted as a function of the diameter of the matrix

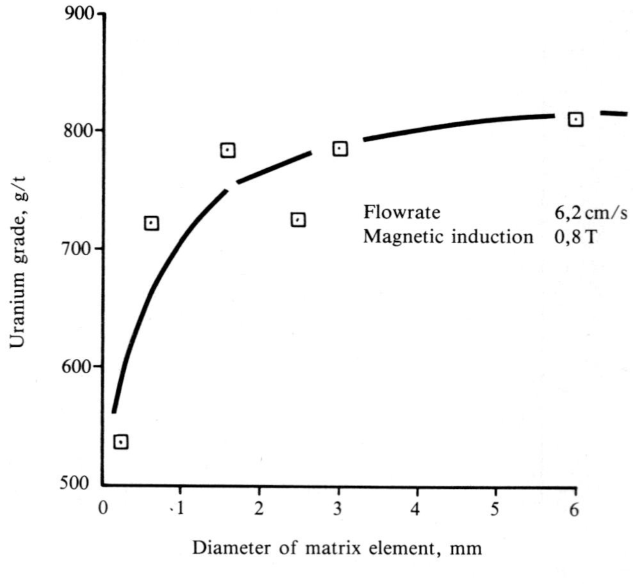

Fig. 4.33 The grade of the magnetic concentrate of uranium-gold leach residues, as a function of the diameter of the matrix element. The mean particle diameter: 16 μm [S29].

element. It can be seen that, for fine matrices, the grade is very low, higher grades being obtained for fairly coarse matrices for which **a/b** is about 350. For these coarse matrices the recovery is also sufficiently high.

The influence of the size of the balls and rods used as a matrix in the recovery of tungsten from the flotation tailings and iron from siderite ore is shown in Figs. 4.34 and 4.35, respectively.

Fig. 4.36 illustrates the effect of the size of loosely packed rods on the metallurgical results and comparison with the expanded metal is also shown.

It can again be seen that with the increasing dimensions of the matrix collectors the recovery decreases and the selectivity of separation increases, although in some cases, for over-large spheres, the grade decreases, as demonstrated by Figs. 4.34 and 4.37.

The influence of the size of the ball collectors on the beneficiation of uranium-gold leaching residues and on copper-lead flotation tailings was thoroughly investigated by Svoboda [S54]. Some of the results are summarized in Figs. 4.37, 4.38 and 4.39.

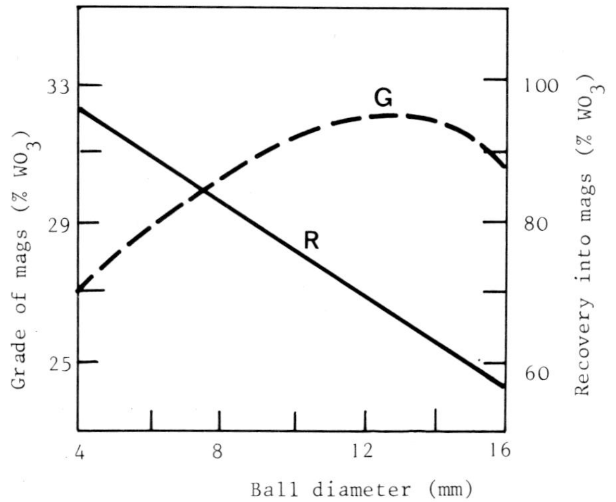

Fig. 4.34 The effect of ball size on magnetic separation of tungsten (after Karmazin et al. [K21]).

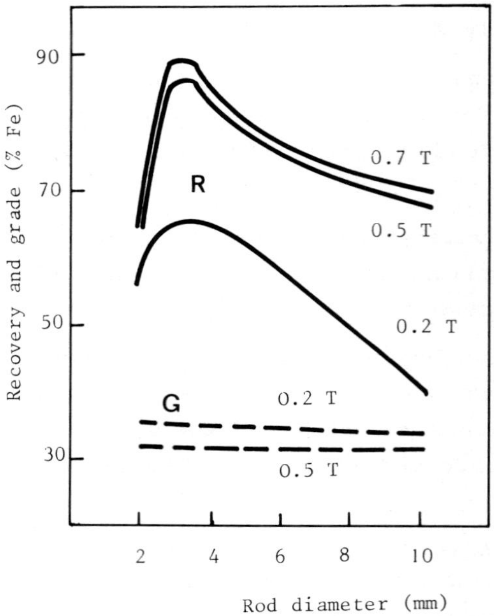

Fig. 4.35 The effect of rod diameter on magnetic separation of siderite ore (after Hencl and Šebor [H35]).

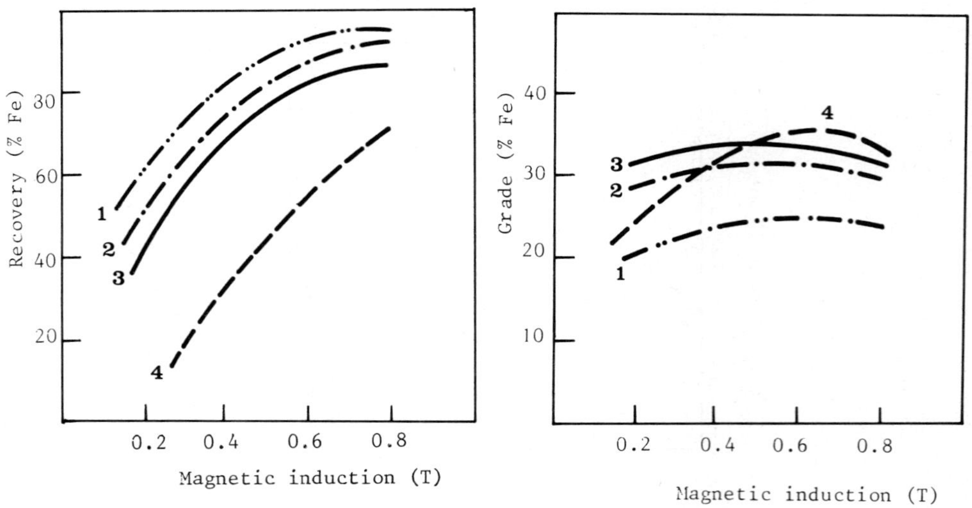

Fig. 4.36 The effect of size and shape of the matrix collectors on separation of geothite ore (Lengede); 1-rods, diameter 8 mm, length 8 mm, 2-rods, diameter 10 mm, length 10 mm, 3-rods, diameter 10 mm, length 18 mm, 4-expanded metal, wire diameter 1 mm (after Hencl [H1]).

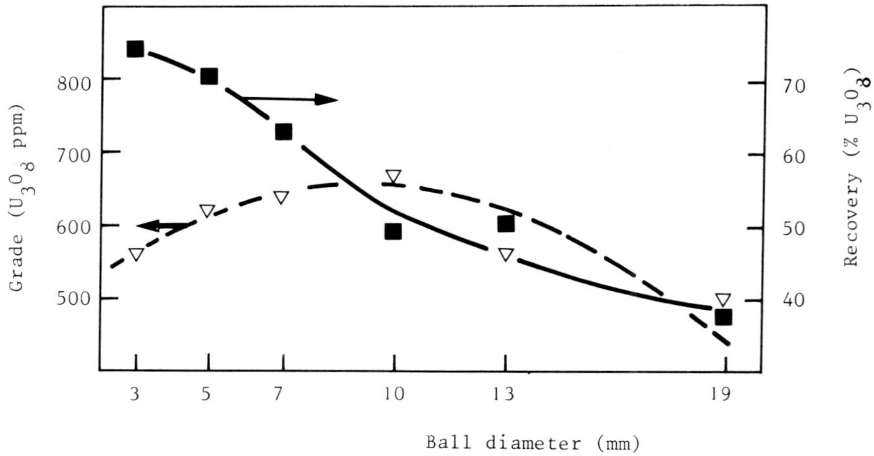

Fig.4.37 The effect of ball diameter on magnetic separation of uranium from cyanidation residues. Magnetic induction 2.2 T, interstitial velocity 0.035 ms^{-1}, L-configuration. [S54].

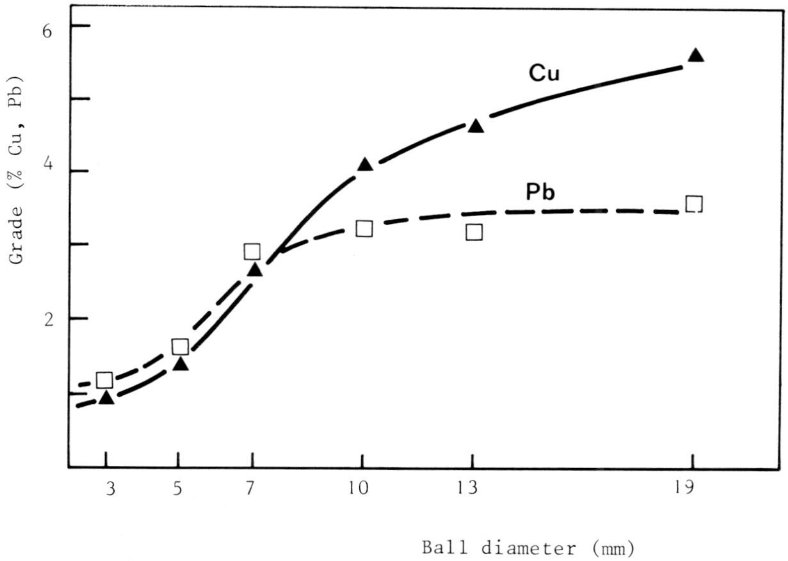

Fig. 4.38 The influence of the ball diameter on the grade of the magnetic concentrate from the Tsumeb flotation tailings, magnetic induction 2.2 T, interstitial velocity 0.035 $m\ s^{-1}$ [S54].

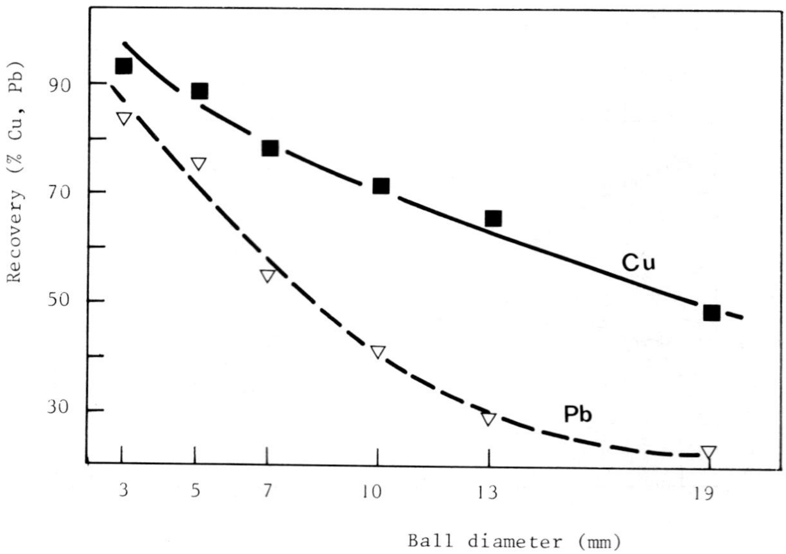

Fig. 4.39 The influence of ball diameter on recovery of copper and lead from the Tsumeb flotation tailings [S54].

Matrices of dimensions similar to those used to treat uranium-gold-leach residues were found to give the best results in pyrite removal from coal by HGMS [D9], with magnetic susceptibility one order of magnitude lower than shown above.

The basic rules for the optimization of the matrix size fail even in simple filtration experiments [H36]. As shown in Fig. 4.40 the filtration efficiency of ferric hydroxide flocs with a mean diameter of 3 μm, using a ball matrix, was comparable to that of fine mesh, the filaments of which have a diameter approximately 20 times smaller than that of the balls.

A fine steel wool matrix is usually used in kaolin brightening where very fine and very feebly magnetic impurities must be removed. However, high efficiency of separation can be achieved even with a much coarser ball matrix, as shown in Fig. 4.20. Owing to a much lower porosity of the ball matrix, the results are not exactly comparable, the interstitial velocity of the slurry through the ball matrix being much higher than that through the steel wool matrix. For the same interstitial velocities, the performance of the ball and steel wool matrices would be even closer.

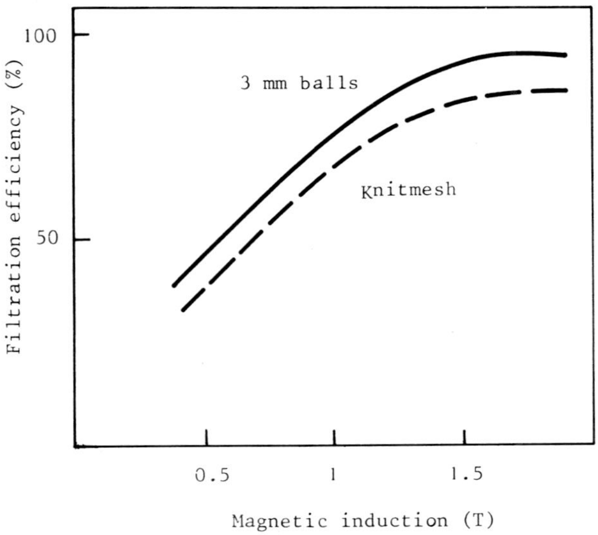

Fig. 4.40 The filtration efficiency of ferric hydroxide flocs as a function of magnetic field for ball and wire matrices (after Harding and Baxter [H36]).

IV.5.2 The optimum shape of a matrix

The criteria for a selection of the optimum shape for matrix collectors are even more vaguely defined than those for the size of the collectors. The choice of the matrix depends on the characteristics of the slurry undergoing the separation process and is often dictated by a personal experience and by the availability of a suitable matrix type. Ideally, the matrix should offer maximum collection efficiency, minimum magnetic reluctance and fluid impedance, together with an acceptable capability for being cleaned. Various shapes of a matrix are used, such as grooved plates, rods, balls, wire mesh, expanded metal and steel wool.

Besides the size of a collector, any matrix is characterized by its filling factor **f** or by porosity (or void fraction) ϵ :

$$f = \frac{\text{volume of matrix}}{\text{volume of separation chamber}}$$

and $\epsilon = 1 - \mathbf{f}$. The porosity or filling factor are expressed either as a fraction of 1 or by percentage. The porosity of the matrix bed changes with time as particle accumulation within the matrix increases in such a way that

$$\epsilon = \epsilon_i - \frac{\sigma_d}{1 - \epsilon_b} \tag{4.6}$$

where σ_d is the specific deposit (volume of deposited matter per unit volume of matrix), ϵ_b is the porosity of a deposit and ϵ_i is the porosity of clean matrix. Some typical values of the filling factor for assorted types of matrices are given in Table 4.6.

TABLE 4.6
Filling factors of assorted matrices

Matrix	Filling factor **f**
Spheres	0.50 to 0.64
Steel wool	0.01 to 0.05
Expanded metal	0.12 to 0.27
Wire mesh	0.15 to 0.25
Grooved plates	0.50 to 0.80

The filling factor is determined mainly by the size of the matrix collectors, by the way the matrix elements are arranged in the separation chamber and by the degree of the wall effect. For instance, if equal spheres are allowed to settle to form a packing, a filling factor of about 0.60 is obtained. If this "loose" random packing is tapped, it gives a 'close' random packing of a filling factor of 0.64 which is impossible to increase [L18].

The performance of various shapes of matrix collectors depends primarily on the particle size distribution and magnetic properties of the solids to be separated. We have seen that fine matrices (e.g. steel wool) are suitable for the removal of very fine and very feebly magnetic particles; in most applications in mineral processing these matrices are not suitable owing to their non-selectivity and susceptibility to clogging.

For the majority of applications in minerals treatment coarse matrices (e.g. balls, expanded metal, grooved plates, etc.) are the best choice but ample experimental work shows that there is no universal type of a matrix that would meet all requirements in all applications. Each type of matrix has its advantages and disadvantages and the attitude of various investigators to a given type of matrix ranges from total rejection to unreserved acceptance. A typical example is a ball matrix that has been used extensively in the magnetic filtration of water in power plants [H37] and in magnetic concentration of minerals [C7, S30, K21]. In other applications, the performance of the ball matrix was found to be unsatisfactory [M11].

Fig. 4.41 compares the metallurgical performance of a ball matrix with the rod matrix applied to the recovery of iron from the hematite ore. It can be seen that the rod matrix gives considerably lower recovery than the ball matrix, while the grade of the concentrate differs only slightly.

A detailed investigation of the performance of various matrices, as applied to recovery of uranium and gold from leaching residues, was conducted by Svoboda [S29]. A list of the matrices used is given in Table 4.7, together with the filling factor **f** for each matrix and the dimensions of the matrix elements.

Several matrices are shown in Figs. 4.42 to 4.44; the selected results are summarized in Figs. 4.29, 4.45 to 4.47.

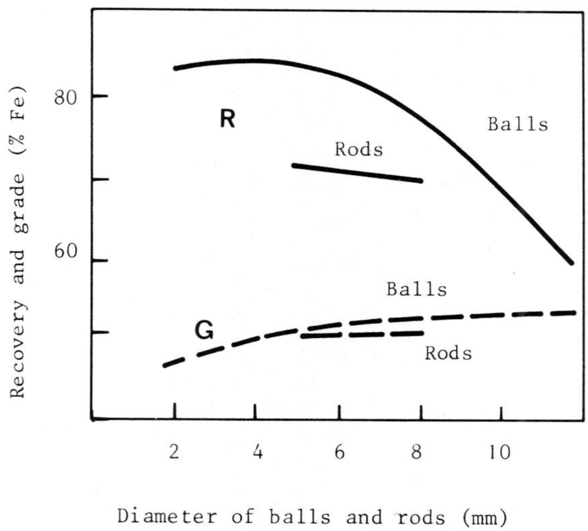

Fig. 4.41 The influence of the size of balls and rods on metallurgical performance of a magnetic separator (after Karmazin et al.[K15]).

TABLE 4.7

List of matrices applied to recovery of uranium and gold from leaching residues [S29]

Matrix	Filling factor, f	Diameter of element, mm	Size of aperture, mm
Balls	0.59	3.2	
Balls	0.55	6	
Steel wool	0.04		
Expanded metal 1	0.27	4	
Expanded metal 2	0.14	1.9	
Expanded metal 3	0.20	1.3	
Woven mesh 8	0.21	0.9	1.25
Woven mesh 9	0.18	2.5	10
Woven mesh 10	0.21	1.6	5
Screens 11	0.42	4.8	2
Screens 12	0.31	2.8	1.5
Screens 13	0.28	1.8	1.0
Woven mesh 15	0.19	0.63	1.0
Woven mesh 16	0.20	1.0	2.0
Knitmesh 26	0.05	0.25	200 stitches per m
Knitmesh 27	0.05	0.25	160 stitches per m
Grooved plates 4R	0.64	4 grooves per inch	
Grooved plates 8R	0.83	8 grooves per inch of the plate width	

Fig. 4.42 Matrices used for magnetic separation (from left to right): grooved plate, expanded metal 2, Knitmesh.

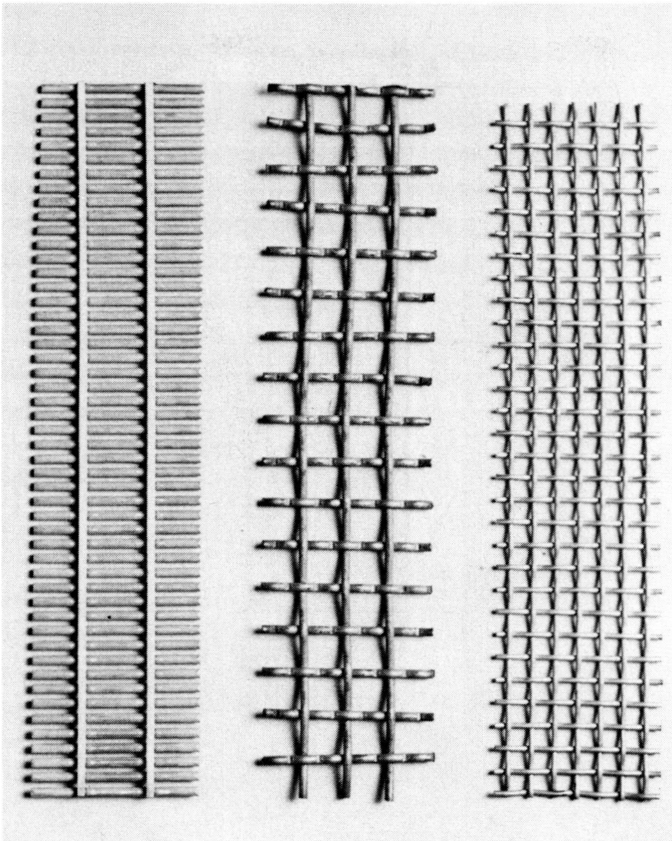

Fig. 4.43 Matrices used for magnetic separation (from left to right): screen 12, mesh 9, mesh 10. Description is given in Table 4.7.

It follows from these diagrams that the highest recoveries were obtained with fine matrices (e.g. steel wool, Knitmesh, mesh 8, expanded metal 3). On the other hand, the grade of concentrate obtained with such matrices was low, owing to their lower degree of selectivity. For this reason, and because of their high degree of mechanical capture and clogging, these fine matrices are unsuitable. Grooved plates, which are often used in the magnetic separation of iron ores, gave generally lower grades and lower recoveries than balls or mesh. Similarly, the metallurgical performance of expanded metal, a popular matrix in some industrial separators has often been inferior to that of other matrices (see Figs. 4.36, 4.45 and 4.29).

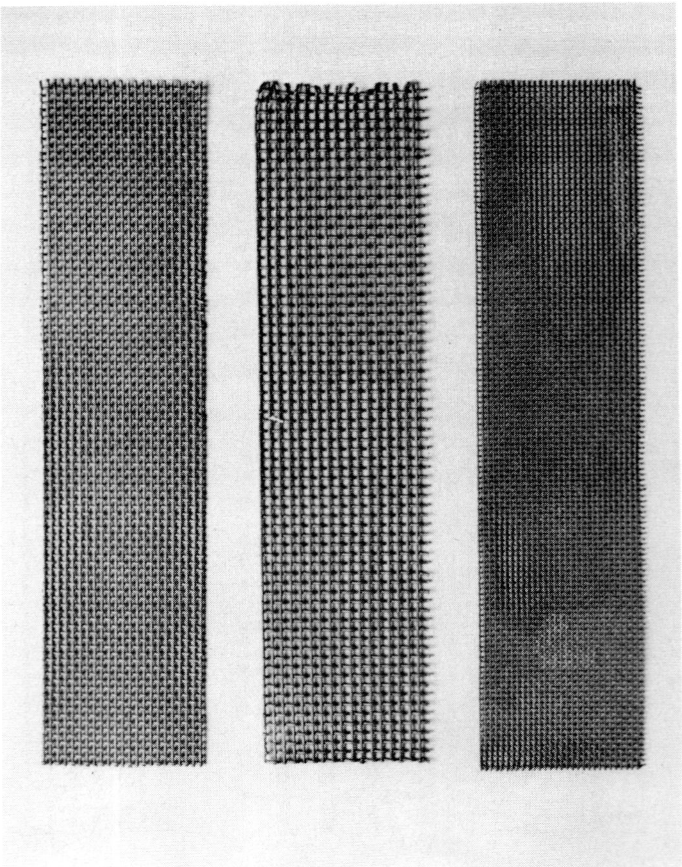

Fig. 4.44 Matrices used for magnetic separation (from left to right): mesh 15, mesh 16, mesh 8. Description is given in Table 4.7.

Steel balls and mesh 10 were found to give the best results, not only in the recovery of uranium and gold, but also in other applications, where the particles are even finer and less magnetic [S30]. The rod matrix was found to give the best results in the beneficiation of iron ores [H1, S31, C8], although in other applications their performance was found to be inferior [M11, K15].

There are many interesting geometries that can be considered for the matrix collectors. Only a few of them were discussed in this paragraph and a great deal of research is necessary in this area.

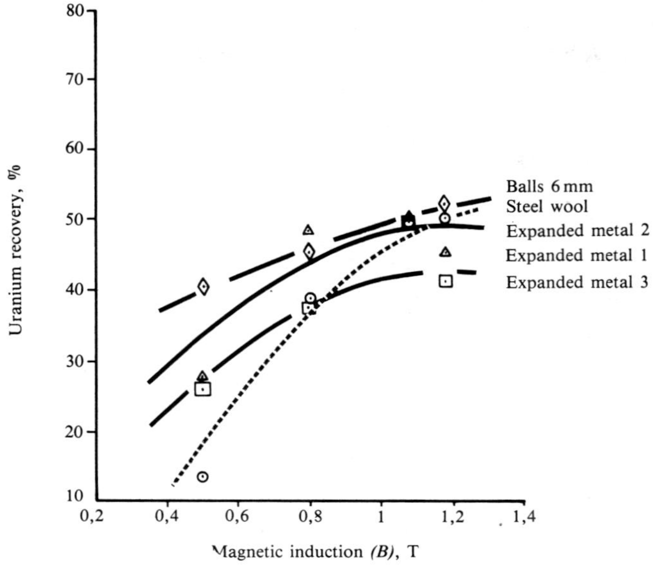

Fig. 4.45 Recovery of uranium from leaching residues by assorted matrices (see Table 4.7); interstitial velocity 0.06 m s^{-1}, T-configuration [S29].

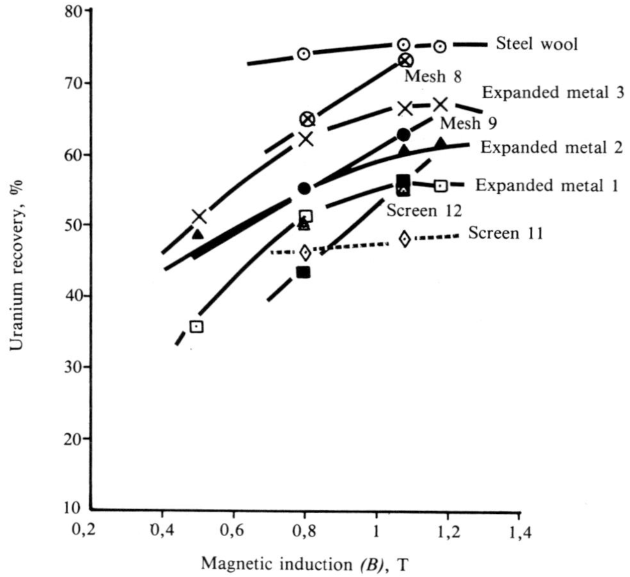

Fig. 4.46 Recovery of uranium from leaching residues by assorted matrices (see Table 4.7); interstitial velocity 0.06 m s^{-1}, T-configuration [S29].

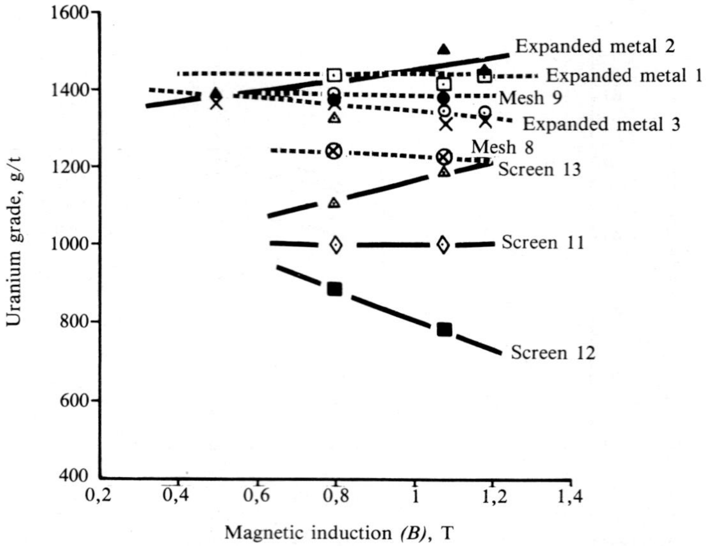

Fig. 4.47 The grade of the magnetic fraction of the uranium leaching residue for different matrices (see Table 4.7); interstitial velocity 0.06 m s^{-1}, T-configuration [S29].

IV.5.3 Mechanical capture

As in any application of deep bed filtration, a certain portion of the particles fed into the magnetic separator will be mechanically captured. Mechanical entrapment increases both for coarse particle sizes and for finer matrices, as can be seen in Fig. 4.48.

The magnetic recovery $\mathbf{R}_{mag}$ can then be evaluated from the equation

$$R_{mag} = \frac{R - R_{mech}}{1 - R_{mech}}$$

where $\mathbf{R}$ is the total recovery and $\mathbf{R}_{mech}$ is the recovery due to mechanical capture.

Fig. 4.49 shows the effect of particle size on mechanical capture for assorted matrices listed in Table 4.7, and Fig. 4.50

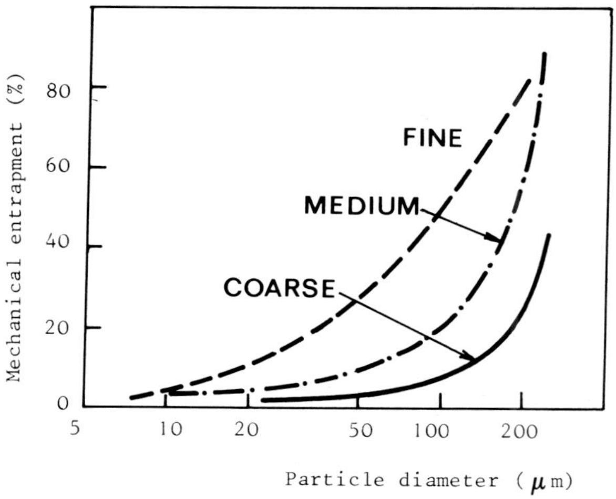

Fig. 4.48 Mechanical entrapment of hematite particles in steel wool matrix; zero applied field, flow velocity 0.06 m s^{-1} (after Clarkson and Kelland [C31]).

depicts the influence of diameter of the matrix element on mechanical capture for various size fractions. The dependence of mechanical capture on the flowrate for selected matrices is shown in Fig. 4.51 where it can be seen that, at a superficial velocity of the order of 0.07 ms^{-1}, which corresponds to the industrially acceptable flowrate for the concentration of valuable minerals, the correction for mechanical entrapment is minimized for all the matrices investigated. As the flowrate decreases, mechanical capture increases only slightly for a matrix of balls, whereas it increases dramatically for a matrix of expanded metal.

For small particles (approximately 8 μm) mechanical capture is practically constant for fine and coarse matrices. With increasing particle size mechanical entrapment rises only slightly for coarser matrices (e.g. balls or mesh 10), whereas the increase is very sharp for fine matrices, as can be seen in Fig. 4.49. For very fine particles (smaller than 17 μm and 25 μm), mechanical entrapment rises only very slightly with decreasing size of the matrix elements; for particles larger than 53 μm, mechanical capture becomes pronounced for fine matrices.

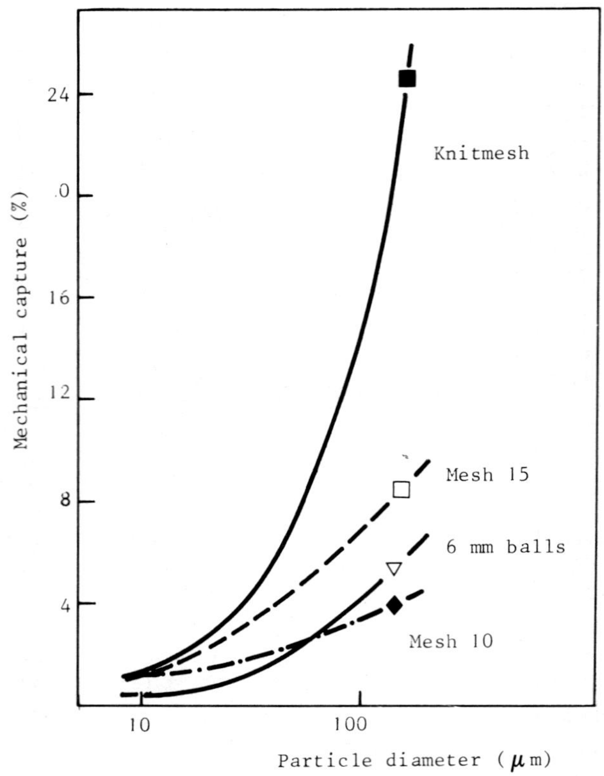

Fig. 4.49 Mechanical capture of uranium-gold leaching residues, versus particle size for various matrices at the superficial velocity of 0.02 ms^{-1} [S29].

Mesh and expanded metal matrices were found to be resistant against the mechanical entrainment in dry separation of coarse pyroxenite ore. Fig. 4.52 shows that for these fine matrices, mechanical entrainment does not exceed 2 per cent and for coarse mesh and 19 mm ball matrices entrapment was well below 1 per cent. Furthermore, mechanical entrapment was found to be independent of flowrate, which indicates that a certain degree of matrix clogging will always take place unless there are other mechanisms of matrix cleaning.

The 7 mm ball matrix, in contrast to wet separation of fine slurries, causes a high degree of mechanical capture, which depends on the air velocity.

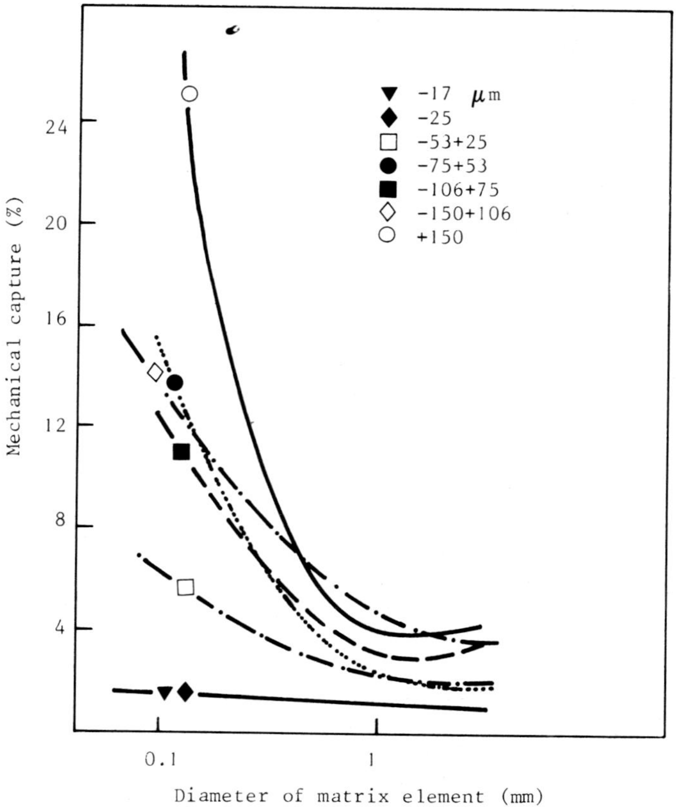

Fig. 4.50 Mechanical capture versus diameter of the matrix elements for various size fractions of uranium-gold leaching residues, at superficial velocity of 0.02 ms^{-1} [S29].

IV.5.4 Matrix height

The height of the matrix is an important parameter that determines the retention time of particles in the magnetic field, the probability of their collision with the matrix collectors and thus the probability of their capture. Similarly, input power and the mass of a magnetic separator, pressure drop and a degree of the blockage of the matrix also depend on the height of the matrix.

In early HGMS machines the height of the matrix was usually 80 to 100 mm. In current cyclic or continuous HGMS separators, the

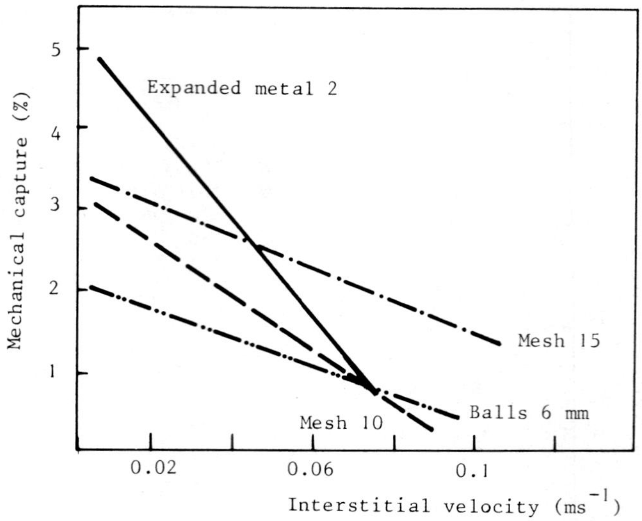

Fig. 4.51 Mechanical entrapment of uranium-gold leaching residues as a function of flowrate for assorted matrices[S29].

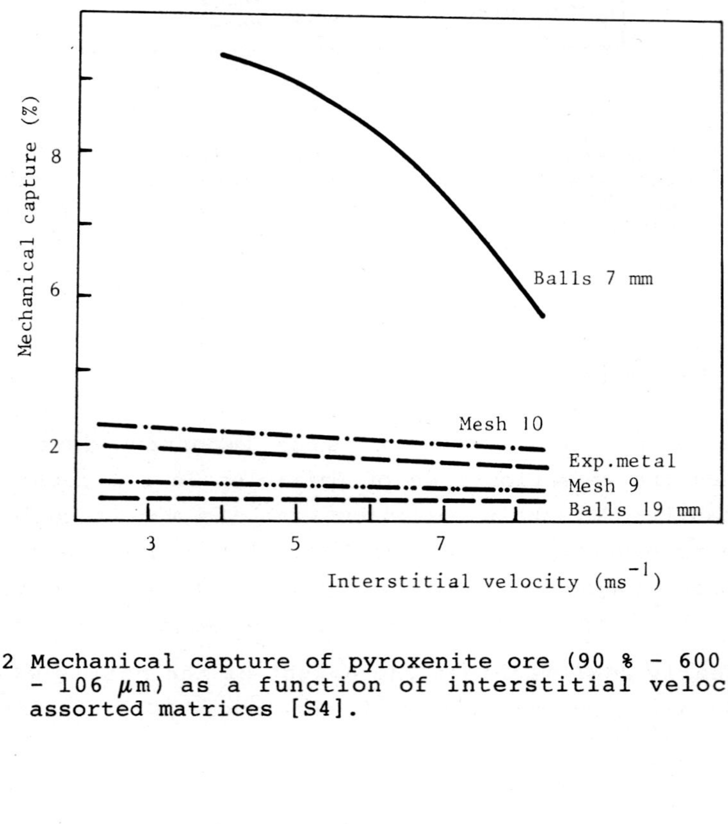

Fig. 4.52 Mechanical capture of pyroxenite ore (90 % - 600 μm, 12 % - 106 μm) as a function of interstitial velocity, for assorted matrices [S4].

matrix height is approximately 200 mm, although in some exceptional cases (e.g. in kaolin brightening) the height can be as high as 500 mm.

For the removal of moderately strongly magnetic particles a matrix height of 100 to 150 mm is usually sufficient, as can be seen in Figs. 4.53 and 4.54; a further increase in matrix height does not increase recovery.

For the recovery of more weakly magnetic minerals, a deeper matrix is often needed. However, an increase in matrix height above 175 mm results in a negligible rise in recovery, as shown in Figs. 4.55 and 4.56. For the treatment of very finely ground minerals, a matrix higher than 200 mm is sometimes used: for instance, in the beneficiation of kaolin a matrix of height of 500 mm is often employed.

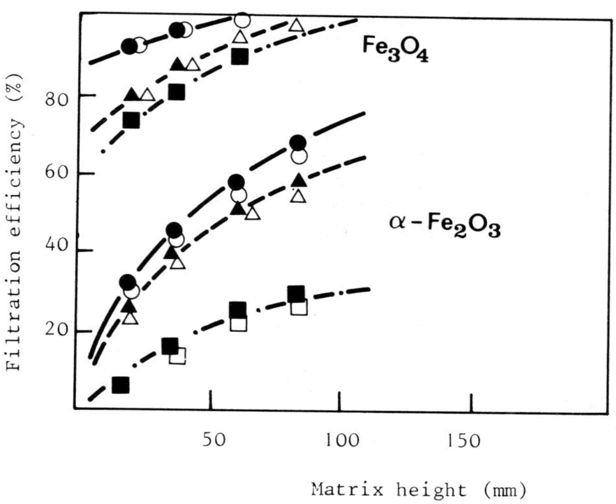

Fig. 4.53 The filtration efficiency of iron oxides from steel mill waste water, as a function of matrix height, for different types of steel wool matrix.
○ - fine wool (2a = 40 μm)
△ - medium wool (2a = 75 μm)
□ - coarse wool (2a = 125 μm)
open symbols - ordinary steel wool
closed symbols - diffusion-bonded wool
(After Soda et al.[S55])

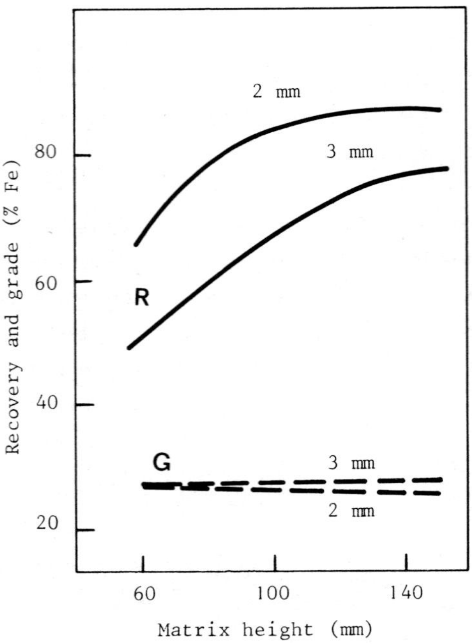

Fig. 4.54 The recovery of iron and the grade of the magnetic concentrate from siderite ore, as a function of matrix height for a matrix of rods of different diameters; magnetic induction 0.9 T, flowrate 0.15 ms^{-1}, rod spacing 1.5 mm (after Hencl and Šebor [H35]).

The grade of the magnetic concentrate does not usually depend on the height of the matrix within the limits mentioned above, although in some cases a decrease in the grade with increasing matrix height was observed (Fig. 4.57).

When interpreting the literature data on the effect of matrix height on the metallurgy one must always satisfy oneself that the tests were conducted under constant experimental conditions. With increasing matrix height the mass of sample fed into the separator must also increase so that matrix loading is constant. Any failure to do so will result in erroneous results since the matrix loading can have a considerable effect on the metallurgical results.

The height of the matrix should be as low as possible, consistent with the acceptable metallurgical results. By reducing the matrix height it would then be possible for the size of the pole-piece or the height of a coil, to be reduced resulting in a decrease in the mass of the steel frame or the iron yoke, as well as in lower input power.

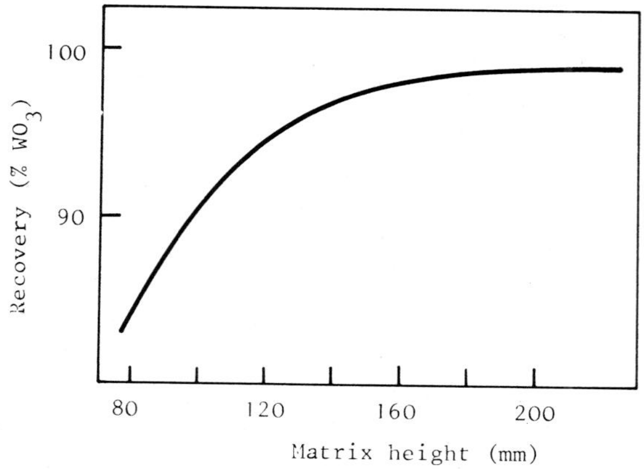

Fig. 4.55 The recovery of WO_3 from the wolframite slimes, as a function of matrix height; matrix: 8 mm balls (after Karmazin et al.[K15]).

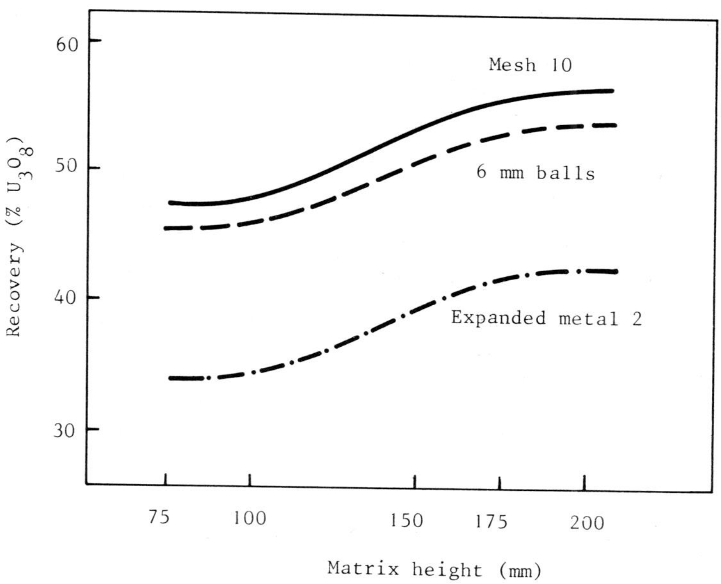

Fig. 4.56 The recovery of uranium from uranium-gold leaching residues, as a function of matrix height for various matrices. Magnetic induction 1.1 T, flow velocity 0.06 m s^{-1} [S29].

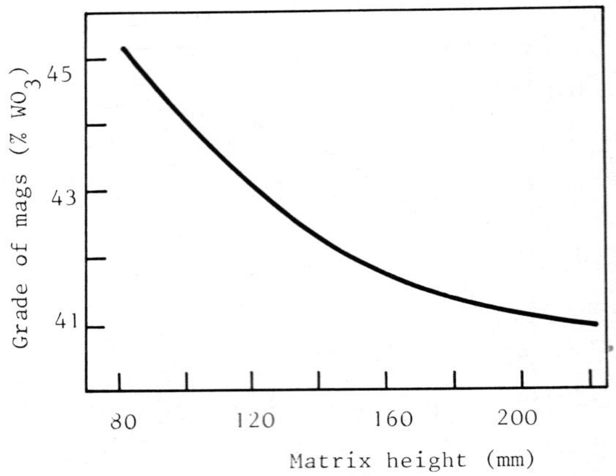

Fig. 4.57 The grade of the magnetic concentrate from wolframite slimes, as a function of matrix height (after Karmazin et al. [K15]).

IV.5.5 Matrix loading

The matrix is able to carry only a limited amount of mineral particles before its capture efficiency begins to deteriorate. Quantitatively, this capability is usually expressed through the matrix loading. There are several ways to embody this quantity, most of them acceptable if used in a consistent manner. Four different definitions of the matrix loading, often used in mineral processing, are given below:

$$L_{F1} = \frac{\text{feed mass}}{\text{matrix mass}} \text{ (dimensionless)} \tag{4.7}$$

$$L_{F2} = \frac{\text{feed mass}}{\text{matrix area}} \text{ (kgm}^{-2} \text{ or gcm}^{-2}\text{)} \tag{4.8}$$

$$L_{F3} = \frac{\text{feed mass}}{\text{matrix volume}} \text{ (kgm}^{-3} \text{ or gcm}^{-3}\text{)} \tag{4.9}$$

$$L_M = \frac{\text{mags mass}}{\text{matrix volume}} \; (\text{kgm}^{-3} \text{ or gcm}^{-3}) \qquad (4.10)$$

Definition L_{F3} is perhaps the most frequent and convenient and describes well the loading capacity of a matrix within a single application, where the mass yield into the magnetic fraction does not change appreciably.

For a more general characterization of a separation process applied to different ores and different matrices the definition of L_M is more appropriate [L13] since it is the mass yield into the magnetic product and not the mass of the feed that determines the degree of loading. The disadvantage is that the mass yield into the mags must be known before this loading factor can be determined.

In kaolin brightening it is customary to express the matrix loading through a number of canister volumes of a slurry passed through a separator. Fig. 4.58 illustrates the concentration of Fe_2O_3 in magnetically processed kaolin versus the amount of kaolin processed. The deterioration of the matrix performance with an increasing number of canister volumes processed is marked.

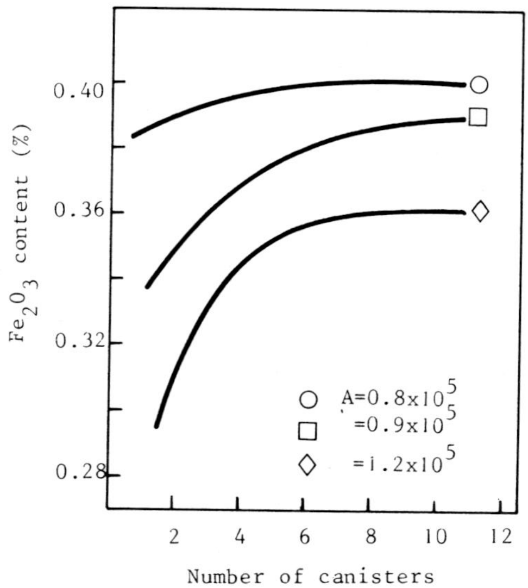

Fig. 4.58 The concentration of Fe_2O_3 in a non-magnetic product of kaolin versus the amount of kaolin processed, for various values of parameter $A = H_o/\eta v_o$ (after Watson et al. [W4]).

In other cases, the effect of matrix loading on kaolin purification is less obvious, as demonstrated in Fig. 4.59.

A typical breakthrough curve that expresses the effect of matrix loading on the filtration efficiency of municipal waste water is shown in Fig. 4.60.

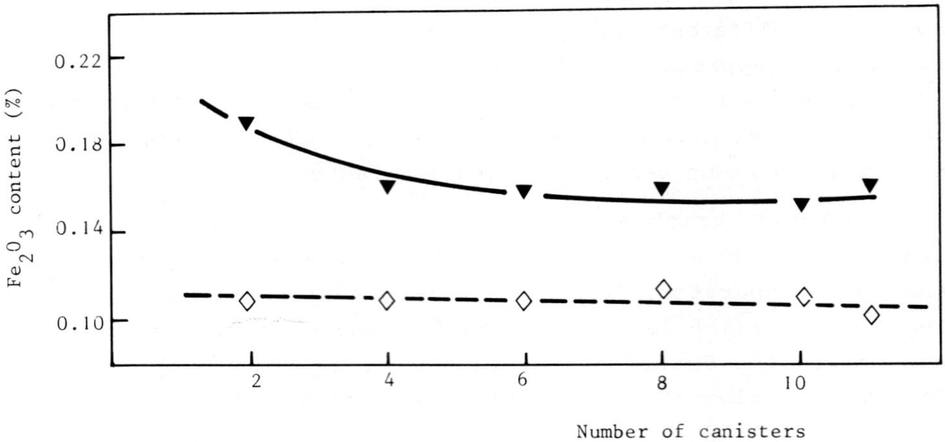

Fig. 4.59 The concentration of iron in the non-magnetic product from South African kaolin, as a function of the matrix loading, for different samples; **B** = 1.9 T, $\mathbf{v}_o = 6 \times 10^{-3}$ m s^{-1} [S4].

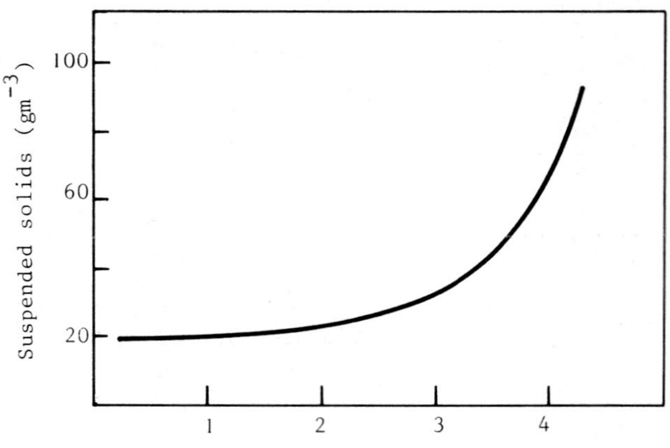

Fig. 4.60 The concentration of suspended solids in the municipal water effluent from HGMS, versus matrix loading (after Allen [A22]).

In the concentration of minerals matrix loading affects not only the recovery, but also the grade of the product. As a rule, with increasing loading recovery decreases and the grade of the magnetic concentrate increases. As demonstrated by Figs. 4.61 and 4.62, the recovery of uranium from cyanidation residues decreases with increasing loading for ball and mesh matrices as low as 50 kg m^{-3} (0.05 g cm^{-3}). On the other hand, the grade of the magnetic concentrate increases substantially as the loading L_M increases from 50 to 280 kg m^{-3} (0.05 to 0.28 g cm^{-3}). This behaviour can be explained by visualizing that at a higher matrix loading, the magnetizable and entrained gangue particles in higher layers of the build-up profile are not sufficiently held and can be stripped off from the matrix surface by fluid shear stress. Although such action decreases recovery of magnetizable particles, the grade of the magnetic fraction is improved. Similar results are usually obtained in the beneficiation of iron ores [H35, M27, H38] and typical dependence is depicted in Fig. 4.63.

If the valuable product reports into the non-magnetic fraction, then, of course, the trend is reversed, as shown in Fig. 4.64. Recovery into the nonmags increases, and the grade of the non-magnetic product decreases with increasing matrix loading.

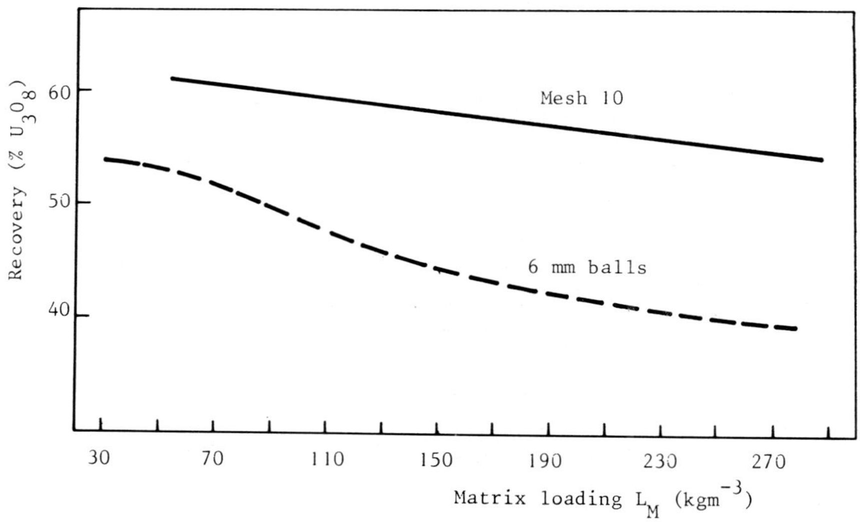

Fig. 4.61 The recovery of U_3O_8 from cyanidation residues as a function of matrix loading L_M. Magnetic induction 1.1 T, flow velocity 0.07 m s^{-1}. [S29].

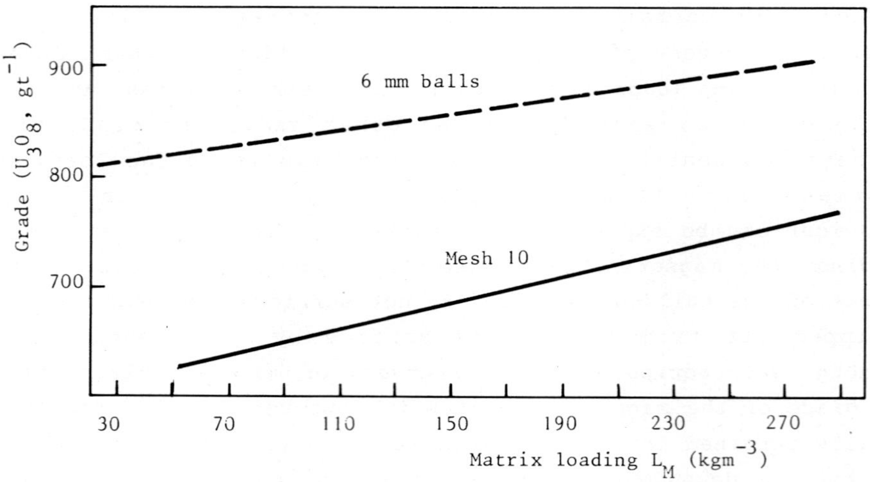

Fig. 4.62 The grade of the magnetic fraction of cyanidation residues as a function of matrix loading L_M. Magnetic induction 1.1 T, flow velocity 0.07 m s^{-1} [S29].

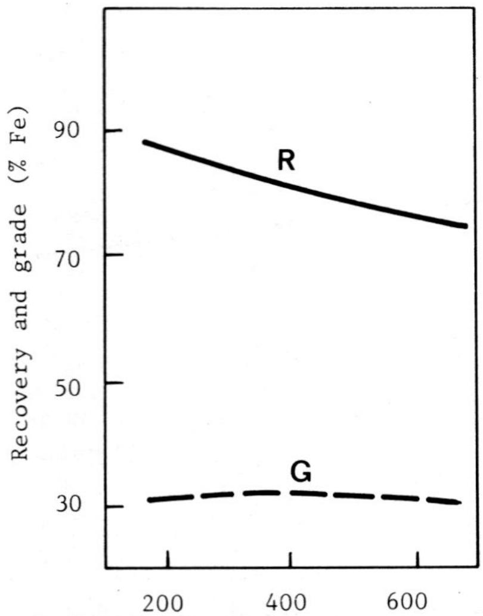

Fig. 4.63 The recovery and grade of the magnetic product of siderite ore as a function of matrix loading L_{F3} (after Hencl and Šebor [H35]).

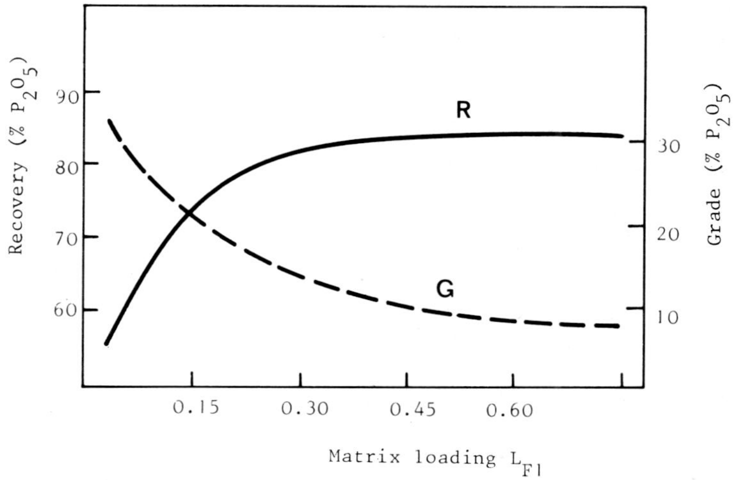

Fig. 4.64 The recovery of apatite by dry HGMS, and the grade of the non-magnetic product from the pyroxenite ore, as a function of matrix loading; **B** = 1.25 T, $\mathbf{v}_o$ = 2.9 m s^{-1}, matrix: mesh 10 [S4].

Typical values of matrix loading for various applications are summarized in Table 4.8.

The effect of matrix loading can be explained when we realize that the overall performance of a matrix magnetic separator is determined by the ratio of the magnetic force to the wash-off erosion force.

In wash-off erosion, the driving force is viscous drag, and the resisting force is a magnetic force that tends to hold the particles on the matrix. The loading limitation that a matrix can sustain is determined by the wash-off process and not by the capture process. At higher loadings, the magnetic force that acts on particles in the more distant layers of a build-up decreases and the shear stress increases, so that the wash-off force will be dominant for very weakly paramagnetic particles, unliberated grains and diamagnetic particles entrained in clusters of paramagnetic particles. Consequently, the selectivity of separation in the higher layers is better and the grade improves.

We have seen that matrix loading is an important parameter that allows a magnetic separator to be operated in a regime of maximum recovery or of maximum grade of a concentrate. And, of course, it determines the throughput of a separator, the angular velocity of a carousel in a continuous machine or the length of a duty cycle in cyclic separators.

TABLE 4.8
Matrix loading in different applications

Application	Matrix	L_{F2} (kgm^{-2})	L_{F3} (kgm^{-3})	L_M (kgm^{-3})	Ref.
Hematite ore	Grooved plates		180-300		[M27]
Steel mill waste water	Steel wool		9000 120		[T12] [T9]
Power plant cooling water	Steel wool		4000		[T12]
	Balls		16 280	14	[M29,H37] [D25]
Siderite ore	Rods		200-600		[H35]
Pyroxenite ore	Mesh		600-6000		[S4]
Uranium leach residues	Mesh or balls		300-3000	30-300	[S29]
Iron ore tailings	Expanded metal		100-1000		[A23]
Oxidized taconites	Balls	60-240			[H38]
Coal desulphurization	Steel wool		50-240		[W25]
Cu-Pb flotation tailings	Balls		800	80	[S30]
Magnetite from coal	Expanded metal		125-275		[D23]
Insoluble fission products	Balls		600		[W23]
Gas filtration	Steel wool		5×10^4		[B49]

IV.5.6 Magnetic properties of matrix material

Models of magnetic separation indicate that the probability of particles being captured on the matrix is directly proportional to the saturation magnetization of the matrix material. However, the limited experimental data given in the literature indicate substantial deviations from this simple theory.

It has been shown [S31] that, at intermediate magnetic fields, a weakly magnetic steel matrix with a magnetic polarization of 0.1 T is only slightly less efficient than a standard steel matrix with a saturation polarization of 2 T.

The effect of the saturation polarization of the matrix on the metallurgical performance of a magnetic separator was therefore investigated, and rods with a diameter of 10 mm and saturation polarization ranging from 0.1 to 2.0 T were used as the matrix. Siderite ore, with particles smaller that 200 μm was used in these tests, which were carried out in transverse and longitudinal configurations and at magnetic inductions varying from 0.2 to 1 T; the results are summarized in Fig. 4.65. For matrices with a saturation polarization greater than 0.5 T, the curves for recovery and grade versus magnetic field were identical within the experimental error. For a weakly magnetic steel matrix with a polarization of 0.1 T, the recovery was strongly dependent on the value of the external magnetic field and, at a background magnetic induction of 1 T, the recovery reached values that were only 10 per cent lower than those for strongly magnetic steels.

For all values of magnetic induction, the grade of the magnetic concentrate was identical to that of ferromagnetic steel.

Similarly, in the purification of kaolin, the amorphous ferromagnetic fibres with a saturation polarization of 1.1 T performed better under identical conditions than ordinary stainless-steel wool, for which 2.1 T was the saturation polarization [Y1].

It appears, therefore, that magnetization of the matrix plays a minor role, which is in contrast to theoretical predictions. Thus, a paramagnetic steel matrix can be used, and acceptable metallurgical performance can be achieved, while advantage is taken of the low remanent magnetization of these steels to facilitate the flushing of the magnetic concentrate from the matrix [S31, S56]. The values of remanent and saturation polarizations of assorted steels are summarized in Table 4.9.

Magnetization curves for matrices of various origins are given in Fig. 4.66.

Very weakly magnetic materials (such as austenitic stainless steel or aluminium) can be used as matrices for magnetic scalping of ferromagnetic impurities from the pulp. The aluminium matrix was used by Riley and Watson [R14] for the filtration of ferromagnetic particles and the beneficial effect of scalping by paramagnetic matrices on the recovery of uranium was demonstrated by Watson et al. [W26] and by Svoboda [S29]; aluminium or 304 stainless steel expanded metal or woven mesh can be used. The magnetic polarization of the aluminium matrix at 8×10^{5} Am^{-1} (10 kOe) was 0.05 T and that of 304 stainless steel mesh was 0.027 T [S29]. The performance of both types of matrices was very similar.

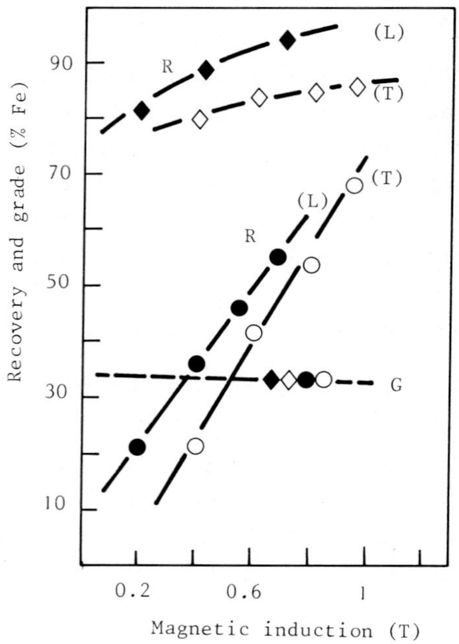

Fig. 4.65 The recovery and grade of the magnetic fraction of siderite ore versus magnetic induction for matrices of ferromagnetic (◇) and paramagnetic steels (o), respectively. Closed symbols - longitudinal configuration, open symbols - transverse configuration [S31, S29].

TABLE 4.9

Remanent polarization J_R and saturation polarization J_S of assorted steels [S31]

USA (AISI)	UK (BS)	Czechoslovakia CSN	J_R(T)	J_S(T)
	420 S45	17 023	0.12	1.5
410	410 S21	17 021	0.08	1.85
		17 113	0.07	1.5
405	403 S17	17 125	0.055	1.6
		17 153	0.04	1.1
321	321 S12	17 248	0.025	
316 Ti	320 S17	17 347	0.008	
1010	045 M10	12 010	0.07	1.7
		19 312	0.12	1.6
		19 712	0.11	1.6
		19 720	0.13	1.4
		14 240	0.10	1.7
		16 720	0.08	1.4
		16 220	0.10	1.6
AISI 444				1.62 [P21]

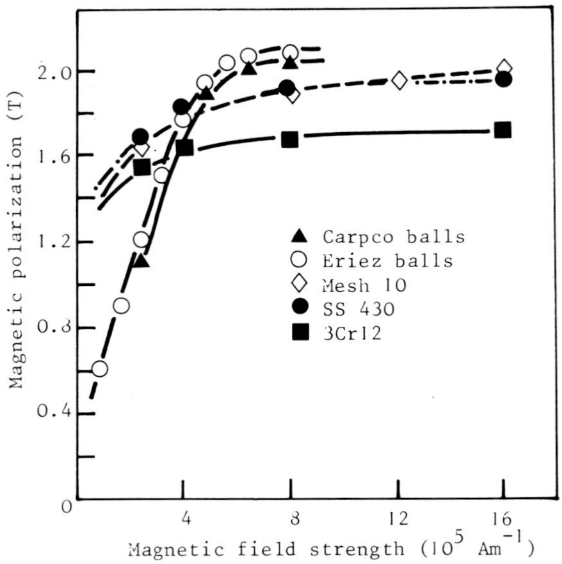

Fig. 4.66 Magnetization curves of different matrices [S29].

IV.5.7 Selection of matrix material

The choice of a matrix material is dictated by its mechanical and magnetic properties. Saturation polarization of the material should be sufficiently large (> 0.5 T, but preferably close to 2 T) to safeguard the acceptable collection efficiency. The remanent polarization should be as low as possible so that the removal of ferromagnetic particles from the matrix is promoted.

The most common materials used for a matrix are mild steel and magnetic stainless steel. Although the mild steel has very good mechanical and magnetic properties, it is not corrosion resistant and its use in long-term applications of magnetic separation is not practical. Stainless steels can be broken down into several basic groups, as shown in Fig. 4.67.

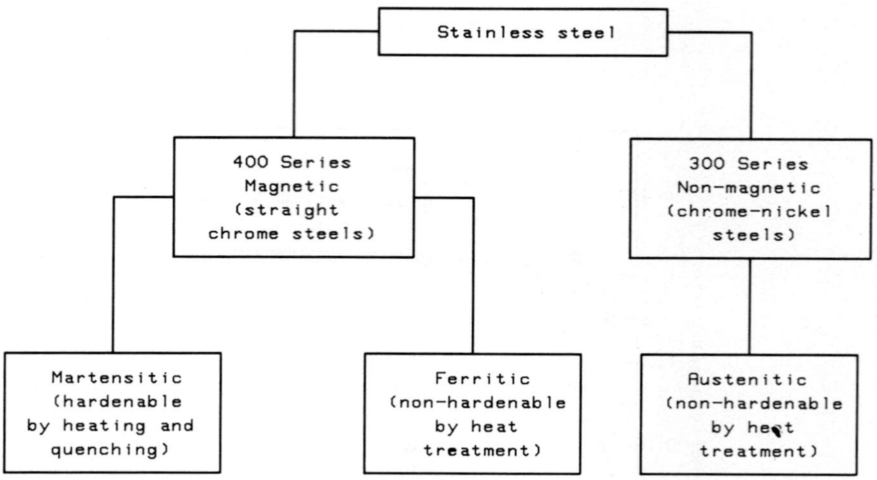

Fig. 4.67 Stainless steel classification.

Martensitic stainless steels (410, 420, 431)

Excellent strength and hardness are the main advantages of martensitic steels. They are hardenable by heat treatment, with the added advantage of being resistant to corrosion. Attainable strength and hardness increase as the carbon content increases. Martensitic steels have a high carbon content (0.2 - 1.2 per cent) with a chromium content of 12 to 18 per cent. Basic properties include moderate corrosion resistance, poor weldability and high strength.

Ferritic stainless steels (409, 430)

These are chromium steels with generally 12 to 18 per cent chromium content, but they have much lower carbon content than the martensitic types. These steels have moderate to good corrosion resistance which increases with the chromium content. They are magnetic, non-hardenable and always used in the annealed condition. SS 430 steel is most widely used as a matrix material.

Austenitic stainless steels (300 series)

If nickel is added to plain chromium stainless steel in sufficient quantity it changes the crystal structure of stainless steel into austenitic steel. Austenitic stainless steels commonly contain 18 per cent chromium and 8 per cent nickel, while steels that are more corrosion resistant also contain 3 per cent molybdenum. They are of low carbon content (0.08 per cent maximum), non-magnetic and highly corrosion-resistant. The basic properties include formability and weldability.

The approximate equivalents of some of the more common types of steels used as a matrix material are given in Table 4.10. The data on the corrosion resistance of selected stainless steels are summarized in Table 4.11. Although corrosion resistance of the matrix material is an important consideration in mineral processing, the requirements are even stricter in those applications of HGMS where the equipment is expected to last for a full lifetime of the plant, for instance in the reprocessing of highly irradiated fuels. Potts et al. [P21] found that ferritic steel AISI 444 was markedly superior to other stainless steels tested, as shown in Fig. 4.68.

TABLE 4.10

Approximate equivalents of some of the more common types of steel [W27]

USA (AISI)	UK (BS)	W.Germany (Stoff Nr.)	France (ANFOR)	USSR (GOST)	ČSSR (CSN)
405	403 S17	1.4724			17 125
410	410 S21	1.4006	Z10C14	12Ch13	17 021
	420 S45	1.4028	Z30C14	30Ch13	17 023
420	420 S37	1.4021			
430	430 S15	1.4016		12Ch17	
304 L	304 S12	1.4306	Z2CN18-10	03Ch18N11	
309	309 S24	1.4828	Z10CN 25-13		
316 Ti	320 S17	1.4571		10Ch17N 13M2T	17 347
321	321 S12	1.4541	Z6CNT 18-11	Ch18N10T	17 248
1010	045 M10	1.1121	XC10	08;10	12 010
Mild steel	4360 43C	1.0483			

TABLE 4.11

Corrosion resistance of stainless steels

	Stainless steel (AISI)		
Substance	316	410	430
Acetone	A	B	A
Hydrochloric acid, <1 %	B	C	C
10 %	B	C	C
Nitric acid	A	A	A
Sodium hydroxide	A	A	A
Sulphuric acid, 5 % 21°C	A		C
10 % boiling	B		C
Water	A	A	A

Legend: A - fully resistant
B - fairly resistant
C - not resistant

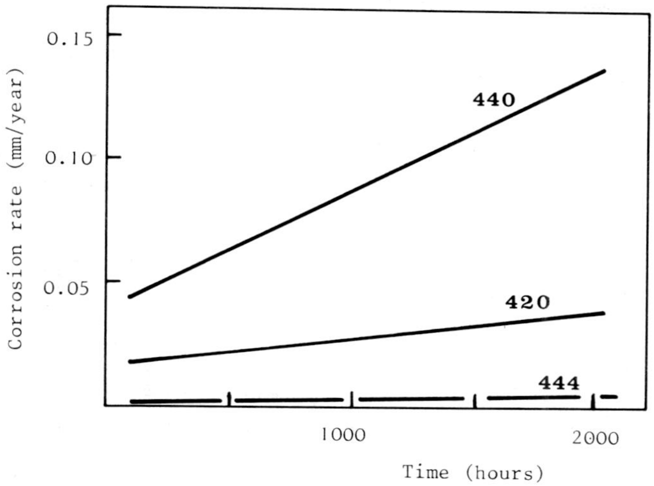

Fig. 4.68 Corrosion of stainless steels in 3M nitric acid at 60°C (after Potts et at. [P21]).

IV.5.8 Wear of matrix material.

There are three major categories of wear in which most situations can be included: abrasive, adhesive and erosive. A matrix in a magnetic separator is exposed predominantly to erosion wear caused by the impact of rapidly moving particles entrained in either a gas or liquid stream. The process of erosion wear is affected by several factors that can be divided into three groups:

a) incidence conditions (angle of incidence, the incidence velocity of particles)
b) characteristics of the slurry (particle size and form, hardness, type of carrying medium)
c) matrix material characteristics (physical and mechanical properties, structure)

The volume erosive wear is proportional to $\mathbf{v}_o^n$, where $\mathbf{v}_o$ is the slurry velocity and the value of the exponent **n** depends mainly on the material. In the case of metals, **n** usually varies in the range 2.05 - 2.43. The volume erosive wear depends linearly on the concentration of particles in the slurry. However, at concentrations greater than some critical concentration the incident and reflected particles may interact and this may lead to a reduction in the rate of erosive wear.

Marked erosive wear is caused in most cases by particles whose size is greater than a threshold size which depends on the incident velocity of particles. For instance at 10 m s^{-1} the threshold particle size is 30 μm, while at 300 m s^{-1} it is 5 μm. The effect of particle size on erosive wear is shown in Fig. 4.69.

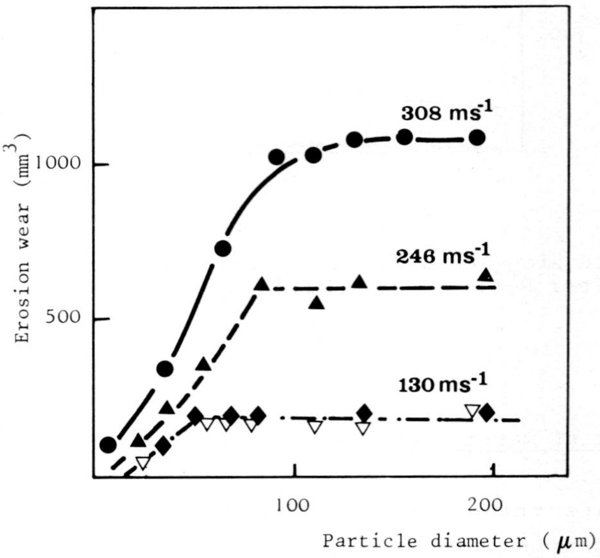

Fig. 4.69 The effect of particle size on erosion wear at various incidence velocities and angles. ● , ▲ , ◆ 90°, ▽ 20°. (After Suchánek et al. [S57]).

The shape of particles is also an important factor that influences the rate of erosive wear. Angular particles require a lower kinetic energy than spherical particles to penetrate into the surface layers of material. A difference in the wear rate of an order of magnitude was observed [S57] for angular and spherical particles.

The higher the strength and hardness of the particles, the greater is the extent of damage to the surface of the material. On the other hand, the increased hardness of the matrix material has a beneficial effect because it increases its resistance to erosive wear. Nevertheless, even the hardest materials are worn down by the extended action of far softer materials.

The structural stiffness of the steel wool matrix was studied by Soda et al. [S55] who found that by diffusion bonding the steel wool, a high resistance to stiffness drop and to rejection of fine filaments is achieved.

Comparison of the erosive wear of assorted steels is given in Table 4.12 while Table 4.13 gives data on the erosion wear of some non-metallic materials. The abrasion wear resistance of several steels suitable as a matrix material is given in Table 4.14.

TABLE 4.12
The comparative resistance to erosive wear ψ_e of selected steels (after Suchánek et al. [S57])

Steel			
Czechoslovakia CSN	W.Germany Stoff Nr.	Hardness (HV)	Erosive wear (ψ_e)
12 014	1.1121	100	1.0
14 209	1.3520	723	2.55
15 261	1.8161	554	1.69
17 246	1.4541	146	1.50
17 618	1.3401	219	1.93
19 436	1.2080	797	3.56

TABLE 4.13
The comparative erosive wear resistance ψ_e of some non-metallic materials (after Suchánek et al. [S57]).

Material	ψ_e
Sintered corundum	1.06
Fused basalt	0.45
Glass	0.28
Vulkolan	1.66 - 8.9 +)

+) Depends on the hardness of Vulkolan, higher ψ_e is obtained for softer Vulkolan

TABLE 4.14
The comparative abrasive wear resistance ψ_a for steels used as a matrix material [S31]

Czechoslovakia CSN	W.Germany Stoff Nr.	Abrasive wear ψ_a
12 014	1.1121	1
17 021	1.4006	1.35
17 125	1.4724	1.38
17 153		1.40
17 248	1.4541	1.39

IV.5.9 Clogging of the matrix

Probably the most important problem encountered in pilot- or production-scale HGMS separators is a permanent retention of particles in the matrix. The effects may range from a slight decrease in metallurgical performance to complete clogging of the system.

The removal of particles that adhere to the matrix determines the efficiency of the subsequent magnetic capture of particles and the length of time between shut-downs for replacement or cleaning of the matrix. Industrial slurries always contain various extraneous material (wood, plastics, tramp iron), oversize grains, and ferromagnetic admixtures. Their presence and frequent shut-downs of the separator for the replacement of a blocked matrix caused by such particles severely impair the economic and metallurgical performance of the unit. The shut-down time required for external matrix washing can be one day or more resulting not only in a decrease of production but also in a deterioration of the mechanical properties of the matrix material [L19, P22].

The first step in eliminating this problem is to match the matrix to the particle size distribution in the feed, so that mechanical straining is eliminated. Permanent retention of particles in the matrix can be further minimized by preliminary screening and low-intensity magnetic scalping, by desliming the feed to eliminate fine ferromagnetic particles and by continuous on-site cleaning of matrix collectors [C6, C7].

Mechanical vibration, air blasting and sonic cleaning of the matrix were used by Boucher and Bulmer [B33] and by Gooding [G14] in the application of HGMS for the removal of particles from the waste gases in various steel-making processes.

The problem can be also partly solved by using a matrix material of low remanent polarization [S31], and by intermittent intense scouring of matrix by air and water jets [B32].

One disconcerting fact is that neither of these measures completely eliminates the clogging of the matrix, although blockage takes place much later. It is usually found that neither flushing with high-pressure water nor with pressurized air completely removes the entrained ferromagnetic particles from the matrix. Moreover, a number of potential users of magnetic separation were discouraged from applying HGMS on a large-scale by the mere prospect of having to remove and clean the matrix of high-capacity machines.

It will thus be seen that each of these techniques can be employed only in special cases, sometimes at the expense of increased complexity of the system or of reduced metallurgical performance of a separator. Although various manufacturers of magnetic separators claim that their matrix is clog-free, all matrix separators suffer to various degree from matrix blockage.

Investigation of the matrix-cleaning process in a high-gradient magnetic separator [R15] showed that, after ten cycles of matrix flushing, the efficiency of particle capture dropped by 30 per cent. Similar deterioration of the capture efficiency and increased build-up of particles that cannot be flushed out, with an increasing number of flush cycles, were observed by Williams et al. [W23] and are illustrated in Fig. 4.70. These experiments, and extensive experience obtained in the operation of production-scale

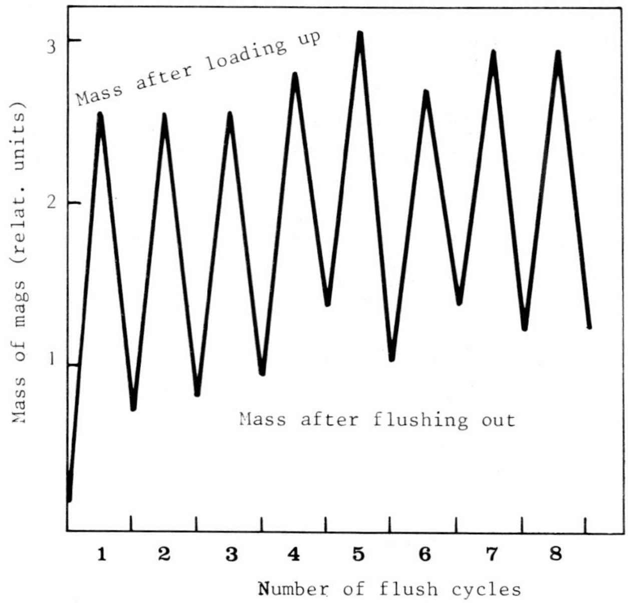

Fig. 4.70 Mass of particles in the matrix as a function of a number of flush cycles (after Williams et at. [W23]).

HGMS separators in kaolin beneficiation [L19] and in other applications, indicate that a certain fraction of particles deposited on the matrix collectors cannot be removed by conventional means and that residual forces obstruct the cleaning process.

The user of HGMS is thus faced with a most frustrating dilemma. He has at his disposal economically attractive magnetic separation technology with a mechanical stumbling block: that of matrix blockage. All preventive measures mentioned above can be regarded only as an interim measure, particularly in large-scale applications of HGMS. The long-term solution can, perhaps, be found in machines using the back-flushing system [A24] or combining the matrix reversal with back-flushing [C8] or employing a flexible self-cleaning matrix [C15, C32].

IV.5.9.1 Cleaning of a ball matrix

Karmazin et al. [K21] developed in 1968 a rotor magnetic separator in which the balls of the matrix were continuously removed from the rotor, externally cleaned and returned to the rotor. The notion of external washing of the matrix collectors was later applied by Colburn [C33] to a linear magnetic separator whereby the ball collectors were removed from the inclined passageway, washed to remove collected paramagnetic particles and extraneous impurities and returned to the magnetic zone by means of a conveyor.

A commercially successful continuous ball-cleaning system was developed by Corrans [C6, C7] and is illustrated in Figs. 4.71 and 4.72.

A horizontal slot near the bottom of the rotor permits some of the balls to be removed for cleaning while the separator continues to operate normally. The height of the rotor is such that the slot can be placed below the level of the magnet pole-faces; an inclined grid at the bottom of the rotor supports the matrix and facilitates the movement of the balls through the slot. A narrow belt surrounding the slot is used to retain the balls but an arrangement of pulleys lifts the belt from the rotor at a desired position to expose a short portion of the slot; the open slot allows some of the balls to be discharged from the rotor (Fig. 4.71). The rate of discharge of the balls can be adjusted by various means so that the amount of entrapped ferromagnetic particles and wood chips can be kept at an acceptable level.

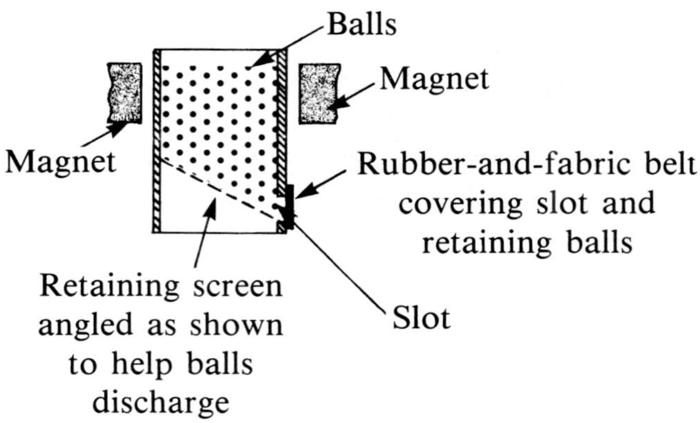

Fig. 4.71 Section through the rotor of the WHIMS machine showing the design for the continuous removal of balls (after Corrans et al. [C7]).

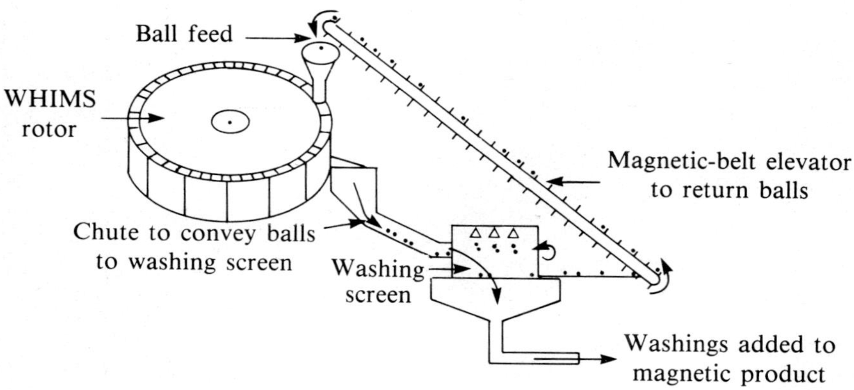

Fig. 4.72 The rotor showing the cleaning and return system for balls (after Corrans et al. [C7]).

Balls discharged through the open area of the slot are conveyed to a washing screen. A rotating-trommel type screen has been found to be the most effective, and this is combined with four high-pressure sprays. Undersize or worn balls are removed by the screen. From the screen the cleaned balls are lifted by a magnetic conveyor into a hopper that choke-feeds the rotating annulus. A constant level in the matrix is thus maintained. Make-up balls are occassionally added to the hopper. The rate of discharge is relatively small. The matrix needs to be turned over once every 3 or 4 hours to prevent blockage, but this period can be varied by modifying the design. A photograph of the ball removal system is shown in Fig. 4.73.

The ball cleaning system installed with the Eriez continuous magnetic separator CF-30 has been tested extensively for the recovery of uranium and gold from cyanide tailings at a gold mine, at a fluorspar operation and at a copper and lead mine. It has been proved that the modified machine can run continuously on a production basis.

Fig. 4.73 Ball removal system.

IV.5.9.2 Removal of particles from the matrix

It has been observed by Svoboda and Corrans [S59] that the removal of hematite particles from a matrix is strongly dependent on the pH value of the flushing water, as shown in Fig. 4.74.

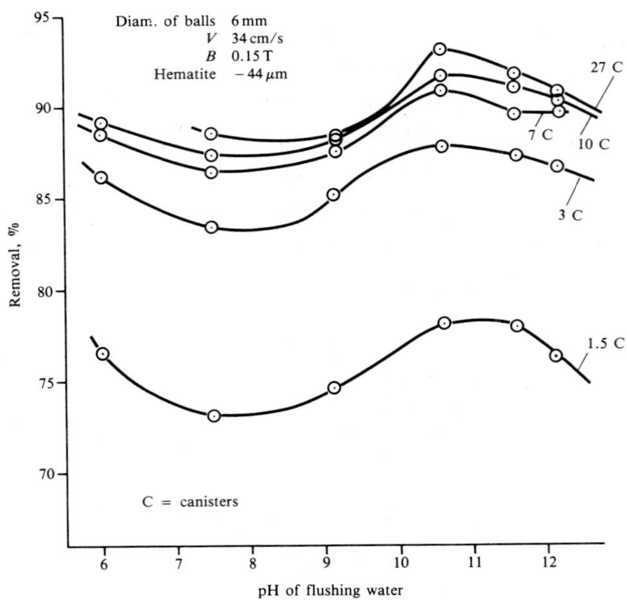

Fig. 4.74 Dependence of removal of hematite particles from a steel ball matrix on the pH value of flushing water for various volumes of flushing water [S59].

It can be seen that removal is not complete, even under optimum conditions, and that the fractions adhering to the matrix after ten canisters of flushing water had been used amounted to 8 per cent of the feed mass. A substantial increase in volume of flushing water resulted in only a mild improvement. Flushing water is usually pH 7 in practice, and an improvement in flushing efficiency of hematite can be achieved by adjustment of the pH value to between 10 and 11. The optimum pH value will, of course, depend on the surface properties of the captured material.

Conversely, for a given efficiency of particle removal, much lower volumes of water can be used. For a removal efficiency of 87 per cent, as achieved with ten canisters of water at pH 7, only three canisters of flushing water at pH 10 are needed.

Fig. 4.75 indicates that, after an initial rapid improvement of removal efficiency with increasing flowrate, particle desorption rises only very little for a velocity of more than 0.2 m s^{-1}. The same diagram implies that an increase in the volume of flushing water brings about only a mild improvement in particle removal.

These experimental results on the desorption of particles from the matrix of a magnetic separator were analysed in a framework of DLVO theory that was extended to include magnetic interaction (see Section III.5.2). The calculations of total interaction energy give a good explanation of the experimental observations, which imply that electric double-layer and London-van der Waals interactions probably represent the residual forces that hamper a complete flushing of the matrix.

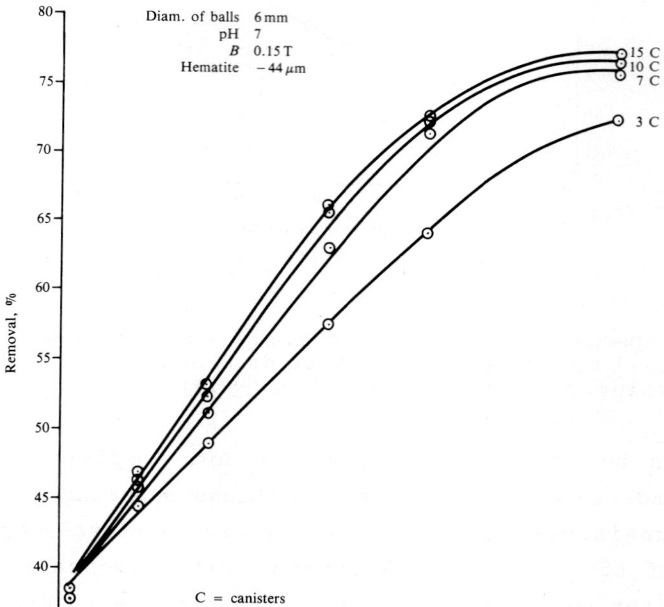

Fig. 4.75 Variation of percentage of particles removed from a matrix with velocity of the flushing water [S59].

It has been shown that, by adjustment of the hydrodynamic conditions, it is possible for a high degree of particle desorption to be achieved. However, a total wash-off of particles cannot be achieved without affecting the surface interactions between the adsorbed particles and the matrix. The adjustment of the pH value of the flushing water can lead either to an increase in the removal of particles from the matrix or to a reduction in the volume of flushing water, as shown in Fig. 4.74. The latter implication can be of particular importance in such applications, in which it is imperative that the value of flushed magnetic fraction should be kept as low as possible as, for example, in the magnetic filtration of steel-mill waste-waters. In cyclic magnetic separators, a reduction in the volume of flushing water can lead to a decrease in the flushing time and therefore to an increase in the length of the duty cycle of the separator.

IV.6 FLOW VELOCITY

IV.6.1 Flow velocity in a matrix

The metallurgical performance of a magnetic separator changes with the velocity of flow of a slurry to such an extent that flow velocity may be ranked with magnetic field strength and matrix as one of the primary factors affecting the efficiency of magnetic separation.

The velocity of the slurry flowing through the matrix bed determines the shear stress acting on the particles deposited on the surface of matrix collectors, and thus controls the recovery and the grade of the magnetic product. Simultaneously, flow velocity controls the retention time of the slurry in the magnetic field and the capacity of a separator.

The superficial velocity $\mathbf{v}_o$ is defined as the average linear velocity the slurry would have in the separation chamber if no matrix was present. Thus, superficial velocity (in m s^{-1}) can be written as the volumetric flow rate $\mathbf{Q}_v$ (in m^3 s^{-1}) divided by the cross-sectional area $\mathbf{S}$ (in m^2) of the matrix

$$v_o = \frac{Q_v}{S} \tag{4.11}$$

The average flow velocity in the cross-section available for flow in a matrix bed is the interstitial velocity **v** which is related to the superficial velocity by

$$v = \frac{v_o}{\epsilon} \tag{4.12}$$

where ϵ is the porosity of the matrix bed.

The effective length of a particle trajectory is greater than the matrix height **h** because the slurry travels a tortuous path $\mathbf{l_t}$, and therefore the velocity along the tortuous path must be correspondingly greater

$$v' = v\,\frac{l_t}{h} = \frac{v_o l_t}{\epsilon h} \tag{4.13}$$

If Karman's assumption $\mathbf{l_t/h} = 2^{1/2}$ is used [S58], we obtain

$$v' = \sqrt{2}\,\frac{v_o}{\epsilon} \tag{4.14}$$

In magnetic separation, a distinction between superficial and interstitial velocities is not often made since most fundamental investigations have been carried out on single wires or fine steel wool matrices. For such matrices, owing to high porosity ($\epsilon \geqslant 0.95$), the difference between superficial and interstitial velocities is negligible. However, for less porous matrices, usually used in mineral processing, the porosity can widely differ (see Table 4.6), and a distinction between superficial and interstitial velocities has to be made, to make the results, obtained for different matrices, comparable.

IV.6.2 The effect of flow velocity on the performance of a magnetic separator

IV.6.2.1 Magnetic filtration

As follows from the expression for hydrodynamic drag (Eq. 1.15), an increase in flow velocity results in an increase of the hydrodynamic drag on a particle and consequently a necessary condition for succesful separation (Eq. (1.1)) may not be met. Thus, by increasing the flowrate the filtration efficiency usually decreases, as demonstrated in Fig. 4.76. A pattern of a decrease of filtration efficiency with flow velocity can vary, and is determined primarily by the magnetic susceptibility of solids and by the applied magnetic field. In some applications the performance of a magnetic filter drops sharply with increasing flowrate and then assumes a constant value as seen in Fig. 4.76, while in other cases filtration efficiency first decreases only very mildly and a sudden reduction occurs at very high flow velocities.

The selection of the operating flowrate is controlled by the balance between the requirements on the purity of the product and the throughput.

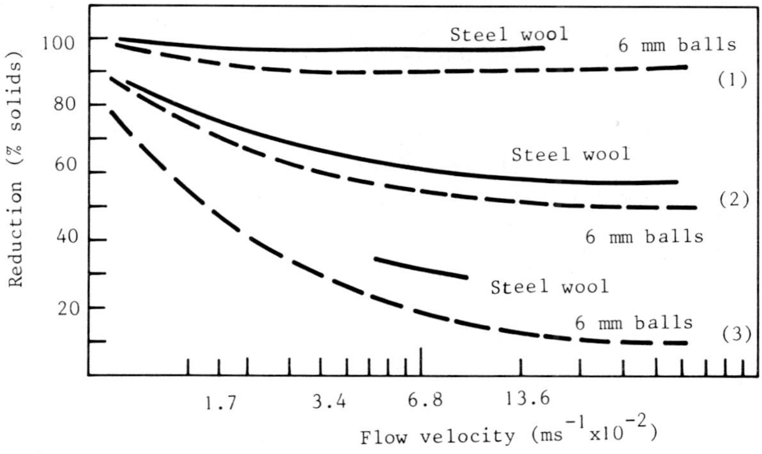

Fig. 4.76 The filtration efficiency of PWR cooling water as a function of flowrate (after Troy et al. [T12]).

IV.6.2.2 Removal of impurities from mixtures

As in magnetic filtration, the lower the slurry flow velocity, the greater the removal of magnetizable particles from a mixture. This effect is depicted for coal desulphurization by wet magnetic separation in Fig. 4.77.

Similarly, in kaolin brightening, the concentration of magnetic impurities in the product increases, and the brightness decreases with increasing flowrate, as shown in Fig. 4.78.

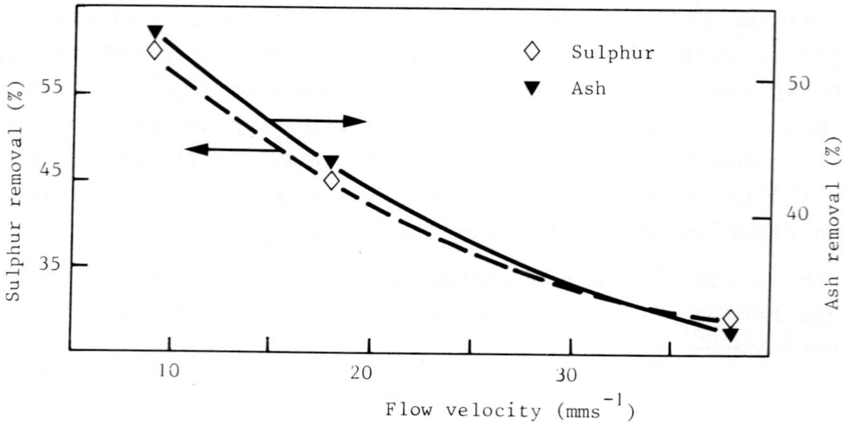

Fig. 4.77 Effect of slurry flow velocity on sulphur and ash removal from coal (after Wechsler et al. [W25]).

The temptation would be, as in magnetic filtration, to employ very low slurry velocities. However, in contrast to magnetic filtration, flow velocity must be specified with due regard for product recovery, and a compromise must be reached between product quality, product recovery and throughput of separator. Fig. 4.79 shows that with decreasing flowrate the recovery of coal decreases, while high flowrates, where the recovery is the highest, produce the poorest product.

A similar effect can usually be observed in kaolin brightening, as depicted in Fig. 4.80. The losses of kaolin into the magnetic product increase with decreasing flowrate owing to insufficiently high fluid shear stress that removes mechanically entrained non-magnetic particles from the matrix. Apart from the loss of a valuable product, this effect also decreases the duty cycle of a cyclic magnetic separator.

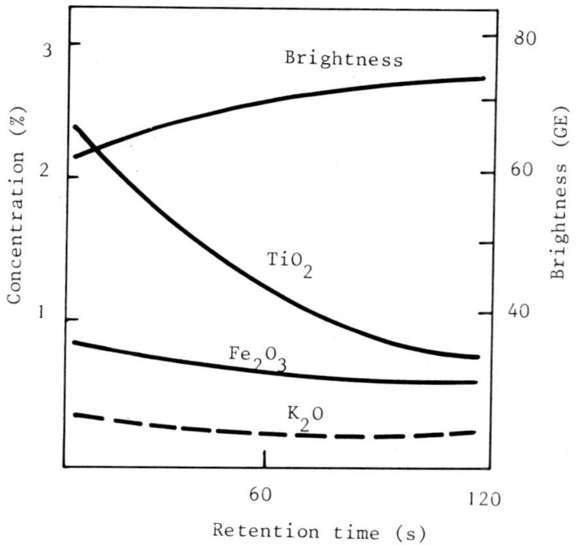

Fig. 4.78 The concentration of impurities and brightness of the kaolin product as a function of the retention time of the slurry in a magnetic field (after Iannicelli [I4]).

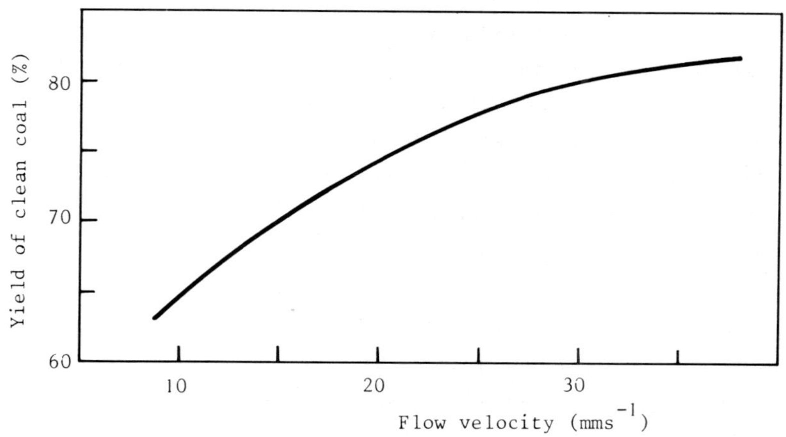

Fig. 4.79 Effect of slurry flow velocity on recovery of clean coal (after Wechsler et al. [W25]).

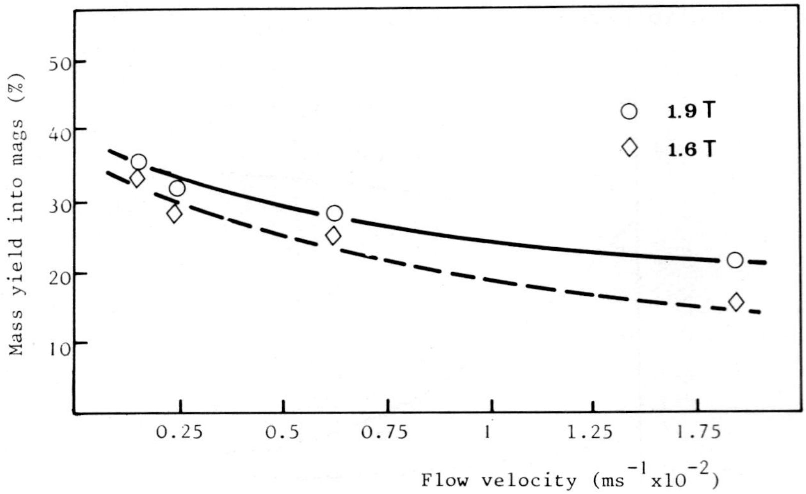

Fig. 4.80 Effect of slurry flow velocity on mass yield of kaolin into the non-magnetic fraction [S4].

IV.6.2.3 Concentration of minerals

The selection of the slurry flow velocity in those applications of high-gradient magnetic separation, in which a valuable mineral is concentrated in a magnetic product, is dictated by the consideration whether the main emphasis is placed on the recovery or on the quality of the magnetic product. Inverse relationship between product grade and product recovery implies that to achieve a high recovery low flowrates must be used, while a high grade can be obtained at high flowrates. Magnetic separators can thus be run in a regime of maximum recovery or in a regime of maximum grade of the concentrate. Typical dependences of recovery and grade on the slurry flow velocity are depicted in Figs. 4.81 to 4.83.

The rate of the drop in recovery and the rate of the increase in grade with increasing flow velocity are determined by the values of magnetic susceptibilities of a valuable mineral and of a gangue, by the magnitude of applied magnetic field, by the size distribution of the ore, and by the geometrical configuration employed in a magnetic separator.

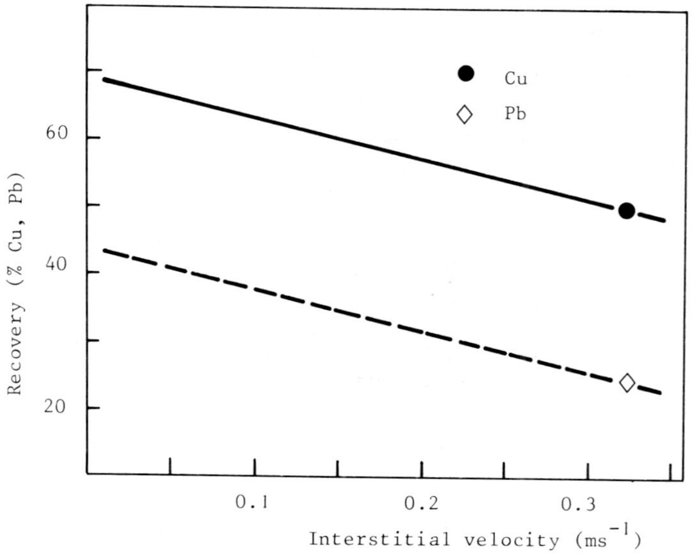

Fig. 4.81 Effect of interstitial slurry velocity on recovery of Cu and Pb from the flotation tailings by HGMS. Matrix: mesh; magnetic induction 1.2 T [S30].

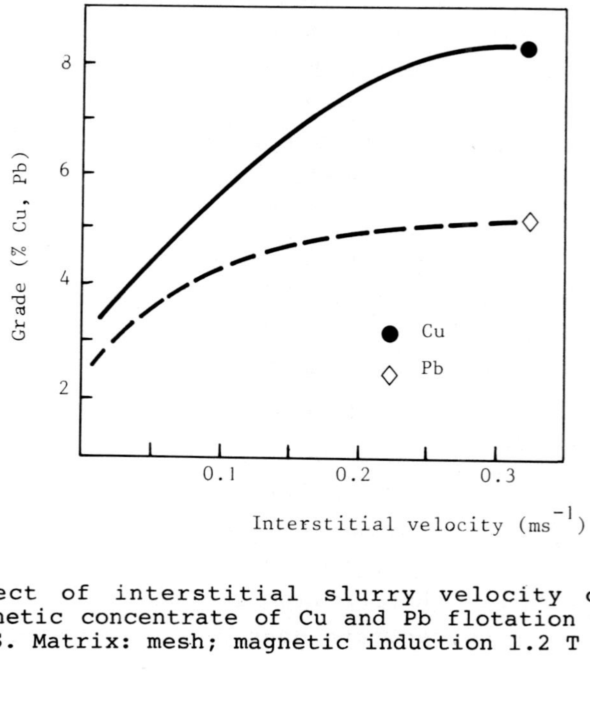

Fig. 4.82 Effect of interstitial slurry velocity on grade of magnetic concentrate of Cu and Pb flotation tailings, by HGMS. Matrix: mesh; magnetic induction 1.2 T [S30].

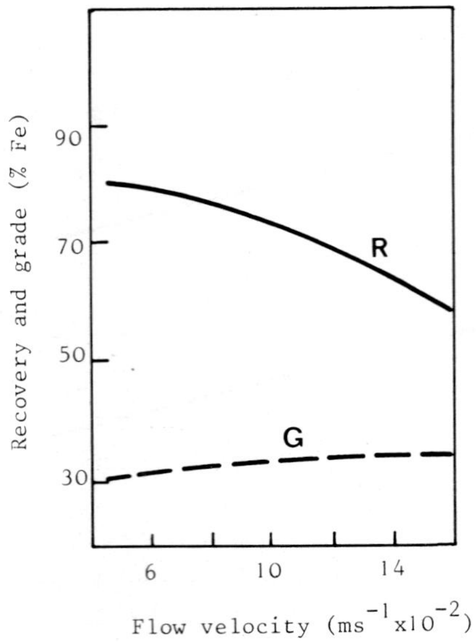

Fig. 4.83 Effect of flow velocity on recovery and grade of magnetic concentrate of siderite ore. Matrix: rods. (After Hencl and Šebor [H35]).

If the difference between magnetic susceptibilities of a useful mineral and the gangue is small, an increase in flow velocity will result in a sharp drop in recovery, while the improvement in the grade will be small. If the difference in susceptibilities is great, an increase in flow velocity can cause only a marginal drop in recovery, while feebly magnetic gangue particles will be swept from the matrix so that the selectivity of separation will increase.

Similarly, if the applied magnetic field is sufficiently strong, more strongly magnetic particles of a valuable mineral will remain attached on the matrix, despite an increase in the flow velocity, while mechanically entrained gangue particles will be stripped from the matrix surface, thereby increasing the grade of the magnetic product.

The presence of excessive amounts of slimes in the feed into a magnetic separator increases the sensitivity of the metallurgical results to flow velocity variations. In such cases, recovery usually drops rather sharply with increasing flow velocity while the grade of the concentrate does not improve appreciably. For efficiently deslimed ores the grade of the magnetic concentrate can be considerably improved by increasing flow velocity, without signally reducing recovery.

The effect of flow velocity differs in longitudinal and transverse configurations of the flow and magnetic field. In a transverse configuration, where the flow and magnetic field directions are mutually perpendicular, the build-up of particles is exposed to a high shear stress by the fluid. Consequently, recovery tends to decrease rapidly with increasing flow velocity, while the grade of the magnetic concentrate often improves dramatically.

In a longitudinal configuration, where flow velocity and magnetic field are parallel the shearing forces exerted by the slurry flow are much lower then in T-configuration, recovery usually does not drop substantially with increasing flow velocity and simultaneously, an increase in the grade is often only modest.

We have seen that flow velocity is an important parameter that considerably affects the performance of a magnetic separator. In many practical applications, however, where continuous units are used, a control over flow velocity, by means other than the natural resistance of the matrix bed, is absent. A potential of such magnetic separators for an optimum regime of operation is thus severely reduced.

IV.6.3 Pressure drop in a matrix

As in any application of deep bed filtration, pressure drop across the matrix is an important parameter in the control of the HGMS process.

If follows from Darcy's law that the superficial velocity $\mathbf{v}_o$ can be expressed as a function of pressure drop $\Delta \mathbf{P}$

$$v_o = -\frac{k\Delta P}{\eta \Delta x} \tag{4.15}$$

where η is the viscosity, $\mathbf{k}$ is the hydraulic permeability and $\Delta \mathbf{P}$ is the pressure difference across thickness $\Delta \mathbf{x}$ of a matrix. Kozeny and Kármán developed a theory that relates the hydraulic permeability to the geometry of the porous solid.

For laminar flow in circular tubes of radius **R** and length **l**, the pressure drop is given as

$$\Delta P = \frac{8\eta l v}{R^2} \tag{4.16}$$

If we imagine now that a matrix bed is just a tube of complicated cross-section with hydraulic radius R_h, then the pressure drop across the clean matrix can be written

$$\Delta P = \frac{8\eta l v}{R_h^2} \tag{4.17}$$

R_h may be expressed in terms of porosity ϵ and then eq. (4.17) yields

$$\Delta P = \frac{38 v_o l \eta (1-\epsilon_i)^2}{a^2 \epsilon_i^3} \tag{4.18}$$

which is the Blake-Kozeny equation. This equation is generally good for porosity below 0.5 and is valid only in the laminar region (i.e. for $a\ \rho_f v_o / [\eta (1 - \epsilon_i)] < 5$).

For turbulent flow a similar equation was derived by Burke and Plummer [B18]

$$\Delta P = 0.88 \frac{\rho v_o^2 (1-\epsilon) l}{a \epsilon^3} \tag{4.19}$$

It can be noted that this equation has a different dependence not only on v_o, **a** and η, but also on ϵ, from that given for laminar flow.

In the course of the separation process the particles are deposited onto the matrix collectors, the effective matrix element radius increases and the porosity decreases

$$a_d = a + t_d \tag{4.20}$$

where t_d is the thickness of a deposit; porosity is given by eq. (4.6). For a loaded matrix, eq. (4.18) can be rewritten as

$$\Delta P_m = \frac{38 v_o \eta l (1-\epsilon)^2}{a_d^2 \epsilon^3} \tag{4.21}$$

and thus, for a relative pressure drop we obtain

$$\frac{\Delta P_m}{\Delta P} = \frac{a^2 \epsilon_i^3 (1-\epsilon)^2}{a_d^2 \epsilon^3 (1-\epsilon_i)^2} \tag{4.22}$$

The empirical correlation of the pressure drop difference to the matrix loading was obtained in [E3, S60], while various experimental observations showed that the pressure drop was a useful monitor of the matrix loading. On the other hand, in the filtration of the corrosion products from the cooling water of thermal power plants, the pressure drop did not rise with the

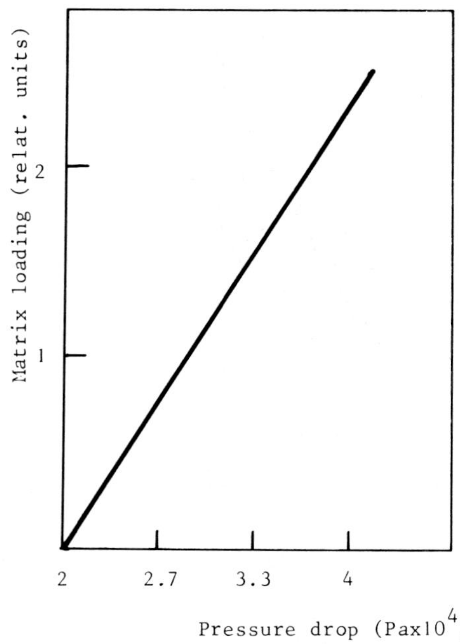

Fig. 4.84 Pressure drop across the matrix and the matrix loading for magnetic separation of chromium particles from a suspension (after Williams et al. [W23]).

increasing loading and the length of a duty cycle had to be determined from the quality of the effluent [M29]. A typical correlation between pressure drop and matrix loading is shown in Fig. 4.84. Matsuzaki et al. [M32] observed that the pressure drop across the matrix did not increase with matrix loading provided the particles were being captured on a clean matrix. A dramatic increase in pressure drop was observed when particles were deposited onto already deposited particles.

For those magnetic separators, where a flow distribution system is necessary to maintain a uniform flow distribution, the pressure drop across such a system will be large compared with the drop across the matrix alone. The effect of the flow distributor and of matrix loading on the pressure drop is shown in Fig. 4.85.

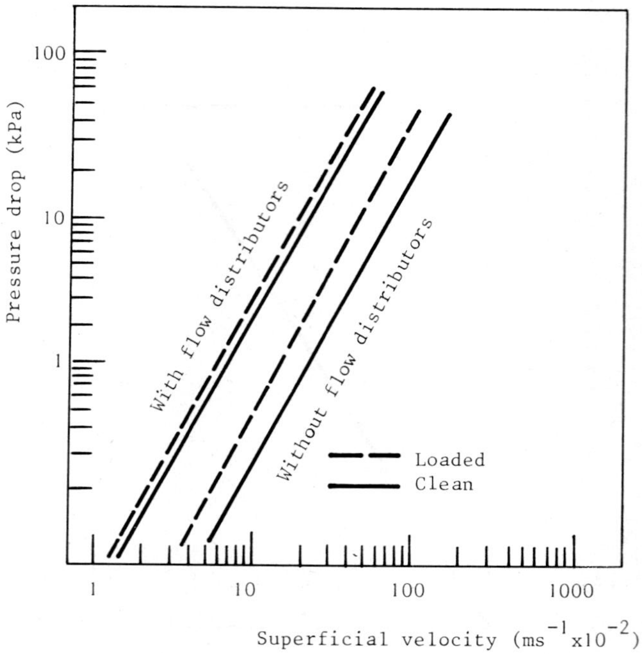

Fig. 4.85 The effect of flow velocity on pressure drop across the matrix (after Troy et al. [T12]).

IV.7 PULP DENSITY

Although the pulp density or the percentage of solids affects the throughput of a separator and matrix loading, its effect on metallurgical results is usually small. Recovery often does not depend on the pulp density in a wide range of densities. In other cases, recovery can increase with increasing pulp density [M11, H1], which is probably a consequence of non-selective capture. This is demonstrated by a frequent drop in the grade of the magnetic concentrate, as shown in Fig. 4.86. In exceptional cases, an increase in pulp density results not only in an increase of recovery, but also in the grade of the magnetic concentrate, as demonstrated in Fig. 4.87.

The effect of pulp density on the recovery of uranium, and on the grade of the magnetic product of uranium-gold leaching residues, for different matrices, is shown in Fig. 4.88.

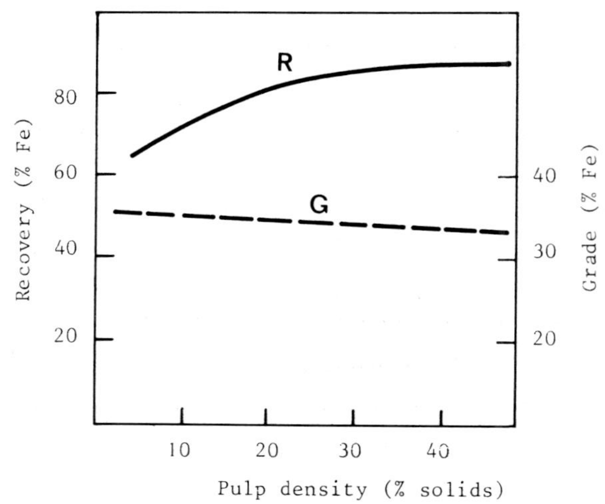

Fig. 4.86 The effect of pulp density on recovery and grade of magnetic concentrate of siderite ore (after Hencl [H1]).

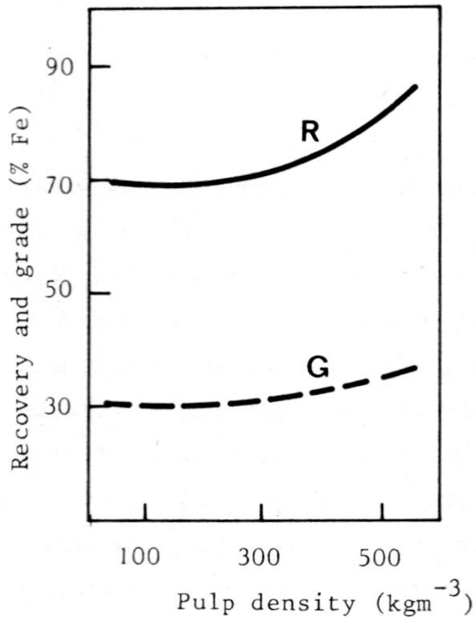

Fig. 4.87 The effect of pulp density on recovery and grade of magnetic concentrate of siderite ore (after Hencl and Šebor [H35]).

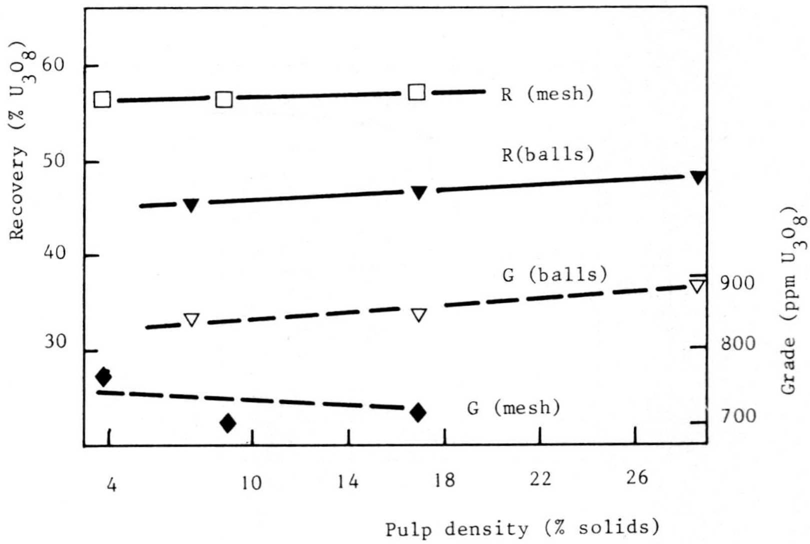

Fig. 4.88 The effect of pulp density on recovery and grade of magnetic concentrate of uranium-gold leach residues, **B** = 1.1 T[S29].

As the effect of the pulp density (within a certain range) on the metallurgical performance of a magnetic separator is small, it is advantageous to use the highest possible pulp densities compatible with permissible matrix loading and feed rate. As a rule, in beneficiation of ores it is possible to vary pulp density between 20 and 55 per cent solids, while with fine clays the upper limit is probably about 30 per cent solids.

IV.8 FEED RATE

The feed rate, which is closely associated with matrix loading, is an important variable in determining the cost-effectiveness of a magnetic separator. The higher the feed rate, the higher the treatable tonnage per machine resulting in lower capital and operating costs per tonne treated. The effect of feed rate on production rate is shown in Fig. 4.89.

At low and moderate feedrates, for which the permissible matrix loading is not exceeded and for which the matrix is not flooded and the continuity of the slurry flow has not yet been established, the interstitial velocity of the slurry is high. In addition the erosion force acting on the deposited particles is large, recovery decreases and the grade of the magnetic concentrate increases with increasing feed rate, as shown in Figs. 4.89 and 4.90.

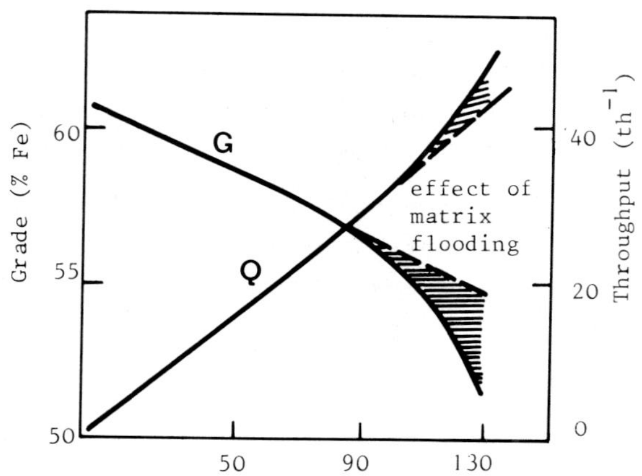

Fig. 4.89 The effect of feed rate on production rate and grade of magnetic concentrate of hematite ore. (Jones DP317 magnetic separator). (After Beech-Jones et al.[B34]).

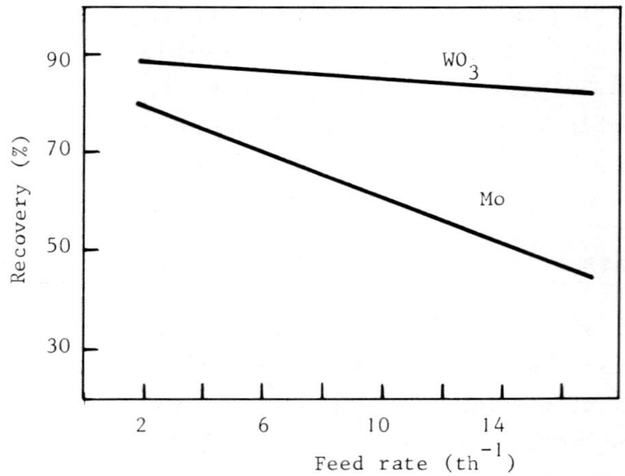

Fig. 4.90 Recovery of non-sulphide molybdenum and tungsten, as a function of feed rate. (Jones separator, matrix: grooved plates).(After Masoner [M11]).

At high feed rates, when the matrix is flooded, interstitial velocity is reduced due to the resistance of the matrix bed, and recovery increases with increasing feed rate. On the other hand, the grade of the concentrate decreases due to the reduced selectivity of the process and to the carry-over of gangue into the magnetic fraction. The effect of matrix flooding is illustrated in Fig. 4.89.

The effect of feed rate on the metallurgical performance of a continuous magnetic separator for reducing the concentration of impurities in fluorspar is illustrated in Fig. 4.91.

The feed rate also affects the metallurgical performance of a low-intensity drum separator, as shown in Fig. 4.92. It can be seen that the concurrent regime of operation is more sensitive to feed rate variations than the countercurrent mode.

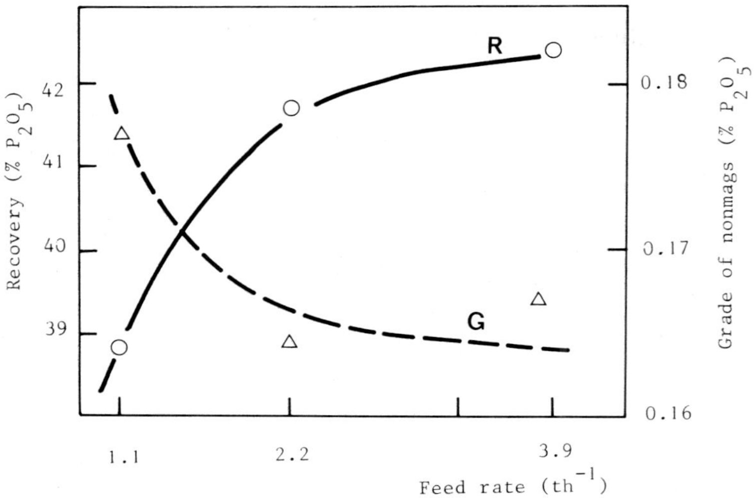

Fig. 4.91 Recovery of P_2O_5 and the grade of the non-magnetic product, as a function of feed rate. (Eriez CF-60 WHIMS separator, matrix: 7 mm balls). (After Jonker [J8]).

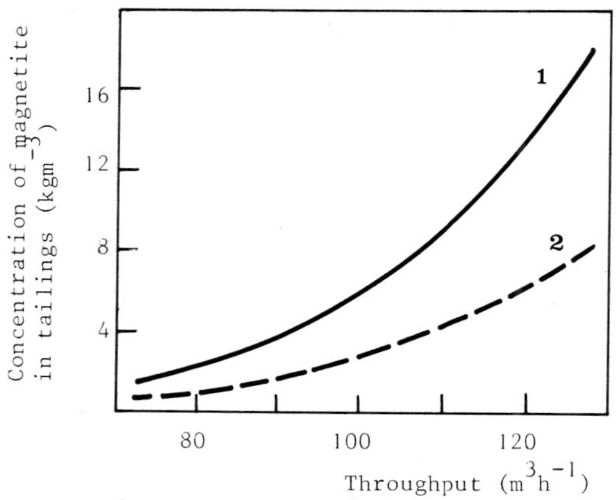

Fig. 4.92 The effect of feed rate onto a magnetic drum separator, on the losses of magnetite in the tailings. Head concentration of magnetite 140 kg m^{-3}, drum diameter 0.9 m, length of separation zone 1.2 m, speed of rotation 9 r/min. 1 - concurrent mode, 2 - countercurrent mode. (After Karmazin et al. [K15]).

IV.9 RINSE AND FLUSH WATER

The metallurgical performance of a matrix magnetic separator can be significantly controlled by varying the volume of rinse and flush water.

Rinse water is added at the end of the magnetic head of a carousel magnetic separator. This rinse water is used to wash out mechanically entrained non-magnetic particles and lightly held unliberated middling particles.

Flush water is used to remove the magnetic product from the matrix. Matrix flushing is performed outside the magnet station in continuous magnetic separators, or when the magnetic field is switched off in cyclic separators.

The quantity and velocity of rinse water that is used can have a very marked influence on the separation obtained. An increase in the volume of rinse water will normally result in a decrease in mass of the magnetic fraction and in recovery. The grade of the magnetic product often increases dramatically with an increasing volume of rinse water, although beyond a certain amount of rinse water this effect becomes negligible.

The effect of the volume of rinse water on the mass yield into the magnetic product is shown in Fig. 4.93 while the effect on recovery and grade of the magnetic product from iron ores is depicted in Fig. 4.94. The effects of the volume of rinse water on the concentrate grade and mass recovery of iron are summarized in Fig. 4.95, where the detrimental effect of an excessive flowrate of water on recovery can be noticed.

The optimum volume of rinse water must be determined experimentally since a wide array of factors, like particle size distribution, magnetic susceptibilities of valuable mineral and gangue and degree of liberation affect the influence of rinse water on the metallurgical performance of a magnetic separator. Some typical values of rinse water volume, employed in commercial matrix magnetic separators are summarized in Table 4.15.

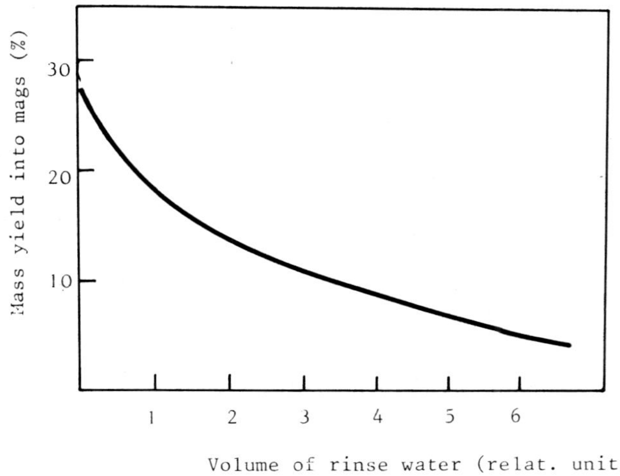

Fig. 4.93 The effect of volume of rinse water on mass yield into the mags from uranium ore (after Stone [S16]).

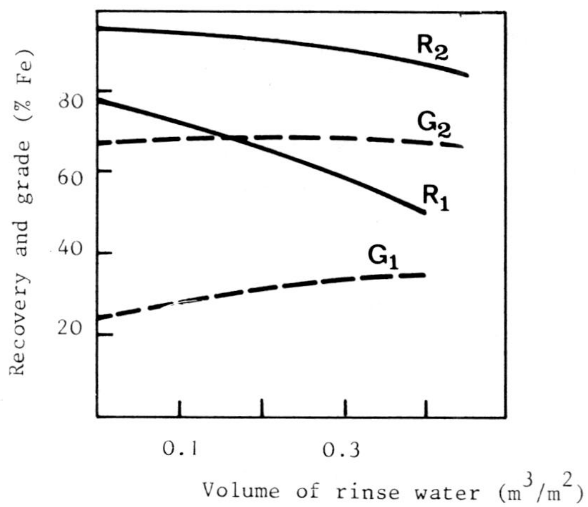

Fig. 4.94 The effect of volume of rinse water on recovery and grade of mags from geothite (1) and hematite (2) ores (after Hencl [H1]).

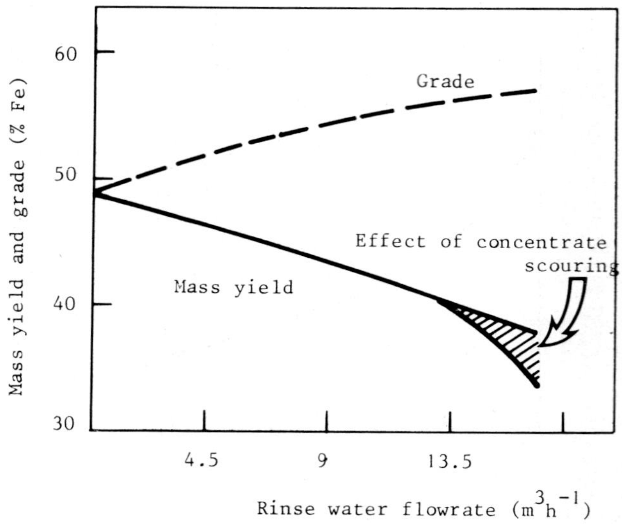

Fig. 4.95 The effect of the flowrate of rinse water on the metallurgical performance of Jones DP 317 magnetic separator (after Beech - Jones et al. [B34]).

TABLE 4.15

Water consumption (per tonne of feed) in applications of commercial continuous HGMS separators

Separator	Rinse water (m^3/t)	Flush water (m^3/t)	Total water (m^3/t)	Reference
Jones			2(8R plates) 2.5(4R plates)	
	0.5	1.1		
	0.7	1.7		[F19]
	1.0			[L20]
Boxmag Rapid	0.8	2.8		
	0.35	1.0		[F19]
MRVK	1.2	1.6		
VMS	0.75	0.80		[C8]
Carpco	1.0			
Eriez	1.2			[S30]
Readings	0.54	1.0		[F19]
SALA HGMS	2-3	1-2.5		[F19]

As mentioned above, the function of flush water is to remove captured magnetic material from the matrix. Flushing is carried out either outside the magnet or with the field switched off, but owing to the presence of a stray magnetic field and of the remanent magnetization of the matrix collectors, the volume and pressure of flush water must be so high, that the matrix can be kept clean for a sufficiently long time. The effect of the volume of flush water on the metallurgical performance of a separator is negligible. As seen from Table 4.15, the consumption of flush water is quite high, and a limited reduction can be achieved with a combination of water and compressed air. Consumption of flushing water in the application of cyclic HGMS is summarized in Table 4.16.

TABLE 4.16
Consumption of flush water in the application of cyclic HGMS

Application	Consumption of flushing water expressed in terms of			
	Canister volumes	Percentage of processed slurry volume	Flow velocity (ms^{-1})	Ref.
Filtration of power plant cooling water	11	0.01 - 0.02		[M29]
Kaolin beneficiation	10	125 - 250	0.06	[L19]

IV.10 THE EFFECT OF PARTICLE SIZE

In Chapter I we saw that the performance of a magnetic separator is determined by the interplay of the forces acting on a particle, namely between the magnetic dipolar force on the one hand, and the fluid drag, inertial and gravitational forces, on the other. All these forces depend, in various ways, on a particle size, and every user of magnetic separation knows very well that the particle size is probably the most important factor that limits the applicability of various types of magnetic separators. The approximate particle-size ranges of current high-intensity magnetic separators are shown in Fig. 4.96. It can be seen that any magnetic separator operates only in a limited particle size range and the upper and lower limits of this range are determined by the relative importance of competing forces.

In matrix magnetic separators the upper size limit of particles that can be trapped is determined mainly by the type of matrix used, although for particles in the upper size range the role of the gravitational force becomes appreciable and capture efficiency can decrease with an increasing particle size as low as 100 μm.

Of great importance is the establishment of the lowest size limit of particles that can be recovered by HGMS, and the effect of fine slimes. Experiments prove that the presence of fine particles in the pulp causes deterioration of the results of magnetic separation, lowering the recovery in the magnetic fraction and the grade of the concentrate. The question "how fine a particle can be recovered by HGMS?" is clearly highly relevant, not only from the theoretical point of view, but also for an engineer.

The lower size limit of particles to be recovered affects not only the recovery and the selectivity of magnetic separation process, but has a signal impact on the efficiency of the overall technological flowsheet, the part of which magnetic separation

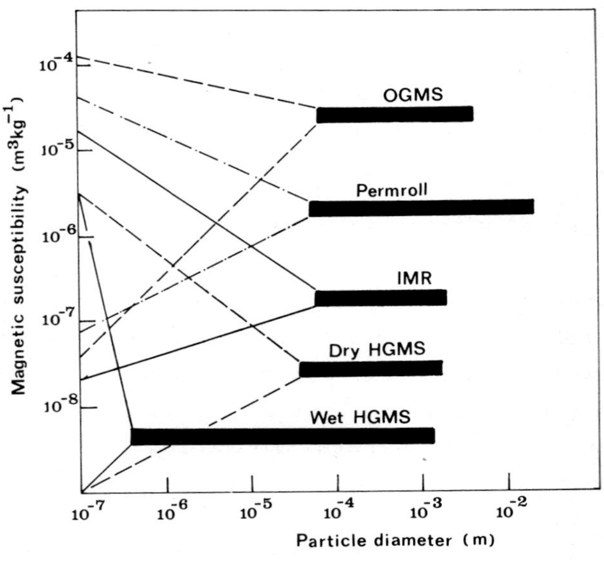

Fig. 4.96 Particle size ranges of high-intensity magnetic separators.

usually is. The excessive amount of slimes in the pulp often requires the pulp flow velocity through the separator to be reduced in order to improve the recovery. Such an approach, however, results in deterioration of the selectivity of magnetic separation and in a reduction of the throughput of a magnetic separator. The other possible solution is to deslime the pulp, although such an approach can result in high losses of a valuable mineral into the slimes, and in the need to install a frequently costly desliming circuit. If desliming is to be avoided, a magnetic separator of a greater throughput is needed, which increases the capital and running costs.

It is thus clear that the knowledge of the separator performance as a function of particle size, and of the lower particle size limit of the recoverable particles is of the utmost importance.

IV.10.1 The experimental investigations into the particle size effects

It has been established in numerous experimental tests that the particle capture by the matrix of a magnetic separator is strongly influenced by the size of the particles. A typical example of the particle size effect is shown in Figs. 4.97 and 4.98. It can be seen that in most cases there is an optimum particle size for which the recovery is the maximum, while for very fine or coarse particles the recovery drops, as a consequence of the adverse effect of competing mechanical forces.

It is of interest to note the insignificant effect of the magnetic induction greater than 1.6 T in the whole range of particle sizes.

The analogous influence of particle size on the performance of a magnetic separator was observed by several other investigators [S63, B36, S62]. The reduced capability of a magnetic separator to recover fine and coarse uranium particles from the leach residues is illustrated in Fig. 4.99. The best average recoveries were obtained for the fraction between 53 and 25 μm, amounting to over 90 per cent for low flowrates and over 80 per cent for high flow rates. Fractions smaller than 25 μm and greater than 106 μm gave lower recoveries.

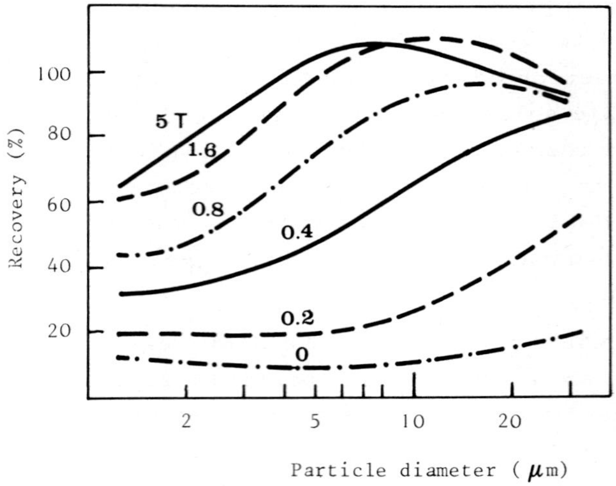

Fig. 4.97 Recovery of $MnCO_3$ versus particle diameter, for different values of background magnetic induction. Flow velocity 0.01 m s^{-1}, matrix: stainless steel wires, diameter 76 μm. (After Maxwell et al. [M28]).

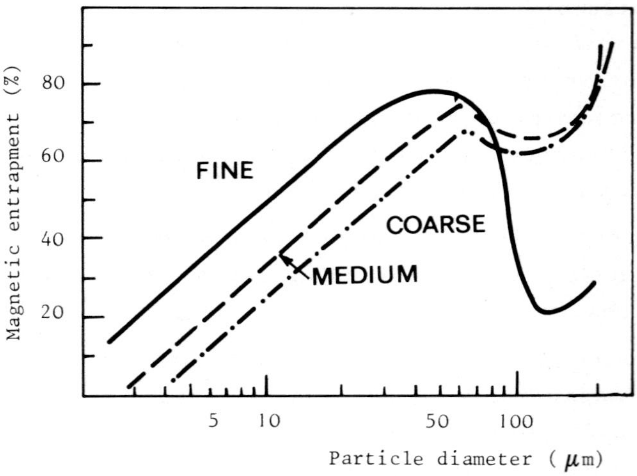

Fig. 4.98 Magnetic capture of hematite particles as a function of particle size, for three different classes of stainless steel wool. Background magnetic induction 0.1 T, slurry flow velocity 0.06 m s^{-1}. (After Clarkson and Kelland [C31]).

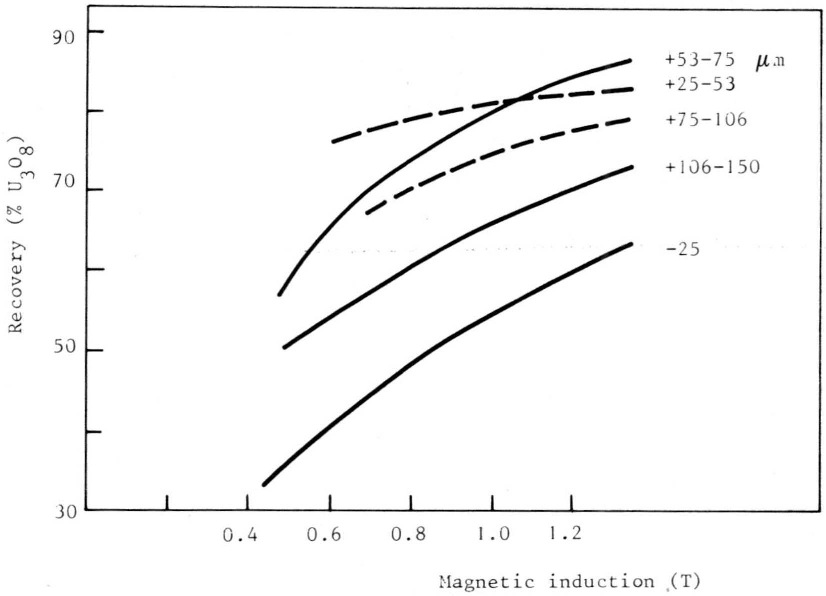

Fig. 4.99 Recovery of U_3O_8 as a function of magnetic induction for various size fractions. Matrix mesh, flow velocity 0.08 m s^{-1}, T-configuration. [S62].

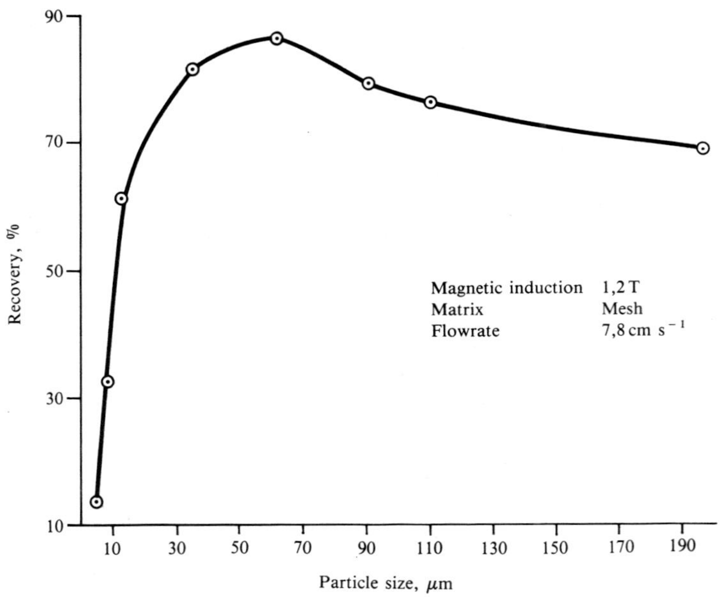

Fig. 4.100 Recovery of U_3O_8 as a function of particle size , the mean particle diameter being taken to represent each of the size fractions [S62].

Figure 4.100 summarizes the effect of particle size on recovery, while Fig. 4.101 shows the influence of particle size on the grade of the magnetic fraction, the distribution of uranium in different size fractions of the feed and the concentration ratio. It is apparent that recovery and grade decrease very sharply for particle size smaller than 15 μm, although the dependence of the grade on particle size is due partly to the distribution of the uranium in various size fractions of the feed.

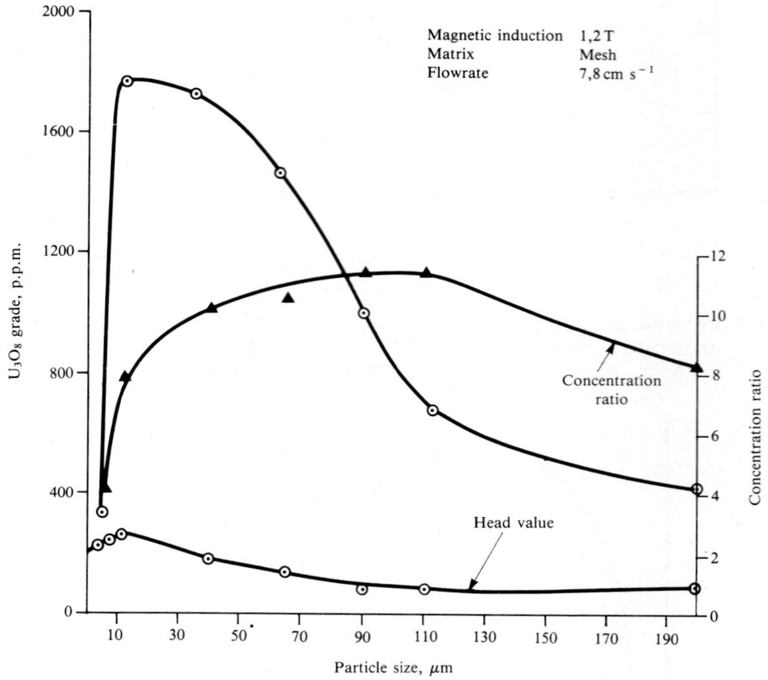

Fig. 4.101 Grade of magnetic concentrate, concentration ratio and head value as a function of particle size [S62].

IV.10.2 The determination of the limiting particle size from theoretical models

Theoretical studies of idealized simple models such as those involving a balance of forces or capture on a single wire provide an understanding of the basic mechanisms involved in the capture process. If these models are to be useful for a selection of initial operating conditions in the separation of industrially treated mineral systems, they must indicate how fine a particle can be recovered by HGMS.

Attempts have been made [S62, S64] to calculate the limiting particle sizes from current models of magnetic separation and to compare their predictive capabilities with experimental results.

Oberteuffer's model [O1]

Oberteuffer's model is based on a simple balance of forces acting on the particles in a magnetic separator. The condition for successful separation of magnetic from non-magnetic particles is that the magnetic force shall be higher or equal to the vector sum of competing forces (eq. (1.1)), a condition simplified here as

$$\vec{F}_m + \vec{F}_d \geqslant 0 \tag{4.23}$$

This model of the balance of forces generally gives higher values for the limiting particle size that can be captured in a magnetic separator than do the other models [S64].

Watson's model [W9]

The dynamics of a magnetizable particle placed in an external magnetic field can be described by an equation for the motion of the particle. Watson [W9] showed that the probability of a particle being captured by a single matrix collector is proportional to the ratio $\mathbf{v}_m/\mathbf{v}_o$ (eq. (3.31)) and the condition of strong capture is then

$$\frac{v_m}{v_o} \geqslant 1 \qquad (4.24)$$

This model gives a better description of the behaviour of a single particle in the vicinity of an isolated matrix element. On the other hand, it does not take into account the probability of the particle collision with matrix collectors, due to "non-magnetic" mechanisms (interception, gravity, inertia etc.), negligible compared to magnetic capture in a single-collector approach, but of great importance in a real multi-collector matrix [S51]. The model also treats magnetic field strength and flow velocity in a symmetrical way, a highly unlikely feature in the magnetic separation of minerals by multi-collector matrices, in view of complicated magnetic behaviour of minerals as a function of magnetic field and of a complex effect of the Reynolds number on the capture efficiency. The applicability of Watson's model to magnetic separation in a real matrix is therefore of limited value, and it predicts the limiting particle size that is almost one order of magnitude larger than that indicated by the experimental results [S64, S62]. The situation is illustrated in Fig. 4.102 where the values of ratio $\mathbf{v}_m/\mathbf{v}_o$ for U_3O_8 particles are shown as a function of particle size.

As a result of the condition of capture (eq.(4.24)), Watson's model predicts the capture of 10 μm particles (5 μm radius) only under extreme conditions (2 T, 0.026 m s^{-1}), as shown in Fig. 4.102. On the other hand, for the experimental conditions used in the tests, the limiting particle size that could be captured at 1 T, 0.078 m s^{-1} was 25 μm. However, the experimental results [S62] show that even for values of $\mathbf{v}_m/\mathbf{v}_o$ as low as 0.2 the recoveries of uranium from the fraction smaller than 25 μm, were high (62 per cent) with very high grades (over 2000 p.p.m.), although theory predicts that strong capture can be expected only for $\mathbf{v}_m/\mathbf{v}_o > 1$. For coarse fraction (between 150 and 106 μm) $\mathbf{v}_m/\mathbf{v}_o$ values as high as 50 were used, but the efficiency of capture did not improve substantially.

In tests performed by Watson et al. [W26, W28] at high magnetic fields (up to 7 T), $\mathbf{v}_m/\mathbf{v}_o$ values amounted to between 5 and 236, so that the probability of capture would be high. Although the mass yield in these tests was high, the recovery of uranite was only moderate (up to 60 per cent) and the quality of the magnetic concentrate was unacceptably low.

The model thus fails to correlate correctly the performance of a magnetic separator with the $\mathbf{v}_m/\mathbf{v}_o$ ratio and its capability to predict the limiting particle size is limited.

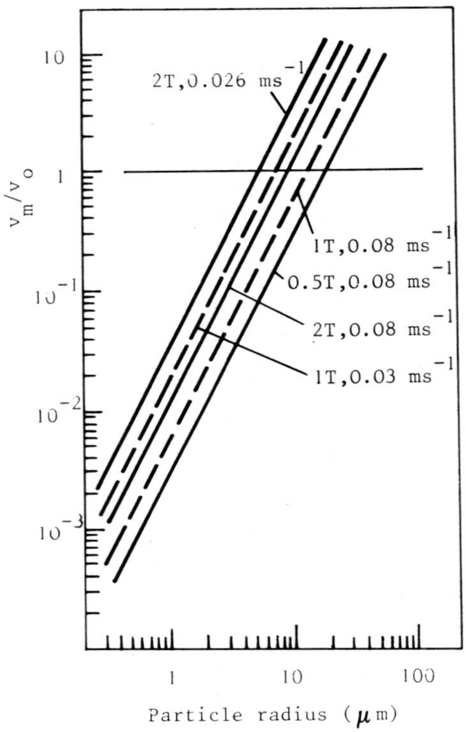

Fig. 4.102 The ratio $\mathbf{v}_m/\mathbf{v}_o$, as a function of the radius of uraninite particles. Matrix: mesh, wire diameter 1.6 mm, saturation polarization 1.7 T [S62].

Nesset and Finch's model [N2]

The model developed by Nesset and Finch [N2] considers the balance between magnetic and Blasius-type drag forces around a matrix element, and shows that the build-up of particles on a matrix is a function of the loading number $\mathbf{N}_L$ given by eq. (3.63). The loading number represents the ratio of the magnetic force to the fluid shear force and permits the determination of limiting particle size. The lower limit corresponds to the condition that the volume loading γ_v (eq. (3.65)) is equal to zero, which is satisfied for $\mathbf{N}_L$ = 2.5. The practical lower limit can be loosely determined from a requirement γ_v = 1, which corresponds to $\mathbf{N}_L$ = 25.

For the same experimental conditions as those used to analyse Watson's model, Nesset and Finch's model gives the practical limiting size of uranite particles as 20 μm, while the lower limit is 2 μm. Again we see that the condition of capture $\mathbf{N}_L$ = 25 gives a particle size much higher than those indicated by the experimental results. The comparison of the limiting diameter of uraninite particles determined from the Watson's model ($\mathbf{v}_m/\mathbf{v}_o$ = 1) and the Nesset and Finch's model ($\mathbf{N}_L$ = 25) is shown in Fig. 4.103.

It can thus be concluded that the theoretical models of particle capture in HGMS do not lead to a correct prediction of the limiting particle size recoverable by a magnetic separator. The particles that can be recovered have a diameter of an order of magnitude smaller than that predicted from the models.

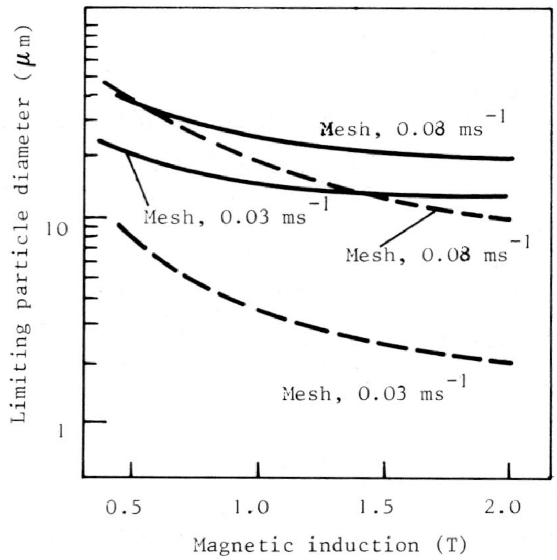

Fig. 4.103 The limiting particle diameter of uraninite particles, as a function of magnetic induction for a mesh matrix, calculated according to the models of Watson and Nesset and Finch. Solid lines $\mathbf{v}_m/\mathbf{v}_o$, dashed lines $\mathbf{N}_L$ [S62].

IV.11 THE SPEED OF ROTATION

IV.11.1 Rotation of a carousel in HGMS

In continuous high-gradient magnetic separators the speed of rotation of a carousel or of a drum influences the retention time of the slurry in a magnetic field and the loading of the matrix. The maximum permissible speed of rotation is determined by the flow velocity of the slurry through the matrix, by the length of the magnetic zone and by the matrix height. As shown in Fig. 4.104 the permitted maximum speed of the carousel can be ascertained from the condition that the particle which enters the matrix at the point A leaves the matrix at point B which is still inside the magnetic zone. The retention time of a particle in the matrix is $\mathbf{t}_R = \mathbf{h}/\mathbf{v}_s$, where **h** is the matrix height and $\mathbf{v}_s$ the slurry flow velocity. The rotation speed of the carousel must satisfy the condition $\mathbf{v}_c \leq \mathbf{l}_m/\mathbf{t}_R$, where $\mathbf{l}_m$ is the length of the magnetic zone. Therefore,

$$v_c \leq \frac{l_m v_s}{h} \qquad (4.25)$$

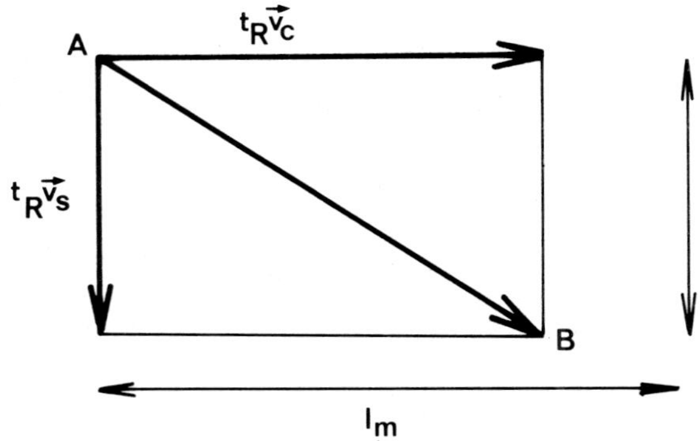

Fig. 4.104 The slurry flow in a continuous magnetic separator.

As an illustration, let us assume that a slurry flows through the matrix with a height of 0.2 m at a velocity of 0.1 m s^{-1}. The retention time of a particle in the matrix is thus 2 s. If the length of the magnetic zone l_m is 1 m, than the maximum rotational velocity 0.5 m s^{-1} of the carousel continues to satisfy condition given by eq. (4.25).

By increasing the speed of rotation the loading of the matrix decreases, at a constant feed rate, or the throughput of the separator increases, at a constant matrix loading. One would thus expect that, at a constant feed rate the recovery would increase and the grade of the magnetic concentrate decrease, as a consequence of a reduced matrix loading.

The enhancement of recovery and the reduction of the grade with increased speed of rotation is actually often observed in practice, as demonstrated by Fig. 4.105.

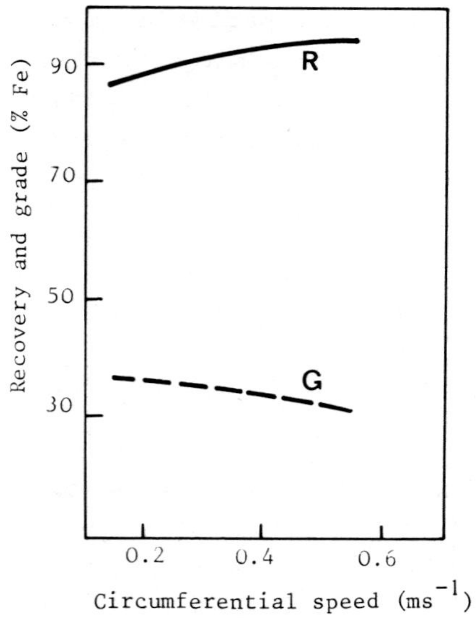

Fig. 4.105 The effect of the peripheral velocity of a drum of a VMS magnetic separator on the metallurgical results of separation of siderite ore (after Hencl and Šebor [H35]).

The grade of the magnetic concentrate can be further impaired, at high carousel speeds, by a carry-over of non-magnetic particles past the tailings launder, into the magnetic product. The elevation of the mechanical entrainment at zero-applied field, as a function of the rotational speed, is shown in Fig. 4.106. The figure also suggests that different types of matrices exhibit a different response to variations in speed of rotation.

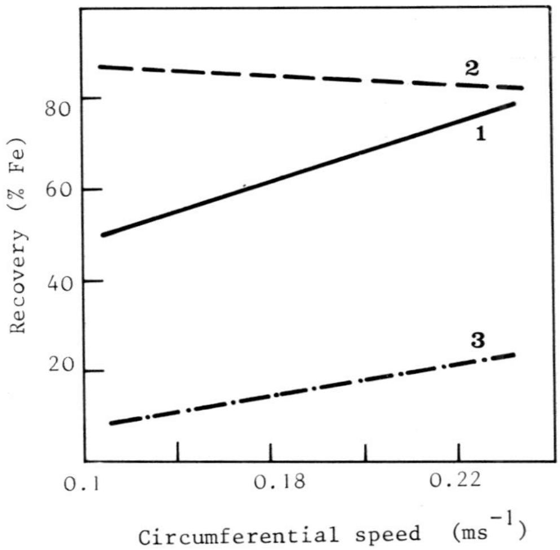

Fig. 4.106 The effect of the speed of rotation of a drum on metallurgical performance of an MRVK separator applied to beneficiation of siderite ore.
1 - recovery of iron; matrix: rods, diameter 12 mm,
2 - recovery of iron; matrix: 12 mm balls
3 - mechanical capture.
(After Hencl [H1]).

IV.11.2 Rotation of a drum in low-intensity magnetic separators

The speed of rotation of the drum affects the performance of a low-intensity magnetic separator. Karmazin et al. [K15] found that the optimum speed of rotation corresponds to the minimum slip of a material, when the surface velocity of a drum is close to the flow velocity of the pulp. The effect of speed of rotation is depicted in Fig. 4.107.

Similarly, the length of the separation zone of a drum magnetic separator has a considerable effect on its performance. Karmazin [K15] established that by doubling the length of the zone (from 600 mm to 1200 mm), the losses of magnetite into the non-

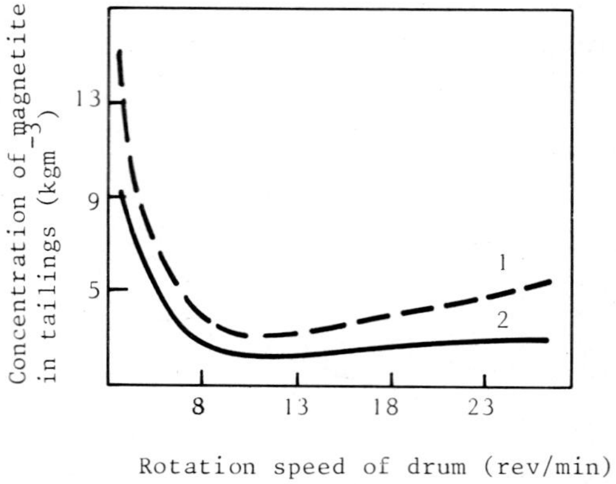

Fig. 4.107 The effect of the speed of rotation of a drum on the losses of magnetite in the tailings. Head concentration of magnetite: 140 kg m^{-3}, drum diameter: 0.9 m, length of the separation zone 1.2 m, feed rate 108 m^3/h. 1 - concurrent mode, 2 - countercurrent mode. (After Karmazin et al. [K15]).

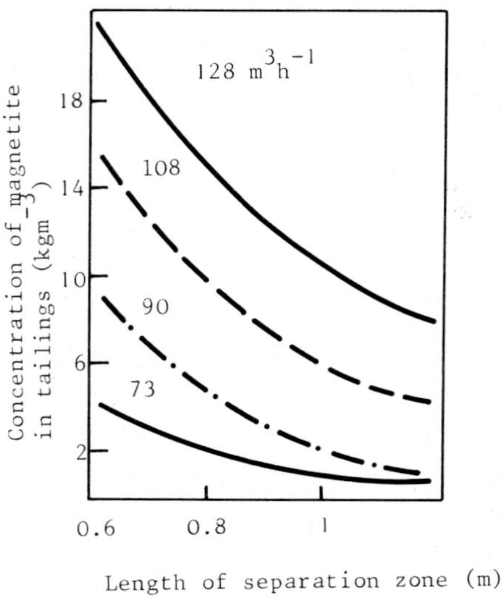

Fig. 4.108 The influence of the length of the separation zone on the concentration of magnetite in the tailings for various feedrates. Head concentration of magnetite 140 kg m^{-3}, drum diameter 0.9 m, speed of rotation 9 r/min, countercurrent operation. (After Karmazin et al. [K15]).

magnetic fraction decreased by a factor of 7 to 10 at the same value of magnetic field strength. It was further found, that an increase in the magnetic field strength cannot compensate for the losses due to insufficient zone length.

The length of the separation zone also affects the sensitivity of a separator to the feed rate, as shown in Fig. 4.108. As the length of the zone increases this sensitivity decreases, while for the greater length investigated (1200 mm) the separator was found to tolerate a wide range of feed rates.

IV.12 PRE-TREATMENT OF PULP

When an ore is to be beneficiated by magnetic separation, it often requires preliminary treatment in order to improve the efficiency of separation . The most frequent operations used to prepare the ore for magnetic separation are magnetizing roasting, magnetic scalping, dispersion, pH adjustment and desliming.

IV.12.1 Magnetizing roasting

Many weakly magnetic minerals can be converted to more strongly magnetic compounds by heating. For instance, weakly magnetic oxides, hydroxides and carbonates of iron can be transformed to strongly magnetic magnetite or maghemite by heating in a reducing atmosphere.

Although the technology of reduction roasting of iron ores, followed by low-intensity magnetic separation and flotation, is well developed [D2, D1, K24] its wider application is hampered by high running and capital costs. Nevertheless the energy consumed in roasting is partially compensated by decreased grinding costs and by the energy recovered in the conversion of magnetite to hematite during pellet induration [H39].

The fuel consumption for the ore heating and reduction and for the evaporation of moisture and crystalline water amounts to six to ten per cent of the mass of the ore roasted. IN CGOK in the USSR, where reduction roasting is used to beneficiate the oxidized quartzites, fuel consumption is as high as 50 m^3 of natural gas per 1 tonne of the roasted ore [K24]. The roasting temperature ranges from 500° C to 800° C, while the duration of heating which is of the order of minutes or hours strongly depends on the type of the ore and particle size. A complicating factor in the reduction roasting of iron ores is usually a high degree of magnetic flocculation of artificial magnetite, as a consequence of increased remanent magnetization and coercive force. High coercive force appears as a result of an abrupt cooling of an ore after its removal from a furnace. The cooling-induced internal tension creates the crystallites of single-domain dimensions and coercive force increases. Such a high coercive force is not, however, an inherent property of the roasting products, but arises as a consequence of amended size distribution [K15].

Karmazin et al. [K24] proposed a new technology of iron ore treatment which could reduce running costs and improve metallurgical parameters. It was suggested that in order to reduce fuel consumption, reduction roasting should be applied to a -0.5 mm fraction from which a large volume of gangue was removed by high-gradient magnetic separation. The technique was applied to oxidized quartzites from Krivoi Rog, which were ground to 80 per cent -71 μm and then upgraded by a high-gradient magnetic separator. The magnetic product, containing 49.9 per cent of iron, at the recovery of 85.5 per cent, was roasted in a natural gas atmosphere at 750° C for 3 minutes, followed by three-stage, low-intensity magnetic separation at the magnetic field strength of 180 kA m^{-1} (2.3 kOe). The grade of the final concentrate was 64.5 per cent of iron at the overall recovery of 79.4 per cent. The metallurgical parameters were better than with conventional flotation technology and fuel consumption was reduced to 20 m^3 per tonne of feed.

The oxidizing roasting of ilmenite from beach sands is practised at Richards Bay Minerals (South Africa). Ilmenite recovered by HGMS is roasted in a fluidized bed at a temperature of 800° C for 40 minutes, thereby increasing magnetic susceptibility from 2.4 x 10^{-6} m^3kg^{-1} (190 x 10^{-6} cm^3g^{-1}) to 15 x 10^{-6} m^3kg^{-1} (1192 x 10^{-6} cm^3g^{-1}). The CO gas derived from the smelter furnace, at a rate of 40 m^3 per tonne of roasted ilmenite, is used to dry and roast the ore, which is further upgraded by a dry low-intensity magnetic separator followed by induced magnetic roll.

The de-coppering of galena and molybdenite concentrates by the removal of chalcopyrite can be significantly enhanced by roasting the chalcopyrite. It was reported by Makarow et al. [M31] that by heating chalcopyrite at the oxidizing atmosphere to a temperature between 250 and 500° C for 10 minutes the magnetic susceptibility of the product increased by three orders of magnitude. The magnetization temperature was found to increase with increasing particle size and dilution of the sample.

Dawson [D18] has shown that reduction roasting of the oxidized ore containing gold and platinum group metals (PGM - platinum, palladium and rhodium) followed by dry HIMS, resulted in a significant upgrading of the PGM values in the magnetic fraction. It was observed that the quantity of the magnetic fraction increased with the temperature of the reducing roast (from 610 to 655° C) and then dropped considerably at temperatures of 745 and 790° C. Increased roasting time (from 20 minutes to 1 hour) also appeared to have an effect similar to that of an increase in temperature between 40 and 50° C.

IV.12.2 Magnetic scalping

Strongly magnetic impurities that are often present in ores cause the blockage of the matrix and of the separation gap in dry high-intensity magnetic separators. The presence of these impurities also impairs the grade of the magnetic concentrate since the gangue and unliberated particles are locked in the clusters of ferromagnetic impurities and report to the magnetic fraction.

The detrimental effect of ferromagnetic impurities and tramp iron can be reduced in a number of ways, as discussed in Section (IV.5.9), although none of these techniques is fully efficient, partly because these impurities are present mainly in the finest fractions, often smaller than 25 μm. In many cases the only solution is magnetic scalping, either with a low-intensity magnetic separator or with a high-gradient magnetic separator, where a matrix made of a paramagnetic metal can be used.

Magnetic scalping with an aluminium expanded metal matrix or a 304 stainless steel mesh, applied to the pretreatment of uranium-gold leaching residues, was found to improve the metallurgical performance of a magnetic separator. As can be seen in Table 4.17,

TABLE 4.17

The effect of scalping on uranium-gold leaching residues on a matrix of SS304 at 1.2 T [S29]
Magnetic induction 1.1 T
Flow velocity 0.06 m s^{-1}

Sample	Matrix	Not scalped			Scalped		
		Mass yield of mag. %	U_3O_8 grade of mag. g/t	U_3O_8 recovery %	Mass yield of mag. %	U_3O_8 grade of mag. g/t	U_3O_8 recovery %
Full sample	Balls	10,0	838	53,8	6,6	1031	46,4
	Mesh 10	11,5	740	56,5	8,4	912	53,7
Fraction <25 μm	Balls	4,0	1140	25,7	4,3	1346	28,9
	Mesh 10	7,4	1012	35,6	5,6	1217	34,7
Fraction >25 μm	Balls	9,1	897	76,2	8,4	911	69,3
	Mesh 10	9,6	824	74,5	9,5	842	76,0

TABLE 4.18

The effect of scalping on size fractions of uranium-gold leaching residues on a matrix of SS304 at 1.2 T [S29]
Magnetic induction 1.1 T
Flow velocity 0.06 m s^{-1}

Fraction μm	Matrix	Not scalped			Scalped		
		Mass yield of mag. %	U_3O_8 grade of mag. g/t	U_3O_8 recovery %	Mass yield of mag. %	U_3O_8 grade of mag. g/t	U_3O_8 recovery %
<25	Balls	4,0	1140	25,7	4,3	1346	28,9
	Mesh 10	7,4	1012	35,6	5,6	1217	34,7
>25<53	Balls	7,6	1523	69,5	7,4	1583	70,8
	Mesh 10	8,5	1439	70,1	9,7	1394	81,5
>53<75	Balls	9,1	1255	79,7	8,5	1319	77,8
	Mesh 10	10,5	1143	78,8	9,2	1388	77,4
>75<106	Balls	9,4	913	80,5	8,7	967	74,2
	Mesh 10	10,2	864	84,5	9,9	887	81,6
>106<150	Balls	8,2	518	73,1	7,6	569	68,1
	Mesh 10	8,2	813	76,7	8,0	527	66,5
>150	Balls	9,6	401	63,0	8,8	436	52,5
	Mesh 10	13,8	417	72,0	8,3	485	54,3

scalping reduces the mass yield into the magnetic fraction from the full sample by about 30 per cent, improves the grade of the magnetic fraction by about 23 per cent and reduces the recovery slightly. The effect of scalping on the fraction smaller than 25 μm follows a similar pattern; the fraction larger than 25 μm is unaffected.

Table 4.18 illustrates the effect of scalping for various size fractions of uranium-gold leaching residues. It can be seen that scalping improves the grade for the fine fractions up to between 75 and 53 μm, whereas the effect is negligible for the coarser fractions.

The results also imply that recovery does not increase for the fine fractions, and that it drops for the coarse fractions. Scalping reduces the mass yield into the magnetic fraction for all size fractions.

IV.12.3 Dispersion

The degree of selective magnetic separation that can be achieved with minerals depends on the liberation of individual mineral particles from each other. Thus, the complete dispersion of all particulate components is an essential prerequisite for the selective magnetic separation that follows. In this way, non-selective flocs originally present in the slurry must be broken down and the colloidal stability of the dispersed state must be established.

An examination of the various particle-particle interactions that were discussed in Sections III.5 and IV.13 suggests the following criteria for selective interactions (dispersion or coagulation) between particles of a valuable mineral in a mixture with other mineral components:

a) Particles of a valuable mineral should carry an electrical charge of the same sign and magnitude as the gangue minerals so that the probability of flocculation between them is minimized.
b) Electrical charge on the particles of the valuable mineral should be such that the repulsive interaction energy between them will be smaller than the attractive energy.

Although the addition of a dispersant is often required, overstabilization should be avoided, because the following step, magnetic separation or flocculation could be impaired.

Dispersion of natural ores is often difficult, owing to interference from dissolved ions and because it is not possible to draw reliable conclusions for selective aggregation or dispersion behaviour of a given slurry from results obtained with pure minerals [S61]. A large variety of chemicals are used as bulk dispersants, e.g. polyphosphates, polysilicates and tannins, which decrease the degree of hydration of the surface layers of particles.

Although dispersion is an inherent part of most flotation and flocculation flowsheets, its application to magnetic separation is not very common. Only with the advent of high-gradient magnetic separation capable of treating quasi-colloidal particles has the importance of a slurry dispersion been recognized. In kaolin beneficiation by HGMS, slurry dispersion by sodium hydroxide and

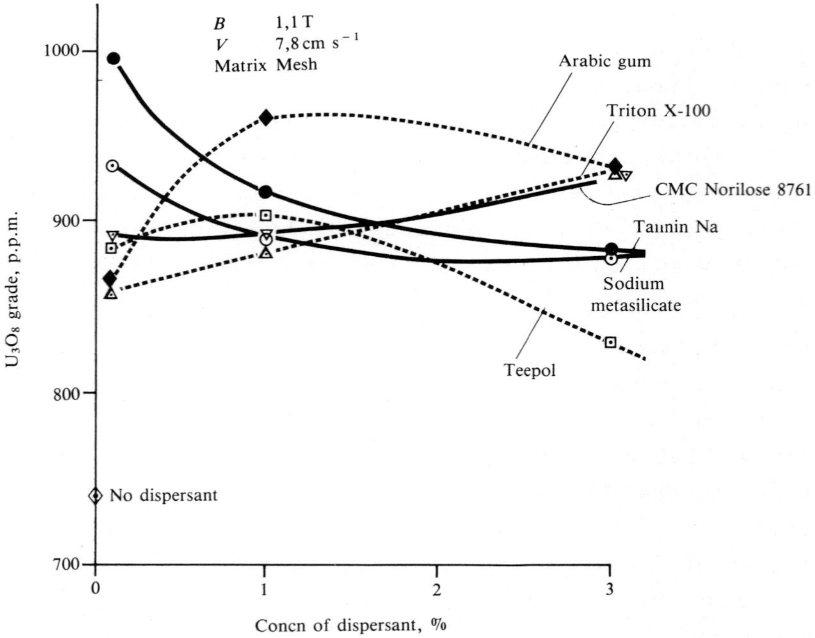

Fig. 4.108 The grade of magnetic concentrate from the whole sample of uranium-gold cyanidation residues as a function of the concentration of dispersant [S62].

sodium silicate is an essential part of pulp preparation. Likewise, dispersion was found remarkably beneficial in the upgrading of uranium-gold leaching residues by HGMS [S62]. As shown in Figs. 4.108 and 4.109, the use of a dispersant substantially increased the grade of the magnetic concentrate, the increase amounting to 35 per cent for the full sample and up to 55 per cent for the fraction smaller than 12 μm. Not all the dispersants were equally efficient and the best performance was observed for tannin Na. The optimum concentration of dispersants varies with the type of dispersant and particle size distribution of the ore. For a full sample, low concentrations (0.1 per cent) were sufficient, while for very fine fractions (smaller than 12 μm) the grade rose with increasing concentration of dispersant to at least 3 per cent by mass. This is because fine particles have a higher specific surface area than coarse particles. The recovery of uranium was not affected by the presence of dispersants.

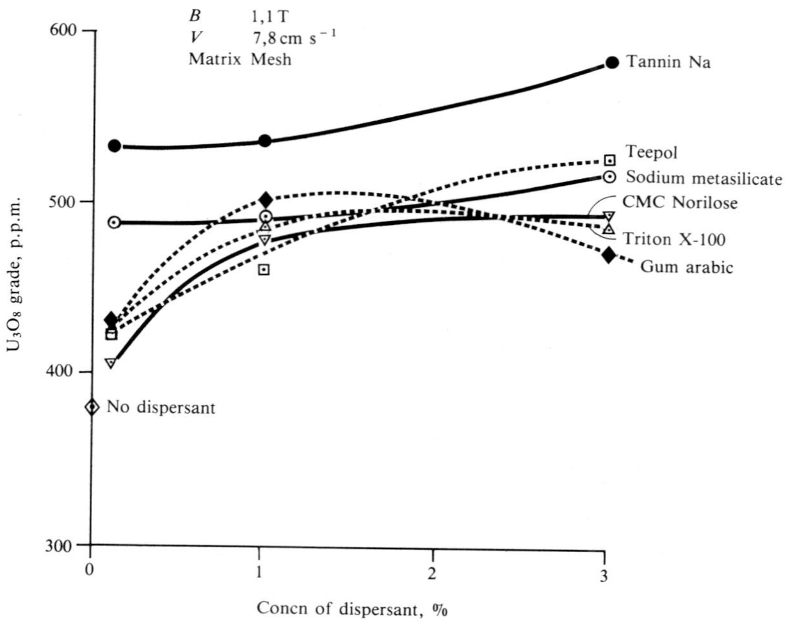

Fig. 4.109 The grade of the magnetic concentrate from the fraction of uranium-gold cyanidation residues, smaller than 12 μm as a function of the concentration of dispersant [S62].

A substantial improvement in the grade of the magnetic concentrate from Cu-Pb flotation tailings was also observed in pilot-plant tests with a carousel magnetic separator by adding dispersant [S30]. Although the type and concentration of a dispersant was not optimized, its presence doubled the grade of the concentrate and a slight increase in the recovery of copper was observed.

IV.12.4 The effect of pH

It has been shown in Section III.5 that for fine particles, treated by high-gradient magnetic separation, volumetric forces in a magnetic separator, viz. magnetic force, force of gravity and hydrodynamic drag, have to compete with surface forces. The question of whether a particle will be captured by a matrix element covered by particles previously deposited is determined, inter alia, by the surface characteristics of the particles. In practice, therefore, particles to be separated should be colloidally unstable, i.e. they should not repel each other and thus prevent contact. On the other hand, particles of gangue minerals must not coagulate with those of the valuable mineral lest they become entrained in the magnetic concentrate.

Electrostatic forces and van der Waals forces, are of comparatively short range and cannot be used to account for the transport of particles to the matrix. Nevertheless, these surface forces are of vital importance in the determination of attachment and strength of adhesion. A combination of flocculation (ionic or magnetic, or both) of the particles, of the transport of the particles to the matrix, and of the promoted or hindered adhesion of the particles on the matrix (depending upon the balance of attractive and repulsive forces), will determine the performance of a magnetic separator.

An experimental study on the effect of pH on magnetic separation of weakly magnetic iron ores [S40, S41] has shown that to achieve high recoveries of iron and high grade of the magnetic concentrate, the pH value of a slurry must be adjusted in such a way as to correspond with the point of zero charge (PZC) of the oxide mineral. The probability increases that magnetic particles will be captured on the matrix and that the volume of the adhered particles will rise. Simultaneously, the concentration of oxides and hydroxides of the metals in the tailings increases, provided that they are diamagnetic and their PZCs are far removed from the working pH value; at this pH value the largest settling velocity and particle sizes were observed. Such behaviour was ascribed to a combination of ionic and magnetic flocculation in the primary and secondary minima, and to a higher retention probability of particles on the matrix [S41]. Thayer and Linkson [T17] observed experimentally that when zeta potentials of hematite and silica had the same sign, the improvement in their separation was effected. Mass yield into the non-magnetic fraction increased and the grade of this product dropped.

The results of experimental investigation of the effect of pH on the recovery of uranium and gold by HGMS were presented by Svoboda et al. [S42]. The dependence of the recovery of uranium and gold on the pH value of the pulp in shown in Fig. 4.110 which reveals that the maximum recovery of uranium is achieved at pH values between 1.2 and 1.8, whereas the recovery of gold is practically independent of pH.

Figure 4.111 depicts the grade of the magnetic concentrate as a function of pH, and indicates that gold follows the same pattern as uranium, exhibiting the highest grade at a pH value of 1.2 and then decreasing continuously to the alkaline region. The secondary maximum that was observed for uranium at pH values from 10 to 11 seems to be absent for gold. The increment in the gold content of

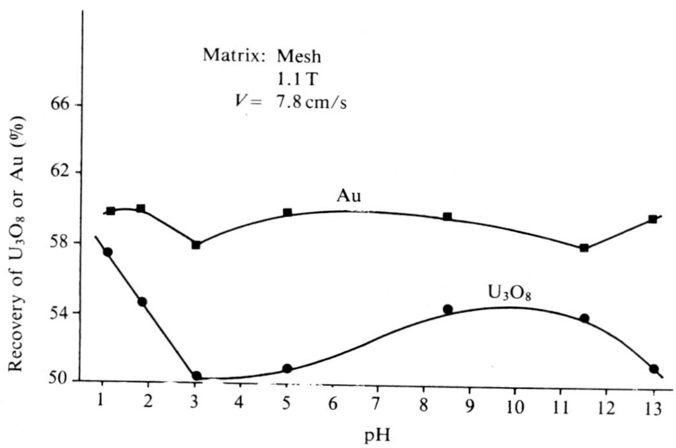

Fig. 4.110 Recovery of uranium and gold as a function of pH[S42].

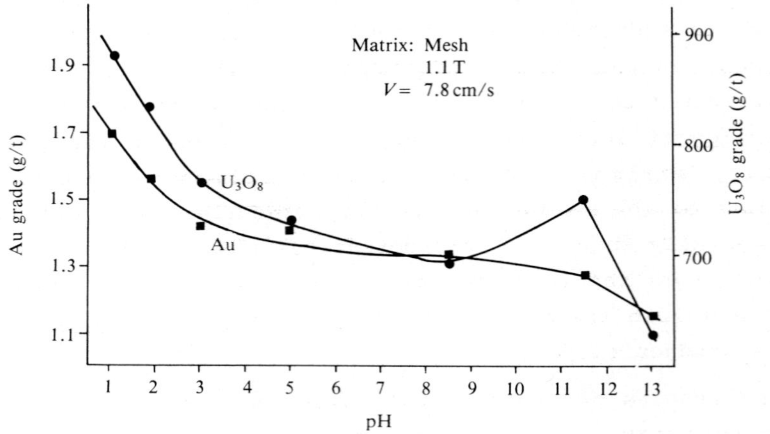

Fig. 4.111 Magnetic concentrate grade as a function of pH [S42].

the concentrate, which results from the adjustment of pH from the value corresponding to the usual operating conditions, viz. 9 to 1.2, is about 25 per cent, which is even higher than that for uranium (20 per cent). The very high selectivity of magnetic separation in the acid region is confirmed by the strong dependence of the mass yield into the magnetic fraction on pH.

Parsonage [P14] has demonstrated that if surface interactions are controlled, the selective coating of mineral and metallic particles with magnetite, followed by magnetic separation, can be used as a method of particle separation.

IV.12.5 Desliming

We have seen in Section IV.10 that the performance of a magnetic separator is strongly influenced by the size distribution of the ore to be treated. We have also seen that there is always a limiting particle size that determines the smallest particle that can still be recovered with acceptable probability. Although capture efficiency decreases also with increasing particle size, the most significant factor that reduces the metallurgical performance of a separator is the presence of slimes.

Although the recovery of slimes can be improved by increasing the magnetic field or reducing the slurry flow velocity, the improvement is often insufficient. Furthermore, these measures cannot always be adopted as most magnetic separators operating in a continuous mode are not equipped for controlling flow velocity by means other than the resistance of the matrix bed, a serious handicap in the treatment of very fine ores. In magnetic separators operating in cyclic mode, flow control is simple and efficient and is successfully implemented in the kaolin industry.

Another possibility is to deslime the slurry by removing those particles smaller than a certain threshold size. The cut-off size is determined by the magnetic susceptibility of particles to be recovered, by the applied magnetic field and by the flow velocity. Efficient desliming very often results in a dramatic increase in the recovery and the selectivity of the process, while inefficient desliming frequently does not improve the metallurgical performance of a magnetic separator but increases the losses of valuable minerals into the slimes and adds to the complexity of the beneficiation circuit. The importance of efficient desliming is illustrated in Fig. 4.112. It can be seen that the effective removal of particles smaller than 10 μm (sample A) not only

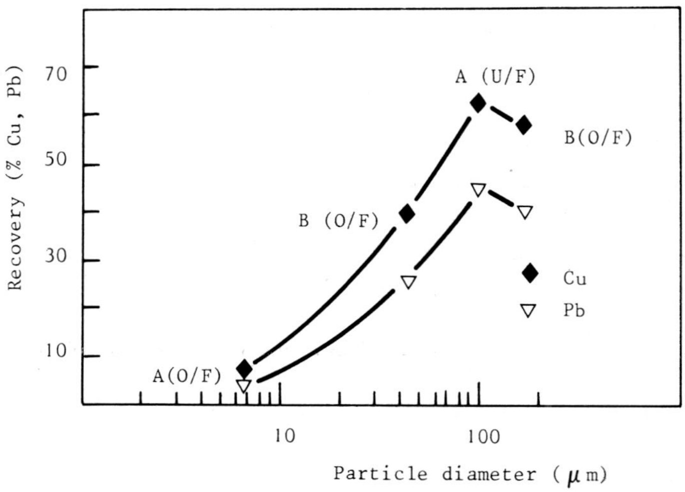

Fig. 4.112 The effect of desliming on recovery of copper and lead from flotation tailings [S30].
A - efficiently deslimed sample (cut-off 10 μm)
B - inefficiently deslimed sample (cut-off 20 μm)
Magnetic induction 1.2 T, matrix: 6 mm balls, interstitial flow velocity 0.12 m s^{-1}.

increases considerably the recovery and the grade of the concentrate, but also creates the overflow fraction that does not practically respond to magnetic separation.

On the other hand, inefficient desliming at a higher cut-off size (sample B in Fig. 4.112) fails to optimize recovery, while the overflow fraction is still susceptible to upgrading by HGMS. Such desliming complicates the circuit and brings about appreciable losses of valuable metals.

Although desliming improves the performance and decreases the volume of the feed into a magnetic separator, it also increases the complexity of the metallurgical circuit and increases the losses of metals into the slimes.

The question whether desliming should be implemented to improve the metallurgical performance of a magnetic separator can thus be answered only after all technical and economical determinants have been taken into account.

IV.12.6 Preliminary magnetization

IV.12.6.1 Magnetic viscosity

Magnetic viscosity is the increase or decrease of magnetization with time in a constant magnetic field. The change of magnetization is generally proportional to the logarithm of time that has elapsed since the last change in the field. The magnetization so produced in an initially demagnetized sample is called viscous magnetization [D14].

It has been shown by Walton [W29] that the magnetic moment produced in an initially non-magnetized sample, placed in the external magnetic field, after a time **t**, is given by

$$\mu_m = \int J_s V n(V) N(V) dV \qquad (4.26)$$

where **V** is the volume of a grain, $\mathbf{J_s}$ its saturation polarization, **N(V)dV** the number of grains between **V** and **V + dV** and **n(V)** the fractional alignment; the latter can be written as

$$n = L(x) + [n_o - L(x)]\exp(-\omega t) \tag{4.27}$$

where **L(x)** is the Langevin function given by eq. (1.34), $\mathbf{n_o}$ the value of **n(V)** at **t** = 0 and ω the rate given by

$$\omega = \nu_r \exp\left(-\frac{K_a V}{kT}\right) \tag{4.28}$$

where ν_r is the fundamental frequency for rotation of the magnetic moment of the grain and $\mathbf{K_a}$ the anisotropy constant. By inserting eq. (4.27) into eq. (4.26) we obtain

$$\mu_m = \int J_s V \left\{ L(x) + [n_o - L(x)]\exp(-\omega t) \right\} N(V) dV \tag{4.29}$$

The rate of increase or decrease of a magnetic moment is determined by the viscosity coefficient

$$S_v = \frac{d\mu_m}{d\,ln t} \tag{4.30}$$

which is strongly dependent on the particle size. It has been shown [D14] that for iron particles

$$S_v^{SD} \cong 300 S_v^{MD}$$

where $\mathbf{S_V^{SD}}$ is the single-domain and $\mathbf{S_V^{MD}}$ the multidomain viscosity coefficient. We can thus anticipate a strong increase in $\mathbf{S_V}$ with a decrease in the grain size.

The viscosity coefficient is also dependent on the magnetic field. At low fields $\mathbf{S_V}$ is usually proportional to the field while at moderate fields $\mathbf{S_V}$ becomes independent of the field.

IV.12.6.2 Experimental investigation of preliminary magnetization

Although magnetic viscosity can play an important role in the magnetic separation of minerals with ordered magnetic structures, the time element has been considered only rarely in magnetic concentration of ores. Fraas [F20] exposed a wide array of magnetic minerals of coarse particle size (0.4 to 0.2 mm) to a homogeneous magnetic field before these minerals were separated in a non-homogeneous magnetic field of a belt separator. Fraas found that the pre-magnetization of samples led to an increased recovery of magnetic minerals compared to the test in which no preliminary magnetization was used. Preliminary magnetization also improved the selectivity of magnetic separation. Typical examples of enhanced recovery, as a result of preliminary magnetization, are shown in Fig. 4.113.

Preliminary magnetization of magnetite ore prior to dry magnetic separation is used at the Magnitogorsk metallurgical plant (USSR) to improve the recovery and selectivity of magnetic separation [R16]. The magnetite ore is exposed to premagnetizing magnetic induction of 0.03 T. In the production-scale operation, the iron recovery increased by 1.42 per cent while the losses of iron into the tailings were reduced by 0.1 per cent.

The improvement in iron recovery by HGMS, and in the grade of the concentrate of weakly magnetic iron ore from Krivoy Rog (USSR) was observed after exposing the ore to a pulse magnetic field [S65]. Premagnetization at a magnetic induction of 1.9 T, of a pulse duration of 1.2×10^{-3} s, resulted in an enhancement of recovery of between 2 and 3 per cent, while the grade of iron concentrate increased by more than 1 per cent. The effect was particularly noticeable with very fine fractions smaller than 10 μm.

Experimental investigations on the effect of preliminary magnetization of hematite ore on high-gradient magnetic separation results [S66] showed that for a fine (-12 μm) fraction the grade of magnetic concentrate increased abruptly with time for very short retention times t_m, it reached the maximum at t_m = 10 s and subsequently did not change (Fig. 4.114).

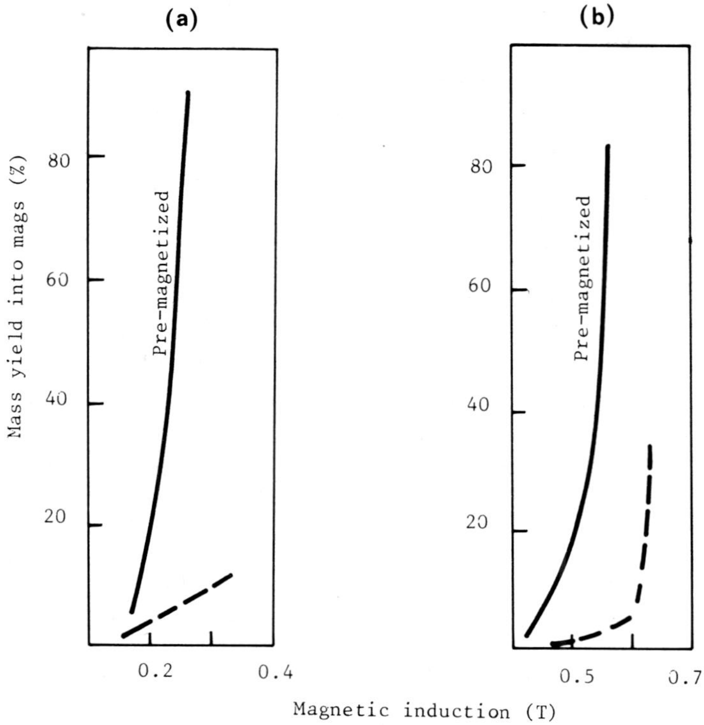

Fig. 4.113 The effect of preliminary magnetization on the recovery of hematite (a) and chromite (b) (after Fraas [F20]).

Recovery of iron into magnetic product increased with increasing premagnetizing time in a similar manner, and then remained constant, as shown in Fig. 4.115.

A higher premagnetizing magnetic field tends to improve recovery, while selectivity of magnetic separation is better at low premagnetizing fields. The improvement in recovery and grade was found to be the greatest for very fine fractions (-12 μm), compared to coarse samples (e.g. -53 +38 μm and -212 μm).

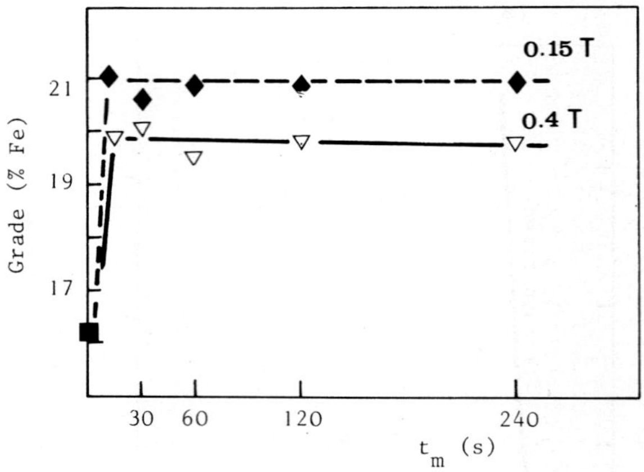

Fig. 4.114 The grade of magnetic concentrate of hematite ore (-12 μm) as a function of premagnetization time t_m for two values of the premagnetizing magnetic field.

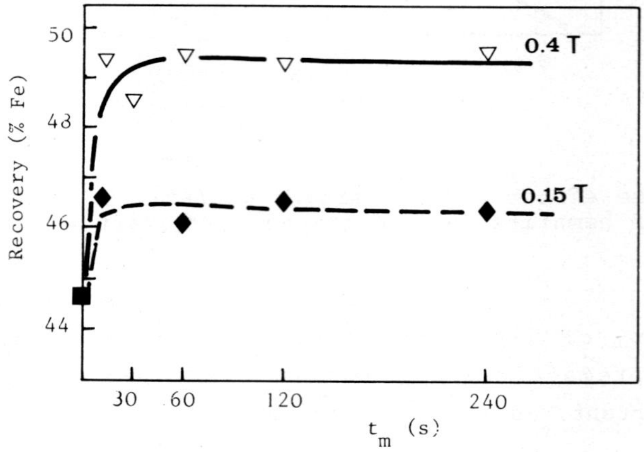

Fig. 4.115 Recovery of iron from hematite ore (-12 μm) as a function of premagnetization time t_m.

IV.13 THE SELECTIVITY OF SEPARATION

As follows from eq. (1.1) the selectivity of magnetic separation of the more magnetic particles from the less magnetic ones is primarily determined by the ratio of magnetic forces acting on the components of the ore to be separated, and by the magnitude of the competing forces $\mathbf{F}_c$.

In a matched system, the magnitude of the magnetic force is proportional to the product $\chi \mathbf{b}^2$, which implies that the role of particle size is more decisive than the difference in magnetic susceptibilities. Particles of different minerals, with $\chi \mathbf{b}^2$ = constant, have the same probability of being captured by the matrix. For example a quartz particle of magnetic susceptibility of -6×10^{-9} m^3kg^{-1} and a diameter of 100 μm has the same probability of being trapped by the matrix as hematite particles with a magnetic susceptibility of 600×10^{-9} m^3kg^{-1} and a diameter of 10 μm. In a more common system where matrix and particle sizes are not matched, $\mathbf{F}_m \sim \chi \mathbf{b}^3$, and the selectivity is even more impaired, as shown in III.1.11. Similarly, intermediate quartz particles with locked hematite will also be retained. The deposition of particles on the matrix is thus dictated by magnetic susceptibility, size distribution and the liberation characteristics of the ore. A clear-cut separation of minerals from gangue is usually not possible due to inadequate liberation, although the selectivity of magnetic separation can be considerably improved by classification.

Although the metallurgical performance of a magnetic separator is strongly dependent on particle size distribution, it is sometimes useful to determine the threshold ratio of magnetic susceptibilities of the components of the ore, for which separation will be efficient, provided that this rule is applied to particles of the same size.

Derkach found [D1] that the magnetic separation of martite from psilomelane had been selective for the ratio of their susceptibilities greater than 5, while for lower values of this ratio the efficiency of separation dropped sharply. Dobby et al. [D10] proposed that the minimum ratio of magnetic susceptibilities was 20, although for wide size ranges this figure is likely to be greater.

In most applications of magnetic separation to mineral treatment little can be done to modify the difference in magnetic susceptibilities, although desliming or classification can alleviate the negative effects of a wide size distribution. On the other hand, the selectivity of separation can be improved by employing higher flow velocities of the slurry and coarse, more selective matrices. Selective separation of either paramagnetic or diamagnetic particles can be accomplished by designing a matrix in which the collecting elements are shielded in such a way that only a part of their surface is available for particle capture [D15, K12, T13]. The selectivity of very finely ground ores can also be enhanced by dispersion and by adjusting the pH value of the slurry, as discussed in paragraph IV.12.4.

IV.13.1 The effect of surface interactions on selectivity of magnetic separation

It has been shown in paragraph IV.12.4 that in order to increase the selectivity of ores, the mineral particles to be separated should be colloidally unstable, i.e. they should not repel each other, while the particles of the gangue minerals must not coagulate with those of the valuable mineral. The highest selectivity can be expected at a pH value at which the surface potentials of the valuable and gangue minerals have the same sign, but which is not too far removed from the PZC of the valuable mineral.

In order to correlate this rule quantitively with the results of experimental investigations, energetic and kinetic criteria were derived [S43] for the selective magnetic separation of a useful mineral from a system containing several mineral species , based on the DLVO theory.

In the plot of total energy of interaction as a function of the distance between particle surfaces, the generalized DLVO theory produces a deep potential well, with the primary minimum at a small separation and a shallower secondary minimum at a larger separation. Between the two minima, there is an energy barrier that obstructs coagulation into the primary minimum. The general form of interaction energy versus separation is shown in Fig. 3.40. The magnitude of the potential barrier V_{max} is usually used as a criterion of stability. Arbitrary values of V_{max}, ranging from 25 to 0 **kT**, where **k** is the Boltzmann constant and **T** is the absolute temperature, so that **kT** is the thermal energy, have been chosen as the basis for the classification of systems as being coagulated or dispersed [W21]. In the presence of an external magnetic field, the value of the threshold of the magnetic field, which indicates the onset of destabilization of a suspension, can be determined, for instance, from the condition $V_{max} = 0$. In concentrated suspensions, where the long-range interactions are taken into account, the magnitude of the energy barrier at the onset of stability rises to about 30 **kT** [B26].

Should the theory be extended to the problem of selective coagulation, the kinetics of the process must be included, in which case information is required on the rates of coagulation of the various combinations of particles in the mixed suspension. As the rates are functions of the concentrations of the relevant particles and their size, simple energy criteria do not provide sufficient information to enable one to distinguish rates in a mixture.

It has been shown in Section III.5.2.3 that the half-time of flocculation given by eq. (3.107) is proportional to the stability ratio **W**, which for a suspension containing **i** and **j** mineral species can be written as

$$W_{ij} = \int_{0}^{\infty} \exp\left(\frac{V_T^{ij}}{kT}\right) \frac{a_i + a_j}{(a_i + a_j + h)^2} dh \tag{4.31}$$

If we assume that the frequency of encounters between **i** and **j** type particles depends on the relative number of such particles in the system, then the overall stability ratio for a system comprising the two species can be written as [H25]

$$\frac{1}{W_T} = \frac{n^2}{W_{11}} + \frac{(1-n)^2}{W_{22}} + \frac{2n(1-n)}{W_{12}} \tag{4.32}$$

where **n** is the relative number of particles of component 1 in the system. The usual criterion for the onset of rapid coagulation in dilute dispersions is that **W** must be 1 and, for moderately rapid coagulation, 10. $\mathbf{W} = 10^3$ might be taken as being typical of a practically stable suspension.

The role of the stability factor and the effect of the coagulation time on magnetic coagulation has been analysed [S35]. It was shown that the value of the threshold magnetic field indicates the onset of destabilization of a suspension, and, determined from the condition $\mathbf{V}_T = 0$, is only a necessary but not a sufficient condition for coagulation. The other condition to be taken into account is that the residence time is $> \mathbf{t}_{1/2}$, which imposes an additional restriction on $\mathbf{W}$ for a given suspension density $\mathbf{N}_o$. Owing to this requirement, the value of the threshold magnetic field increases, particularly for coarse particles.

Furthermore, the use of the magnitude of $\mathbf{V}_{max}$ as indicating the degree of instability of a suspension is unrealistic where a secondary minimum of significant depth precedes the energy barrier. It has been shown [S36] that rapid magnetic coagulation into the secondary minimum can occur in systems for which the interaction energy curves show large energy barriers and a deep secondary minimum. Undoubtedly, it can also occur in systems for which the secondary minimum is still deep but the energy barrier is small. The presence of the external magnetic field allows us to create sufficiently deep secondary minima that will substantially change the stability of the system, although this destabilization would not be reflected in the energetic criteria.

It is therefore clear that the onset of destabilization of a suspension and, consequently, the critical conditions for selective magnetic separation determined from the energy conditions $\mathbf{V}_T = 0$ represents just a loose necessary condition, whereas sufficient conditions can be determined by including the time factor.

If a particle of a useful mineral is denoted as 1, and the gangue particle as 2, high recoveries will be achieved if $\mathbf{V}_T^{11} < 0$ and $\mathbf{W}_{11} < 1$. Simultaneously, for the grade of the magnetic concentrate to be high, one must ensure that valuable and gangue particles do not coagulate mutually, so $\mathbf{V}_T^{12} > 0$ and $\mathbf{W}_{12} >> 1$. Furthermore, to avoid the mechanical entrainment of large clusters of gangue particles, it would be preferable for one to work at $\mathbf{V}_T^{22} > 0$ and $\mathbf{W}_{22} > 1$. The conditions for selective magnetic separation of a useful mineral from a mixture of mineral species can be written, in terms of the stability ratio, as

$$W_{11} \leq 1$$
$$W_{12} \gtrless 1 \qquad (4.33)$$
$$W_{22} > 1$$

where 1 denotes the useful mineral and 2 denotes the gangue.

The importance of these phenomena will be illustrated by reference to magnetic separation of gold-uranium ore.

The curves of the total potential energy of interaction of two uraninite particles, plotted as a function of a normalized inter-particle distance $\mathbf{s} = \mathbf{R}/\bar{\mathbf{a}}$, where $\bar{\mathbf{a}} = (\mathbf{a}_i + \mathbf{a}_j)/2$ are shown in Fig. 4.116.

The magnetic dipole interaction is dominant at almost all values of pH and the particles will coagulate into the primary minimum. At high values of surface potential (pH 1.2 or pH 10) the potential barrier will prevent mutual coagulation, which can, however, be compensated for by the presence of a secondary minimum, if it is sufficiently deep.

In order to evaluate the energetic criteria of stability of a suspension and of selective magnetic separation one can plot the height of a potential barrier as a function of pH for assorted uraninite particles for $\mathbf{B} = 1$ T and $\mathbf{X} = 10^6$ m^{-1}. Fig. 4.117, which illustrates the results for the interaction U_3O_8 - U_3O_8, shows that rapid coagulation can be expected at a pH value of 3, while the potential barrier rises sharply to the acid and alkaline regions. It would therefore be desirable for magnetic separation to be carried out at around a pH value of 3, since the uraninite particles create bigger clusters that are recovered more easily by HGMS. A similar picture is presented by the interaction U_3O_8 - SiO_2, as can be seen in Figure 4.118. Heterocoagulation is again most probable at a pH value of 3, at least for very fine particles. Aggregation of these two mineral species is undesirable because it reduces the grade of the magnetic concentrate. Therefore, one should try to work at pH values that are far from 3 to guarantee high-quality concentrates. Uranium dissolution and a subsequent precipitation does not seriously affect the results [S42].

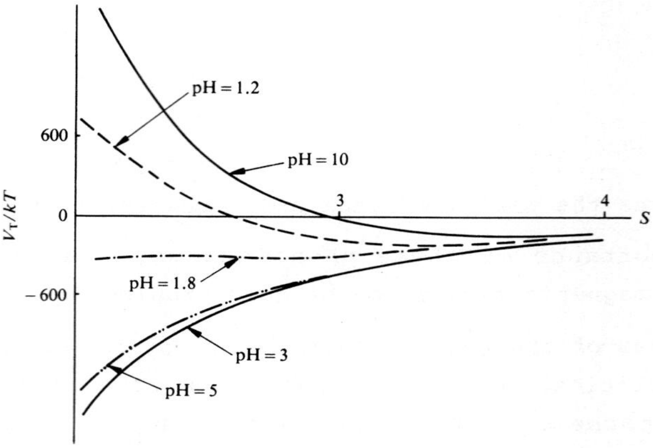

Fig. 4.116 Curves of total potential energy $\mathbf{V_T}$ of two interacting uraninite particles of diameter 5 μm for various pH values of the suspension, where **B** = 1 T, **X** = 10^6 m^{-1}, **A** = 2 x 10^{-20} J [S43].

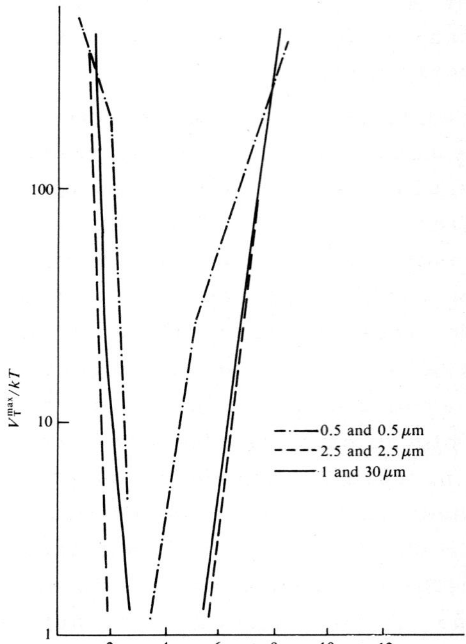

Fig. 4.117 Variation in the height of the potential barrier as a function of pH for the interaction of uraninite particles when **B** = 1T, **X** = 10^6 m^{-1}, **A** = 2 x 10^{-20}J [S43].

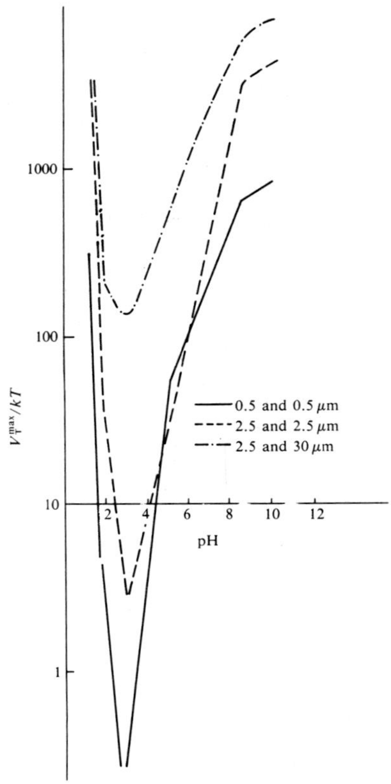

Fig. 4.118 Variation of the height of the potential barrier as a function of pH for the interaction of uraninite and quartz particles when **B** = 1T, X = 10^6 m^{-1}, **A** = 2 x 10^{-20}J [S43].

The effect of pH on the stability ratio for a system of uraninite particles of various sizes is shown in Fig. 4.119, which indicates that rapid coagulation of small particles (radius ≃ 0.5 μm) will take place at the pH value close to the PZC of uraninite (pH = 3.5). For larger particles (2.5 μm), magnetic interaction becomes predominant, and the stability ratio drops to less than 10^{-1} for a wide range of pH values. When the zeta potential exceeds 30 mV, electrostatic repulsion becomes decisive, and coagulation is hindered. The interaction of a small particle (2.5 μm) with a large one (30 μm) is solely attractive, and coagulation becomes rapid throughout the range of pH values. These conclusions are similar to those drawn from the energetic criteria (Fig. 4.117).

The variation of the stability ratio with pH for a system of uraninite-quartz at **B** = 1 T, **X** = 10^6 m^{-1} is illustrated in Fig. 4.120. It can be seen that rapid coagulation particularly for very fine particles, will take place in the range between the PZCs for uraninite and quartz.

For coarse particles a potential barrier increases the stability ratio to above 1, but slow coagulation can be expected in the pH range between 2 and 3.5.

It follows from the theoretical analysis of the potential barrier and of the stability ratio that the heterocoagulation between uraninite and quartz is most probable in the vicinity of a pH value of 3 and, consequently, a drop in the grade of the

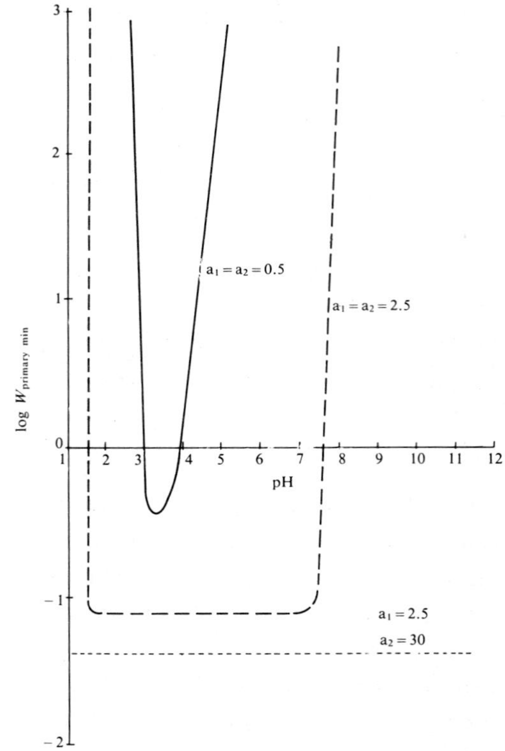

Fig. 4.119 Variation of the stability ratio with pH for a system of uraninite particles at **B** = 1 T, **X** = 10^6 m^{-1}, A = 2 x 10^{-20}J [S43].

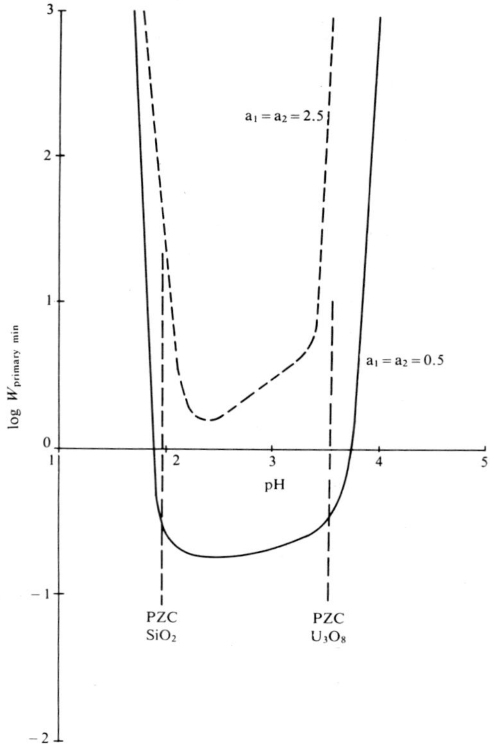

Fig. 4.120 Variation of the stability factor with pH for a system U_3O_8-SiO_2 at **B** = 1 T, **X** = 10^6 m^{-1}, **A** = 2 x 10^{-20}J [S34].

magnetic concentrate should be expected. The stability of the uraninite-quartz system will be expected at pH < 2 or a pH > 4. The high selectivity of magnetic separation at pH < 2 is confirmed by the experimental results (Fig. 4.111), whereas the predicted selectivity at pH > 4 is cancelled by a promoted coagulation in a system U(Al), U(Fe)-SiO_2, where U(Al) and U(Fe) are the uranium oxides with adsorbed Al and Fe cations. It has been shown [M19] that these systems are present in pulp, either as a result of contamination or as a result of deliberate additions during previous treatment of the ore.

Thus, at pH 3 the stability of the uraninite-quartz system is reduced and rapid coagulation of the uraninite particles should also be expected at a pH value of 3, which is actually confirmed by experiment (Fig. 4.111). The combination of these two phenomena leads to a rapid increase of the mass yield into the magnetic fraction at a pH value of 3. Nevertheless, the poor grade of the magnetic fraction at a pH value of 3 causes a drop in uranium and gold recovery at this value of pH. The interaction of U(Fe)-SiO_2 and U(Al)-SiO_2 particles is attractive in the pH region 3 to 9, and the stability ratio lower than 1 guarantees a high degree of colloidal instability in this region. Consequently, magnetic separation will be non-selective in this interval of pH, which is experimentally demonstrated by a rapid decrease in the grade of the magnetic fraction (Fig. 4.111) and by an increase in the mass yield into the magnetic fraction. However, for a pH value higher than 9, separation should be selective, since the stability ratio for the U(Al)-SiO_2 system rises sharply and experimentally we observe a secondary maximum in the grade of the magnetic fraction at a pH value of about 10 to 11.

As a function of pH, the gold grade in the magnetic fraction follows a pattern similar to that of the uranium grade, confirming the assumption that gold and uranium are intimately associated in cyanidation residues. However, a certain portion of the gold is associated with minerals that do not respond to magnetic separation and to colloidal interactions with the magnetic species, as can be inferred from the finding that gold recovery is not dependent on pH in the acid region.

The investigation of total interaction between various mineral species thus provides very good confirmation of the experimental results. The required selectivity of magnetic separation is achieved by control of the surface properties of various mineral components and by adjustment of the magnitude of the external magnetic field. Theoretical analysis [S34] implies that the best metallurgical performance of magnetic separation can be expected at the value of pH at which the valuable and gangue minerals have the same sign of the surface potential. On the other hand, this pH value should not be too far from the PZC of the valuable mineral since, at PZC, one would expect a preferential aggregation of the valuable mineral particles and, consequently, more efficient recovery by HGMS. In addition, magnetic separation must be performed at carefully selected values of the external magnetic field because, when the magnetic field is increased, interaction between dissimilar particles becomes repulsive, and the recovery of uranium decreases. This phenomenon is usually observed experimentally, and confirms the incorrectness of the concept that, in mineral processing, one can improve the efficiency of magnetic separation by increasing the magnetic field, and increase the thoughput simply by increasing the magnetic field and flowrate simultaneously in such a way that their ratio is constant.

Furthermore, the magnetic field can induce a deep secondary minimum in the interaction energy of useful mineral particles with one another or of useful mineral particles with gangue particles. Coagulation into this secondary minimum can affect the recovery and grade of magnetic concentrates.

The behaviour of a system of magnetizable quasi-colloidal particles is therefore a function of the magnetic field, and a complex interplay between various interactions is very sensitive to the value of the magnetic field.

IV.13.2 Carrier magnetic separation

Since carrier (or piggy-back) magnetic separation uses coarse particles as a carrier for fine particles (as a consequence of their mutual interaction), it is possible for the latter to be recovered. Carrier magnetic separation makes use of the fact that the frequency of particle collisions is much greater when there are particles of different sizes, which are the result of different trajectories in the accelerating fluid. The rate of adhesion of fine particles ($\mathbf{b} \simeq 2$ μm) to coarse ones ($\mathbf{b} \simeq 30$ μm) is from 10^3 to 10^4 as high as the rate of adhesion between fine particles [S44].

The results of the investigation of the possible effect of carrier magnetic separation on the overall performance of a magnetic separator [S62] are summarized in Table 4.19. Fraction smaller than 17 μm of the uranium-gold leach residues were mixed with fractions between 25 and 53 μm, between 53 and 75 μm and between 106 and 150 μm in the ratio 5 : 1 by mass. The results of magnetic separation of these mixtures indicated that the presence of fractions between 53 and 75 μm, and between 53 and 75 μm considerably enhanced the recovery of uranium from fine (-17 μm) fraction and the grade of the magnetic fraction.

TABLE 4.19

Magnetic separation of mixtures of size fractions of uranium-gold leach residues
Magnetic induction 1.1 T, interstitial velocity 0.08 m s^{-1}, matrix: mesh, pH 1.2

Mixture 5:1 (μm)	Head U_3O_8 (ppm)	Theoretical Grade of mags U_3O_8 (ppm)	Theoretical Recovery into mags U_3O_8 (%)	Experimental Grade of mags U_3O_8 (ppm)	Experimental Recovery into mags U_3O_8 (%)
-17+ (-53+25)	185	820	30	1033	52
-17+ (-75+53)	173	700	29	850	40
-17+ (-150+106)	159	616	27	600	22

Although recovery and grade become strongly dependent on the pH value of the slurry, no single component of the mixture exhibited such a dependence. This is again a consequence of the fact that pH can influence the attachement of fine particles to the coarse ones. On the other hand, the presence of the fraction between 106 and 150 μm neither improved the selectivity of the process nor increased the recovery of uranium from the fine fraction. Furthermore, no dependence of recovery and grade on pH was observed.

These results indicate that carrier separation (in which fine particles coagulate onto the coarse ones and the cluster is recovered by the matrix), plays an important role in the capture of slimes. This two-stage mechanism of magnetic separation also explains the failure of the theoretical models to predict the minimum particle size recoverable by HGMS (see paragraph IV.10.2). At the same time, these conclusions imply that the behaviour of very fine particles in a magnetic separator should always be studied in the presence of other particle sizes comprising the ore. The conclusions drawn from the results obtained with isolated short-cut size fractions can be incorrect and can contribute to the inaccurate assessment of the performance of a magnetic separator.

IV.13.3 Selective separation by magnetic susceptibility gradient

It has been shown by Takayasu and co-workers [T14, H40] that mixtures of particles with a wide range of magnetic susceptibilities can be separated by utilizing a colloidal magnetic fluid with a magnetic susceptibility gradient.

A magnetic field gradient creates a concentration gradient (and hence a density gradient) when applied to a colloidal magnetic fluid in which particles are subject to thermal motion [T5]. When gravitational and magnetic forces are orthogonal, this gradient causes particles to be separated by moving to a position where the magnetic susceptibilities of particles (κ_p) and fluid (κ_f) are in equilibrium.

Takayasu et al. [T5] found that the steady-state solution of the equation of motion (eq. (3.25)) of a particle exposed to a mutually orthogonal magnetic force (eq.(3.8)) and gravitational force (eq.(1.14)), immersed in a magnetic fluid, can be expressed as

$$\kappa_p = \frac{\kappa_f - \left(\frac{H_o}{H_x}\right)^2 \kappa_{fo}}{1 - \left(\frac{H_o}{H_x}\right)^2} \tag{4.34}$$

where $\mathbf{H}_x$ is the magnetic field strength as a function of a position $\mathbf{x}$, $\mathbf{H}_o$ being $\mathbf{H}_x$ at $\mathbf{x} = \mathbf{x}_o$.

It follows from eq. (4.34) that the position of a particle in the force field depends only on the volume magnetic susceptibility of a particle, and does not depend on other factors such as density, size and shape. Since κ_f is an exponential function of magnetic field $\mathbf{H}_x$ [T14], a very wide range of values of magnetic susceptibility κ_f as a function of $\mathbf{x}$ can be obtained when the field gradient is applied. Although different minerals, ranging from ferromagnetic to diamagnetic, can be separated by this technique, one disadvantage is that it takes a long time to establish the susceptibility gradient.

Takayasu and Kelland [T15] increased efficiency by feeding a family of streams with different fluid magnetic susceptibilities into a separation chamber so as to form the susceptibility gradient; the basic idea of the method is illustrated in Fig. 4.121. Two magnetic fluids of different magnetic susceptibilities κ_1 and κ_2 ($\kappa_1 < \kappa_2$) are fed into the chamber while particles **A**, **B** and **C** are simultaneously introduced into the fluid stream. Particle **A**, whose magnetic susceptibility κ_A is smaller than κ_1, moves toward the left side of the chamber (diamagnetic capture), while particle **C** whose susceptibility κ_C is greater than both κ_1 and κ_2, moves to the right side of the chamber (paramagnetic capture). Particle **B**, with a susceptibility $\kappa_1 < \kappa_B < \kappa_2$ moves to the right side of the layer of fluid 1 until the particle (by the paramagnetic capture process) reaches the position where $\kappa_B < \kappa_2$. As a result of this process, particles **A**, **B** and **C** can be obtained from outlets 1,2 and 3 respectively. By using such a spatial susceptibility distribution selective separation can be achieved and both paramagnetic and diamagnetic fluids, as well as a ferrofluid, can be used to adjust fluid susceptibility.

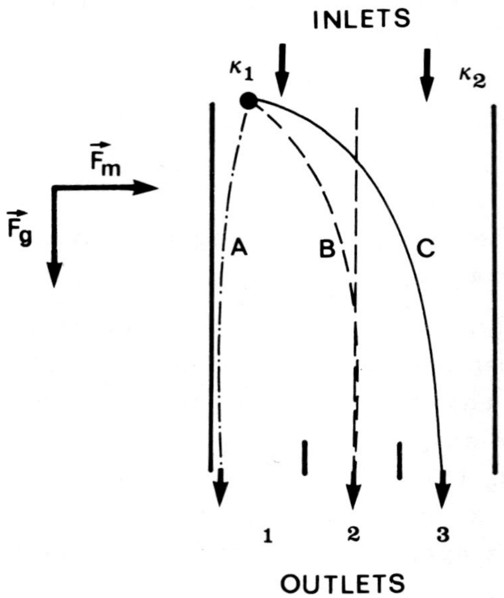

Fig. 4.121 The schematic diagram of a susceptibility gradient magnetic separator (after Takayasu and Kelland [T15]).

IV.13.4 The vibration of a matrix

The vibration of a matrix, either by mechanical or electromagnetic means, or by the combination of both, was proposed by several investigators, to assist in the flushing of the matrix [K3] or to propel the ore through the matrix in dry high-gradient magnetic separation [L17]. A typical experimental arrangement of vibration-assisted dry magnetic separation is shown in Fig. 4.122. Lateral mechanical vibration can be provided by an eccentric rotating shaft driven by a variable speed motor or by the regular rapping of the separation chamber by a vibrating feeder [S4]. Longitudinal electromagnetic vibration is provided by a coil on the chamber that is energized with alternating current, at a frequency in the sonic range. A vibratory force is exerted on the coil by the interaction of the dc field generated by the magnet and of the ac field in the coil; amplitude of vibration can be adjusted by changing either the dc or ac fields.

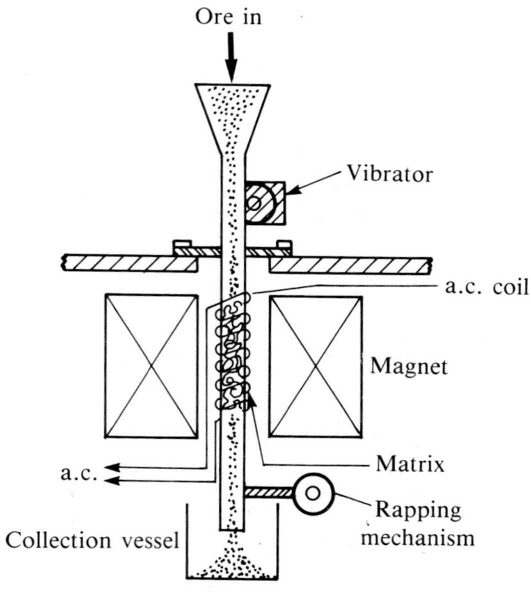

Fig. 4.122 Apparatus for vibration-aided magnetic separation.

The electromagnetic vibration of a matrix was also found to improve appreciably the selectivity of wet high-gradient magnetic separation [S67, S68]. The effect of vibration on the recovery of WO_3 as well as the grade of the magnetic concentrate from the wolframite-quartz mixture, is illustrated in Fig. 4.123. It can be seen that the grade of the magnetic product increased with increasing ac current from 7.9 per cent WO_3 at zero current to 35,4 per cent at I = 4 A; the effect on recovery was much less pronounced.

Matrix vibration was also found to significantly improve the selectivity of magnetic separation of wolframite-cassiterite slimes. Although the magnetic susceptibilities of wolframite and cassiterite are similar (5×10^{-7} m^3kg^{-1} and 3.5×10^{-8} m^3kg^{-1} respectively), with a wide size distribution (90 per cent - 74 μm and 54 per cent - 37 μm), the selectivity of their separation was further enhanced by employing the electromagnetic vibration of the matrix, as shown in Table 4.20.

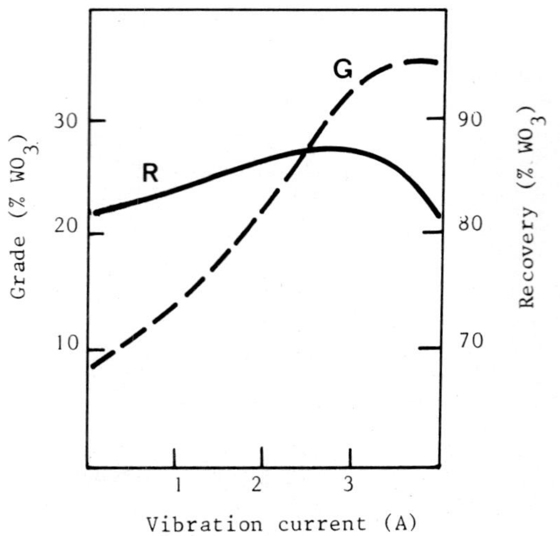

Fig. 4.123 Magnetic separation of wolframite-quartz mixture as a function of the vibration current.
Magnetic induction 1.5 T, flow velocity 0.014 m s^{-1}, matrix: steel wool.
(After Sun and Li [S67, S68]).

The vibration of a matrix reduces the mechanical entrainment of gangue particles and unliberated metal particles in the magnetic product and facilitates their desorption from the matrix by fluid flow. For coarse matrices which exhibit a much lower level of mechanical entrainment, the beneficial effect of matrix vibration should diminish, as demonstrated by Table 4.21.

TABLE 4.20

The effect of matrix vibration on magnetic separation of wolframite and cassiterite. Laboratory tests. Matrix: steel wool. (After Sun and Li [S67])

No vibration

Product	Mass yield (%)	Grade (%)		Recovery (%)	
		WO_3	Sn	WO_3	Sn
Mags	65.0	50.0	2.5	95.3	39.2
Non-mags	35.0	4.5	7.1	4.7	60.8
Feed	100.0	34.1	4.1	100.0	100.0

Vibration

Mags	52.6	60.0	0.5	92.1	6.2
Non-mags	47.4	5.7	7.9	7.9	93.8
Feed	100.0	34.3	4.0	100.0	100.0

TABLE 4.21

The effect of matrix vibration on magnetic separation of wolframite and cassiterite. Pilot plant tests. Matrix: mesh

No vibration

Product	Mass yield (%)	Grade (%)		Recovery (%)	
		WO_3	Sn	WO_3	Sn
Mags	56.9	59.4	1.2	88.9	12.4
Non-mags	43.1	9.8	11.2	11.1	87.6
Feed	100.0	38.0	5.5	100.0	100.0

Vibration

Mags	53.2	63.0	0.5	88.3	4.6
Non-mags	46.8	9.5	11.3	11.7	95.4
Feed	100.0	37.9	5.6	100.0	100.0

IV.14 COMPARISON OF WET HIGH-GRADIENT MAGNETIC SEPARATORS

As discussed in Chapter III, the dramatic development of high-gradient magnetic separation as a commercially viable technique for mineral beneficiation resulted in the design and manufacture of a wide range of continuous, wet high-gradient separators that use either iron-yoke electromagnets or solenoids to generate the magnetic field.

Since each type of magnetic separator was often designed with a particular application in mind, it would be naive to attempt to directly compare the performances of these separators with the use of a standard ore. All commercial HGMS separators differ widely in fundamental operational characteristics that significantly affect their performance. Although background magnetic induction is probably very similar in all types of separators, the numerical values given in the commercial literature often vary either because magnetic induction inside the matrix or the background magnetic induction is stated, as discussed in Section IV.4.1.

The selection of a type and height of matrix was usually dictated by the particular application for which the separator was designed. Experience has shown there is no universally acceptable type of matrix that would meet all requirements and the selectivity of the matrix and its resistance to clogging is strongly dependent on the type of ore treated and on the way the ore was mined and treated prior to magnetic separation.

A straightforward comparison of various types of magnetic separators should also be avoided unless a reliable economical and technical assessment of the operation has been made. Magnetic separators that use conventional iron-yoke magnetic circuits are usually metallurgically efficient and very cost-effective, provided the capacity is several tens of tonnes per hour; the commercial success of such separators proves their efficiency beyond any doubt. On the other hand, continuous HGMS separators that use iron-yoke magnets generate only a modest magnetic induction in a small volume, they have a limited capacity and it is unlikely that larger and more powerful machines could be built. Furthermore, the cost-effectiveness of large machines cannot be improved substantially; any large operation could require a fleet of machines, while problems associated with feed distribution, pumping of tailings and concentrates, maintenance and floor requirements, become enormous. However, cost-effective high capacity separators can be built that use the iron-clad solenoid. Such separators can be easily scaled-up, while still preserving a modest input power to generate a magnetic field.

Any comparison of magnetic separators should thus take into consideration the size of the operation and the capability of a machine to be scaled-up if the need arises.

IV.14.1 Comparative pilot-plant tests with HGMS separators

Comparative tests of several wet HGMS separators, on the pilot-plant scale, were described by Forssberg and Kostkevicius [F19]. Although the conditions of the operation inevitably championed some separators at the expense of others, some interesting metallurgical and economical comparisons were made.

An artificial mixture of hematite concentrate (73 per cent) and quartzite (27 per cent) was used as a test material; the mixture also contained magnetite and quartz. The sample contained 48 per cent of particles smaller than 64 μm and 21 per cent smaller than 15 μm. The mean magnetic susceptibility of the mixture was equal to approximately 2.5×10^{-6} m^3kg^{-1} (200×10^{-6} cm^3g^{-1}).

The technical data of separators that were used in the tests are summarized in Table 4.22, while Table 4.23 shows the optimum operational parameters. Fig. 4.124 depicts the grade-recovery curves, including the theoretical curve that was obtained with the laboratory batch magnetic separator. The effect of particle size on iron recovery is illustrated in Fig. 4.125. Of all the separators tested, the best metallurgical results were obtained with the SALA HGMS, which exhibited the best recovery, selectivity and sensitivity to fine particles. The results obtained with the Jones and Readings separators were comparable, the Jones separator being more efficient with fine particles while Readings performed better with coarse fractions.

The SOL separator was found to be insensitive to fine particles (smaller than 50 μm) and its selectivity was lower than

TABLE 4.22
Specifications of magnetic separators used in tests

	Jones P71	Boxmag Rapid SHW1	Readings 16-pole	Krupp-SOL 24/14	SALA HGMS Mk II
Matrix	Grooved plates 8R gap 2.3mm	Triangular bars 2x3mm	Salient plates gap 2.5mm	Grooved plates 6R gap 2mm	Expanded metal diam. 1mm
Rotor diameter (mm)	710	800	1800	2400	1720
Nominal capacity (t/h)	1 - 2.5	1.5 - 6	15 - 30	20 - 40	1 - 3
Number of separation zones	2	1	8	4	1
Matrix width (mm)	75	25	66	180	120
Matrix height (mm)	220	150	205	170	140
Circumferential velocity of rotor (ms^{-1})	0.11	0.19	0.48	0.37	0.11
Mass (t)	7.1	3.8	7.1	10	6

TABLE 4.23
Optimum operational parameters

	Jones P71	Boxmag-Rapid SHW1	Readings 16-pole	Krupp-Sol 24/14	Sala HGMS Mk II
Capacity (t/h)	1.8	1.7	22	20	1.0
Energy consumption (kWh/t)	0.50	2.6	0.84	3.45	24
Pulp density (% solids by mass)	53	50	55	45	25
Rinse water (m^3/t)	0.70	0.35	0.54		18
Flush water (m^3/t)	1.7	1.0	1.0		6
Matrix loading L_{F3} (kgm^{-3})	200	1150	400	600	200

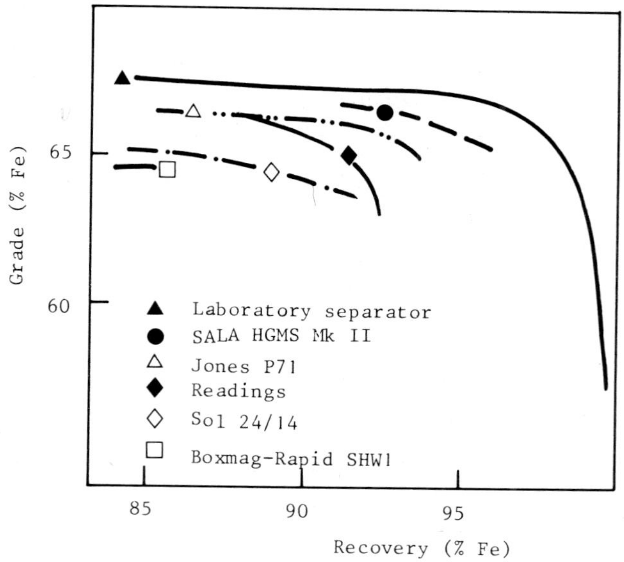

Fig. 4.124 Relationship grade vs recovery for various continuous HGMS separators [F19].

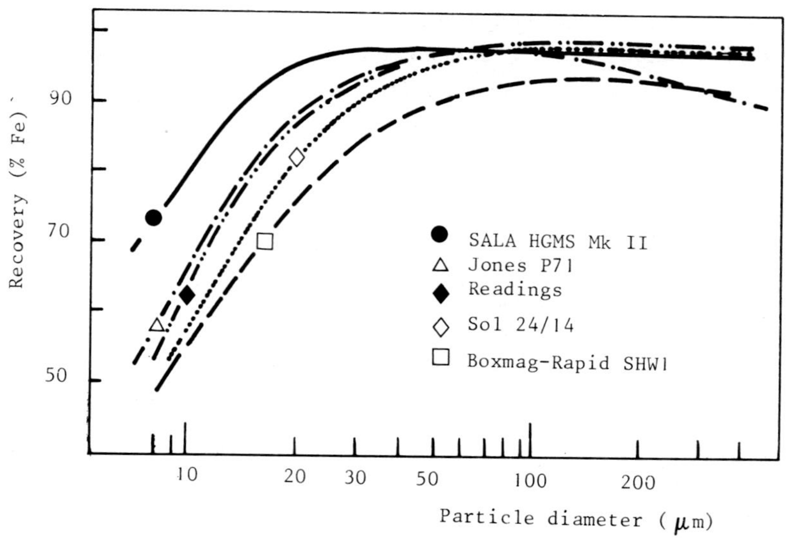

Fig. 4.125 Recovery of iron as a function of particle size [F19].

TABLE 4.24

Representative metallurgical results obtained with different HGMS machines

Separator	Fe grade (%)	Fe recovery (%)
Sala HGMS Mk II	66.0	94.0
Jones P 71	66.0	92.0
Readings WHIMS-16	65.0	91.5
Krupp-SOL 24/14	63.5	91.5
Boxmag-Rapid SHWI	64.5	85.5

that of the other machines. The optimum metallurgical results are summarized in Table 4.24.

Although these results indicate the potential of various continuous HGMS machines, their direct translation into large-scale applications must be done with caution. As discussed above, the metallurgical performance is strongly influenced by the choice of matrix and its height, and by the background magnetic field, which cannot be consistently assessed when measured in the matrix. Furthermore, specific energy consumption strongly varies with the size of the machines, which have a widely differing potential for a scale-up. The assessment of the technological and economic potential of each machine must be based on a detailed analysis of their performance with a wide range of ores of different magnetic susceptibilities and size distributions, and taking into account the anticipated magnitude of the operation.

IV.14.2 Capacity of a cyclic HGMS separator

It was shown by Watson [W19] that the performance of a cyclic magnetic separator can be described by the equation

$$ln(1 - R_m) = K_1(1 - K_2 n)H_o t_R \tag{4.35}$$

where $\mathbf{R_m}$ is the fraction of material retained in the separator, $\mathbf{K_1}$ and $\mathbf{K_2}$ the parameters which depend on $\mathbf{v_m}/\mathbf{v_o}$, $\mathbf{n}$ the volume of slurry passed in units of canister volume, $\mathbf{H_o}$ the applied magnetic field and t_R the time taken to pass one canister volume of slurry. Thus

$$t_R = \frac{h}{v} = \frac{h(1-f)}{v_o} \tag{4.36}$$

where **h** is the matrix height.

After a certain number n_o of canister volumes have passed, the factor $(1 - R_m)$ will become unacceptably large and the process must be stopped. At this stage it is necessary to displace the remaining canister volume of slurry from the canister, at the velocity v_o and at the applied magnetic field, with n_1 canister volumes of fluid. The applied field is then switched off and the captured magnetic product is removed by flushing. With the fast field decay accessory, the field can drop to zero in 15 to 30 seconds. The field is restored to its operating value H_o and slurry processing is continued.

The throughput of a separator (in m^3s^{-1}) is then written as

$$Q = v_o S \rho_b \delta_d \tag{4.37}$$

where ρ_b is the slurry density (% of solids), **S** is the cross-sectional area of the canister and δ_d the duty cycle, which is defined as the fraction of the total time that the slurry is processed in the separator

$$\delta_d = \frac{n_o}{n_o + n_1 + t_D/t_R} \tag{4.38}$$

In eq. (4.38), t_D is the dead time equal to the sum of the time taken to switch the field on and off and the time taken to clean the matrix. Similarly, the throughput in tonnes per hour is given by

$$Q = 3.6\ v_o S f L_{F3} \tag{4.39}$$

where **f** is the filling factor of the matrix and L_{F3} is the matrix loading given by eq. (4.9) (in kg m^{-3}). The dependence of the throughput of a cyclic HGMS separator on a number of canister volumes of processed slurry, at two different values of dead time, is shown in Fig. 4.126. For comparison, the throughput of the reciprocating canister separator is also shown.

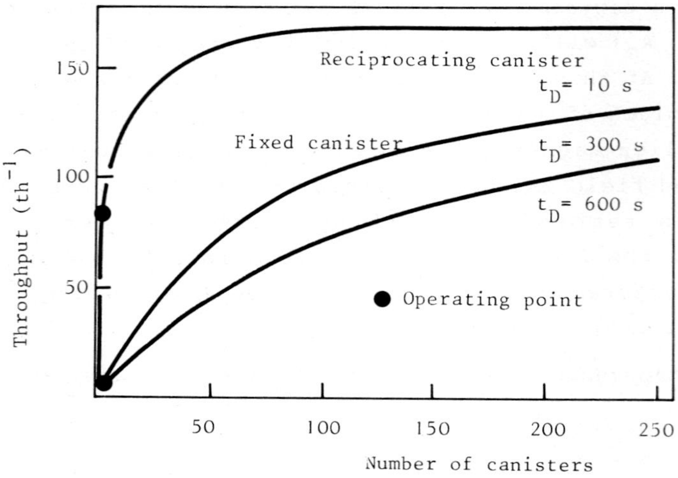

Fig. 4.126 The throughput of a cyclic HGMS separator as a function of number of canister volumes of slurry processed, for different values of dead time $\mathbf{t}_D$.
Magnetic induction 4 T, slurry density 33 per cent of solids, flow velocity 0.2 m s^{-1}, canister diameter 0.85 m.
(After Watson [W30]).

The comparison of the throughput of cyclic HGMS machines that use a superconducting (5 T) and conventional magnets (2 T), is illustrated in Fig. 4.127. The curve was calculated under the assumption that the performance of a separator is proportional to $\mathbf{v}_m/\mathbf{v}_o$. In industrial conditions the cyclic separator usually operates at four canister volumes which implies that the throughput of a superconducting separator would be one order of magnitude greater.

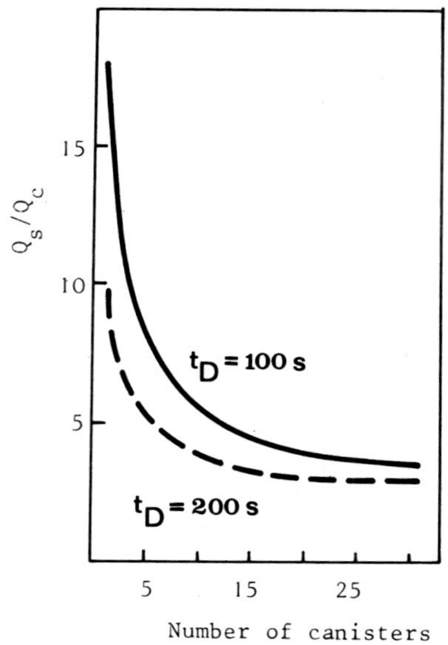

Fig. 4.127 The ratio of throughputs for superconducting (Q_S) and conventional (Q_C) separators of equal size, as a function of number of canister volumes of slurry processed (after Watson [W19]).

IV.14.3 Capacity of a continuous HGMS separator

The duty cycle of a continuous, high-gradient magnetic separator is equal to unity and the throughput is determined by the matrix width, flow velocity of the slurry, velocity of the matrix movement and by the matrix loading. The situation is schematically shown in Fig. 4.128.

The residence time of a slurry in the matrix is given by eq.(4.36). Since the slurry must leave the matrix while in the magnetic field the maximum permissible velocity of the carousel is

$$v_c \leq l_m \frac{v}{h} \tag{4.40}$$

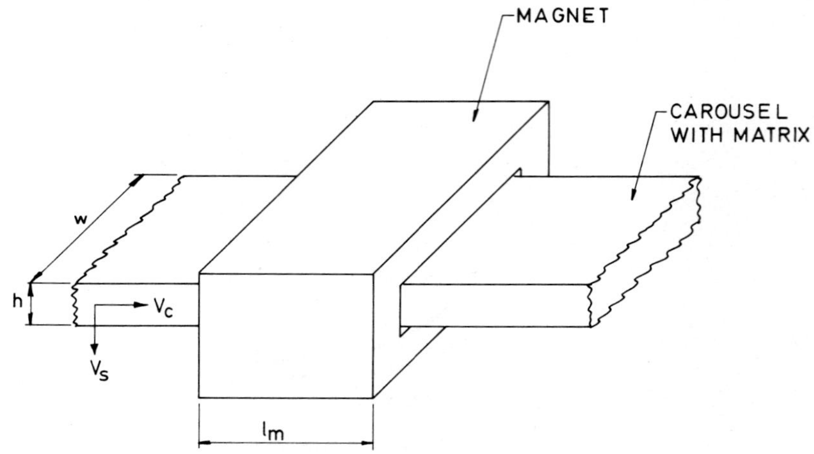

Fig. 4.128 Schematic diagram of a continuous HGMS separator.

where $\mathbf{l}_m$ is the length of the magnetic zone, **h** is the matrix height and **v** the interstitial velocity of the slurry in the matrix.

The capacity of a separator is given by

$$Q = 3.6 \frac{hwv_c L_M f}{R_M} \tag{4.41}$$

or

$$Q = 3.6 hwv_c L_{F3} f \tag{4.42}$$

where **Q** is the capacity (in tonnes per hour), **h** the matrix height (in m), **w** the matrix width (in m), $\mathbf{v}_c$ the velocity of matrix motion (in m s^{-1}), **f** the filling factor of a matrix, $\mathbf{L}_M$ the matrix loading given by eq. (4.10), $\mathbf{L}_{F3}$ the matrix loading given by eq. (4.9) (in kg m^{-3}) and $\mathbf{R}_M$ the mass yield into the magnetic fraction.

IV.14.4 Longitudinal and transverse configurations

As shown in Fig. 3.3, the directions of magnetic field and particle flow velocity can assume, with respect to the matrix collector, three orientations, all of which are employed in commercial high-gradient magnetic separators. Most separators with electromagnets fitted with iron yokes use the transverse (T) arrangement, with the exception of the Jones separator, in which the grooved plates assume the axial configuration. On the other hand, in separators employing iron-clad solenoids, the directions of the field and slurry flow are parallel and thus L-configurations is effectual.

The patterns of particle build-up on the matrix differ in these three modes of operation and result in a dissimilar metallurgical performance of a magnetic separator. Different theoretical models, either microscopic single-wire description, or phenomenological models, often give contradicting predictions about the performance of separators in T- and L- configurations. For instance, it follows from a single-wire approach [W9] that recovery of magnetizable particles is the largest for T-configuration, and the smallest for axial configuration, namely

$$R_T > R_L, \quad R_L > R_A$$

A similar conclusion was reached by Moyer et al. [M15] for a system of spherical collectors. It was found that $\mathbf{R}_T > \mathbf{R}_L$ for paramagnetic particles, while $\mathbf{R}_T < \mathbf{R}_L$ when applied to diamagnetic particles. In contrast, the phenomenological description [B20] predicts less efficient performance in T-configuration as compared to L-configuration, i.e.

$$R_L > R_T, \quad R_A \simeq R_T$$

as follows from eqs. (3.79) and (3.80).

Although systematic experimental investigations into the effect of geometrical configuration on metallurgy are rather scarce, the majority of them indicate that recovery of a mineral into the magnetic fraction is more effective in L-configuration. The difference can be quite considerable as is demonstrated by Figs. 4.23, 4.25, 4.65 and 4.129.

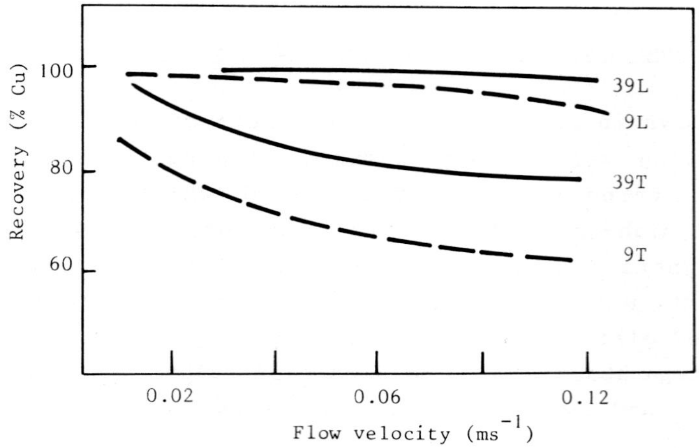

Fig. 4.129 Comparison of the effect of longitudinal (L) and transverse (T) configurations on recovery of copper by HGMS, for assorted particle diameters (in μm) (after Kim et al. [K27]).

It has also been observed (Figs. 4.23 and 4.25) that in T-configuration, recovery tends to increase with an increasing magnetic field up to higher magnetic fields than in L-configuration, where the recovery is almost field-independent, starting with a very modest magnetic field. In contrast to recovery, the grade of the magnetic fraction is usually higher in T-configuration than in L-configuration, as shown in Figs. 4.23 and 4.27.

As follows from Fig. 3.3, in L-configuration the particles are captured on the upstream side, and to a lesser extent on the downstream side of the matrix collectors. The particle build-up, being in the proximity of the stagnation points, is exposed to moderate shear stress by the fluid flow, which favours high retention probability but low selectivity of the process.

On the other hand, in T-configuration, the particles are captured on the sides of the matrix collectors (where fluid flow velocity is high) and are thus subjected to high shear stress. Mechanically entrained gangue particles, together with unliberated mixed grains, are thus stripped from the matrix so that the selectivity of magnetic separation increases, at the expense of recovery. In practice, this means that the metallurgical performance of high-gradient magnetic separators is governed by the following inequalities

$R_L > R_T$

$G_L < G_T$

Significantly, contrasting metallurgical performances of magnetic separators, using either transverse or longitudinal geometrical configurations, imply that when scaling-up the laboratory results, the same geometrical arrangements must be used in a laboratory separator and in a production-size machine.

IV.15 MAGNETIC FLOCCULATION

IV.15.1 Magnetic flocculation of ferromagnetic minerals

As discussed in Section III.5.1 magnetic flocculation of strongly magnetic materials is a process commonly encountered in wet low-intensity magnetic separation of finely ground ferromagnetic minerals. Although flocculation can improve the efficiency of separation by recovering very fine particles, the grade of the magnetic product can decrease if magnetic flocculation is non-selective.

Magnetic separation is not the only branch of the mining industry where magnetic flocculation is used to enhance the efficiency of the process. Magnetic flocculators are frequently used in the following applications:

a) to increase the settling rate of solids and reduce the size and costs of dewatering equipment
b) to produce cleaner water and thus reduce maintenance and pollution problems
c) to increase the difference in settling rates between flocculated magnetite and unflocculated gangue slimes
d) to improve the performance of centrifuges that remove solids from steel plant effluents
e) to improve filtering rates of ferromagnetic products and reduce the blinding of filter cloth
f) to reduce the addition of flocculating reagents.

Magnetic flocculation of strongly magnetic materials is largely an empirical technology with very few supporting meaningful theoretical models. Owing to their simplicity, magnetic flocculators are generally used without any extensive research, although several basic rules common to all applications must be obeyed. The effect of such variables like magnetic field strength, particle-size distribution, content of solids and degree of stirring were investigated by several researchers and some basic findings will be summarized in this Section.

IV.15.1.1 Magnetic field strength

It has been reported by Bartnik and Shively [B22] that agglomerates of natural magnetite started to form at the magnetic induction of 10 mT, but 40 mT (400 G) was necessary to obtain stable flocs. The most stable agglomerates of natural magnetite were formed at 0.14 T, and with increasing magnetic induction above this value no improvement was observed.

The investigation of the Soviet magnetite ores [K15] has shown that the threshold magnetic induction required for a full flocculation is a function of the origin of the ore and ranges from 8 mT (80 G) to 80 mT (800 G), as seen in Fig. 4.130. It was found that complete magnetic flocculation occurs before magnetic saturation of the material is reached. Furthermore, for the same mineral, the degree of magnetic flocculation is proportional to its remanent magnetization $\mathbf{J}_R$, although the magnitude of remanent magnetization does not unambiguously characterize the degree of flocculation.

The threshold magnetic field required for a complete flocculation is a function of coercive force $\mathbf{H}_C$ of the sample and the empirical formula describing this relation was found by Karmazin et al. [K15].

$$H_M = 123 H_c^{\frac{1}{2}} \tag{4.43}$$

where $\mathbf{H}_M$ is the magnetic field strength at which a complete magnetization occurs.

In order to avoid the entrainment of the gangue particles in the flocs of the ferromagnetic grains, the optimum magnitude of the external magnetic field must be found. Since magnetic permeability μ and magnetic susceptibility χ are strongly dependent upon the magnetic field and reach their maxima at some intermediate value of the applied magnetic field, the force among the particles in the flocs, and consequently the length of the floc, reach a maximum value at the magnetic field, corresponding to the maximum of the magnetic permeability. This is demonstrated by the experimental results shown in Fig. 4.131.

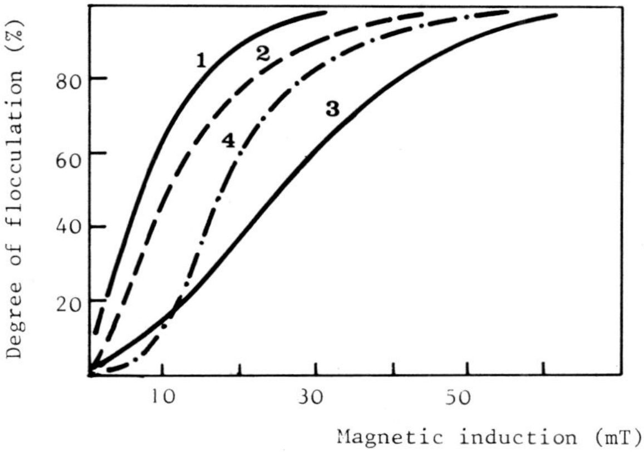

Fig. 4.130 Effect of magnetic field on the degree of magnetic flocculation of assorted Soviet magnetite ores.
(1) Sokolovsko-Sarbaisk ore
(2) Kachkanarskii ore
(3) Krivoi Rog ore
(4) Kerchenskii ore
(After Karmazin et al.[K15])

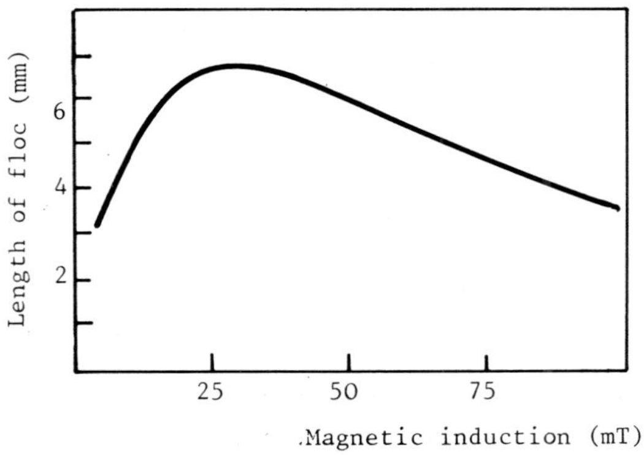

Fig. 4.131 The length of the floc as a function of the applied magnetic field (after Berdichevskii et al. [B21]).

Furthermore, a greater amount of non-magnetic material is entrained in flocs formed from more finely ground ore than in coarse flocs formed from relatively fewer particles [L15], which in turn leads to weakening and shortening of the flocs. It was shown by Lomovtsev and Shugol [L14], that the shape of the flocs is a function of the character of the magnetic field and of the properties of an ore. The form of the flocs ranges from long chains to massive irregular clusters. The effect of the magnetic field strength upon the tensile strength of the magnetite flocs is shown in Fig. 4.132.

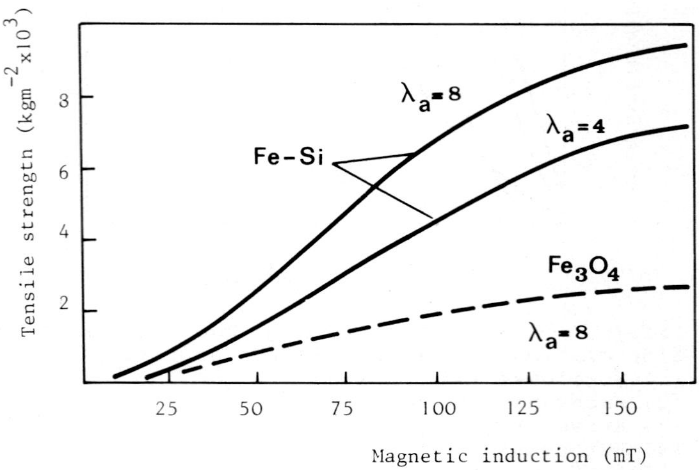

Fig. 4.132 Effect of magnetic field on the tensile strength of the flocs of ferrosilicon (aspect ratio 4 and 8) and magnetite (aspect ratio 8) (after Karmazin et al. [K15]).

IV.15.1.2 Velocity of flow

Very few experimental data on the effect of the residence time of particles in the magnetic field upon flocculation efficiency are available. Benson et al. [B23] state that for effective flocculation, slurries with more that 1 per cent of natural magnetite should be exposed to a 60 mT (600 G) magnetic field for a period of not less than 0.07 seconds. Although good flocculation was obtained after exposure to a magnetic field as short as 0.03 s the best results were obtained after the residence time of 0.2 seconds with a flow of 1.8 m s^{-1} [B22]. Natural magnetite flocs also show good stability to high turbulence.

IV.15.1.3 Temperature

The operating temperature can have a diverse effect upon a process of magnetic flocculation. As a result of lower fluid viscosity at increased temperatures formation rate and stability of the flocs improved [B22]; on the other hand, a decrease in the coercive force at higher temperatures led to a reduction in breaking strength of the aggregates, while an increase in temperature from 20° C to 90° C reduced the strength of the flocs by up to 10 per cent [K15]. Lantho [L15] observed that the magnetite content of the titanoferrous magnetite concentrate decreased steadily with increasing temperature in the interval from 200 to 500° C, while the TiO_2 content dropped only between 400° C and 450° C, which corresponds to the temperatures exceeding the Curie temperature of the titanoferrous magnetite. Since the Curie temperature of titanoferous magnetite falls with increasing ilmenite content, selective concentration can be achieved by adjusting the temperature.

IV.15.1.4 Concentration of suspension and particle size

As the concentration of solids in a suspension increases, the half-time of flocculation decreases (eq. (3.107)) while the probability of particle collision and adhesion increases. If the retention time of particles in the magnetic field is sufficiently longer than the half-time of flocculation, which in turn is determined by the concentration of solids, then a high degree of agglomeration can be expected. The conclusion is that with suitably selected operating conditions suspensions of very low solids concentration can be agglomerated. An enhanced degree of flocculation of magnetite under common industrial conditions has been observed with suspensions containing as little as 1 per cent of solids [B22].

The density of a suspension also affects the tensile strength of the flocs, as shown in Fig. 4.133. An effect of particle size on tensile strength is depicted in Fig. 4.134. It can be seen that the character of this dependence is similar to the dependence of the magnetic susceptibility χ on particle diameter **d**, $\chi = f(d)$. The strength of the magnetite flocs containing particles of the micrometer dimensions is significantly reduced, reflected in the increased losses of fine magnetite particles into the tailings.

The tensile strength of the agglomerates is also affected by the shape of the particles, as illustrated in Fig. 4.135.

Since the magnetic interaction of ferromagnetic particles is many orders of magnitude larger than the surface interactions the effect of pH of the slurry is negligible.

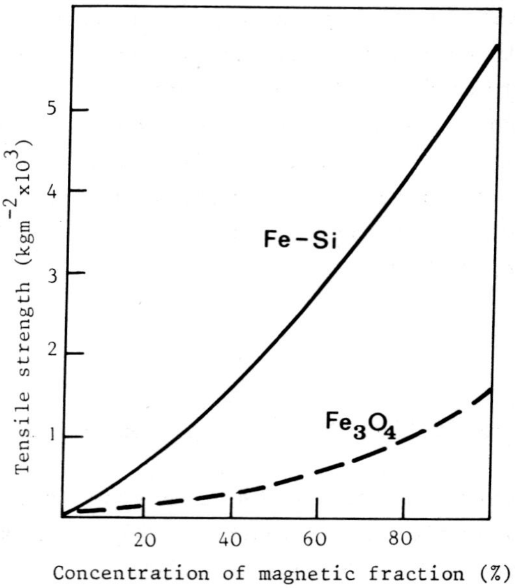

Fig. 4.133 Effect of the concentration of a magnetic fraction in a suspension on the tensile strength of flocs of ferrosilicon and magnetite with the aspect ratio 4 (after Karmazin et al. [K15]).

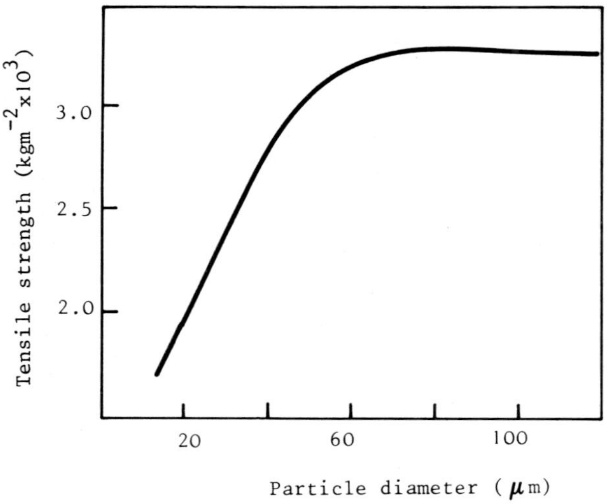

Fig. 4.134 The effect of the particle size on the tensile strength of the magnetite flocs at 0.16 T, aspect ratio 8 (after Karmazin et al. [K15]).

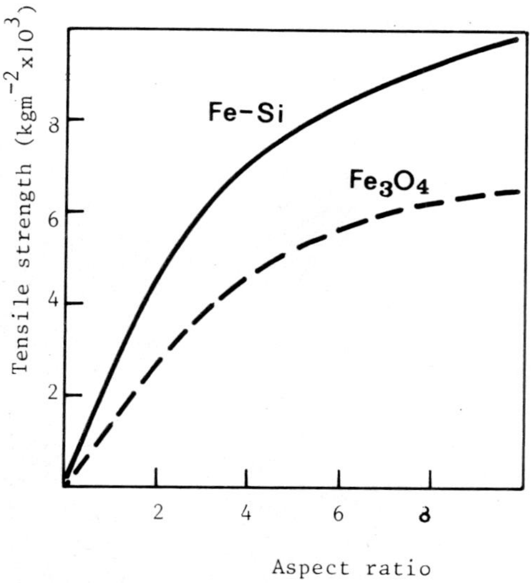

Fig. 4.135 The effect of the aspect ratio on the tensile strength of the ferrosilicon and magnetite flocs, at 0.16 T (after Karmazin et al. [K15]).

IV.15.1.5 Magnetic flocculators

Commercially available magnetic flocculators (Fig. 4.136) are commonly used in the mining and steel producing industries to flocculate ferromagnetic particles and thus increase the settling rate in ore and heavy media slurries. Permanent magnets are used to generate the magnetic field, which ranges from 0.06 T to 0.1 T in the air gap.

Magnetic flocculators should be placed over pipelines in a vertical position close to the feed entry into the thickener or hydrosizer. The pipe section in the magnetic flocculator is made of non-magnetic material and its diameter ranges from 50 mm to 600 mm.

The standard flocculators are made in lengths of 0.3 m to 0.5 m and widths from 0.2 m to 1.2 m. The openings between the poles are 0.1 m, 0.18 m and 0.25 m. For pipes with diameters larger than 0.25 m, rectangular pipe sections are recommended for economic reasons [B23].

Laminar flow of the slurry helps to improve flocculation because excessive turbulence as a consequence of vigorous pumping can break the flocs.

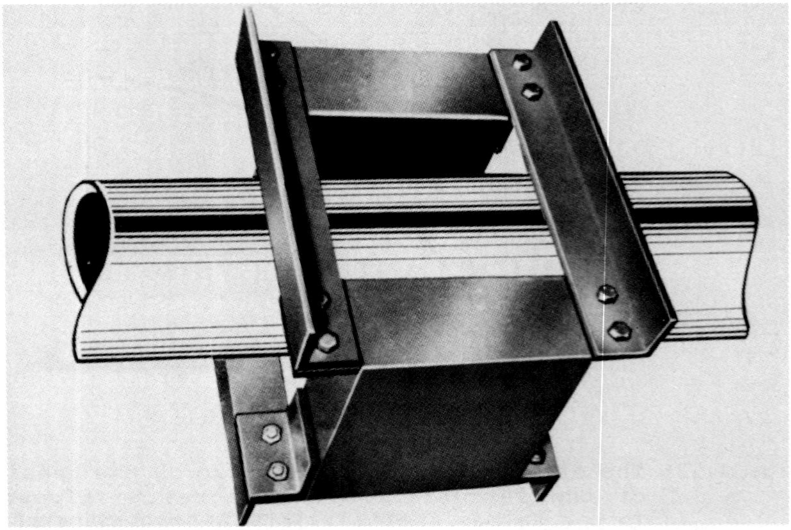

Fig. 4.136 Magnetic flocculator (courtesy of Eriez Magnetics, Inc.).

In selecting a correct magnetic flocculator the following information should be provided:

a) type of suspension treated
b) magnetic susceptibility and coercive force of the magnetic fraction
c) density of the suspension
d) flowrate of the suspension
e) temperature of the suspension.

IV.15.2 Magnetic flocculation of weakly magnetic minerals

In contrast to magnetic flocculation of strongly magnetic minerals, the application of this process to feebly magnetic matter has been limited to a small number of experimental laboratory tests. These invariably suffered from a lack of information about the sample properties, and a serious interpretation of the experimental data in terms of current theoretical models is difficult.

The first experimental observation of the magnetic flocculation of weakly magnetic minerals was reported by Hencl and Svoboda [H23] in 1979. A sample of siderite ore, with a mean susceptibility of 0.75×10^{-6} m^3kg^{-1} (60×10^{-6} cm^3g^{-1}), was passed through a separation device shown in Fig. 4.137 [H26]. The device consisted of a chamber of circular cross-section with an outer diameter 100 mm made of non-magnetic material; a collecting trough was placed in its upper part. The lower section of the chamber which was tapered terminated in a discharge nozzle. Slurry was fed from the top through a pipe, coaxial with the separation chamber, the outlet of which was positioned at the zone of the maximum magnetic field. The device was inserted into the bore of a solenoid magnet.

The magnetizable particles experience a magnetic dipolar interaction which induces their agglomeration and the flocs thus formed settle under the force of gravity and can be collected at the bottom outlet. The diamagnetic gangue particles which remain dispersed, are carried away by the rising flow and leave the device via an overflow trough. Fig. 4.138 shows the experimentally determined dependence of the overflow density on the magnitude of the magnetic field for different flow velocities.

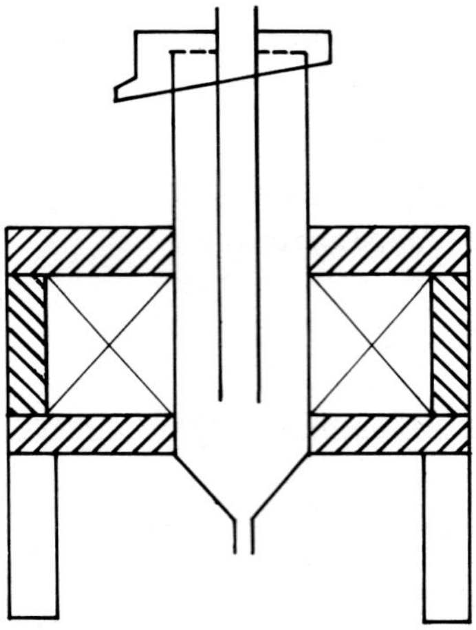

Fig. 4.137 Schematic diagram of a magnetic flocculation device (after Hencl et al. [H26]).

It can be seen that the concentration of solids in the overflow decreases with increasing **B** and the aggregates of the magnetic particles report to the underflow.

The effect of the magnetic field on the grade of the underflow and overflow is illustrated in Fig. 4.139. As the magnetic field increases so the Fe content in underflow increases, while the grade of overflow decreases.

Owing to a low concentration of solids in the slurry (10 per cent by mass) and to large particle size, the retention time of the slurry in the magnetic field was too short compared to the half-time of flocculation. It was observed, however, that magnetic flocculation of feebly magnetic particles occurred at moderate magnetic fields.

The settling velocity of magnetic flocs was enhanced by exposing the slurry to the gradient of magnetic field present in a solenoid. Van Kleef et al. [V6] placed the separation device in a magnetic field in such a way that the discharge end of the chamber was above the point of maximum field. Experiments carried out on mixtures of paramagnetic $MnCO_3$ and diamagnetic quartz showed that even for a very dilute suspension magnetically-enhanced settling

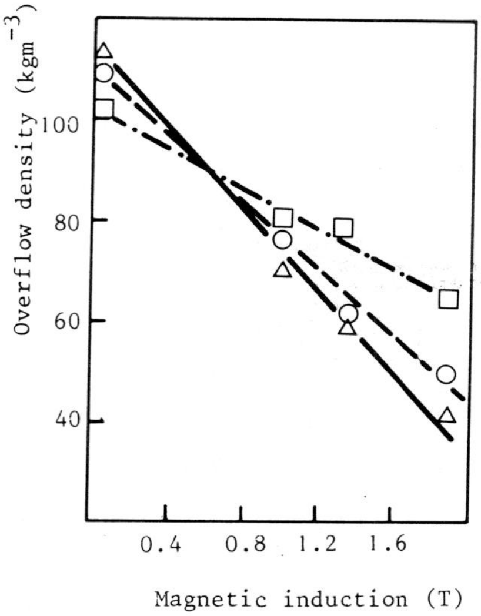

Fig. 4.138 The dependence of the overflow density on magnetic induction for different flowrates. △ - **v** = 5 x 10^{-3} m s^{-1}, ○ - 12 x 10^{-3} m s^{-1}, □ - 26 x 10^{-3} m s^{-1} (after Hencl and Svoboda [H23]).

rates can be used to separate fine paramagnetic particles from a gangue. The experimental conditions used in [V6] were not suitable for efficient magnetic flocculation to be observed, namely, that the concentration of a magnetizable mineral was too low and the characteristic time of flocculation had to be made sufficiently short to be comparable to the retention time of the slurry in the magnetic field by employing a very high magnetic field. It can be expected that at higher slurry densities, the magnetic field that was needed to achieve acceptable metallurgical results would be much lower, as demonstrated in [H23].

Fundamental experimental investigations into the magnetic flocculation of paramagnetic and diamagnetic particles were carried out by Watson [W20] and Parker et al. [P11], who observed that the field-induced particle aggregation, other than binary-pair formation, was not observed in fields up to 15 T. The information about the sample properties was incomplete [P11] and a serious interpretation of their experimental data was difficult. Curves of a settling velocity versus magnetic field were plotted for the suspension of several minerals irrespective of differing particle

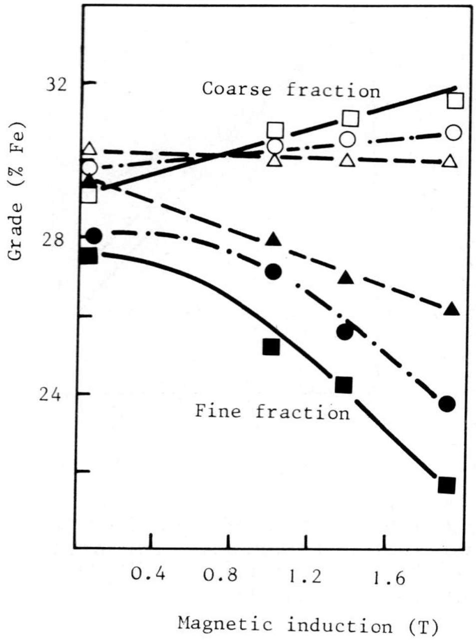

Fig. 4.139 Grade of overflow as a function of magnetic field (after Hencl and Svoboda [H23]).

particle size, surface potential and width of a diffuse layer. The valency of the ions was not taken into account and the concentration of potential-determining ions was not included in the analysis. In some instances, the authors [P11] observed a minimum in the settling velocity of particles in low magnetic fields. For example, for a suspensions of Mn_2O_3 in water, this minimum was observed at **B** = 0.7 T, which corresponds to the disappearance of a secondary minimum in a potential energy curve for a given system [S36]. Therefore, for a field greater than 0.7 T flocculation in a primary minimum can be expected, while the increasing settling rate for **B** smaller than 0.7 T can be attributed to flocculation in a secondary minimum [S36].

IV.15.2.1 The possibility of practical applications of magnetic flocculation

Selective chemical flocculation of the mineral slimes provides an efficient means of recovering particles of such small dimensions that other techniques fail to concentrate them economically. This method, however, has a number of disadvantages, viz. that a high consumption of chemicals increases running costs and causes ecological problems and that the time of chemical flocculation can be influenced only to a limited extent, which means that the flocculation of coarse fractions is hindered because of long flocculation time.

On the other hand, no chemical agents are required in magnetic flocculation and flocculation time can be substantially reduced by increasing the magnetic field. In practice, however, the reduction in flocculation time is limited by the maximum magnetic field of, say, 6 T, available in large superconducting magnets. Another important limitation of magnetic flocculation is that in practice it can be used only for paramagnetic (and of course, ferromagnetic) minerals. Although the magnetic flocculation of diamagnetic minerals is possible [S34] in principle, the required magnetic fields are impractically high.

Nevertheless, it is probable that magnetic flocculation can play an important role in the beneficiation of weakly magnetic finely ground iron ores [S46]. Magnetic flocculation allows us to create clusters of particles several times greater than individual particles, and these flocs can then be easily recovered in a high-gradient magnetic separator. As follows from Table 4.25, hematite or siderite particles smaller that 10 μm can be flocculated at reasonably low magnetic fields, while large particles will be unaffected. It is thus possible to obtain a dimensionally homogeneous slurry free from -10 μm fractions.

The application of magnetic flocculation to the beneficiation of fine weakly magnetic ores is depicted schematically in Fig. 4.140. Fig. 4.140a shows the application of magnetic flocculation alone to the separation of magnetic and non-magnetic fractions. The clusters of paramagnetic particles created in a magnetic field settle rapidly and separate from the non-magnetic gangue, which can be carried out of the system by rising fluid flow. The advantage of such a mode of operation is the high throughput and the absence of a matrix that forms an inherent part of all high-gradient magnetic separators.

TABLE 4.25
The threshold magnetic field $\mathbf{B_F}$ calculated from the condition $\mathbf{W} = 1$, and a true flocculating magnetic field $\mathbf{B_F}^+$ determined from the condition $t_{1/2} = 1$ s; for a suspension containing 10 per cent by (volume) of solids [S46]

Particle diameter (μm)	Hematite $\mathbf{B_F}$ (T)	Hematite $\mathbf{B_F^+}$ (T)	Siderite $\mathbf{B_F}$ (T)	Siderite $\mathbf{B_F^+}$ (T)
1	0.1	0.1	0.45	0.45
5	0.05	0.5	0.23	2.0
10	0.03	3.0	0.12	12.0
20	0.015	10.0	0.05	300.0
40	0.004	20.0	0.017	700.0

However, the separation of both fractions will probably not be very high since particles greater than 15 μm will not flocculate readily at commonly-used magnetic fields. In order to achieve an acceptable recovery, the overflow will have to contain only the finest fraction of the gangue. Therefore, the magnetic fraction will contain coarse non-magnetic particles, or conversely, in order to guarantee sufficient selectivity of the process, a part of the magnetic fraction will have to report to the overflow.

The incorporation of magnetic flocculation into a flowsheet containing wet high-gradient magnetic separation seems to be more advantageous; the situation is depicted in Figs. 4.140b and 4.140c. If magnetic flocculation precedes magnetic separation, the flocculation eliminates slimes smaller than 10 μm and the underflow can subsequently be treated by HGMS; if magnetic flocculation follows the HGMS, it can treat the tailings which contain, aside from non-magnetic gangue, magnetic particles smaller than 10 μm. However, since the concentration of the magnetic fraction in the tailings is low, flocculation time will be long and the efficiency of flocculation low.

Fig. 4.140d illustrates the situation when only a magnetic separator is used. Magnetic particles smaller than, say, 10 μm are usually lost in the tailings.

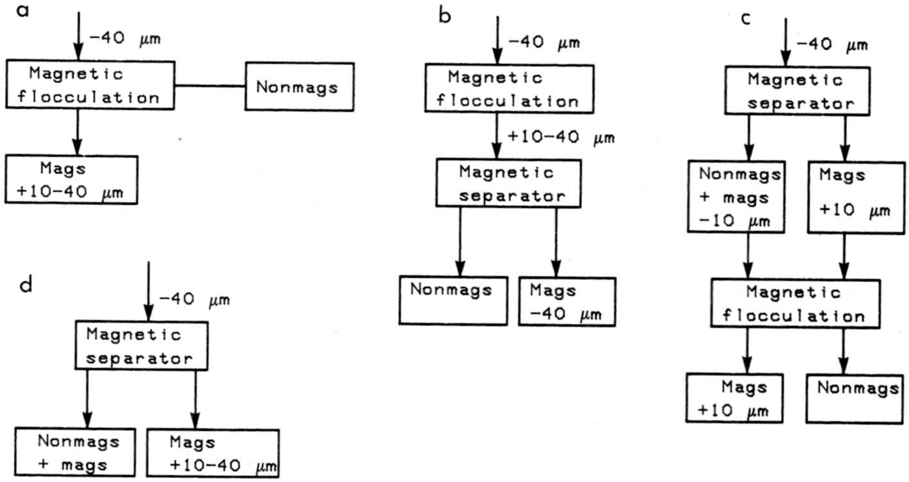

Fig. 4.140 Application of magnetic flocculation to the beneficiation of weakly magnetic ores.

Let us estimate the technical parameters of a device for the magnetic flocculation of weakly magnetic minerals. Let us assume a solenoid magnet of internal diameter 2 m, length 0.5 m and background magnetic induction 2 T, as used in the purification of kaolin. A hematite particle of 1 μm diameter flocculates in a magnetic field 0.1 T with a characteristic time of 1 s. In a magnet with a length of 0.5 m, the slurry flowrate of 0.2 m s^{-1} can be used (the retention time of a particle in a magnetic field will be 2.5 s, i.e. 2.5 times longer than the flocculation time). At this flowrate and with a slurry density of 33 per cent (by mass), the device will be able to treat 1000 tonnes of ore per hour.

The flocculation of 5 μm particles is less favourable. If a magnetic field of 0.05 T is used, the characteristic flocculation time will be about 100 s and the required slurry flowrate must be 2×10^{-3} m s^{-1}. Under these conditions the capacity of the device will be a mere 10 t/h. The other possibility is to reduce the time of flocculation by increasing the magnetic field. At a magnetic field of 0.5 T the half-time of flocculation will be 1 s, slurry velocity 0.2 m s^{-1} and the throughput will be again 1000 t/h.

The hematite particles of 10 μm diameter will flocculate with even greater difficulty: at 0.015 T the half-time of flocculation is 1000 s and the throughput will thus be very low. On the other hand at 3 T the characteristic time of flocculation will be again 1 s and the throughput will approach 1000 t/h. Because the magnetic induction of 3 T is not economically available with conventional magnets it is possible either to work at a lower field (e.g. 2 T) which will imply a longer flocculation time and lower throughput, or to use a superconducting magnet capable of generating high magnetic fields in large volumes at low energy consumption.

IV.15.3 Demagnetization

The demagnetization of an ore during beneficiation represents a more complicated problem than that of magnetization. This is particularly true about finely pulverized magnetite and all magnetically roasted iron minerals that have a high coercive force. The passage of ore containing these substances results in the formation of magnetic flocs that interfere with classification and reduce the selectivity of separation.

There are three main methods of demagnetization: a gradually reduced alternating field, a single application of a carefully adjusted demagnetizing field and thermal demagnetization. The first mentioned method is the most satisfactory and the most widely applicable.

In the alternating field method the magnetizing force must start at a value appreciably greater than the coercive force and be gradually reduced with each alternation until it reaches a very low value. For a rough demagnetization 3 or 4 alternations may be sufficient, but when the specification for demagnetization is very strict, as many as a hundred alterations may be needed [M1].

For demagnetization to be efficient several conditions are to be met [K15]:

a) alternating magnetic field of sufficiently large strength has to be generated
b) the flow of the slurry through the demagnetizing device should be laminar

c) the minimum residence time of the slurry in the magnetic field should be of the order of 0.24 - 0.28 s, which corresponds to 12 to 14 cycles of the ac magnetic field.

The demagnetizing unit may be a solenoid or an electromagnet with a laminated core of transformer steel. The sample to be demagnetized may be gradually withdrawn from the solenoid, or the sample can be left in a position and the current slowly reduced by means of a variable voltage transformer.

Experience indicates that ac magnetic induction greater than 0.03 T (300 G) at 50 Hz is usually sufficient for the demagnetization of natural magnetite or other ferromagnetic material of low coercive force [K15, B23, W24, D2].

It was shown by Karmazin [K15] that magnetically soft magnetite ore from the Sokolovsk-Sarbay region ($\mathbf{H}_c$ = 2.7 kA m^{-1} or 34 Oe) can be easily demagnetized. Although for complete demagnetization a magnetic induction of 0.04 T (400 G) is needed, a high degree of demagnetization can be achieved already at 0.01 T (100 G). A higher magnetic induction (0.07 T) is needed to demagnetize magnetically harder titano-magnetite ore from the Kachkanarskiy region ($\mathbf{H}_c$ = 7.5 kA m^{-1}, or 94 Oe) while magnetically roasted iron ores appeared to be difficult to demagnetize. As seen in Fig. 4.141 the demagnetization of Krivoi Rog magnetite ore ($\mathbf{H}_c$ = 15.6 kA m^{-1} or 196 Oe) was far from complete even at 0.1 T (1000 G) while the Kerchenskaya magnetite ore could not be demagnetized even at 0.18 T (1800 G). The demagnetization of very fine magnetite particles, close to single-domain size is also increasingly difficult.

The magnetic properties of the ores investigated by Karmazin et al. [K15] are summarized in Table 4.26.

Fig. 4.142 summarizes the results of the measurements of remanent polarization of the magnetite ores as a function of the demagnetizing field. A comparison of the results shown in Figs. 4.141 and 4.142 indicates that a complete magnetic deflocculation can be accomplished only when the remanent polarization of the ore is reduced to zero and that, for a given ore, the degree of magnetic deflocculation is inversely proportional to remanent polarization.

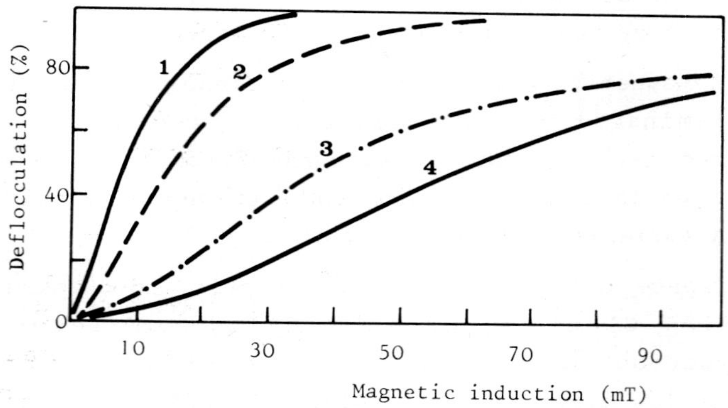

Fig. 4.141 Effect of the applied magnetic field on the deflocculation of the magnetite ores.
1 - Sokolovsko-Sarbay ore
2 - Kachkanarskaya ore
3 - Krivoi Rog ore
4 - Kerchenskaya ore
(After Karmazin et al. [K15])

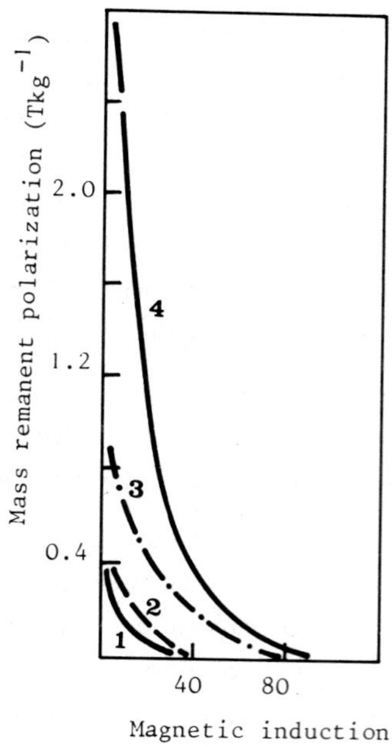

Fig. 4.142 Effect of applied magnetic field on the mass remanent polarization of flocs during demagnetization.
1 - Sokolovsko-Sarbay ore
2 - Kachkanarskaya ore
3 - Krivoi Rog ore
4 - Kerchenskaya ore
(After Karmazin et al. [K15]).

TABLE 4.26

Magnetic properties of the magnetite ores (after Karmazin et al. [K15])

Ore	Content % Fe	FeO	Specific magnetic suscep. (m^3kg^{-1})	Coercive force (Am^{-1})	Mass remanent polarization (Tkg^{-1})
Kerchenskaya	51.8	23.7	$5.9x10^{-4}$	$1.49x10^4$	2.82
Krivoi Rog	21.3	9.0	$2.5x10^{-4}$	$1.56x10^4$	0.92
Sokolovsk-Sarbay	40.4	18.4	$7.0x10^{-4}$	$2.7x10^3$	0.40
Kachkanarskaya	19.3		$2.0x10^{-4}$	$7.5x10^3$	0.38

Demagnetizing coils

For the demagnetization of ferromagnetic slurries in the mining industry, air-core ac coils are frequently used. The air cores are usually circular in shape so that round slurry pipe can be accommodated. Little or no advantage has been found in rectangular laminated iron core which is often used in the demagnetization of steel parts [B23].

The pipe section through the demagnetizing coil is made from non-metallic material for the purpose of eliminating the shielding of the slurry from the magnetic field and from eddy current generation. Semi-empirical equations based on eq. (1.69) were developed [B23] to calculate the design parameters of a demagnetizing coil:

$$H=\frac{1.03NI}{[(a+c)^2+b^2]^{\frac{1}{2}}} \tag{4.44}$$

$$L_e=\frac{7.88N^2(a+c)^2}{(3a+9b+13c)}\times 10^{-6} \tag{4.45}$$

where **H** is the magnetic field strength (in A m^{-1}),

$\mathbf{L_e}$ is the inductance (in Henry)

a - inside diameter of winding (in m)

b - length of coil (in m)

c - radial depth of winding (in m)

N - number of turns

I - current in the coil (in ampères).

Demagnetizing coils vary widely in size, shape, construction etc; typical configurations are shown in Fig. 4.143.

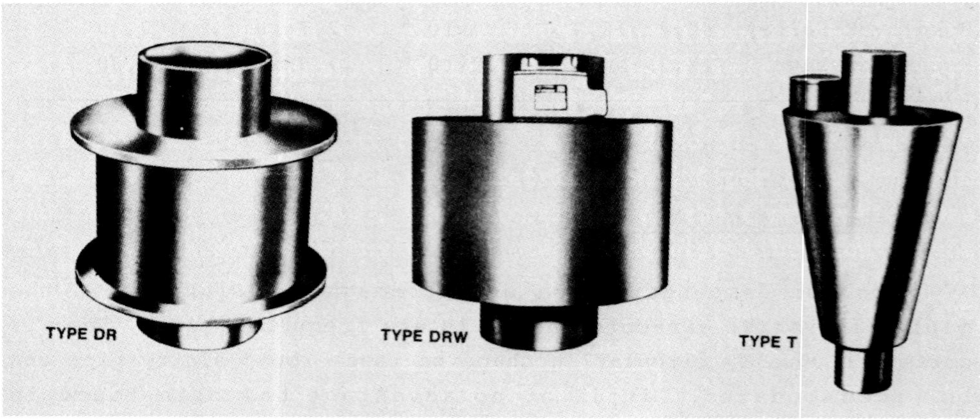

Fig. 4.143 Demagnetizing coils (courtesy of Eriez Magnetics, Inc.)

CHAPTER V

INDUSTRIAL APPLICATIONS OF HIGH-GRADIENT MAGNETIC SEPARATION

Although magnetic separation constitutes an old and established technique for the beneficiation of strongly magnetic ores and for the removal of ferromagnetic impurities from mixtures, the advent of high-gradient magnetic separation (HGMS) has been one of the most significant advances in the handling of quasi-colloidal particles in the last 20 years. Although HGMS has been applied in numerous branches of industry and science for the solving of various problems, its greatest potential application technologically and economically lies in the beneficiation of weakly magnetic ores. Although a large-scale application of HGMS to the treatment of minerals has been hampered by a lack of a cost-effective versatile magnetic separator and by an insufficient understanding of the fundamental principles of HGMS, as applied to complex systems of minerals, high-intensity and high-gradient magnetic separation can create great values in terms of high quality concentrates for a large spectrum of minerals.

V.1 Minerals treatment

V.1.1 Iron ores

Caue beneficiation plant, Brazil

High-gradient magnetic separation offers its greatest application as an alternative to flotation in the concentration of low grade weakly magnetic iron ores.

The world's largest production-scale plant using wet high-gradient magnetic separation for the beneficiation of hematite, is the Caue mine belonging to the Brazilian State Enterprise CVRD at Itabira. The overall plant has a capacity of 46 Mt/year, of which 18 Mt are high-grade hematite requiring only crushing and screening; the remaining 28 Mt/year is itabirite, the name given to finely disseminated hematite and quartz mixture requiring fine grinding for liberation. Wet, high-gradient magnetic separation is used for concentration and by 1981 28 Jones magnetic separators (D-317) were installed at Caue. A maximum throughput of 4000 tonnes of itabirite ore per hour was achieved [C34, D17].

The present Caue flowsheet is shown in Fig. 5.1. The itabirite ore, smaller than 1 mm, is treated at a rate of about 120 t/h per machine and a magnetic concentrate of a grade of 67 per cent Fe, middlings of 52 to 54 per cent Fe for recirculation and tailings between 8 and 15 per cent Fe are obtained.

The upgrading of itabirite was also tested with a carousel SALA HGMS separator, on a pilot-plant scale and the representative results ore shown in Table 5.1.

The Conceicao Mine of CVRD uses a very similar flowsheet to that of Caue and 15 Jones DP 317 separators have been set up in the plant.

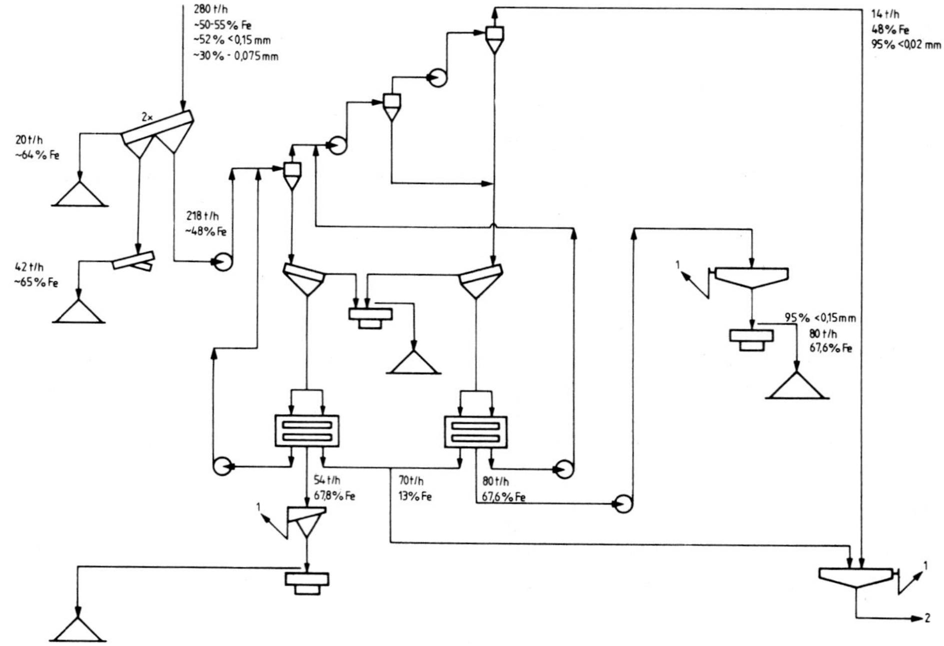

Fig. 5.1 Flowsheet of Caue beneficiation plant (CVRD Itabira) (after Cordes, Erzmetall [C34]).

TABLE 5.1

Metallurgical results of 2-stage beneficiation of itabirite by Sala HGMS [D19]
Matrix: expanded metal, magnetic induction 0.3 T, flow velocity 0.075 m s^{-1} (laboratory), 0.135 m s^{-1} (pilot plant)

	Grade		Recovery
	% Fe	% SiO_2	% Fe
Laboratory test:			
Feed	48.8	25.0	100
Mags	68.6	2.0	97.6
Non mags	3.0		2.4
Pilot-plant tests:			
Feed	48.8		100
Mags	68.6		95.8
Non-mags	3.7		4.2

Buchanan beneficiation plant, Liberia

Five Jones DP 317 separators were installed at the Buchanan plant in Liberia [L20] to substitute flotation in the pellet feed preparation. Most of the ore treated is the hematite with a grade high enough for it to be used directly as blast furnace burden.

However, owing to strongly fluctuating ore composition, which features a high portion of iron and aluminium hydroxide a flotation facility was installed, which, however, was unable to cope with the heterogenity of the deposit. The Jones magnetic separators evened out variations in the feed almost completely while increasing the grade of the concentrate to a level that cannot be achieved by the flotation process. Comparison of the results obtained by flotation and HGMS is given in Table 5.2, the flowsheet for beneficiation is shown in Fig. 5.2.

Experience has shown that optimum results can be obtained when the separator is run at 1.1 T, with distance of 2.5 mm between grooved plates used as a matrix and 55 to 60 per cent solids. The pressure of rinse water turned out to be critical variable, while the pH of the feed had very little effect on metallurgy. The iron recovery in the +45 μm fraction was 95 per cent or more, while in the fine fraction -45 μm recovery dropped to between 70 and 75 per cent.

TABLE 5.2
Comparison of results obtained by flotation and HGMS at Buchanan (1977) [L20]

		% Fe	
	Mass %	Assay	Recovery
Flotation process:			
Flotation concentrate	56.4	66.9	65.8
Hydrocyclone overflow to PF	30.7	53.0	28.3
Total pellet feed	87.1	62.0	94.1
Flotation tailings	8.9	14.4	2.2
Hydrocyclone overflow to tails	4.0	53.0	3.7
Total tailings	12.9	25.9	5.9
Classifier overflow = feed	100.0	57.3	100.0
Magnetic separation:			
Nimba Jones conc. = PF	73.0	65.1	82.5
Nimba Jones tailings	27.0	37.3	17.5
Classifier overflow = feed	100.0	57.6	100.0

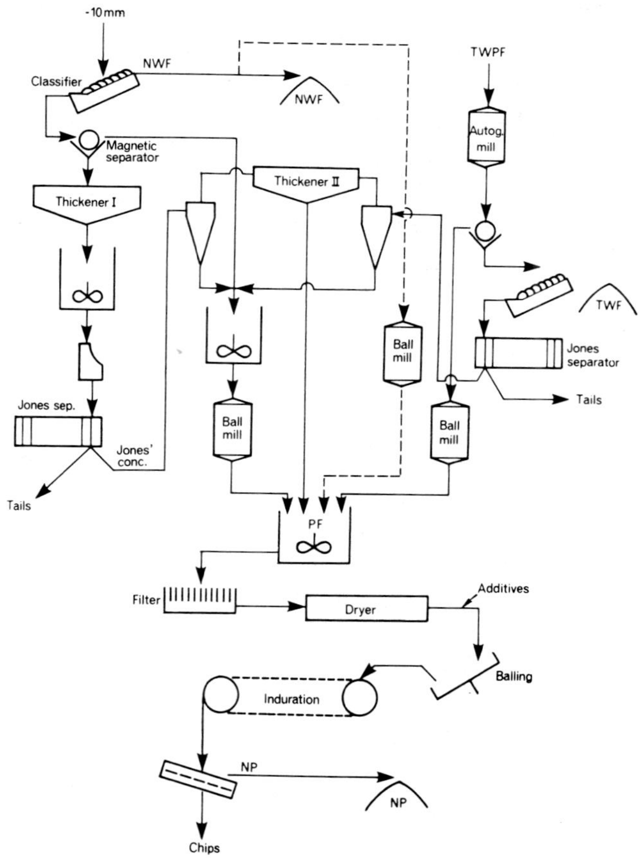

Fig. 5.2 Flowsheet for beneficiation plant at Buchanan [L20].

Wabush Mines, Canada

At Wabush Mines (Canada) Jones magnetic separators were installed to recover fine hematite from tailings [B34]. Wabush ore is specular hematite averaging 36 per cent iron, with some hematite, limonite and goethite. The bulk of fine hematite (-75 μm) concentrate by spirals, is lost to the tailings. Flotation of tailings was not successful and the environmental considerations, combined with rising costs of the reagents, were also unfavourable. Magnetic separation was found to be successful in upgrading the tailings and two Jones DP 317 machines were installed on the mine for secondary cleaning of about 190 t/h of spiral tailings, at the same time providing a secondary cleaning of some 50 t/h of primary concentrate. The separators were operated at 0.95 T, and grooved plates, used as a matrix, had a gap of 2.5 mm. The fine magnetite caused the clogging of the plates and consequently a reduction in unit capacity. When the concentration of magnetite exceeded 4 per cent, the plates clogged completely. The magnetic circuit produces 350 kt per year of an acceptable-grade product and mechanical availability of the circuit exceeds 90 per cent [B34]. A diagram of the magnetic circuit is given in Fig. 5.3 and the metallurgical performance of HGMS is illustrated in Fig. 5.4.

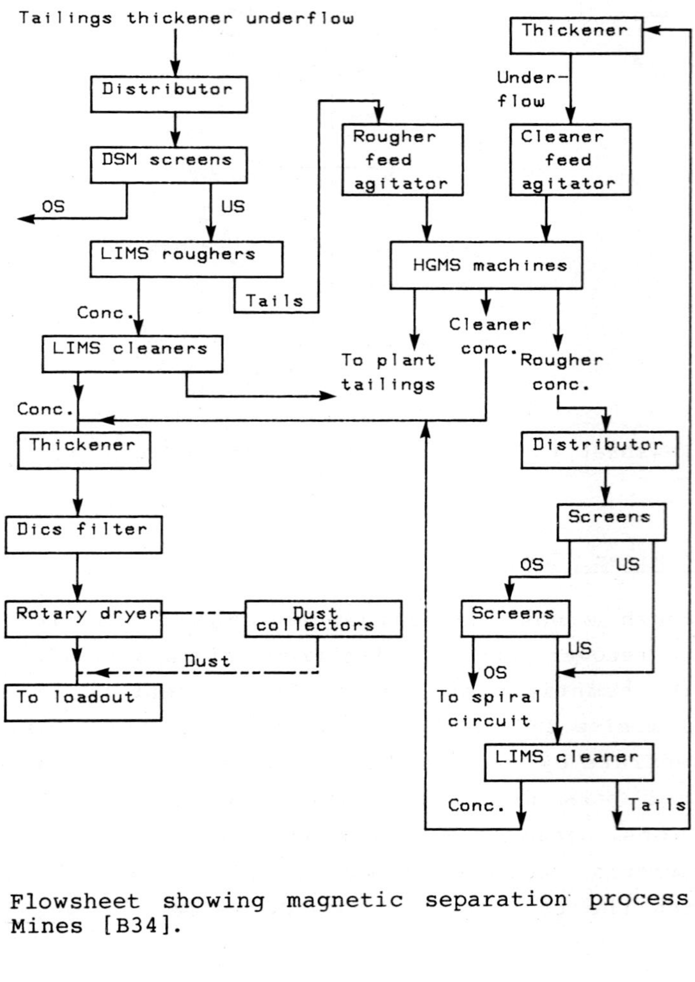

Fig. 5.3 Flowsheet showing magnetic separation process at Wabush Mines [B34].

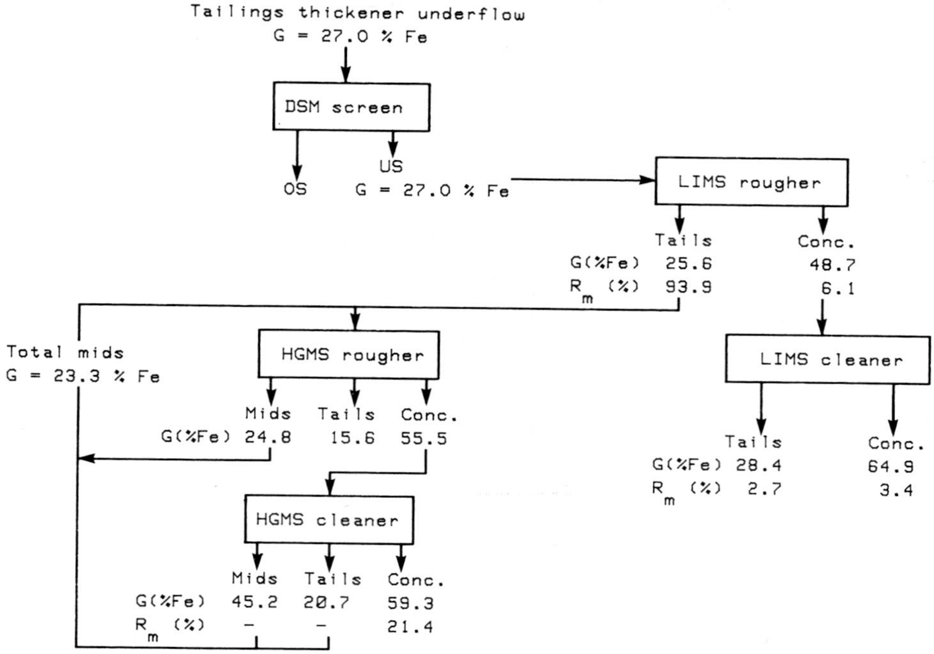

Fig. 5.4 Metallurgical performance of a magnetic separation circuit at Wabush Mine [B34].

Las Truchas beneficiation plant, Mexico

The treatment of martite iron ores by Jones HGMS machines employing the standard 2.5 mm gap between the grooved plates is often hampered by a reduced throughput, due to the presence of an excessive concentration of strongly magnetic components that had remained after low-intensity magnetic separation. A two-stage HGMS process was developed [B37, B38] in which the non-magnetic fraction from LIMS was upgraded in the first stage with Jones HGMS, with a large gap of about 5 mm at low magnetic field strength, adjusted according to martite content. The non-magnetic product is subsequently re-treated by a Jones machine using the gap of 2 mm at high magnetic field strength. The results of the representative tests are summarized in Table 5.3 and the two-stage flowsheet is shown in Fig. 5.5.

This flowsheet is used in the beneficiation plant in Las Truchas (Mexico) to preconcentrate magnetite-martite-hematite ore by two Jones machines [A27].

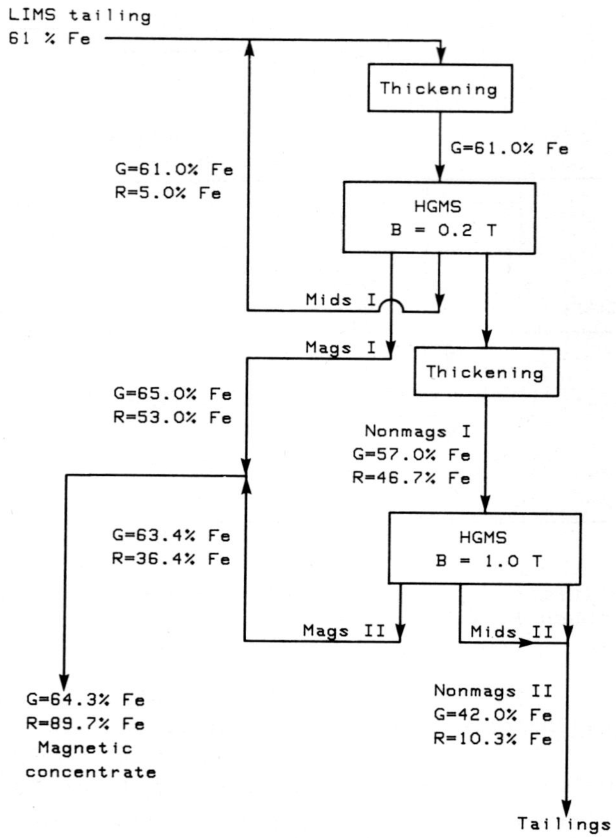

Fig. 5.5 Flowsheet showing two-stage magnetic separation treatment of martitised iron ore [B38].

Minnesota oxidized taconites

Wet, high-gradient magnetic separation, either alone, or in combination with flotation, is a technically feasible method of processing the enormous resources of oxidized taconite in the Lake Superior region (USA): HGMS was found to be less costly than either reduction roasting or selective flocculation-flotation [H38, H39], while the cost advantages of HGMS operation over flotation are reduced reagent, grinding and water treatment costs.

The application of HGMS to oxidized taconites was evaluated on a pilot-plant scale with Carpco and Sala carousel magnetic separators [L21, L22, L23, K25]. It was shown that from the ore ground to 90 per cent -37 μm, a concentrate of between 49 and 54 per cent Fe can be obtained, at an iron recovery ranging from 68 to 79 per cent, in a single pass [K25]. The flowsheet and material balance for the laboratory HGMS tests are shown in Fig. 5.6.

Since the HGMS operation is significantly more expensive than flotation from the standpoint of capital costs, the high cost of investment is a significant barrier to its application while progress is hindered by the high cost of production-size machines [H38].

TABLE 5.3

Results of two-stage HGMS tests with a Jones P40 separator on a martitised ore [B37, B38]

Gap width (mm)	Grooved plates Type	B (T)	Q (t/h)		R_M (%)	G (% Fe)	R (% Fe)
I 4.0	I 8R	I 0.22	I 249	Mags I	28.8	68.0	35.5
II 1.8	II 8R	II 1.0	II 120	Mags II	46.9	62.9	53.5
				Mags I+II	75.7	64.85	89.0
				Nonmags	24.3	25.1	11.0
I 5.0	I 8R	I 0.35	I 291	Mags I	29.8	68.0	36.0
II 1.8	II 8R	II 1.0	II 104	Mags II	49.0	62.8	54.7
				Mags I+II	78.8	64.75	90.3
				Nonmags	21.2	24.6	9.3
I 5.0	I 4R	I 0.35	I 308	Mags I	24.5	67.1	29.9
II 1.8	II 8R	II 1.0	II 121	Mags II	48.4	65.0	57.3
				Mags I+II	72.9	65.7	87.2
				Nonmags	27.1	25.9	12.8
I 5.0	I 4R	I 0.7	I 200	Mags I	40.0	67.8	48.6
II 1.8	II 8R	II 1.0	II 129	Mags II	33.4	66.1	39.5
				Mags I+II	73.4	67.0	88.1
				Nonmags	26.6	25.0	11.9

Strassa beneficiation plant, Sweden

A Sala 480 high-gradient magnetic separator was used at the Strassa mine (Sweden) to recover iron oxides from plant tailings [A23]. The Strassa crude ore contains approximately 28 per cent iron with a magnetite to hematite ratio of 3 to 2. The ore is beneficiated with low-intensity magnetic separators, the magnetic concentrate is sent to the pelletizing plant and the tailings are further upgraded by Humphrey spirals and shaking tables. The tailings are thickened and the thickener underflow, with an iron content of up to 10 per cent, is processed by HGMS.

The Sala HGMS machine, with a capacity of 180 t/h, was operated at 40 to 80 t/h. The maximum background magnetic induction employed was 0.55 T, expanded metal was used as a matrix and the superficial flow velocity of the feed ranged from 0.175 m s^{-1} to 0.45 m s^{-1}. The machine produced a concentrate that contained 50 per cent Fe at 42 per cent recovery and which can be included in the feed to the existing pelletizing plant.

Rudňany beneficiation plant, Czechoslovakia

The beneficiation of a complex siderite ore from the Rudňany deposit (Czechoslovakia) by magnetic separation has been thoroughly investigated in the last twenty years. The ore, the composition of which is shown in Table 5.4, is beneficiated by a combination of magnetic separation and flotation, high-intensity wet roll magnetic separators being used to recover siderite [H41].

An effort to improve the technological parameters of the magnetic circuit was induced by the decreasing quality of the deposit, by low sensitivity of the roll separators to particles smaller than 100 μm and by the need to operate a great number of low-performance machines that require classified feed. MRVK high-gradient magnetic separators were found to simplify the operation and to improve the recovery of iron from the slimes [H1] and it is anticipated that in 1988 all siderite ore in Rudňany will be treated by a recently developed VMS high-gradient magnetic separator [K26]. The proposed flowsheet for the operation is shown in Fig. 5.7.

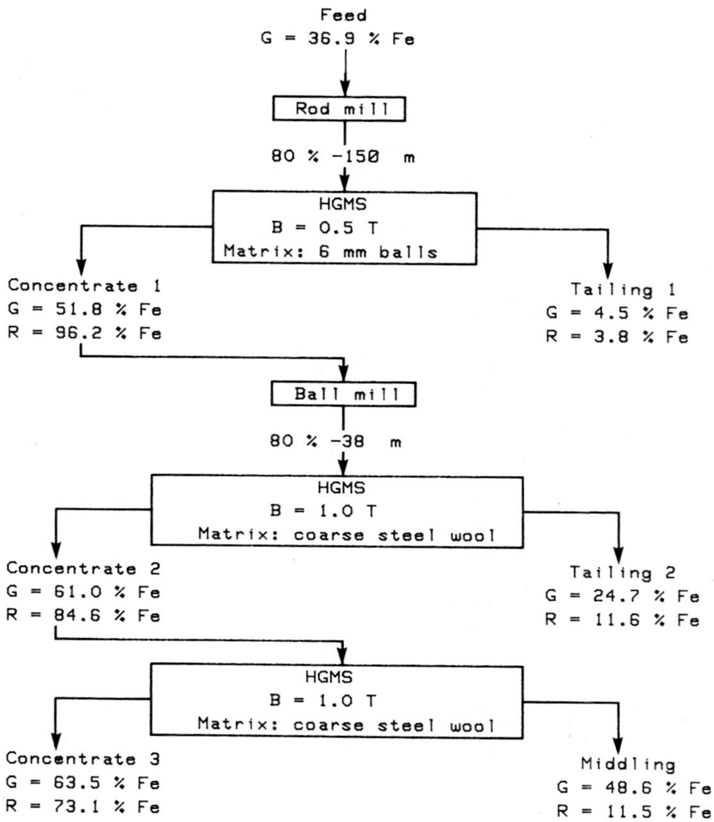

Fig. 5.6 Flowsheet and material balance for laboratory beneficiation of oxidized taconite by wet high-gradient magnetic separation [H38].

TABLE 5.4

Composition of the Rudňany siderite ore and magnetic susceptibilities of the components [K26]

Mineral	Content (mass %)		Mass magnetic susceptib.	
	Fe	SiO_2	$m^3kg^{-1}x10^{-8}$	$cm^3g^{-1}x10^{-6}$
Siderite	40.62	0.20	62 - 75	49 - 60
Baryte	-	1.32	0.05	0.03
Fuchsitic quartzites	20.17	49.48	26.7	21.2
Sericitic shales	17.32	40.97	20.2	16.1
Chloritic shales	7.43	47.98	10.8	8.6
Graphitic shales	6.82	59.30	4.2	3.3

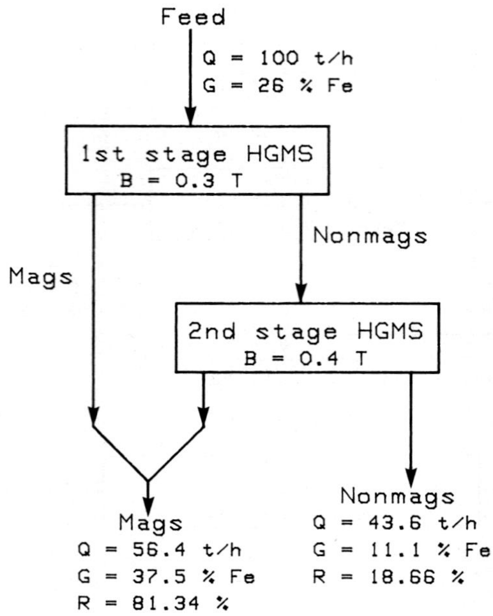

Fig. 5.7 Proposed flowsheet for the magnetic beneficiation of siderite Rudňany ore [H35].

Krivoi Rog concentrator, Ukraine

At the present time, almost one half of the iron ore mined in the USSR originates from the Krivoi Rog District in the Ukraine. Moreover, by 1990 beneficiation plants will be commisioned that treat iron ore from vast deposits of finely intergrown weakly magnetic oxidized quartzites from the Krivoi Rog District and Kursk Magnetic Anomaly. The proposed capacity of the plants is 30 Mt/year and it is expected that at least forty high-gradient magnetic separators, each with a capacity of 100 t/h will be required.

The iron minerals are mainly martite, hematite, hydroxides and, to a lesser degree, magnetite. The minerals are very fine, over 80 per cent being smaller than 45 μm. In order to meet the basic technological parameters shown in Table 5.5, a number of continuous high-gradient magnetic separators were tested [M27, C8, K24]. It was observed that by increasing the magnetic induction in

TABLE 5.5

The target technological parameters of beneficiation of oxidized iron quartzites by HGMS at Novo-Krivoi Rog plant [M27]

	HGMS Separator Jones	SALA
Mass yield into mags (%)	39	42
Recovery of iron into mags (% Fe)	64	70
Grade of mags (% Fe)	60	61

a Jones separator from 1 T to 1.5 T, the concentration of iron in the tailings decreased by 2 to 2.5 per cent. A reduction in the matrix height from 220 mm to 150 mm resulted in a drop of recovery of 10 per cent, the grade being constant. The investigation also revealed that for feed size distribution 75 per cent smaller than 74 μm, the gap between the grooved plates should range from 1.5 to 2 mm, while for a finer grind, 100 per cent smaller than 45 μm, the gap should be as small as 1 mm. The minimum particle size of recovered particles amounted to 15 μm for a 2 mm gap, and 8 to 10 μm for a 1 mm gap. For narrow gaps however, matrix clogging was observed [M27].

The flowsheet for the treatment of oxidized quartzites by Jones HGMS is shown in Fig 5.8. The concentrate obtained in this flowsheet contained 60.4 per cent Fe, in a recovery of 72.4 per cent Fe.

The flowsheet incorporating the high-gradient magnetic separator using solenoid magnets with the expanded metal matrix is depicted in Fig. 5.9. Although background magnetic induction was equal to only 0.5 T, the recovery of 75.6 per cent at the grade of 62.2 per cent Fe was achieved. The pilot-plant tests for the beneficiation of oxidized quartzites from Krivoi Rog and Kursk Magnetic Anomaly were also performed using the VMS high-gradient magnetic separator (see Section II.4.3.6). In a two-stage flowsheet, similar to that shown in Fig. 5.9 background magnetic induction of 0.5 T was used. In the first stage HGMS, rods of 3 mm diameter and of 2.5 mm spacing were employed, while in the second stage woven mesh of 1.6 mm wire diameter with an aperture of 2.5 mm, was used. Typical metallurgical results are shown in Table 5.6.

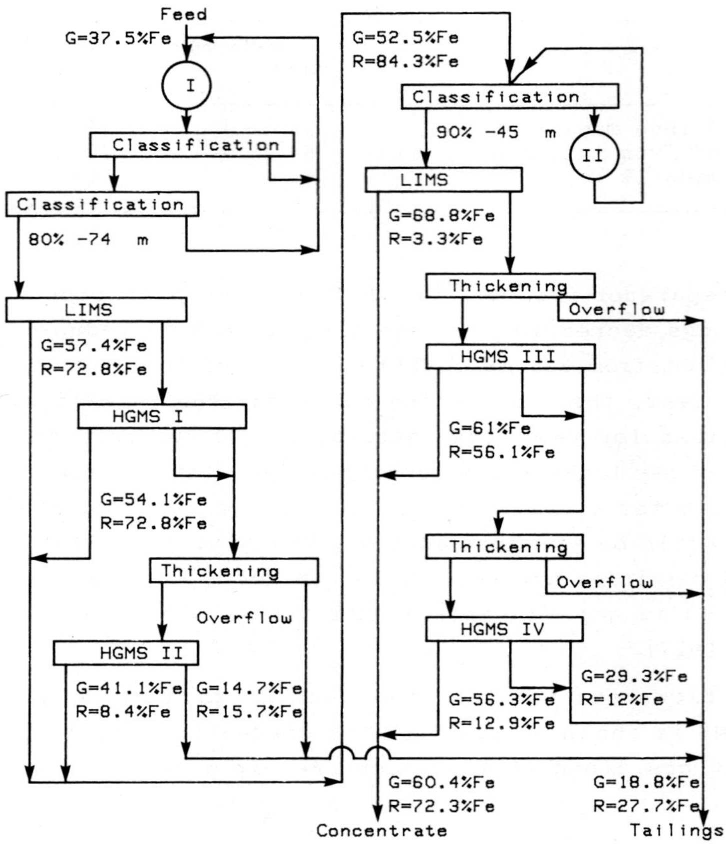

Fig. 5.8 Flowsheet showing recovery of oxidized quartzites (Krivoi Rog) by HGMS using electromagnet with iron yoke. R_m - mass yield (%), G - grade (% Fe), R -recovery (% Fe). (After Maliy et al. [M27]).

TABLE 5.6

Two-stage magnetic separation of oxidized quartzites with a VMS separator [C8]

	Krivoi Rog		Kursk Magnetic Anomaly	
Product	Grade (%)	Recovery (%)	Grade (%)	Recovery (%)
Stage I:				
Feed	38.3	100.0	39.9	100.0
Magnetics	50.7	90.5	50.1	90.5
Non-magnetics	11.4	9.5	15.0	9.5
Stage II:				
Magnetics	61.7	71.9	57.5	78.3
Non-magnetics	31.5	19.6	27.4	12.2
Total non-magnetics	20.0	29.1	20.1	21.7

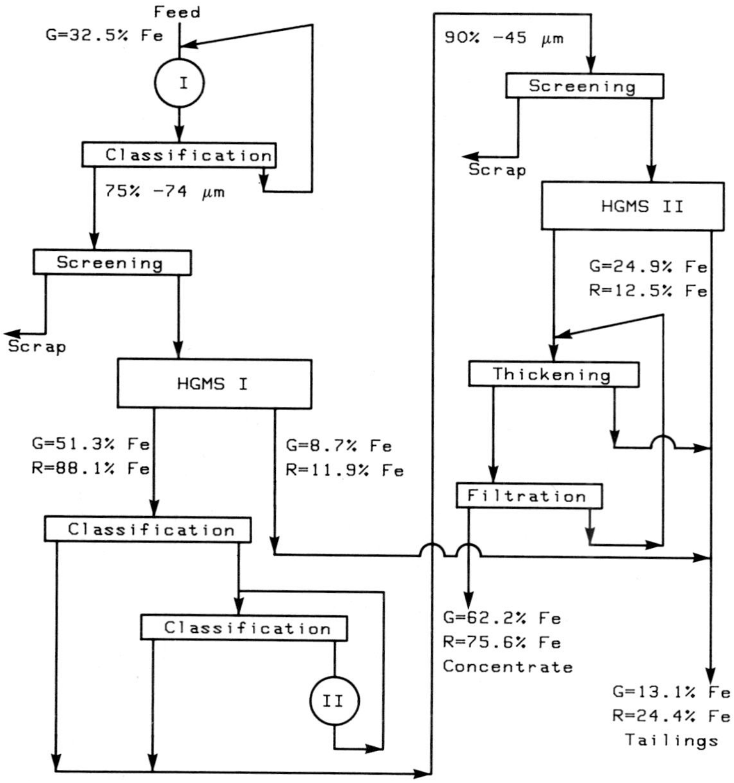

Fig. 5.9 Flowsheet showing recovery of oxidized quartzites (Krivoi Rog) by HGMS using solenoid magnet (matrix: expanded metal, magnetic induction 0.5 T) (after Malyi et al. [M27]).

V.1.2 Manganese ores

The Soviet Union is a most important producer of manganese, the most extensive reserves being found in the Nikopol and Chiatursk basins. Wet high-intensity magnetic separation with four-roll separators with focused magnetic field, is used to beneficiate ore smaller than 3 mm or products from log washers and jigs. The flowsheet for the beneficiation of low-grade manganese concentrate in the Chiatursk beneficiation plant is shown in Fig. 5.10.

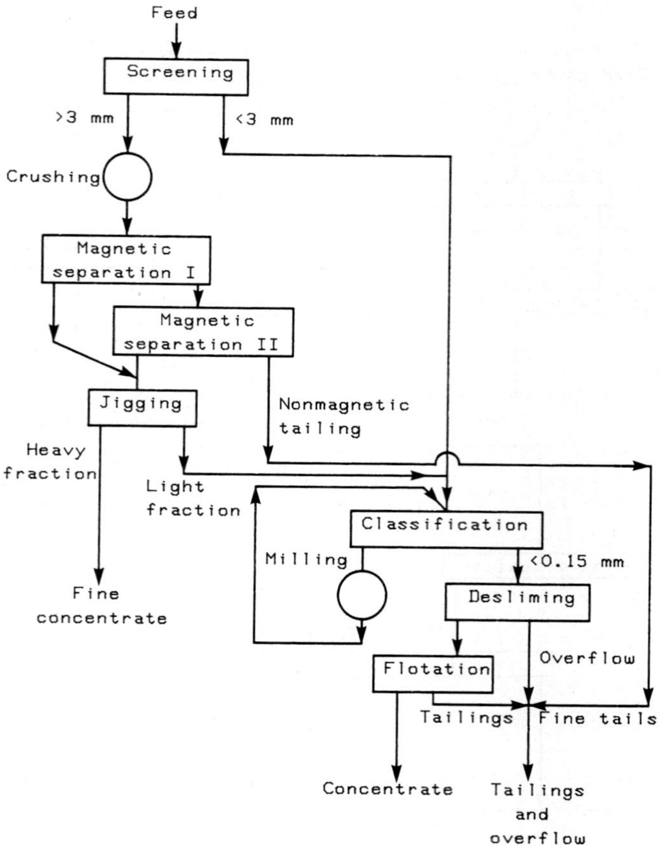

Fig. 5.10 Flowsheet showing beneficiation of manganese concentrate in the Chiatursk beneficiation plant [D1].

During the beneficiation process of the manganese ore, a large quantity of tailings, containing between 10 and 14 per cent of manganese is generated , with manganese minerals usually smaller than 20 μm. Since the upgrading of manganese slimes by anionic flotation is inefficient, extensive research at the Bogdanovsk beneficiation plant of the Ordzhonikidze Concentrator has revealed that, by using high-gradient magnetic separation (Jones P-40 and Sala HGMS 120-15-00) a concentrate containing up to 33 per cent of manganese [O9] can be produced.

With the use of Jones separator, with a gap of 2.5 mm between the grooved plates, the recovery of manganese from a sample containing particles smaller than 0,5 mm amounted to 55 per cent, fractions smaller than 40 μm (30 per cent recovery) being responsible for a drop in recovery compared to deslimed samples. For a fraction -500 + 160 μm recovery of manganese of 90 per cent was obtained, while for a fraction -160 + 40 μm the recovery amounted to 82 per cent [T16].

When the gap between the grooved plates was reduced to 1.5 mm, recovery rose to 93 per cent for -500 + 160 μm, to 88 per cent for -160 + 40 μm and 43 per cent for -40 μm fractions. Magnetic induction significantly affected the results: at 1 T the recovery of manganese from -500 + 40 μm fraction was 57.3 per cent, increasing to 88.4 per cent at 1.4 T.

Pilot-plant testwork with the Jones P-40 separator has shown that in a two-stage process, depicted in Fig. 5.11 a concentrate containing up to 33 per cent of manganese, at the recovery of 60 per cent, can be obtained. The grade of the magnetic concentrate is, however, lower than that achieved by flotation (36 per cent Mn at the recovery of 32 per cent), due to a simultaneous recovery of iron oxides, and it is anticipated that a combined HGMS-flotation scheme would be employed to upgrade the manganese slimes [T16].

The effectivenes of HGMS in the upgrading of different types of manganiferous iron ores in the Cuyuna Range, Minnesota, USA was studied by Sabelin et al. [S69]. Of the four samples selected as representative of Cuyuna manganese ore types, only hematite ore responded positively to HGMS. The recovery-grade curves are shown in Fig. 5.12. The carbonate ore responded reasonably well to preconcentration by HGMS, followed by anionic flotation of the carbonate mineral. Other extensively oxidized ores were extremely slimy and very little upgrading by HGMS was obtained.

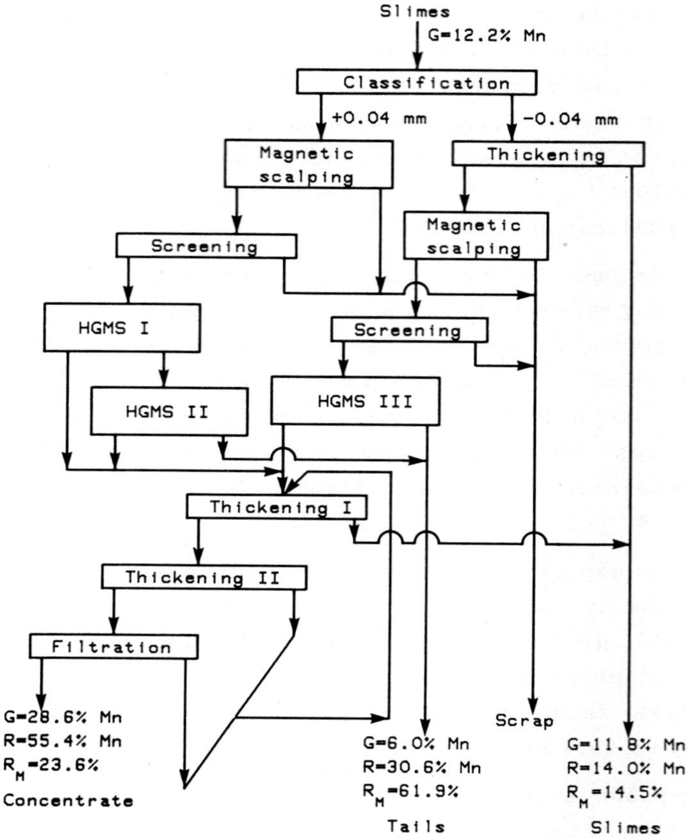

Fig. 5.11 Flowsheet showing beneficiation of manganese slimes by HGMS [T16].

Manganese ore from the Obrotchishte deposit (Bulgaria) after being crushed below 50 mm was screened into three fractions: the -50 + 6 mm fraction was upgraded by heavy media separation, the -6 + 0.5 mm fraction by jigging and the -0.5 mm fraction, after desliming, was beneficiated by pinched sluices. Recovery of manganese from the fine fraction does not exceed 32 per cent at 31 per cent grade and alternative techniques were sought to improve the metallurgy. Barzev et al. [B39] have shown that high-gradient magnetic separation, as applied to -0.5 mm fractions, is far more effective than pinched sluices and that the recovery of manganese exceeding 79 per cent, at 30 per cent grade, can be obtained in a three-stage HGMS, as shown in Fig. 5.13a. Desliming at 45 μm was found to be not beneficial, and although selectivity improved marginally, a significant drop in recovery (to 63 per cent Mn) was observed in a deslimed sample, as shown in Fig. 5.13b.

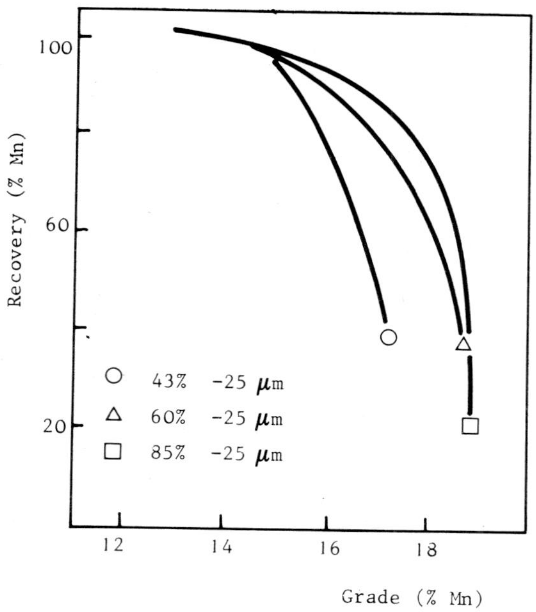

Fig. 5.12 Chart showing the relationship between recovery of Mn and the grade of magnetic fraction of hematite ore (Cuyuna Range) [S69].

V.1.3 Tin and tungsten ores

Cassiterite (SnO_2) is the only tin mineral in sufficient abundance in the earths's crust to have any commercial value (stannite is occasionally won). Although its high density in relation to its gangue constituents makes it an ideal mineral for the application of gravity-separation techniques, fine cassiterite responds poorly to conventional beneficiation techniques, and its upgrading, by the removal of more magnetic impurities, including tourmaline was attempted by HGMS [A16, A21].

Wolframite, a tungstate of iron and manganese, $(Fe,Mn)WO_4$, is often associated with cassiterite and the difference in their magnetic susceptibilities is often used to separate the minerals

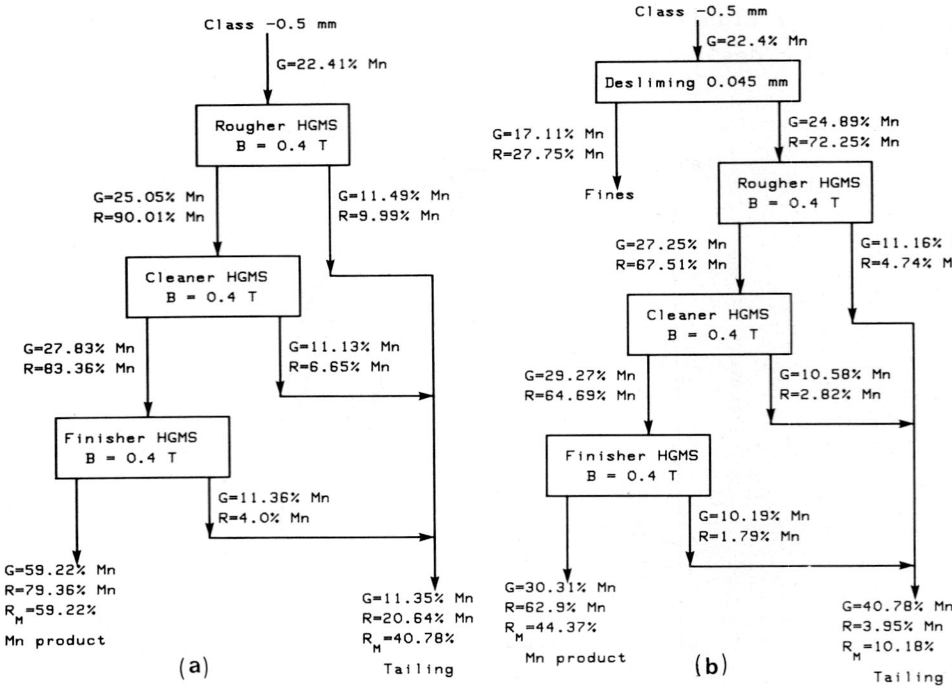

Fig. 5.13 Flowsheet showing beneficiation of the manganese ore (Obrotchischte, Bulgaria) by HGMS (after Barzov et al. [B39]).

magnetically. The fine fractions that are generated either in dry magnetic separation (scrubber dust) or in a gravity-flotation process, are not usually reprocessed, although several efficient flowsheets involving HGMS were developed over the years.

Karmazin et al. [K21] investigated the upgrading of the dust from dry magnetic separation and the slimes obtained from gravity-flotation treatment of intermediate tin products. As follows from Fig. 5.14, in a two-stage HGMS, using steel balls as a matrix and with a background magnetic induction of 0.2 T, the scrubber dust assaying 28 per cent WO_3 and 6.2 per cent Sn was upgraded to a magnetic concentrate containing 36.4 per cent WO_3 at a recovery of 96 per cent, while cassiterite was concentrated into the non-magnetic fraction, having a grade of 14.9 per cent Sn at a recovery of 82.8 per cent. The effect of the magnetic induction on the

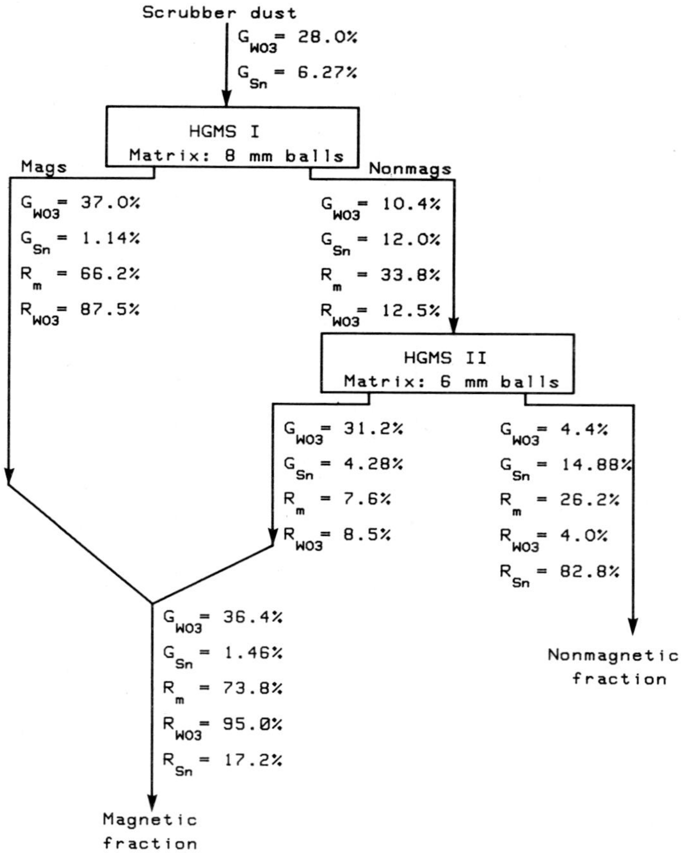

Fig. 5.14 Flowsheet showing recovery of wolframite and cassiterite from scrubber dust [K21].

metallurgical results is shown in Fig. 4.28, while the influence of the size of the ball matrix is illustrated in Fig. 4.34.

Hencl [H1] has shown that by a three-stage HGMS (expanded metal as a matrix, background magnetic induction up to 0.9 T) the wolframite ore from Mongolia (-0.5 mm, 3.6 per cent WO_3) can be upgraded to such an extent that the magnetic concentrate contains 39 per cent WO_3 at a recovery of 67 per cent.

The recovery of wolframite from the slimes by the combination of flotation and magnetic separation is schematically depicted in Fig. 5.15. As shown in Table 5.7, a product containing 21 per cent WO_3, at a recovery of 79 per cent, was obtained [G28].

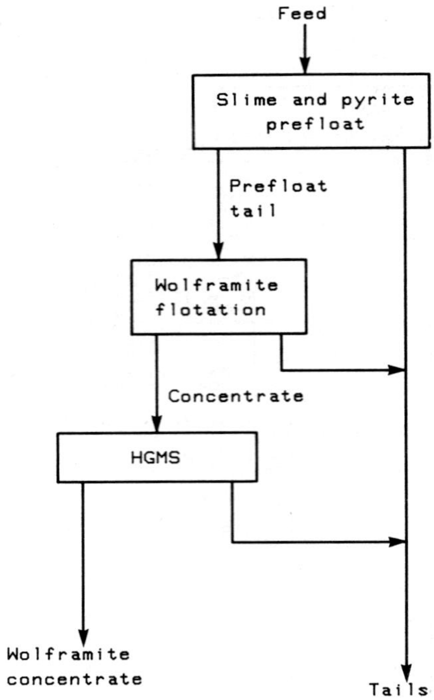

Fig. 5.15 Flowsheet showing beneficiation of wolframite slimes [G28].

TABLE 5.7

Magnetic separation of wolframite concentrate [G28]

Product	Mass yield (%)	Grade (% WO_3)	Recovery in HGMS (% WO_3)	Overall recovery (% WO_3)
Feed	100	1.90	100	35.0
Mag. fraction	6.95	21.70	79.5	27.8
Nonmag. fraction	93.05	0.42	20.5	7.2

The improvements in the beneficiation of Chinese wolframite slimes by the inclusion of magnetic separation, were reported by Sun et al. [S70]. Wolframite, having a magnetic susceptibility of 5 x 10^{-7} m^3 kg^{-1} (40 x 10^{-6} cm^3 g^{-1}) can be effectively separated from cassiterite and other associated non-magnetic minerals. The recovery of tungsten depends chiefly on the metal distribution in the size range smaller than 10 μm, which was found to be the lower recoverable limit, as shown in Table 5.8. The concentration ratio is dependent on the proportion of other weakly magnetic minerals, such as tourmaline, limonite and siderite.

Results of laboratory magnetic separation tests in Table 5.9 show that recoveries of up to 87 per cent of WO_3 and a concentration ratio from 4 to 20, were obtained. Pilot plant tests conducted at mine F yielded a concentrate grade of 8.4 per cent WO_3, with a recovery of 76 per cent and a concentration ratio of 19.

TABLE 5.8
Recovery of tungsten as a function of particle size [S70]

Particle size range (μm)	Recovery (% WO_3)	Concentration ratio
-150 + 104	75.2	14.0
-104 + 76	79.3	36.8
-76 + 66	86.0	43.9
-66 + 50	95.8	38.5
-50 + 40	93.4	31.6
-40 + 30	89.4	47.7
-30 + 20	84.7	31.5
-20 + 10	61.9	20.3
-10	18.9	45.0

TABLE 5.9
Two-stage magnetic separation of wolframite slimes (China) [S70]

		Mass yield (%)	Grade (% WO_3)	Recovery (% WO_3)	Concentration ratio
Mine C	Feed	100	0.53	100	
	Mags	8.05	5.2	79.2	9.8
	Non-mags	91.95	0.12	20.8	
Mine D	Feed	100	0.46	100	
	Mags	15.8	2.38	82.5	5.2
	Non-mags	84.2	0.095	17.5	
Mine F	Feed	100	0.64	100	
	Mags	3.9	12.84	79.0	20.1
	Non-mags	96.1	0.14	21.0	
Mine G	Feed	100	0.68	100	
	Mags	18.6	2.58	70.6	3.8
	Non-mags	81.4	0.25	29.4	

Wolframite products from HGMS require further cleaning to meet market specifications. It was found that at the mine F, where hübnerite occurs as the main tungsten mineral and the magnetic concentrate has a grade of 6.9 per cent WO_3, the subsequent arsenic acid flotation gives the final concentrate grade of 51 per cent WO_3 at a recovery of 95 per cent. The flowsheet for the process is shown in Fig. 5.16 and the performances of the pilot-plant and production circuits are summarized in Table 5.10.

TABLE 5.10
Metallurgical performance of the HGMS circuit at Mine F (China) [S70]

	Grade (% WO_3)	Recovery (% WO_3)
Pilot-plant tests		
Feed	0.51	100
Concentrate	47.7	72.8
Tail	0.12	27.2
Production (1983)		
Feed	0.42	100
Concentrate	35.4	73.8
Tail	0.12	26.2

Four Boxmag-Rapid high-gradient magnetic separators were installed at three different plants in Peru to recover wolframite from accumulated or current plant tailings [W36]. A total of one million tonnes of tailings, with a grading of 0.15 to 0.20 per cent WO_3 were identified at Compania Minera Puquio Cocha. After the pilot-plant tests, a flowsheet incorporating high-gradient magnetic separators, shaking tables, bulk sulphide flotation and high intensity dry separators, was selected.

A flowsheet for a full-scale 14 tonnes/h plant is shown in Fig. 5.17. Tramp iron and magnetite is removed by a wet low-intensity drum separator and the non-magnetic product is upgraded by a Boxmag-Rapid HIW-4 carousel separator from 0.15 per cent WO_3 to 2 per cent WO_3 at a recovery of 80 per cent and mass yield into the magnetic fraction of 10 per cent. The concentrate is classified and treated on shaking tables. The table concentrate is fed to the flotation circuit to remove the sulphides while the table tailings are re-treated in a carousel high-gradient separator SHW-1. The magnetic concentrate from SHW-1 is upgraded by gravity separation and flotation. The flotation tailings are upgraded by a dry belt magnetic separator to obtain the final concentrate with 66 per cent WO_3 at the overall recovery of 55 per cent.

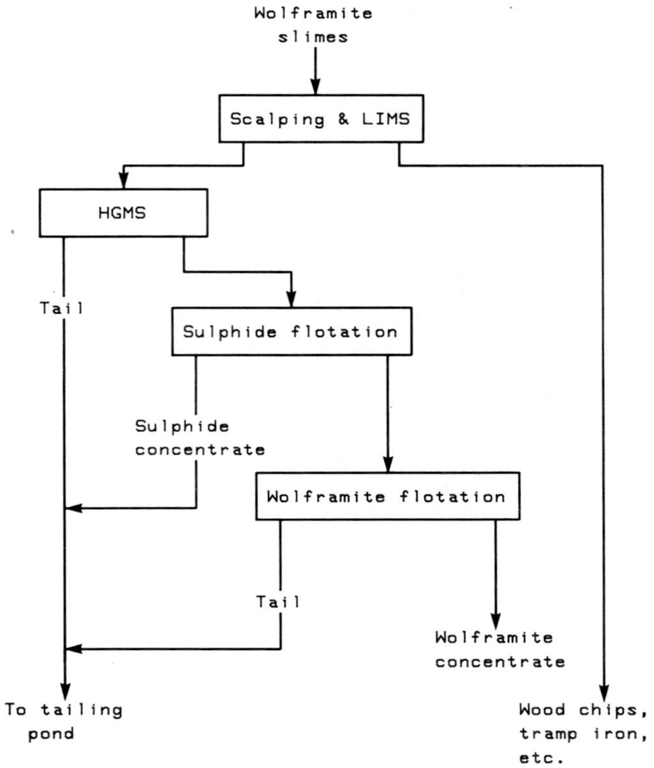

Fig. 5.16 Flowsheet showing beneficiation of wolframite slimes (China) [S70].

V.1.4 The Witwatersrand uranium - gold ores

The ability of high-gradient magnetic separation to recover uranium has been known for some time [S71]. Numerous investigations have shown that recoveries of uranium exceeding 95 per cent can be obtained [M32, S71], while the often unsatisfactory separation of uranium minerals by HGMS is related to ore-specific properties [M32], as well as the unoptimised operation of a magnetic separator [B40].

When Levin [L24] showed that uranium and gold can be recovered by magnetic separation from Witwatersrand ores, a possibility of using HGMS as an upgrading procedure for uranium and gold had been extensively investigated by the Council for Mineral Technology (Mintek), Randburg, in close collaboration with the gold-uranium industry in South Africa. It was shown [C7] that HGMS can recover over 60 per cent of the gold and uranium from many cyanidation residues and flotation tailings, into a concentrate that usually varies between 6 and 12 per cent by mass of the original feed. Of further interest was the finding that HGMS can also recover additional gold and uranium from the tailings of a flotation process that uses cyanidation residues as feed, i.e. that HGMS and flotation can complement each other. Recoveries of gold and uranium are closely related because most of the refractory gold remaining in the residues is intimately dispersed in pyrite or locked in silicates. Flotation recovers some of the former and HGMS some of the latter.

The paper by Corrans and Levin [C7] described how the HGMS process was tested on a continuous pilot plant basis at two gold

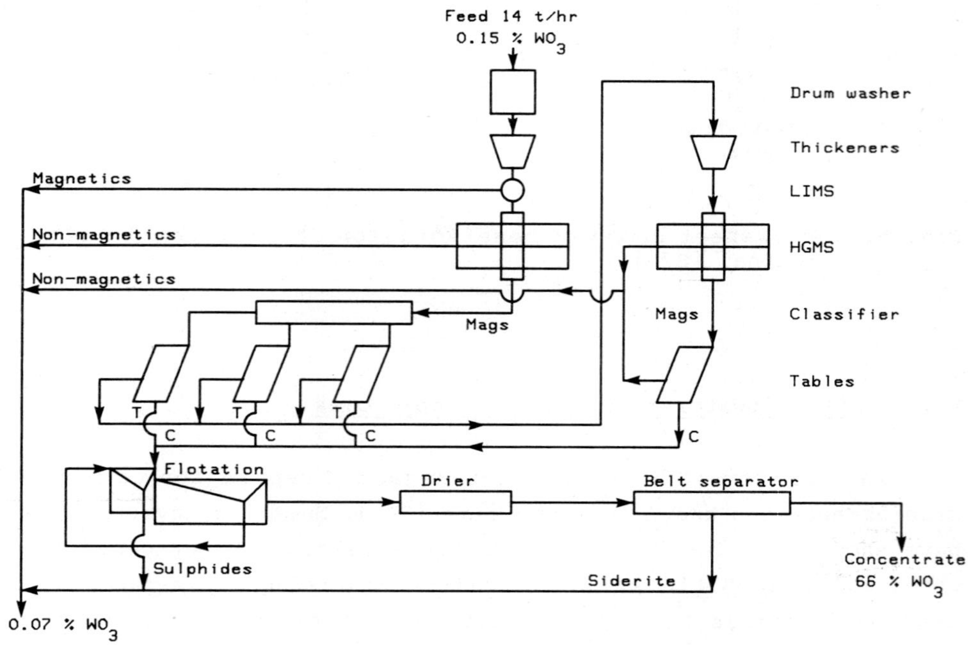

Fig. 5.17 Recovery of wolframite at Puquio Cocha plant (Peru) [W36].

mines. During the tests, a major mechanical problem was brought to light, namely the tendency for the matrix of iron balls (6 mm in diameter) to become progressively blocked with ferromagnetic material and fine wood fibres. This problem was solved by the development of a washing system that allowed the balls to be continuously removed from the machine and, after being washed in a trommel screen, to be returned to the separator [C6, C7, C15, C35].

With the washing system installed, several successful pilot-plant operations were carried out on cyanidation tailings from a mine in the Klerksdorp area. Typically, 55 per cent of the gold and 45 per cent of uranium was recovered into a magnetic concentrate that was 14 per cent by mass of the original feed.

On the strength of the pilot-plant results in 1979 the production-size Eriez CF-60 HGMS machine with the ball-cleaning system was installed as a prototype at the Stilfontein Gold Mine. The flowsheet for the HGMS installation is given in Fig. 5.18. The machine operated at the background magnetic induction of 0.75 T, although up to 1 T could be produced in the air gap. The CF-60 was designed to treat 30 t of dry solids per hour in a pulp of 40 per cent solids.

The machine which ran for over three years, was successful both mechanically and metallurgically, and the results of the work were published by Corrans et al. [C23]. The mass balance for the continuous HGMS machine is given in Table 5.11 while the batch test results are summarized in Table 5.12. It can be seen that recoveries and tailings grades for the batch tests and the large machine are similar, but that the concentrate grades obtained on

TABLE 5.11
Material mass balance for the CF-60 HGMS machine [C23]

	Mass yield (%)	Grade U_3O_8 (ppm)	Grade Au (g/t)	Grade S (%)	Recovery U_3O_8 (%)	Recovery Au (%)	Recovery S (%)
Feed	100	164	0.36	1.2			
Mags	15.2	633	1.32	2.07	56.4	51.4	25.1
Non-mags	84.8	84	0.21	1.08	43.6	48.6	74.9

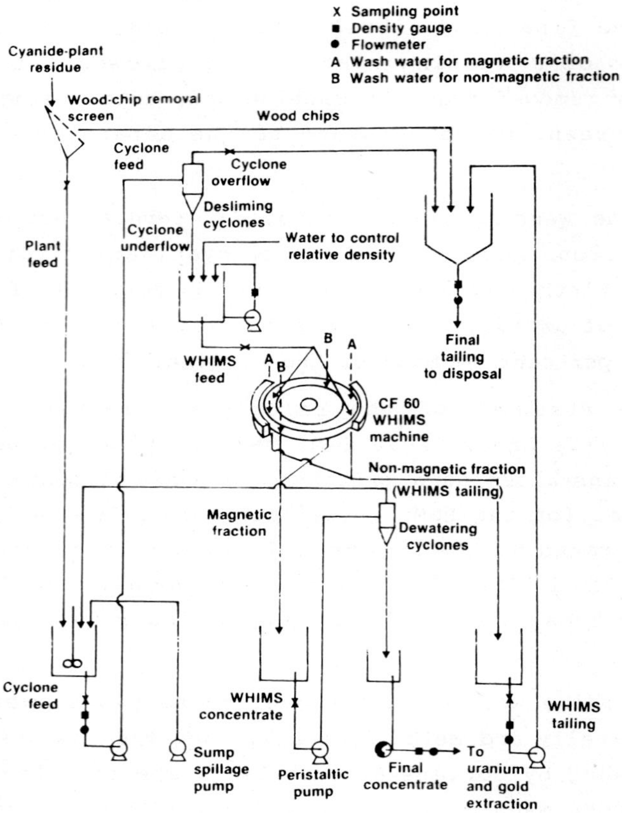

Fig. 5.18 Flowsheet showing the HGMS installation at a gold mine [C23].

TABLE 5.12

Material mass balance for HGMS batch tests [C23]

	Mass yield (%)	Grade		Recovery	
		U_3O_8 (ppm)	Au (g/t)	U_3O_8 (%)	Au (%)
Feed	100	170	0.38		
Mags	9.9	1187	2.25	58.8	59.8
Non-mags	90.1	90	0.17	41.2	40.2

the large machine are lower than those obtained by batch separation.

The tailings assays and mass percentage for the concentrate indicate substantial dilution of the magnetic concentrate by the tailings during operation of the continuous machine.

Good recoveries of gold and U_3O_8 indicate that HGMS installations can be highly profitable. The direct operating costs were as low as 30 cents per tonne of feed for the production of a magnetic concentrate with recoveries up to 55 per cent of gold and 45 per cent of the uranium.

Re-treatment of sand dumps

The treatment of tailings dams and sand dumps, emanating from past gold recovery operations, is receiving ever-increasing attention from major mining groups in South Africa. Tonnages available for re-treatment are enormous and a number of large-scale operations have been established.

Sand dumps are the result of metallurgical operations carried out during the early part of this century. Typical values are: gold (sands) 0.7 to 1.2 g/t, gold (slimes) 0.2 to 0.7 g/t, sulphur 0.8 to 1.2 per cent, and uranium oxide 0.04 to 0.20 kg/t [R17]. In the late 1970s, two of the world's largest dump-retreatment operations were commissioned: the East Rand Gold and Uranium Company (Ergo) in the Transvaal, and the Joint Metallurgical Scheme (JMS) in the Orange Free State. Each plant treats about 1.5 million tonnes per month of slime material, their economic viability being based on the production of three products: gold, uranium and sulphuric acid, from the recovered pyrite.

In the choice of a re-treatment process a complex scenario can develop, based on the grade of the material with respect to gold, uranium and sulphur; the market conditions and prices, scale of operation, capital and operating costs, etc. The simplest process would be direct cyanidation of the reclaimed material. The big operators have introduced a concentration step, flotation being, at present, the only concentration technique in use [C36]. Sulphide flotation can efficiently recover the pyrite (and, hence, the gold locked in, or associated with the sulphide), as well as varying amounts of the free gold. The flotation tailing is discarded to a tailings dam, so the recovery, by flotation, of gold, uranium and sulphur determines to a large degree their ultimate recovery. The pyrite concentrate is leached with acid, and roasted and leached with a cyanide solution, as shown in Fig. 5.19.

With this type of flowsheet, sulphur recoveries are good, viz., up to 85 per cent. The uranium recoveries, however, are not good, being usually 20 to 40 per cent because of the poor flotability of the uranium-mineral species. Gold recoveries are moderate, i.e. 40 to 50 per cent [C36].

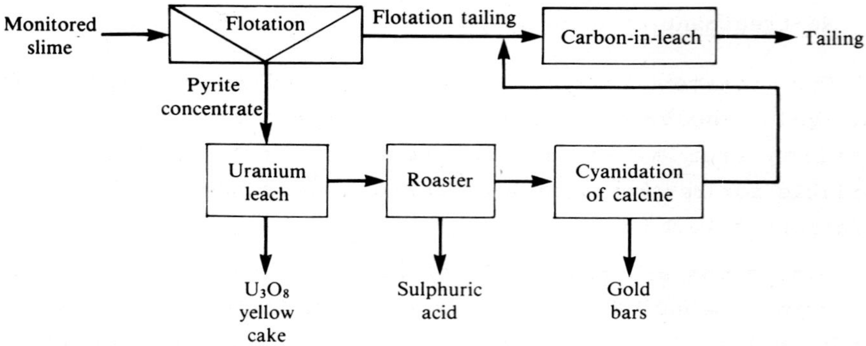

Fig. 5.19 Flowsheet for Ergo and JMS (after Corrans and Dunne [C36]).

Gold locked in oxides, mainly quartz or other silicates can only be recovered by fine milling to liberate the gold, followed by cyanide leaching, the procedure being uneconomical for slime residues. The only possibility for the economic recovery of part of this gold is preconcentration and intensive milling of the reduced mass of the concentrate.

Extensive pilot-plant tests were conducted at Ergo to demonstrate more fully the economic potential of HGMS for the recovery of gold in the context of a large, integrated re-treatment operation [C37, C36]. On average, the gold content of the Ergo flotation feed is 0.57 g/t and that of flotation tailings 0.32 g/t. The carbon-in-leach (CIL) tests showed that, on average, an additional amount of gold (0.1 g/t) could be recovered, thus effectively reducing the gold (essentially all gold locked in oxides) in the tailings to 0.22 g/t.

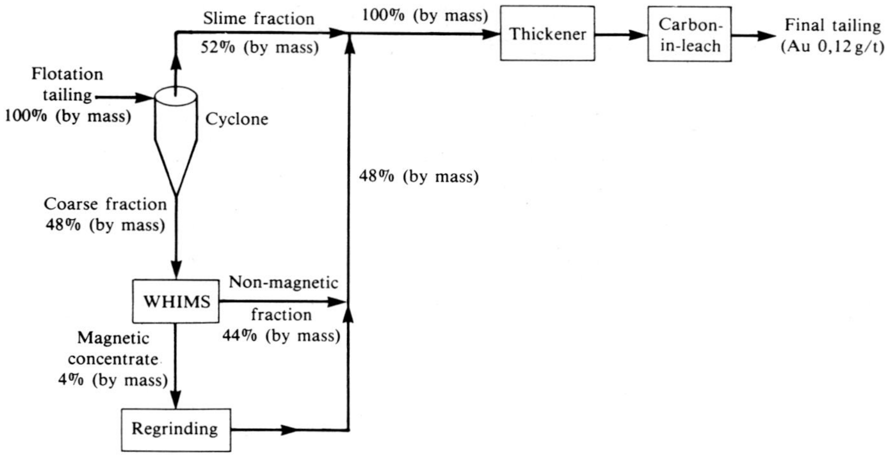

Fig. 5.20 Optimum HGMS flowsheet for Ergo [C36].

TABLE 5.13

Typical HGMS results from pilot plant at Ergo [C37, C36]

Feed to HGMS	Magnetic concentrate: Mass yield (%) a	Mass yield (%) b	Grade Au(g/t)	Recovery (% Au) a	Recovery (% Au) b	Net recovery, Au (g/t)
Deslimed CIL tail	6.4	11.2	1.34	37.3	50.0	0.08(0.09)
Deslimed flotn. tail	4.0	8.3	1.28	20.5	39.3	0.07(0.09)
CIL tail	6.5	-	0.86	27.1	-	0.06(0.06)
Flotn. tail	7.8	-	1.13	33.1	-	0.10(0.11)

a - feed to desliming cyclone, i.e. float or CIL tailings,
b - cyclone underflow, or feed to HGMS.
Note: The values in parenthesis are the average results for batch tests on feeds from different residue dumps at different times.

Four HGMS flowsheet configurations were tested: before and after CIL, each with and without desliming. An Eriez CF-30 HGMS separator, with a capacity of 1 t/h, was used, and the typical results summarized in Table 5.13 demonstrate the effectiveness of HGMS in concentrating the gold in the larger than 23 μm fraction of the CIL tailings. Of the four flowsheets investigated at Ergo, that shown in Fig. 5.20 is considered the optimum and the further implementation of HGMS in re-treatment operations depends to a large degree on the availability of a suitable machine. Ideally, the separator should be able to treat at least 250 t/h, to keep the installation simple and the costs low. For Ergo, five such machines would be needed at low energy consumption, as power constitutes about one-half of the operating costs of the magnetic separator. Equally importantly, the machine must be so designed that clogging of the matrix by strongly magnetic matter and extraneous material is avoided.

V.1.5 Copper and lead ores

Some important copper minerals are weakly magnetic and several attempts were made to recover copper by HGMS in those applications where flotation was inefficient.

Pb flotation concentrate (Japan)

In the flotation process of black ore, Cu and Pb sulphide minerals are floated together to form a bulk concentrate by blowing sulphur dioxide into the pulp, after which the floated minerals are separated by hot flotation. The Pb flotation concentrate as a sink product contains a small amount of unfloatable Cu minerals. Kim et al. [K27] found that the Pb flotation concentrate from Dowa Mining Co. (Japan) contains 5 to 8 per cent of Cu, mainly bornite and chalcopyrite. Magnetic susceptibilities of the minerals found in the Pb flotation concentrate are shown in Table 5.14.

TABLE 5.14
Volume magnetic susceptibility of sulphide minerals (after Kim et al. [K27])

Mineral	Volume magnetic susceptibility (SI)x10^{-4}
Chalcopyrite $CuFeS_2$	1.6
Bornite Cu_5FeS_4	0.4
Chalcocite Cu_2S	0.03
Covellite CuS	0.03 - 0.05
Galena PbS	-0.05

The removal of copper minerals from the Pb concentrate is facilitated by relatively high values of magnetic susceptibility of copper minerals compared to diamagnetic galena and by the fact that the mean size of Cu minerals was larger. Typical results of the removal of copper by HGMS are shown in Fig. 5.21, while Fig. 5.22 illustrates the recovery of Cu, the grade of the magnetic fraction and the losses of Pb into the magnetic fraction as a function of the matrix loading.

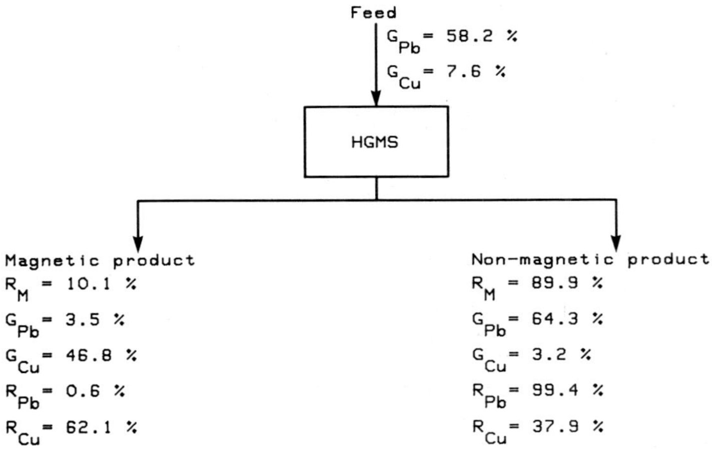

Fig. 5.21 HGMS of the Pb concentrate. Magnetic induction 1.8 T, matrix: expanded metal, flow velocity 0.054 m s^{-1}. (After Kim et al. [K27]).

It can be seen that the Cu grade in the non-magnetic Pb concentrate decreased from 7.6 to 3.2 per cent while the losses of lead into the magnetic fraction were as low as 0.6 per cent, partly due to dispersion of the sample.

Fig. 5.22 demonstrates that by increasing the matrix loading the recovery of copper into the magnetic fraction slightly decreases while the selectivity of separation increases. Simultaneously, with increasing matrix loading, the losses of Pb into the magnetic fraction drop.

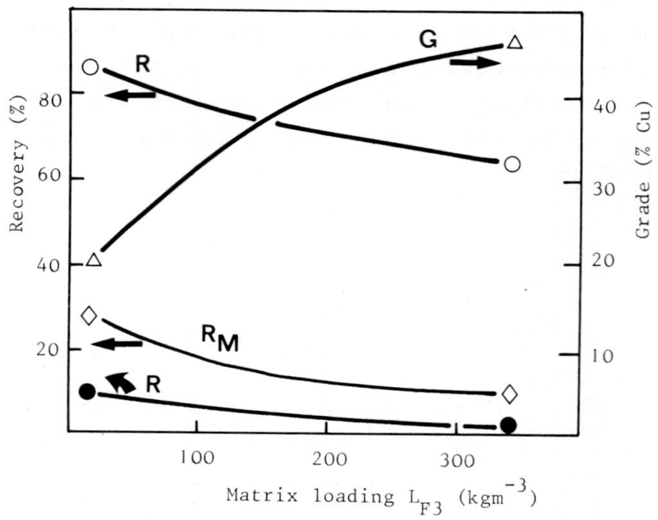

Fig. 5.22 The effect of matrix loading on Cu recovery from Pb flotation concentrate by HGMS (after Kim et al. [K27]).
○ - recovery of Cu,
△ - grade of concentrate (Cu),
● - recovery of Pb,
◇ - mass yield into magnetic product.

Upgrading of sulphide minerals

Some copper concentrates contain excessive amounts of pyrite. El Tawil and Morales [E5] have shown that HGMS can produce a high-grade copper concentrate, as illustrated in Table 5.15. A moderate copper recovery can be further improved by re-treating the non-magnetic fraction by sizing the feed into the HGMS and optimizing the matrix for each size fraction.

By-product molybdenum concentrate obtained from processing porphyly copper ores sometimes contains unacceptably high copper values. The use of HGMS to remove chalcopyrite from molybdenum concentrates offers an alternative to leaching. Laboratory tests have shown [E5] that in a two-pass HGMS, the Cu content can be reduced from 0.8 to 0.5 per cent while recovering over 97 per cent of the molybdenum in the non-magnetic product.

TABLE 5.15

Chalcopyrite-pyrite separation by HGMS [E5]
Magnetic induction: 2 T, matrix: expanded metal

Product	Mass yield (%)	Grade (%)		Recovery (%)	
		Cu	Fe	Cu	Fe
Head (calc.)	100	23.8	31.6	100	100
Magnetics	24.6	31.0	28.6	32.0	22.2
Non-magnetics	75.4	21.5	32.6	68.0	77.8

Cu-Pb flotation tailings, Namibia

The concentrator at the Tsumeb mine treats the ore containing sulphide and oxide minerals of copper, lead and zinc. The recent move of the mining operations into increasingly oxidized zones has resulted in a significant reduction in the sulphide-oxide ratio of the ore entering the flotation plant. Concurrent with the increase in the abundance of the oxide minerals in the ore has been a decline in the overall recovery of the copper and lead values. The drop in recovery is due to poor flotability of complex copper-lead arsenic-vanadate minerals while pilot-plant tests established that high-gradient magnetic separation and gravity separation, and a combination thereof, can remove a high proportion of copper and lead from the flotation tailings [S30].

Table 5.16 summarizes the values of magnetic susceptibilities of the Tsumeb minerals. The typical laboratory metallurgical results of upgrading the flotation tailings by HGMS are given in Table 5.17. It can be seen that in spite of very low values of magnetic susceptibilities and small particle size (43 per cent smaller than 38 μm for non-deslimed sample and 12 per cent smaller than 38 μm for deslimed sample), sizeable portions of copper and lead were recovered. It is particularly surprising for lead, as the only lead-containing minerals with positive magnetic susceptibilities are mottramite and duftite. The remaining lead compounds are diamagnetic.

Mineralogical analysis has shown that malachite and mottramite tend to report to the magnetic fraction while duftite and conichalite, which were expected to report to the mags, were not detected.

TABLE 5.16
Magnetic susceptiblities of Tsumeb minerals [S30]

Mineral	Mass magnetic susceptibility	
	$cm^3g^{-1}x10^{-6}$	$m^3kg^{-1}x10^{-8}$
Azurite $2CuCo_3Cu(OH)_2$	13.7	17.2
Descloizite $PbZn(VO_4)OH$	0.4	0.50
Bornite Cu_5FeS_4	5.3	6.7
Tennantite $5Cu_2S.2(Cu,Fe)S.2As_2S_3$	6.5	8.2
Malachite $CuSO_3.Cu(OH)_2$	6.8	8.5
Conichalite $CaCu(A_sO_4)(OH)$	8.6	10.8
Mimetite $Pb_5(Cl,AsO_4)_3$	-1.7	-2.1
Mottramite $PbCu(VO_4)(OH)$	3.0	3.8
Cerussite $PbCO_2$	-1.9	-2.4
Duftite $PbCu(AsO_4)(OH)$	4.2	5.3

Magnetic susceptibility of the non-magnetic fraction was, in most cases, negative (of the order of -1.3×10^{-8} m^3 kg^{-1} or -1×10^{-6} cm^3 g^{-1}) which indicates that all paramagnetic minerals were recovered into the magnetic product, even at modest magnetic field strength, as demonstrated by Fig. 4.25.

The pilot-plant tests conducted at Tsumeb, using the Eriez CF-30 carousel magnetic separator, confirmed the potential of HGMS to recover up to 70 per cent of copper and over 50 per cent of lead from the tailings, at, however, a combined grade of less than 15 per copper and lead. Since the minimum grade of the feed to the smelter is 20 per cent combined metal, gravity separation was used to clean the magnetic concentrate. The flowsheet of the pilot HGMS-gravity separation plant at Tsumeb is shown in Fig. 5.23. The results of the pilot-plant runs are shown in Table 5.18, in comparison with the laboratory tests.

TABLE 5.17

Metallurgical laboratory results of upgrading of the Tsumeb flotation tailings by HGMS. Magnetic induction: 1.2 T, interstitial flow velocity: 0.123 m s^{-1}, matrix: 6 mm balls, head: 0.50 % Cu, 1 % Pb [S30]

Sample	Mass yield into mags	Grade of mags (%)					Recovery into mags (%)				
	(%)	Cu	Pb	Zn	As	Ag (ppm)	Cu	Pb	Zn	As	Ag
Non-deslimed flotation tailings	8.9	4.1	5.5	1.9	3.3	37	56.7	47.3	32.2	73.8	18.4
Deslimed flotation tailings	10.3	4.1	5.6	2.0	3.8	42	62.7	56.1	39.7	79.5	21.2

TABLE 5.18

Metallurgical results of the HGMS-gravity separation or gravity-separation-only tests at Tsumeb [S30]

Test	Preconcentration	Recovery at 20 % combined grade	
		Cu	Pb
Pilot-plant	HGMS	44.9	33.7
	Spiral	37.1	54.5
Laboratory tests	HGMS	41.8	31.8
	Spiral	33.3	32.1

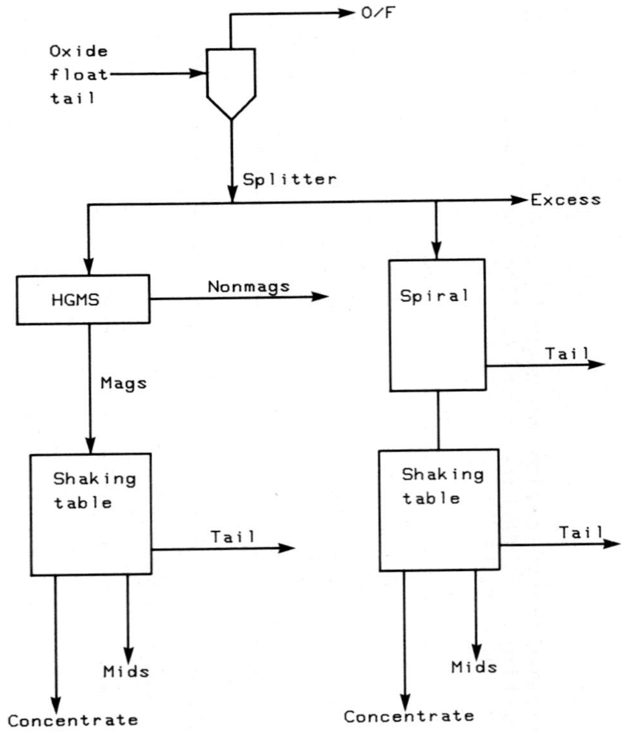

Fig. 5.23 Flowsheet showing the Tsumeb pilot-plant [S30].

Results showed that the use of primary magnetic or gravity separation, with efficient cleaning of the concentrate on a shaking table, is likely to recover about 25 per cent of the copper and lead from the final oxide flotation tailings at a combined metal grade of 20 per cent copper, plus lead. Magnetic separation was found to recover copper minerals more efficiently than lead minerals and it can recover much finer particles than standard gravity separators. Some of the finer particles recovered by HGMS were lost during the upgrading process on the shaking table.

V.1.6 Recovery of other metals

Baddeleyite (ZrO_2)

Ore from the Phalaborwa complex in South Africa contains small amounts of the zirconium oxide mineral baddeleyite. ZrO_2 is recovered from the gravity separation concentrate with induced magnetic roll separators. Typical results from Foskor are given in Table 5.19; these were obtained with the feed 50 per cent smaller than 150 μm.

TABLE 5.19
Recovery of baddelyite by IMR

Product	TiO_2	Fe_2O_3	ZrO_2
Feed	11.06	4.53	75.0
Non-magnetic fraction	0.29	0.06	>98

Platinum-group metals

The Merensky Reef and the chromitite reefs of the Bushveld Complex in South Africa have long been major sources of platinum-group metals (PGM), nickel, copper and of chromium ores for the manufacture of ferrochromium. The Merensky Reef is low in chromitite (0.1 per cent) and HGMS was found to give high recoveries, although at poor selectivities [C15].

Dawson [D18] has shown that an economically feasible separation of PGM and gold from the oxide ore from the

Potgietersrus Platinum Prospects (South Africa) can be achieved by magnetic separation. The tests have shown that the heating of mixtures of ore and carbon, followed by dry magnetic separation resulted in a significant upgrading of the PGM values in the magnetic fraction. The roasting was performed at a temperature of 600 to 700° C for 20 to 40 minutes. Figure 5.24 illustrates the relative behaviour of some of the individual PGM and gold; average grades and recoveries obtained when the quantity of magnetic material was 4.8 per cent by mass of the feed are given in Table 5.20.

TABLE 5.20

Recovery of PGM after a reducing roast and magnetic separation [D18]

	Pt	Pd	Rh	Au
Concentrate grade, g/t	14	16	2	3
Recovery, %	29	39	59	20
Feed, g/t	2.34	1.90	0.15	0.70

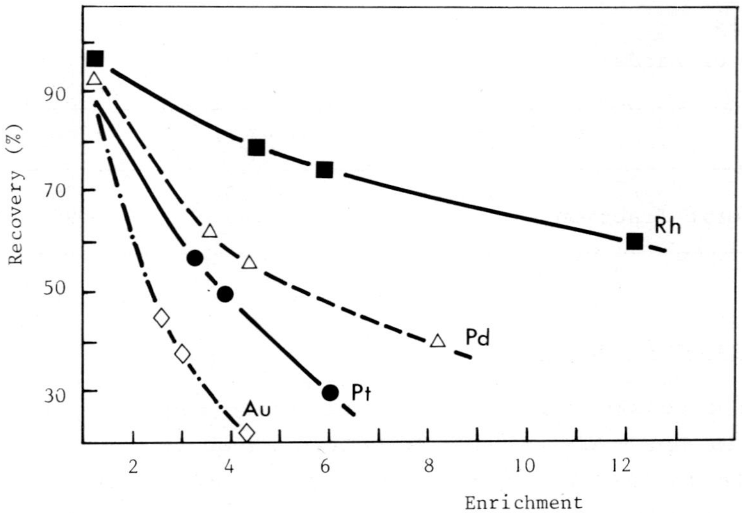

Fig. 5.24 Recovery of platinum, palladium, rhodium and gold by magnetic separation, versus enrichment. Enrichment is defined as assay of cummulative magnetic fraction divided by calculated assay of feed to magnetic separation (after Dawson [D18]).

Ilmenite

Ilmenite ore from the complex anorthite deposits at Tellues in Norway is extracted by Titania A/S using open-pit mining techniques. Until recently, the principal method of beneficiation was flotation, but a decision was taken to investigate the means of improving the process and of reducing the cost of flotation reagents by introducing gravity and magnetic separation circuits.

Three Boxmag-Rapid HIW separators were installed in 1983 and trials showed that they could accommodate between 70 and 75 t/h of feed, yielding a magnetic concentrate grade of 32 per cent TiO_2 at a recovery of 85 per cent from the head value of 15 per cent TiO_2 [W36].

V.1.7 Beach sands

Beach sands are the source of titanium, zirconium, thorium and rare earths. The principal producers of these minerals are South Africa, Australia, USA and USSR. At the Richards Bay Minerals operation in Zululand, South Africa, heavy minerals are extracted from beach sands by conventional dredging and gravity separation (Figs 5.25 and 5.26) to produce five saleable products: high-grade titania slag, low-manganese iron, zircon, rutile and monazite. The major heavy minerals are given in Table 5.21.

TABLE 5.21
Heavy mineral sands

Mineral	Composition	Magnetic susceptibility [A2] $m^3kg^{-1} \times 10^{-7}$	$cm^3g^{-1} \times 10^{-6}$
Ilmenite	$FeTiO_3$	11.3	90
Leucoxene	Altered ilmenite	6.2	50
Rutile	TiO_2	0.16	1
Zircon	$ZrSiO_4$	-0.03	-0.2
Monazite	$(Ce,La,Y,Th)(PO_4)$	1.9	15

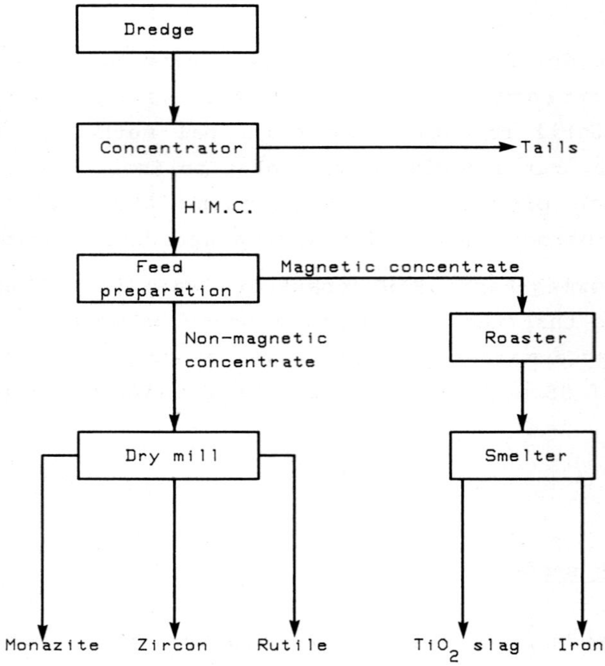

Fig. 5.25 Beach sand production method at Richards Bay Minerals [W33].

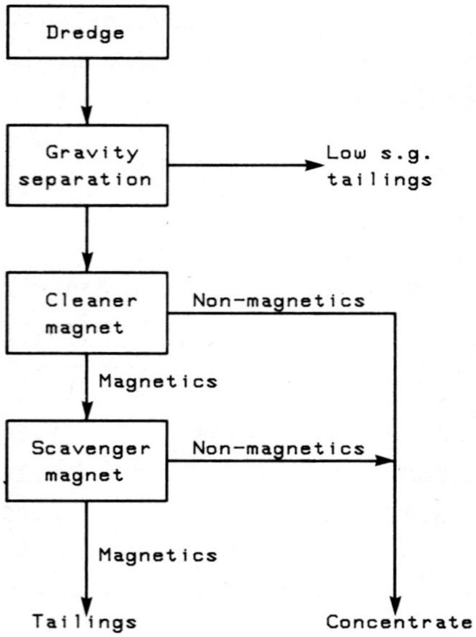

Fig. 5.26 The concentrator circuit at Richards Bay Minerals [W33].

The concentrate which contains 85 per cent of heavy minerals, is first passed through a low-intensity wet drum magnetic separator to remove magnetite. The non-magnetics from this stage are re-treated with Readings wet high-gradient magnetic separators in a complex rougher-cleaner circuit to produce a concentrate that consists largely of ilmenite containing 46 per cent TiO_2.

The non-magnetics from HGMS are again gravity concentrated and then further upgraded by the use of HGMS to remove feebly magnetic material. The resulting concentrate is dried and fed to the "dry mill", where high-tension roll separators and induced magnetic roll separators are used to produce saleable rutile and zircon products.

A diagram of the feed preparation circuit is shown in Fig. 5.27. A total of 31 Readings high-gradient magnetic separators are used at RBM in the wet treatment plant. Each separator consists of 16 oil-cooled coils, rotor diameter is 1750 mm and wedge-wire screens with a gap of 2.5 mm are used as a matrix. The machines operate at a rotor speed of 4.3 r/min at a slurry density ranging from 30 to 40 per cent solids, the feed rate is 25 to 35 t/h.

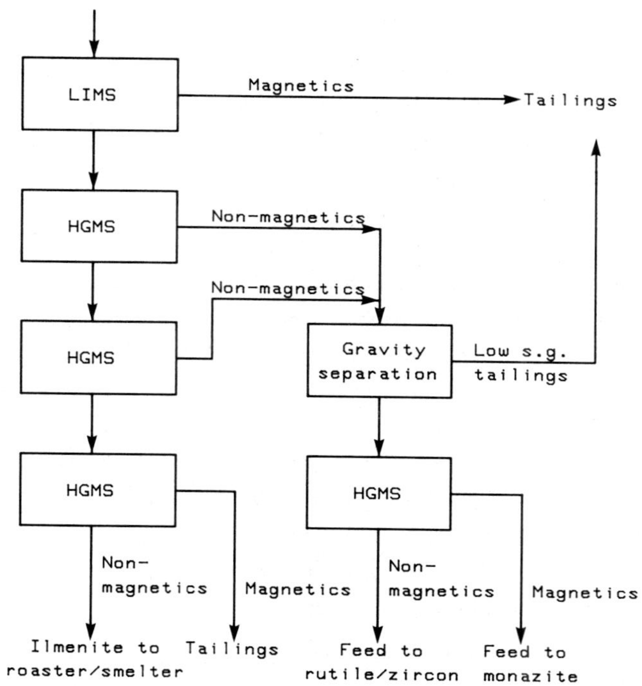

Fig. 5.27 Feed preparation circuit at RBM [W33].

The magnetic fraction from the HGMS machines contains in addition to ilmenite, small quantities of weakly magnetic impurities. These are removed prior to feeding the material to the smelter by dry magnetic separation,following magnetizing roasting in a fluidized bed.

V.1.8 Industrial minerals

V.1.8.1 Kaolin beneficiation

The most successful application of HGMS technology to date is in the brightening of kaolin clay. Kaolin is a white aluminium-silicate mineral that is principally employed as a filling and coating agent in the manufacture of paper products. While kaolin is ubiquitous in nature, high-purity reserves of economic value constitute only a small fraction of the kaolin content of the earth's crust. All economic deposits of kaolin are contaminated by small amounts (generally between 0.5 and 3 per cent) of ferruginous minerals, including oxides of iron, anatase, rutile, siderite, pyrite, mica and tourmaline.

Since iron contaminants are deleterious to clay brightness, considerable effort has been expended to remove them by physical and chemical means. Under the most favourable conditions chemical bleaching usually removes less than half of the iron present in clays. Physical separation techniques such as flotation are generally even less effective for the removal of iron than chemical bleaching [I3]. Although discolouring contaminants are usually very weakly magnetic and very fine (e.g. 80 per cent smaller than 2 μm in Cornwall china clay), they can be efficiently removed by HGMS if a sufficiently large magnetic field strength and low flow velocity of the slurry are used [I5, L19, B42]. The high-gradient magnetic separation of kaolin clay became a practical economic process that has been thoroughly proven on a large-scale in Georgia (USA) [I4], Cornwall (UK) [L19], and Czechoslovakia [B43]. A photograph of the production-scale HGMS for kaolin treatment is shown in Fig. 5.28.

The use of a magnetic separator in a clay processing flowsheet is illustrated in Fig. 5.29. Iron-clad solenoid cyclic magnetic separators are usually used, since these can generate background

Fig. 5.28 High-gradient magnetic separator for kaolin purification (courtesy of Sala Magnetics, Inc.)

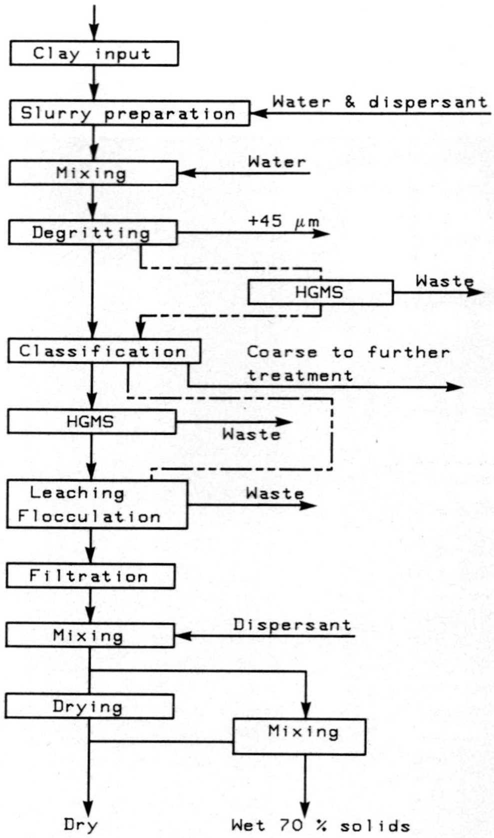

Fig. 5.29 The use of HGMS in kaolin beneficiation.

magnetic induction up to 2 T and have a facility to control the flow velocity of the slurry. A photograph of one of the largest unit is shown in Fig. 2.43. The machine generates 2 T in the canister which has a diameter of 2.1 m, and a height of 0.5 m. The separator itself weighs 340 tonnes and operates at 3500 A and 400 kW.

The operating parameters of such a separator, as applied to the beneficiation of Cornish kaolin clay [L19], are summarized in Table 5.22; typical results from the HGMS separator are shown in Table 5.23.

The effect of magnetic field and flow velocity on the efficiency of extraction of iron oxides is illustrated in Figs.

4.19, 4.20, 4.78 and 4.80. The importance of a correct choice of appropriate dispersant and its optimum dose is shown in Fig. 5.30.

As described in paragraph IV.14.2, process variables also influence the throughput of a separator. Babůrek et al. [B43] developed throughput diagrams that correlate the capacity of a separator with process variables like matrix loading, dead time and concentration of solids in the slurry. Fig. 5.31 shows such a diagram, computed for a cyclic separator of a canister with a diameter of 1 m, and a height of 0.3 m, operating at 1.9 T and at a flow velocity of 6 mm s^{-1}.

TABLE 5.22

Operating parameters of a cyclic HGMS separator (2.1 m diameter)[L19]

Background magnetic induction	2 T
Canister	2 m (diameter) 0.5 m (height)
Matrix	Stainless steel wool
Filling factor of matrix	6 per cent
Feed: pH	6 to 7
Solid content	12 to 15 per cent (mass)
Flow velocity	4 to 12 mm s^{-1}
Flowrate	5 to 15 t/h (dry) 0.7 to 2.1 m^3/h (slurry)
Volume per cycle	4 to 8 canister volumes
Feed time	250 s
Slurry displacement:	
Flow velocity	As for feed
Volume	1 canister volume
Magnetics flush:	
Flow velocity	50 to 60 mm s^{-1}
Volume	10 canister volumes
Flush time	90 s
Total duty cycle	480 s

TABLE 5.23

Process results from the HGMS separator [L19]. (B = 2 T, volume of slurry processed: 4 canister volumes).

	Mass yield (%)	Fe_2O_3 (%)	Fe_2O_3 extraction (%)	Fired at 1180° C % brightness
Feed	100	0.81		83.3
Magnetics	12	2.72		89.1
Nonmagnetics	88	0.55	40	

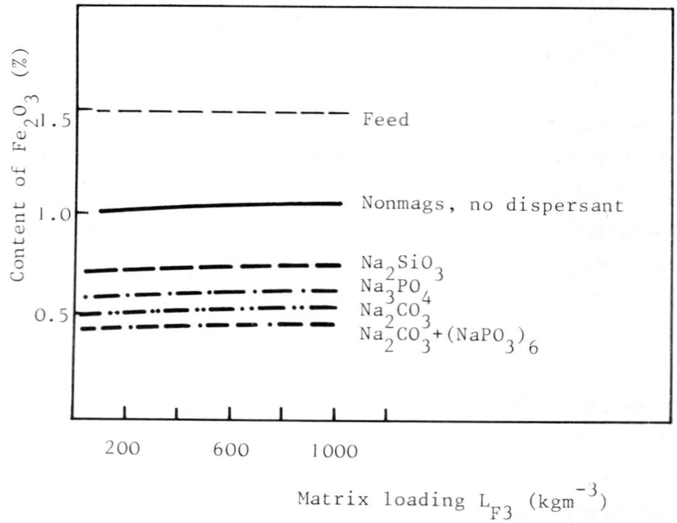

Fig. 5.30 Effect of dispersants on the efficiency of magnetic separation of kaolin (Rumania)(after Dobrescu et al. [D21]).

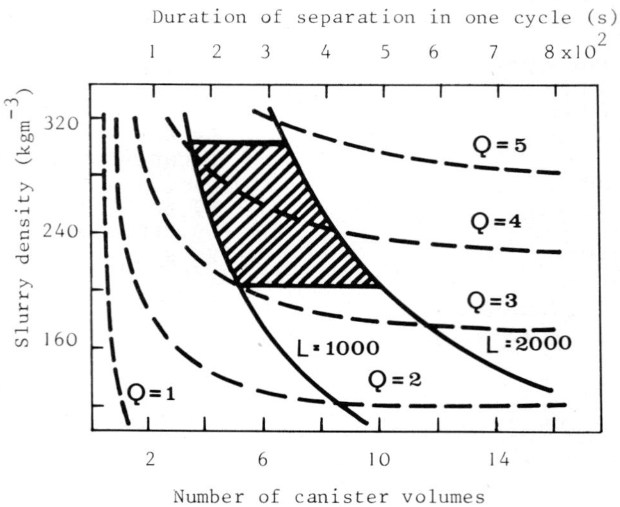

Fig. 5.31 Throughput diagram for a cyclic HGMS (canister diameter 1 m, height 0.3 m) operating at 1.9 T and with a flow velocity 6 mm s^{-1}.
L - matrix loading (in kg m^{-3})
Q - throughput of a separator (in t h^{-1})
(After Babůrek et al.[B43])

The target of a 5 per cent increase in kaolin brightness with Czechoslovak kaolin could have been achieved at 1.9 T and 6 mm s^{-1} for L_{F3} = 1000 to 2000 kg m^{-3} and ρ = 200 to 300 kg m^{-3}. The operating regime of a magnetic separator is represented by the parallelpiped in Fig. 5.31 so that the corresponding throughput is 3 to 4.8 t/h at the feed time from 165 to 500 s and the overall duty cycle 230 to 570 s.

V.1.8.2 The beneficiation of fluorspar

Fluorspar, the commercial name for fluorite, is a mineral composed of calcium fluoride CaF_2, and is consumed mainly in the production of hydrofluoric acid and in the steel and ceramics industry.

At the Buffalo Fluorspar mine in South Africa, a carousel high-gradient separator is used to reduce the phosphate content in the final concentrate within the specification limits of less than 0.15 per cent P_2O_5. The phosphate in the concentrate which is mainly due to monazite in the ore, is concentrated during the heavy media and flotation processes. The P_2O_5 content of the flotation concentrate varies from 0.05 to 0.45 per cent and HGMS was found to be the only technique for reducing P_2O_5 content to 0.025 to 0.20 per cent.

The CF-60 Eriez machine (Fig. 5.32), with a 7 mm steel-ball matrix, employs the continuous ball removal system and has been in operation since 1979; its metallurgical performance is illustrated by Tables 5.24 and 5.25.

In another fluorspar mine in the Transvaal (South Africa), the Jones carousel separator is used to remove the iron oxide and arsenic contaminants from the concentrate.

TABLE 5.24
Effect of feed rate on the removal of P_2O_5 from fluorspar (after Jonker [J8])

Product	Concentration of P_2O_5 (%)		
Tonnes/h	11	22	39
Feed	0.413	0.365	0.369
Magnetics	2.74	2.96	3.25
Non-magnetics	0.176	0.164	0.167
Removal (% P_2O_5)	38.7	41.7	42.3

TABLE 5.25
Influence of pulp density on the removal of P_2O_5 from fluorspar (after Jonker [J8])

	Concentration of P_2O_5 (%)			
Pulp density	1.25	1.30	1.40	1.51
Feed	0.132	0.130	0.127	0.122
Magnetics	0.95	1.06	0.92	0.80
Non-magnetics	0.077	0.083	0.064	0.062
Mass yield into non-magnetics	93.3	95.2	92.6	91.9
Removal (% P_2O_5)	54.4	60.7	46.6	46.7

Fig. 5.32 The CF-60 Eriez carousel HGMS on South African fluorspar mine (with permission of Transvaal Mining and Finance Co. Ltd.).

V.1.8.3 The beneficiation of bauxite

Bauxite is used as an aluminium ore in the manufacture of refractory products and aluminous chemicals. It is composed of hydrated aluminium oxide minerals while silica, iron oxides and titania are major impurities always present in deposits of bauxitic raw minerals. In order to meet technical specifications (Table 5.26) the iron and titania content must be reduced. Iannicelli [I6] showed that the non-magnetic fraction from HGMS of bauxite from Arkansas contains less than 3 per cent Fe_2O_3 and 1.6 per cent TiO_2, as shown in Table 5.27.

TABLE 5.26
National Stockpile specifications (USA) for bauxite [I6]

		Per cent by mass	
		Hydrous bauxite	Calcined bauxite
Alumina (Al_2O_3)	min.	56.2	86.5
Silica (SiO_2)	max.	4.55	7.00
Iron oxide (Fe_2O_3)	max.	1.63	2.5
Titania (TiO_2)	max.	2.44	3.75

TABLE 5.27
HGMS of bauxite (Arkansas) [I6]

	Al_2O_3	SiO_2	Fe_2O_3	TiO_2
Head-hydrous	56.6	2.60	8.65	2.46
Head-calcined	82.4	2.98	10.69	3.51
Non-magnetic product-calcined	92.1	3.12	2.93	1.61

Bolsaitis et al. [B44] investigated the possibility of using HGMS for the removal of impurities from ferruginous bauxite from Venezuela, while Tatsienko et al. [T21] described the beneficiation of bauxite with high iron content using a combined method of magnetizing roasting and magnetic separation. The optimum conditions for pretreatment of bauxites containing 18 to 22 per cent Fe were obtained at temperatures ranging from 920 to 1020 K, roasting time 20 to 40 minutes, addition of solid reductant into the feed 2.5 to 3 per cent, and optimum FeO content 9 to 10 per cent when applying low-intensity magnetic separators (0.1 T). The non-magnetic product was suitable for aluminium production by means of conventional production methods.

The high chromium content of the North-Onega bauxites (USSR) complicates the treatment of the raw material, owing to the toxicity of volatile compounds of chromium. It was found by Favorskaya and Naumchik [F21] that it was possible to reduce the chromium oxide content by matrix magnetic separation, with a magnetic fraction representing 5 to 10 per cent of mass and containing 3 per cent of Cr_2O_3.

Semin et al. [S73] investigated the possibility of using low-grade aluminosilicate ore in the developing aluminium industry in Siberia. Several technological schemes have been worked out for the concentration of the ore, one of them being magnetic separation. The requirements imposed on the method were a high degree of extraction of the alumina into the non-magnetic product, the considerable reduction of iron in the product and low losses of alumina into the magnetic fraction.

The results of magnetic separation tests on nepheline and bereshite ores are summarized in Table 5.28. It can be seen that the degree of concentration of nepheline ore is small while the reduction of the iron content in the concentrate is considerable.

TABLE 5.28

Magnetic separation of aluminosilicate ores from the Krasnoyarsk region (USSR) [S73]

Product	Mass yield (%)	Grade (%)		Recovery (%)	
		Al_2O_3	Fe_2O_3	Al_2O_3	Fe_2O_3
Nepheline ore					
Nonmags	91.8	25.4	4.0	97.6	54.1
Mags	8.2	6.9	37.7	2.4	45.9
Feed	100.0	23.8	6.8	100.0	100.0
Bereshite ore					
Nonmags	88.8	23.1	6.9	91.9	79.0
Mags I	1.5	13.0	27.4	0.9	5.4
Mags II	9.7	16.7	12.4	7.2	15.6
Feed	100.0	22.3	7.7	100.0	100.0

Furthermore, the losses of alumina into the magnetic fraction are very small. Nepheline ores with such an iron oxide content can be successfully processed to alumina by the baking process.

With bereshite ore, the results are not satisfactory and further methods must be sought to reduce the iron remaining in the product and thus produce a concentrate suitable for processing by the usual baking method.

V.1.8.4 The beneficiation of asbestos ore

Asbestos is the name given to a group of fibrous mineral silicates that can be divided into two main groups: one contains the fibrous serpentine called chrysolite and comprises about 94 per cent of the world production of asbestos [L26], the second contains the minerals in the amphibole series (crocidolite, amosite et al.)

Martinez [M34] has shown that the ore particles containing chrysolite fibres can be recovered by dry magnetic separation, while the non-magnetic particles contain little or no fibre. The pilot-plant and production-scale tests with asbestos ore from Quebec (Canada) revealed that magnetic separation could be used for the preconcentration of raw ore with the benefit of increasing the life of the mine and the value of the feed to the fibre recovery section. The results obtained with an electromagnetic pulley are given in Table 5.29.

TABLE 5.29

Magnetic separation of asbestos ore (Quebec), (-125 mm + 12 mm) [M34]

Product	Mass yield (%)	% fibre
Feed	100	1.74
Magnetics	25.4	4.34
Non-magnetics	74.6	0.86

Magnetic separation also increased the recovery of the fibres from the tailings after the screening and aspiration of free fibre, as can be seen in Table 5.30.

The magnetic filtration of airborne respirable asbestos fibres was investigated by Timbrell [T10]. Fibres of amosite (Transvaal) and crocidolite (Cape Province, South Africa), smaller than 200 μm are of two magnetic types: P fibres and N fibres that align with their long axes parallel and normal, respectively, to the direction of the applied magnetic field. The fibres exhibited a high degree of retention in a steel wool matrix even at zero field. At the residual field of 0.03 T, the retention of P fibres amounted to 42 per cent. At 1.5 T, 97 per cent of P fibres and 89 per cent of N fibres were captured, as can be seen in Fig. 4.13.

TABLE 5.30
Magnetic separation of asbestos tailings (-25 + 6 mm) with the use of electromagnetic pulley [M34]

	High field		Low field	
	Mass yield (%)	Fibre (%)	Mass yield (%)	Fibre (%)
Feed	100	0.18	100	0.22
Magnetics	22.6	0.56	3.9	2.55
Non-magnetics	77.4	0.06	96.1	0.13

V.1.8.5 The beneficiation of andalusite

Andalusite is a refractory mineral used for the manufacture of high-duty refractory bricks, electrical insulators and spark-plug bodies. Some South African andalusite contains iron oxides in excess of 2 per cent, which is highly objectionable when used in the above applications, and considerable work has been done to reduce the iron content to a level as low as 0.5 per cent. Tests at Mintek, in which the performance of the cross-belt, induced magnetic roll and permanent magnetic roll separators were compared favoured the permanent roll separator. South Africa's leading producers of andalusite are using these separators to re-treat the concentrate from heavy media separation, sized into four fractions; the results are shown in Table 5.31. Table 5.32 illustrates the effect of sizing on the efficiency of magnetic separation.

Figs. 5.33 and 5.34 show the operation of a permanent magnetic roll separator on an andalusite mine, while Fig. 5.35 shows two plugs made from two grades of andalusite.

Fig. 5.33 Permanent magnetic roll separator (on the extreme right) on South African andalusite mine.

TABLE 5.31

Magnetic separation of the andalusite HMS concentrate by permanent magnetic roll separator [O10]

Product	Mass		Andalusite		Grade	
	(%)	% of HMS feed	Content (%)	Recovery (%)	Al_2O_3 (%)	Fe_2O_3 (%)
Non-mags	74.9	26.1	99.5	85.6	59.5	1.11
Magnetics	25.1	8.7	5.0	14.4	-	-
Head(calc.)	100.0	34.8	87.0	100.0	57.0	3.10

Fig. 5.34 Separation of andalusite by Permroll.

TABLE 5.32

The effect of sizing on magnetic separation of andalusite [010]

Product	Mass (%)	Al_2O_3 (%)	Fe_2O_3 (%)
> 3.5 mm non-magnetics	20.0	58.7	0.90
< 3.5 mm non-magnetics	6.1	59.4	0.76
Combined non-magnetics	26.1	58.9	0.87
Magnetics	8.7	-	-
Head (calc.)	34.8	57.0	3.10

Fig. 5.35 Plugs made from two grades of andalusite. Left: andalusite without upgrading by magnetic separation, right: andalusite upgraded by permanent magnetic roll separator. (Courtesy of Ore Sorters (Pty.) Ltd.)

V.1.8.6 The beneficiation of pyroxenite ore

The Phosphate Development Corporation (Foskor), S.A., extracts rock phosphate (apatite) from the Phalaborwa Igneous Complex in the Northern Transvaal. Currently, the apatite is concentrated by grinding and by fatty-acid flotation. In future years, Foskor will exploit other areas of the orebody, where the apatite occurs in a pyroxenite gangue rock.

Apatite can be separated from pyroxenite by dry high-gradient or high-intensity magnetic separation, as shown in Fig. 5.36. A dry beneficiation process can offer substantial cost-savings by eliminating the use of water, pumps, slime dams, thickeners, flotation reagents and fuel for drying. Essentially, the apatite is diamagnetic, whereas the minerals in pyroxenite rock vary from weakly paramagnetic to strongly paramagnetic. Liberation of the apatite from pyroxenite occurs at relatively coarse sizes (up to 500 μm) which are ideally suited to dry separation.

Foskor has carried out an extensive programme of testwork with various types of magnetic separators, e.g. induced magnetic rolls, permanent-magnet rolls, matrix carousel separators and superconducting open-gradient separators. Foskor thus faces the same problem that many other mining companies have encountered in their work, namely, that although it possesses a technology that is metallurgically suited to the process, the equipment is uneconomical, being too small in capacity, too costly or too unreliable. A comparison of the performance of different magnetic separators is summarized in Section IV.2.2 and illustrated in Figs 4.4, 4.5 and in Table 4.1. A flowsheet for a carousel matrix HGMS is shown in Fig. 5.37.

V.2 The magnetic beneficiation of coal

The environmental impact of acid rain is rapidly becoming a world-wide problem and a great deal of attention is focusing on the reducing or eliminating of acid emissions. The emission levels of sulphur dioxide and ash particles for coal burning facilities are now being regulated by stringent environmental standards.

Coals generally contain 1 to 5 per cent of sulphur, of which over a third is carbonaceous in form and is referred to as organic sulphur. The rest is mineral sulphur consisting of weakly paramagnetic pyrite or marcasite FeS_2.

However, pyrite found in coal often has a much higher magnetic susceptibility than the pure mineral; this is presumably due to impurities or to the partial alteration to pyrrhotite. The conversion of only a small fraction of pyrite to pyrrhotite or sulphate greatly enhances its net magnetic susceptiblity [M35].

Organic sulphur is closely bonded to coal so that its removal by concentration methods is not possible. Consequently, virtually all work on the desulphurization of coal has been directed towards the removal of inorganic sulphur [M36]. Various techniques have

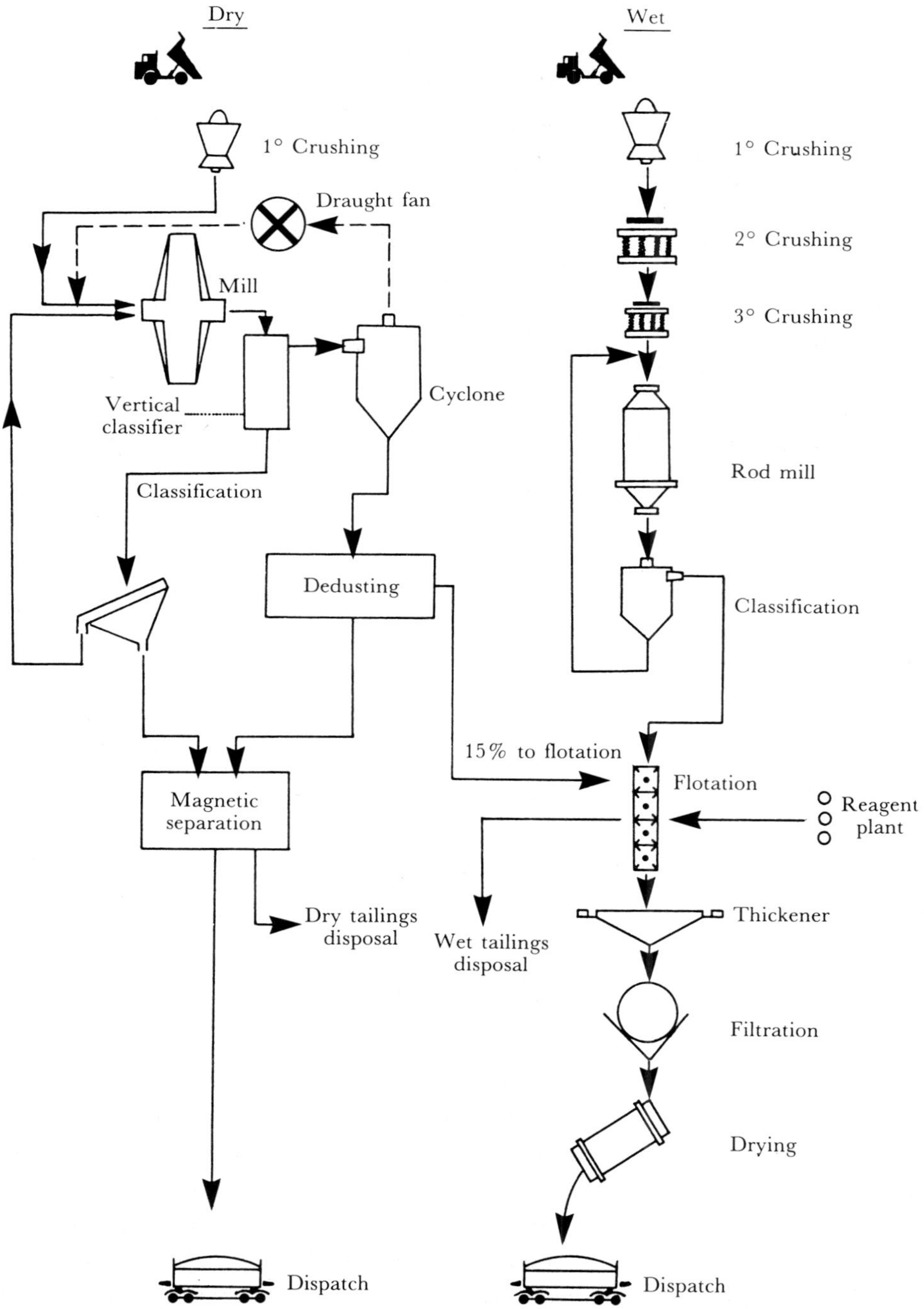

Fig. 5.36 Schematic flowsheet for existing wet and proposed dry beneficiation routes at Foskor [R3].

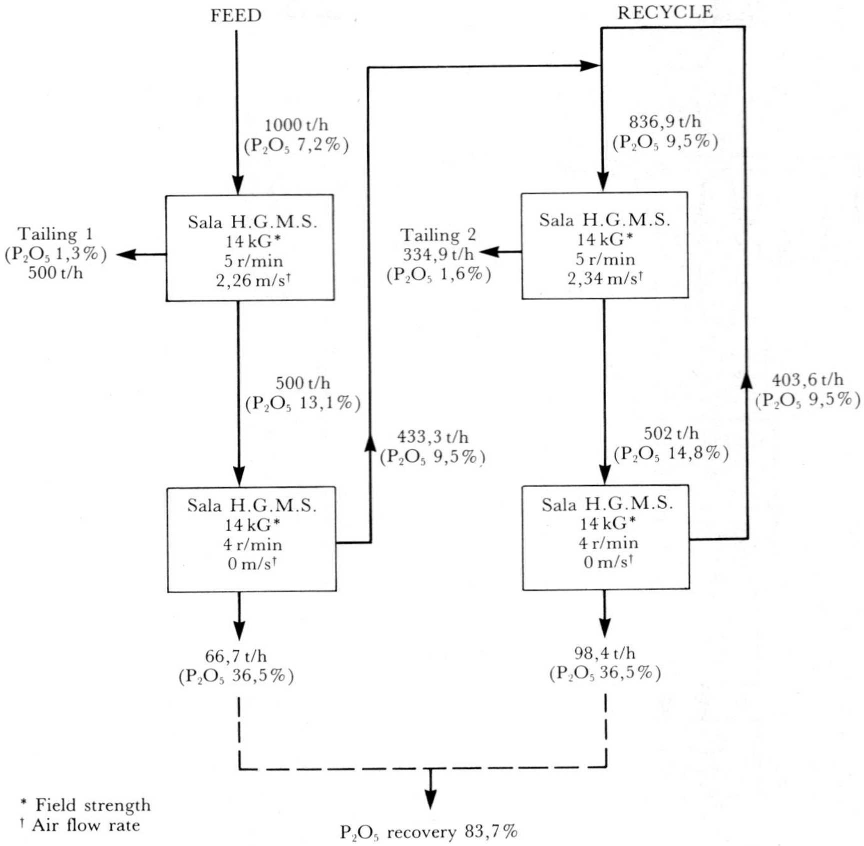

Fig. 5.37 A schematic Foskor flowsheet for Sala HGMS plant based on Model 480-15-15-15 separators [R3].

been investigated for the extraction of pyrite, ranging from froth-flotation to electrostatic separation, with varying degree of success. Magnetic separation is another option. Although recognized as a potentially cost-effective method of desulphurization since the late fifties, it was not until the early seventies, with the advent of HGMS, that its technical and commercial feasibility was demonstrated. Magnetic susceptibilities of constituents of coal, as compiled by Male [M36], are given in Table 5.33.

TABLE 5.33

Magnetic properties of the constituents of coal [M36]

Compound	Specific magnetic susceptibility ($m^3kg^{-1}x10^{-8}$)
Pyrite	0.34
Pyrrhotite	ferrimagnetic
Limonite	70
Siderite	400
Ferrous sulphate	126
Calcite	1.2
Coal	-0.5

V.2.1 The objectives of the magnetic beneficiation of coal

Male [M36] summarized the fundamental pre-requisites that a magnetic separator must satisfy before it becomes economically viable:

1. The pyrite and ash must be removed with as high an efficiency as possible, (even with 100 per cent removal, organic sulphur would still be present). Simultaneously, the amount of combustible matter lost into the magnetic fraction should be low.
2. The capacity of a magnetic separator should be 100 t h^{-1} or more, in view of high throughputs in most power stations (for instance, a 2000 MW power station burns about 750 tonnes of coal per hour).
3. Dry operation is essential for the desulphurization of pulverized coal because the dewatering of coal is a difficult and expensive process.

V.2.2 Magnetic enhancement

The first attempt at the magnetic separation of pyrite from coal occurred in the late fifties. These were experiments in which simple magnetic separators were used. Since the magnetic forces that could be exerted on pyrite in these devices were weak, attempts were made to convert pyrite to the more magnetic pyrrhotite or sulphate by thermal or mechanical means, such as steam treating or crushing. Yurovski and Remesnikov [Y4] steam-air-treated selected Soviet coals at temperatures ranging from 120 to 360° C for periods between 15 seconds and 5 minutes. By using an induced magnetic roll separator, it was possible to reduce the sulphur content from 2.30 per cent to 1.36 per cent. These results were later supported by Taits and Andreeva [T18], who observed that an increase in the temperature of heat treatment resulted not only in an increase of the mass yield into the magnetic fraction but also in a reduction of the sulphur content in the non-magnetic product, as shown in Fig. 5.38.

In other publications the heat treatment of coal was found to be not beneficial [K28, E6]. Ergun and Bean [E6] proposed the use of radio frequency (rf) field for selective heating of the pyrite without affecting the coal so as to avoid a decomposition of coal and excessive energy consumption. This depends on the fact that the

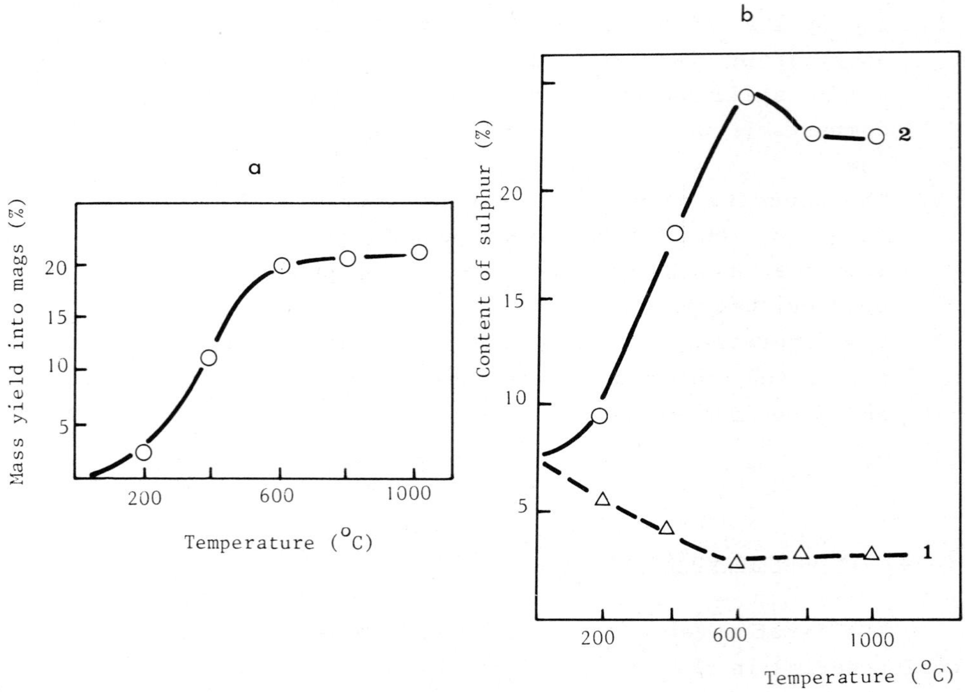

Fig. 5.38 (a) Mass yield of coal into the magnetic fraction and (b) the content of sulphur in non-magnetic (1) and magnetic (2) fractions, as a function of temperature of heat treatment. (B = 0.9 T) [T18].

conductivity of pyrite is larger compared to that of coal and that a particle of pyrite in a rf field will become heated much more rapidly than a coal particle. They demonstrated that the pyrite particles became strongly magnetic after treatment at a frequency of 10 GHz. Bluhm et al. [B45] found that mass magnetic susceptibility of dielectrically heated pyrite samples was approximately proportional to the time and power of pretreatment and that the pyrite and mineral matter in coal were heated 2 to 3 times faster than the carbonaceous coal at 2.45 GHz. Although a subsequent dry magnetic separation showed a slight reduction of sulphur in the product, magnetic enhancement was not significant. In contrast, Taits and Andreeva [T18] observed that by rf heating the coal to 500° C, magnetic separation could reduce the sulphur content from 7.55 to 2.9 per cent, with a mass recovery into the non-magnetic product of 87.6 per cent.

A different approach to pyrite conversion has been described by Kinding [K29], i.e. the so-called Hazen Magnex process. In this process, dry pulverized coal is heated to 170° C and then subjected to iron pentacarbonyl vapour. The iron carbonyl treatment provides the magnetic enhancement of the pyrite and ash-forming minerals. Subsequently, the pyrite and ash-forming minerals are removed as a magnetic refuse. In the pilot-plant operation [P24], reductions of up to 85 per cent of pyrite and 26 per cent of the ash were achieved, as compared to 11 and 40 per cent, respectively, by gravity separation techniques.

Hall et al. [H28] have shown that the oxidative alkaline pressure leach can increase the magnetic susceptibility of pyrite particles. The leach was successfully exploited [H44] as a pretreatment in the magnetic desulphurization of an eastern Canadian coal. Rejection higher than 50 per cent total sulphur (65 per cent pyritic sulphur) at 85 per cent clean coal yield, was achieved using HGMS at a magnetic induction of 1.3 T and a flow velocity of 5.3×10^{-3} m s^{-1}, after pretreatment at 0.5 mol NaOH, 85° C, 930 kPa air pressure, 15 minutes of retention time. Without pretreatment, sulphur rejection under the identical conditions was less than 5 per cent. On the other hand, the ash reduction by this process was low: leaching removed about 10 per cent of the ash, and HGMS a further 5 to 10 per cent. The high flow velocities did not allow capture of the ash minerals, since their magnetic susceptibility was not enhanced by leach treatment.

V.2.3 Wet HGMS of coal

The first investigation into the beneficiation of coal by HGMS was carried out at MIT. Trindade [T19, T20] studied desulphurization and de-ashing of a Brazilian coal by HGMS at 2 T with a steel wool matrix. The results showed that the HGMS reduced the total sulphur by 20 to 50 per cent and the other mineral matter by 14 to 21 per cent. American coals were investigated by Murray [M37], Lin et al. [L27] and Luborsky [L17]. Results comparable to Trindade's were obtained as can be seen in Figs. 4.16, 4.17 and 4.18.

A comprehensive analysis of wet HGMS as applied to coal cleaning was published by EPRI [W25, S72]. The primary objective of this work was to conduct an engineering assessment of the feasibility of magnetic coal cleaning for the electrical utilities. Good correlation was found between performance of the batch and carousel separators. Up to 82 per cent of pyritic sulphur and up to 60 per cent of ash were removed, albeit at unacceptably high losses of coal into the magnetic product, as illustrated by Figs. 4.14 and 4.15.

Oder [O12] investigated the beneficiation of the micronized Upper Freeport coal. The results show that the ash can be reduced from 16.3 to 6.5 per cent at 2 T, or further reduced to 4 per cent in processing at 15 T. Excellent removal of the pyrite sulphur (99 per cent) had already been achieved at low field strengths.

A number of studies in Europe [V8, D9, M39] have shown that a large fraction of pyrite (up to 90 per cent), and a significant amount of ash (up to 50 per cent) of ash can be removed from German coals. Even though the results suggest that desulphurization by wet HGMS is technically possible, there are major problems that must be overcome before a commercially viable system can be realized. The first is the cost of preparation of the coal and the subsequent dewatering and handling of the product. The second is the non-selectivity of separation, causing an excessive amount of coal to be lost in the magnetic product. Because of the high cost of the auxiliary equipment and loss of coal, wet HGMS does not appear economically feasible at present [M36].

To circumvent the difficulty of wet processing the use of methanol [M35] and liquid nitrogen [D22] has been proposed. Although this approach offers several advantages, further research and economic analysis are required.

V.2.4 Dry HGMS of coal

Studies of dry HGMS of unclassified coal have generally shown that the performance is poor [L17, M37], although classification and removal of fines considerably improved the performance of a magnetic separator [H42, H43]; representative results are given in Table 5.34 and Figs. 5.39 and 5.40.

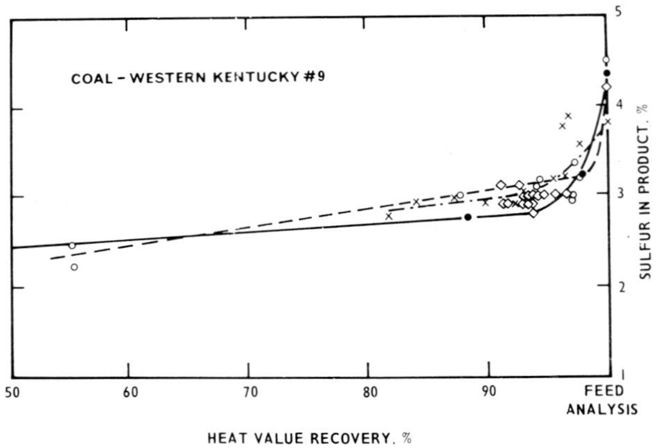

Fig. 5.39 Removal of sulphur by HGMS and gravity separation.
○ specific gravity (-600 + 150 μm)
● specific gravity (-1400 μm)
× batch HGMS
◇ carousel HGMS (-700 + 150 μm)
(After [H42, H43, M36]. Reprinted from [M36], by courtesy of Gordon and Breach Science Publishers.)

TABLE 5.34

Coal benefication by HGMS, B = 2 T, expanded metal matrix [H42]

Coal	Flow velocity (ms^{-1})	Reduction (%) S	Reduction (%) Ash	Thermal recovery (%)
Lower Freeport	0.25	63.3	56.3	94.7
(-150 + 75 μm)	0.46	61.7	62.0	95.6
Kentucky no. 9				
(-150 + 75 μm)	0.25	23.0	45.6	89.2
	0.46	23.0	51.9	92.4
(-600 + 150 μm)	0.25	19.9	49.1	93.0
	0.46	18.4	40.8	93.9

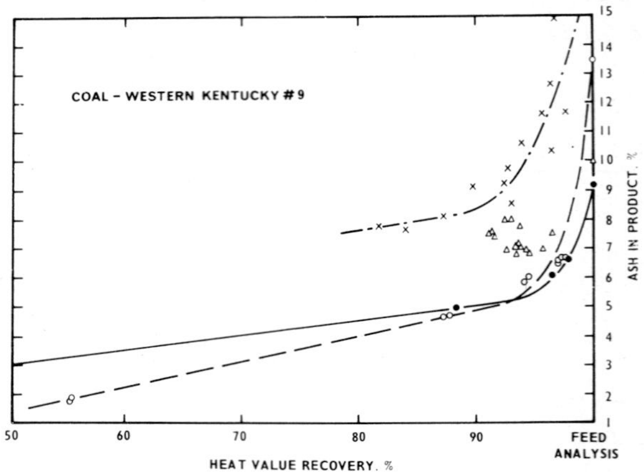

Fig. 5.40 Removal of ash by HGMS and by gravity separation. Notation is the same as in Fig. 5.39. [H42, H43, M36]. (Reprinted from [M36], by courtesy of Gordon and Breach Science Publishers.)

V.2.5 Open-gradient magnetic separation of coal

The application of a split-coil superconducting magnet to the beneficiation of coal was reported by Hise et al. [H17, H18] and by Chovanec et al. [C28]. Tests performed by Hise et al. on a coal sample -600 + 150 μm (Figs. 5.41 and 5.42) indicate that results approaching those obtained with heavy media method can be achieved.

Chovanec et al. [C28] investigated desulphurization and deashing of lignite coal with high sulphur (11 to 12.5 per cent) and ash (20 to 22 per cent) content. The specific magnetic susceptibility of coal was 1.4×10^{-8} m^3 kg^{-1} (1.1×10^{-6} cm^3 g^{-1}), while the susceptibility of pyrite particles was 2×10^{-7} m^3 kg^{-1} (16×10^{-6} cm^3 g^{-1}). The beneficiation of size fractions -500 + 100 μm and -100 + 40 μm was investigated and it was observed that sulphur concentration can be reduced by 43 to 68 per cent, and that of ash by 38 to 62 per cent. As in any application of dry HGMS, the selectivity of separation was a problem that resulted in low recoveries of the heat value, amounting to 33 to 58 per cent only.

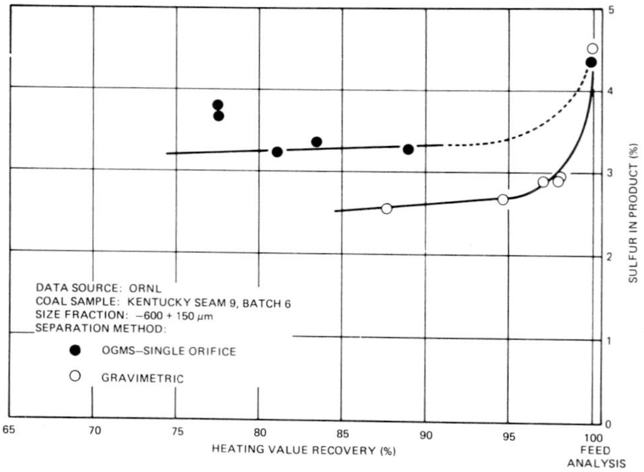

Fig. 5.41 Removal of sulphur from Kentucky coal (-600 + 150 μm) by a split-coil OGMS and by gravimetric separation [H17].

V.2.6 Magnetic separation of liquified coal

In the solvent-refining process, coal is liquified by dissolution of the carbonaceous part by a solvent at high temperature and pressure. The mineral matter in the coal does not dissolve and must be separated from coal solution to produce a clean product. The pyrite is converted to a pyrrhotite-like material during the reaction process and can be removed by magnetic separation. The separation step should be carried out at a temperature compatible with the overall process, preferably in the

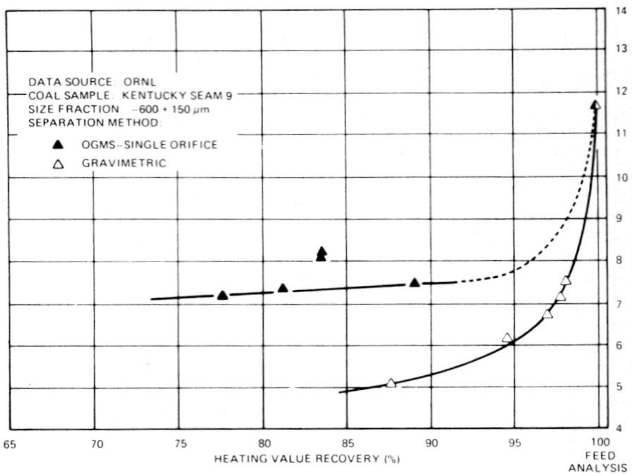

Fig. 5.42 Removal of ash from Kentucky coal (-600 +150 μm) by a split-coil OGMS and by gravimetric separation [H17].

150° C to 300° C range. A typical particle size of the SRC (solvent-refined-coal) filter feed is in the 5 to 19 μm range [L27].

Work by Jacobs and Levinson [J9, O11] in which commercial HGMS with a steel-wool matrix was used removed up to 90 per cent of inorganic sulphur and up to 30 per cent of ash from the SRC filter feed. Similar results were obtained by Liu et al. [L28, L29]. Maxwell and Kelland [M35, M38] managed to remove almost all the inorganic sulphur and up to 40 per cent of ash. Although at 2 T the optimum temperature was 200° C, at higher fields (up to 8 T) there was little improvement in removal. However, temperature dependence was not critical.

Petrakis et al. [P25] have also shown that HGMS is capable of removing over 90 per cent of the inorganic sulphur, solids and ash from a diluted SRC filter feed at flow velocities less than 1.7 x 10^{-3} m s^{-1}, applied field 2 T and temperature 190° C or greater. If flow velocity is less than 4 x 10^{-4} m s^{-1}, then removal is over 97 per cent for ash and 95 per cent for inorganic sulphur. Fig. 5.43 illustrates the effect of applied magnetic induction of ash removal while Fig. 5.44 depicts the relative efficiency of the removal of ash, sulphur and solids as a function of matrix loading.

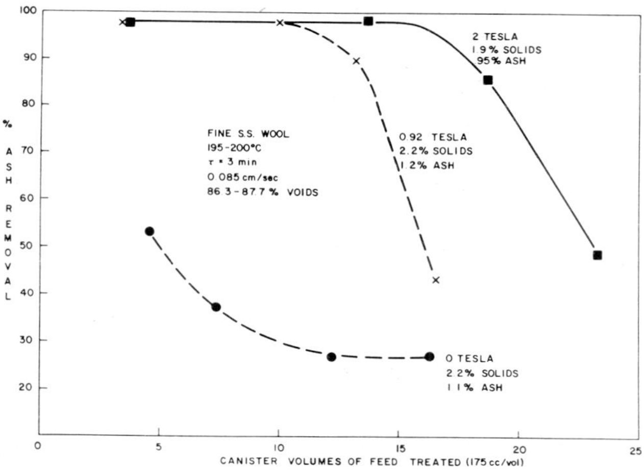

Fig. 5.43 Removal of ash as a function of matrix loading at various field strengths. (Reprinted from [P25], by courtesy of Marcel Dekker, Inc.)

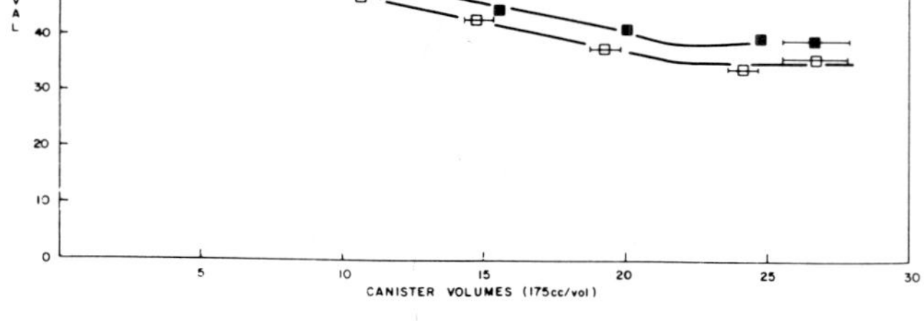

Fig. 5.44 Removal of ash, sulphur and solids as a function of matrix loading. (Reprinted from [P25], by courtesy of Marcel Dekker, Inc.)

V.2.7 Recovery of magnetite from coal by HGMS

Fine magnetite suspended in water is used to clean coal by heavy-media separation. Low-intensity drum separators are used to reclaim the magnetite and multipole units are required to circumvent the limitations imposed by coal entrainment in the magnetic fraction and by insufficient capture of fine particles. Kelland et al. [K30, D23, K33] proposed a high-gradient separator to improve the recovery of small magnetite particles at very high flow velocities. To avoid clogging of the matrix by strongly magnetic magnetite, Kelland et al. [K31] designed a matrix consisting of expanded metal screens arranged with their plane parallel to the flow direction and field. The captured magnetizable particles fill the voids of the screen without extending beyond the screen thickness. The sides of the screen, being regions of diamagnetic capture, avoid the bridging to other screens. The distance between the screens is determined primarily by the maximum size of coal particles.

The application of HGMS to magnetite recovery has several advantages [K32]. First, the rate at which magnetite can be collected is very high and high throughputs per machine can be obtained. The pilot-plant rates can be projected to 300 t h^{-1} per square metre of matrix area. Second, HGMS is insensitive to the slurry density and no difference in performance was found over a range of ratios Fe_3O_4 : coal from 4 : 1 to 1 : 3.

As a result of high flow velocities the selectivity of HGMS is very good. Fig. 5.45 shows how the entrainment of coal can be as low as 5 per cent when the recovery of magnetite exceeds 99 per cent, compared to coal entrainment of 30 per cent sometimes found in drum separators.

Typical results obtained at high flow velocities in pilot-plant tests with the Sala carousel magnetic separator are summarized in Table 5.35 while Table 5.36 shows the effect of the magnetite-to-coal ratio on the metallurgy.

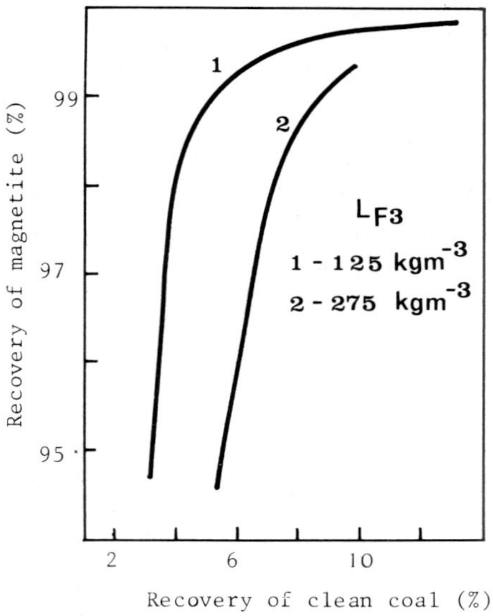

Fig. 5.45 Magnetite recovery versus coal recovery. B = 0.6 T, flow velocity 0.095 to 0.52 m s^{-1}. [K32]

TABLE 5.35

Recovery of magnetite by Sala carousel HGMS, 0.9 T, expanded metal matrix, screen spacing 2.4 mm [K33]

Flow velocity (ms^{-1})	Slurry density (% solids)	Magnetite to coal ratio	Loss of coal into mags (%)	Recovery of magnetite (%)
0.50	17.9	1.91	4.9	99.82
0.58	28.4	1.79	8.1	99.36
0.70	16.2	1.77	4.4	99.65

TABLE 5.36

Recovery of magnetite and loss of coal into the magnetic fraction as a function of magnetite-to-coal ratio [D23]

Fe_3O_4 : coal	Recovery of magnetite (%)	Loss of coal (%)
3.9 : 1	96.3	3.3
1.7 : 1	96.4	3.3
1 : 1	96.4	3.0
0.5 : 1	96.7	2.4
0.33 : 1	95.6	2.4

V.3 THE MAGNETIC BENEFICIATION OF FLY ASH

The treatment of fly ash from pulverized coal power plants for the recovery of constituent metals has been proposed as an alternative to disposal. The utilization of waste helps conserve natural resources by using coal wastes in place of some raw materials while decreasing the tonnages of coal waste requiring disposal.

Fly ash contains many valuable minerals as major and minor constituents. As can be seen from Table 5.37 aluminium is the primary product although titanium, iron and some of the trace metals may also be recovered. Although the percentages of metals found in fly ash are lower than those in higher grade ores, the large amounts of fly ash produced make it a potentially valuable source of minerals. Table 5.38 shows the demand in the USA during 1979 for aluminium, titanium and iron and the quantity that coal fly ash can supply of these metals.

TABLE 5.37
Composition of coal fly ashes [C38]

Constituent	Composition (%)	
	Bituminous	Subbituminous
SiO_2	48.0	30.0
Al_2O_3	20.6	19.7
Fe_2O_3	6.1	5.4
CaO	5.5	28.8
TiO_2	0.8	1.5

Recently, several processes for the recovery of metals from coal fly ash have been proposed. They are based on the following methods of treatment : physical beneficiation, leaching, sintering and gas-solid reactions.

Magnetic separation has been used to separate fly ash into magnetic and non-magnetic fractions. Direct use for the magnetic

TABLE 5.38

Demand in the USA for aluminium, TiO_2 and Fe_2O_3 [C38]

Material	Demand (Mt)	Equivalent material recoverable from fly ash (Mt)	(%)
Aluminium	4.6	3.6	80
Titanium dioxide	0.78	0.36	47
Iron ore	105.0	5.5	5

fraction has included heavy-media material for coal washing and filler for high-density cement. Although its use as an iron ore has been suggested, attempts to market the material as such have not been successful [D24]. The use of magnetic fly ash as iron ore is an attractive possibility since the iron-ore market, unlike others that have had to be penetrated, is capable of absorbing an ash-based product with minimal displacement of conventional sources.

Dobbins and Burnet [D24] have shown that as much as 50 to 60 per cent of the iron can be recovered in the magnetic fraction. Table 5.39 gives a comparison of partial composition of a typical commercial iron-ore concentrate and of two fly ash iron-rich fractions.

TABLE 5.39

Composition (%) of iron ore concentrate and magnetic fractions of Missouri fly ash

Product	Fe	SiO_2	Al_2O_3	CaO
Iron concentrate	65.4	4.72	0.33	0.45
Missouri I	46.1	22.0	9.39	1.73
Missouri II	51.5	16.6	7.0	1.12

Because magnetic fractions of fly ash are lower in iron and higher in gangue components than commercial iron concentrate it becomes necessary to beneficiate the ash fractions so that the concentrate will meet market expectations. A hydrothermal caustic extraction followed by a mild acid wash gives a residue containing iron oxides as high as 95 per cent [D24].

Gilliam et al. [G29] have shown that direct acid leach of ash with HCl and a pressure digestion-acid leach have the greatest economic potential.

Although the application of magnetic separation to the beneficiation of fly ash has been studied in a number of laboratories, the published results are very scarce. Roy et al. [R18] used moving-field dry separation to obtain a magnetic fraction usable as a heavy medium material; the selected results are given in Table 5.40. The recovery of iron oxide amounted to between 35 and 70 per cent.

Michaliková and Špaldon [M40] reported magnetic separation of the Czechoslovak coal fly ash by several belt and roll dry and wet magnetic separators. The values of specific magnetic susceptibility of the Czechoslovak coal fly ash are summarized in Table 5.41. Tables 5.42 and 5.43 show the results of magnetic separation of the Vetřní fly ash containing 3.9 per cent iron.

The application of dry permanent roll magnetic separators to the upgrading of the South African coal fly ash is illustrated in Table 5.44.

TABLE 5.40
Magnetic separation of US coal fly ash (-100 μm) [R18]

Constituent	Feed	Grade (%)		
		Non-mags	Middling	Mags
Kentucky				
SiO_2	47.2	51.8	44.1	20.0
Al_2O_3	17.4	18.9	17.4	8.3
Fe_2O_3	21.3	13.6	27.7	65.9
Missouri				
SiO_2	45.7	51.2	42.9	22.2
Al_2O_3	16.6	18.2	17.1	9.4
Fe_2O_3	25.0	16.3	29.9	62.6

TABLE 5.41
Specific magnetic susceptibilities (at 300 A m^{-1}) of Czechoslovak coal fly ash [S4]

Source	Specific magnetic susceptiblity	
	(m^3kg^{-1} x 10^{-6})	(cm^3g^{-1} x 10^{-6})
NHKG	14.0	1114
Počerady	17.0	1350
Vetřní	24.0	1910
Tisová	11.5	915

TABLE 5.42

Low-intensity wet magnetic separation of Vetřní fly ash [M40]

	Mass yield (%)	Fe content (%)
Feed		3.9
Mags	22.3	14.9
Non mags	77.7	0.91

Kerkdijk et al. [K20] applied HGMS with magnetic induction up to 6 T to the beneficiation of Australian and US fly ashes. Results showed that the excessively high magnetic field and excessively fine matrix caused unacceptable non-selective separation and proved that a less sophisticated and cheaper separator would have given more encouraging results.

TABLE 5.43

Dry magnetic separation of Vetřní fly ash [M40]

	Mass yield (%)	Fe content (%)
Feed	100	3.9
Mags	9.1	22.65
Non-mags	90.9	1.24

TABLE 5.44

Magnetic separation of coal fly ash (South Africa) by permanent roll magnetic separator

Content (%)	Magnetics	Non-magnetics
Fe	34.0	1.9
Al	6.7	12.5
Ti	2.9	3.2
Ca	3.6	6.2

V.4 THE APPLICATION OF HGMS TO WASTE WATER TREATMENT

V.4.1 Steel industry

In modern integrated steel plants, the volume of water applied to the metal itself, or to the steelmaking equipment, varies from 130 to 170 m^3 per tonne of raw steel produced [N7]. Of the total amount of water used, almost one half is for steelmaking operations (coke plant, blast furnaces, BOFs), somewhat over one quarter for hot mills, and slightly under one quarter for steel-finishing. About one half of the total volume of dirty water produced originates in the hot mills, the principal contaminants being millscale and oils. The quantity of scale produced is usually estimated at between two and three per cent of the mass output from the mill [N7].

Because many of the fine suspended particles in these waters are magnetic, high-gradient magnetic separation offers an alternative to conventional sedimentation and flocculation in clarifiers. The early studies by Oberteuffer et al. [O13], Mc Nallan et al. [M41] and Harland et al. [H34] have shown that HGMS provides comparable efficiency at substantially higher flowrates, lower space requirements and considerably lower running costs. A schematic diagram of HGMS in steelmaking applications is shown in Fig. 5.46.

The first full scale HGMS device for steel mill water treatment was installed in 1977 at the Chiba Plant of the Kawasaki Steel Corporation to treat the gas scrubber waste water from the vacuum degassing process. The major constituents of the water were iron and manganese oxide at an average concentration of 100 mg l^{-1}. Particles were smaller than 100 μm, the majority (85 per cent) being in the range 5 to 20 μm. The cyclic magnetic separator, shown in Fig. 5.47, had a diameter of 2.1 m and generated the maximum background magnetic induction of 0.5 T. Table 5.45 summarizes the operating conditions of the separator while Table 5.46 shows typical results of filtration.

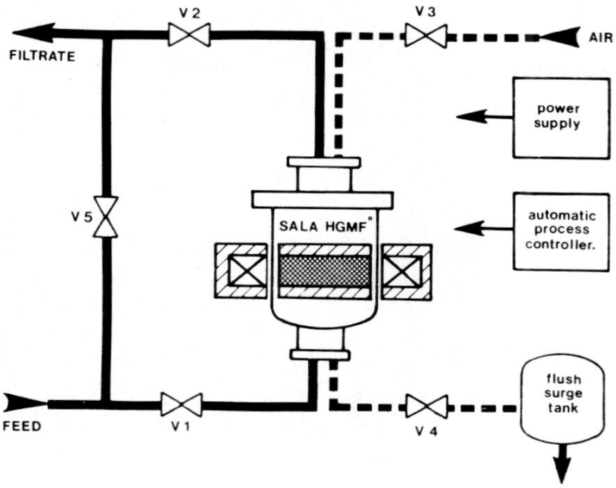

Fig. 5.46 High-gradient magnetic separator in steelmaking applications.
A - gas scrubber water from furnaces
B - scrubber water from steel degassing operations
C - streams from coking, casting and plating processes
D - scale and pit overflow from hot forming operations.
(Courtesy of Sala Magnetics, Inc.)

TABLE 5.45

Operating conditions of HGMS as applied to gas scrubber waste water treatment at Chiba [T9]

Magnetic induction	0.3 T
Flow velocity	max. 0.08 m s^{-1}
Matrix	steel wool, 100 μm diameter
Matrix height	0.15 m
Matrix loading L_{F3}	120 kg m^{-3}
Throughput	900 m^3/h
Feed time	40 to 90 mins
Flush time	1 min
Back-flush flowrate	0.4 m s^{-1}
Input power	20 kW
Mass	40 tonnes

TABLE 5.46

Magnetic filtration results at Chiba Works [T9]

Concentration of solids:	
Feed	100 mg/l
HGMS effluent	< 20 mg/l
Back flushed sludge	1000 to 3000 mg/l
Filtration efficiency	> 80 per cent

Fig. 5.47 SALA-HGMS 214-15-5 magnetic separator installed at a steel plant (courtesy of Sala Magnetics, Inc.).

A detailed investigation of the performance of the HGMS as applied to the filtration of steel-mill waste water with a varying content of ferromagnetic impurities, was carried out by Yano and Eguchi [Y5]. They observed that for a water containing solids of which more than 60 per cent are ferromagnetic, removal efficiency is greater than 99 per cent, at 0.3 T and flow velocity 0.06 m s^{-1}; efficiency decreases only slightly with increasing matrix loading. The comparison between HGMS filter and a sand filter for this type of water is shown in Table 5.47. HGMS has almost the same installation cost as that of a sand filter, but surpasses the latter with regard to operating costs and space requirements.

TABLE 5.47

Comparison between HGMS and sand filter for waste water from the scrubber for an oxygen converter [Y5]

Influent suspended solids 150 mg/l
Target effluent suspended solids < 15 mg/l
Influent volume 20 m^3/h

	HGMS	Sand filter
Flowrate	6×10^{-2} ms^{-1}	8×10^{-3} ms^{-1}
Feed time	20 minutes	8 hours
Flush time	1 minute	15 minutes
Back-flush water volume	16 m^3/day	30 m^3/day
Space requirement	4.5 m^2	8 m^2
Relative operating cost	0.7	1
Relative installation cost	1.1	1
Magnet input power	5 kW	-

For water containing solids that are more than 30 per cent ferromagnetic (waste water from a hot rolling mill) the removal efficiency of solids still exceeds 90 per cent, while the removal efficiency of oil ranges between 50 and 90 per cent. The magnet again operates at 0.3 T and flow velocity is equal to 0.17 m s^{-1}. Comparison of HGMS with other filtration methods is given in Table 5.48.

HGMS surpasses other methods in almost all performance indices. The removal efficiency of solids more than 90 per cent weakly magnetic (e.g. vacuum degassing installation) does not exceed 60 per cent at 0.3 T, while an increase in the field to 0.6 T improves removal to almost 90 per cent. A similar efficiency can be obtained by operating at 0.3 T and adding a small quantity (0.1 ppm) of a polymeric flocculant. Table 5.49 shows a comparison

TABLE 5.48

Comparison between HGMS and other filtration methods for waste water from a hot rolling mill [Y5]
Influent suspended solids 150 mg/l
Target effluent suspended solids <15 mg/l
Influent volume 1100 m^3/h

	HGMS	Clarifier-sand filter	Clarifier
Flowrate	0.17 ms^{-1}	0.01 ms^{-1}	0.005 ms^{-1}
Feed time	20 minutes	24 hours	continuous
Flush time	2 minutes	15 minutes	-
Back-flush water volume	340 m^3/day	300 m^3/day	-
Magnet input power	30 kW	-	-
Space requirements	40 m^2	500 m^2	900 m^2
Relative operating cost	0.4	1	2.6
Relative installation cost	0.4	1	0.9

TABLE 5.49

Comparison between HGMS and clarifier for water from vacuum degassing instalation [Y5]
Influent suspended solids 80 mg/l
Target effluent suspended solids < 25 mg/l
Influent volume 720 m^3/h

	HGMS	Clarifier
Flowrate	0.06 ms^{-1}	6×10^{-4} ms^{-1}
Feed time	60 minutes	continuous
Flush time	4 minutes	-
Back-flush water volume	270 m^3/day	-
Magnet input power	40 kW	-
Space requirements	200 m^2	700 m^2
Operating cost	0.3	1
Installation cost	1.5	1

between HGMS and a clarifier. HGMS has a higher installation cost than a clarifier, but surpasses the latter with respect to operating cost and space requirements.

Table 5.50 shows the selected applications of magnetic filters in the steel industry (as at 1985); Fig. 5.48 shows the flowchart of a water treatment process in a hot-rolling mill.

TABLE 5.50

The selected applications of magnetic filters (SALA HGMS) in the steel industry

Year installed	Location	Application	Influent (mg/l)	Effluent (mg/l)	Magnet diameter (m)	Flowrate (m^3/h)
1977	Kawasaki Steel Co. Chiba Works, Japan	Vacuum degassing scrubber water	100	20	2.1	800
1978	Nippon Steel Co. Yawata Works, Japan	Vacuum degassing scrubber water	80	20	2.1	800
1982	Nippon Steel Co. Kimitsu Works, Japan	Continuous caster cooling spray water	60	12	2.1	1500
1982	Kishiwada Steel Co. Kishiwada Works, Japan	Continuous caster cooling spray water	100	20	1.5	800
1982	Wheeling-Pittsburg Steel Co. Monessen, PA, USA	Continuous caster cooling spray water	80	10	1.0	220
1982	Wheeling-Pittsburg Steel Co. Steubenville, OH, USA	Continuous caster cooling spray water	80	10	2.1	1200

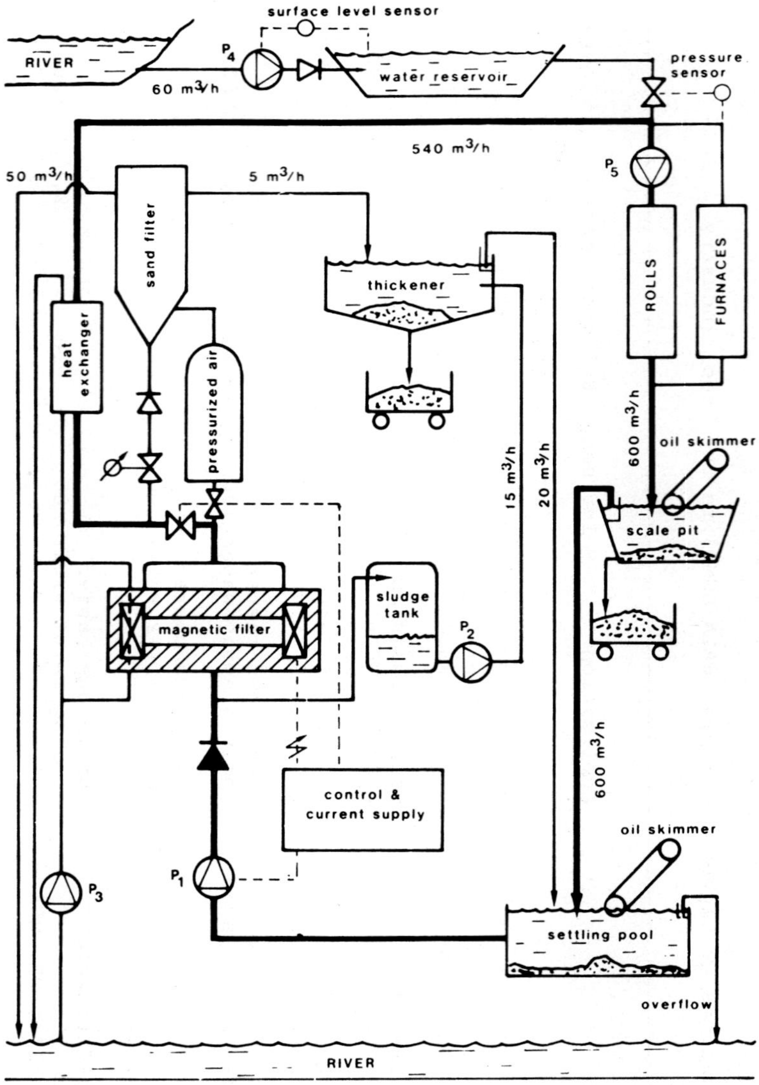

Fig. 5.48 Flow chart for a water-treatment process in a hot rolling mill (after Kokkala [K34]).

V.4.2 Thermal power plants

Corrosion products (crud), in the primary coolant of thermal power plants are deposited on the surface of circuits, where they can interfere with fluid flow and heat transfer. Corrosion products can also become radioactive in nuclear reactors, thereby leading to the irradiation of personnel and the restriction of access. HGMS was found to be effective for the removal of these fine, predominantly ferromagnetic particulates.

Heitmann [H21, H37] described the operational experience with magnetic filters in conventional thermal power plants. Magnetic filters at Kraftwerk Union AG have been in use for the purification of boiler feed-water since 1970. Since then, a large number of filtration plants have been installed in the thermal power industry, e.g. for cleaning turbine condensates. The removal of iron oxide is typically between 90 and 99 per cent. Table 5.51 summarizes some technical data of these magnetic separators, while Table 5.52 shows the efficiency of the filter. During the seventies, extensive studies were carried out to investigate the applicability of HGMS for the purification of water in nuclear power plants.

In nuclear power plants, crud composition varies between pressurized water reactors (PWR) and boiling water reactors (BWR).

TABLE 5.51

HGMS for water filtration in conventional power plants [H37, M29]

Magnetic induction	0.2 T
Flow velocity	0.3 m s^{-1}
Feed time	3 to 7 days
Flush flow velocity	0.22 m s^{-1}
Volume of flush water	10 canister volumes
Matrix	6 mm balls
Matrix height	1000 mm
Matrix loading L_{Fl}	2 x 10^{-3}
Water temperature	50 to 180° C
Working pressure	10 to 40 bar
Pressure drop	0.7 bar
Throughput	100 to 1000 m^3/h
Energy consumption	35 kW at 700 m^3/h
Diameter of magnet	1.5 m at 700 m^3/h

TABLE 5.52

Efficiency of HGMS as applied to water filtration in a conventional thermal power plant [H21, M29]

Inlet iron content	Efficiency of removal (%)	
(ppm)	Total iron	Iron oxides
0 to 0.02	89	93
0.02 to 0.1	90	93.5
> 0.1	86	94
0 to max	89	93.5

However, the principal dose-associated radionuclides are Co^{58}, Co^{60}, Fe^{59} and Mn^{54}. Under normal PWR operating conditions of 300° C, the only stable oxides are magnetite and other substituted spinel ferrites of Co, Ni and Mn which are ferromagnetic.

In BWRs, system conditions vary from approximately 300° C at the core to 50 - 100° C at the feed-water purification system. Under these conditions, about 30 per cent of the oxides are hematite, the balance being formed by Fe_3O_4 spinels found in the PWR [E7].

In the USA, the application of magnetic filtration to the primary and secondary sides of PWRs has been studied jointly by EPRI and Westinghouse [T12]. The results of these studies indicate that the treatment of 10 per cent of the steam generator secondary flow will reduce sludge accumulation by 70 to 80 per cent, and that treatment of 0.5 per cent of the primary coolant has the potential to reduce radioactivity by a factor of 2 to 5.

Heitmann et al. [H45] studied the performance of various types of magnetic filters in the purification of feedwater in PWR; the results are summarized in Table 5.53. It has been observed that all types of filters and matrices (6 mm balls, steel wool and the combination of both) have the same filtration efficiency. It was estimated that the maximum time of operation of the steel wool matrix was between 100 and 200 days, while for a combination of steel balls and steel wool matrix, feed time as long as 650 days can be expected. The ball matrix was found to give a higher efficiency of back flush.

An extensive investigation by Dollè et al. [D25] has shown that a magnetic filter with a steel ball matrix applied to the primary circuit of PWR can extract radioactive corrosion products, reduce stationary radioactivity in the circuit and capture the increased amount of radioactive crud after a transient release at reactor power cycling or shut-down. In the secondary circuit the magnetic filter can be used preferentially at a high temperature in the steam generator blowdown line.

TABLE 5.53

Performance of magnetic filters as applied to filtration of feed-water in PWR [H45]

Matrix	B (T)	Concentration of particles (ppb) Influent	Effluent	Efficiency of filtration (%)	Filtration time (days)
6 mm balls	0.2	7.3	1.1	85	33
Steel wool	0.2	6.9	0.4	94	26
	5			99	.
Balls + wool	0.2	1.9	0.27	86	140

V.4.3 Magnetic seeding

High-gradient magnetic separation may be extended to the filtration of non-magnetic substances by adding magnetic particles and coagulant to the stream. The magnetic particles coagulate with the impurities, thereby allowing the flocs to be recovered by a magnetic separator. Although the technique of magnetic seeding had been proposed as early as 1969 by Mamula et al. [M30], only the availability of high-gradient magnetic separators in the seventies made this concept feasible for large-scale applications.

In practice a coagulant (e.g. alum) is added to the feed stream, in a concentration ranging from 10 to 200 mg l^{-1}, followed by the addition of the magnetite seed in a concentration from 50 to 5000 mgl^{-1}. A small amount of polyelectrolyte may be added to strengthen the flocs prior to magnetic filtration. The flowsheet of this process is shown in Fig. 5.49. Magnetic seeding technique can also be used for selective separation of a mixture of diamagnetic minerals [H46]. In the mixture of gibbsite and quartz, gibbsite was selectively flocculated with magnetite seeds and removed by magnetic separation.

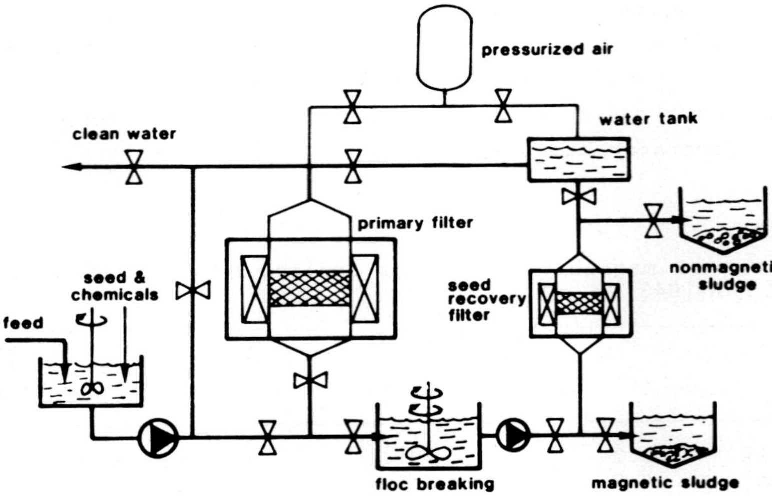

Fig. 5.49 Magnetic filtration process with a magnetic seeding technique (after Kokkala [K34]).

V.4.3.1 Filtration of river water

An extensive study of the magnetic filtration of non-magnetic solids with the aid of the seeding technique was conducted by de Latour et al. [D16, D26]. In their tests on water from the Charles River, the conditions of which are shown in Table 5.54 they were able to improve the colour as well as the quality of the filtered water, as illustrated by Table 5.55.

Bitton and Mitchell [B46] investigated this technique for the virus removal. 95 per cent of viruses were removed by HGMS, following ten minutes of contact time with magnetite. A concentration of 250 ppm of magnetite was sufficient to yield maximum removal. The addition of calcium chloride as a bridge between the virus and magnetite improved separation.

Algae can be removed by HGMS as efficiently as bacteria. Bitton et al. [B47] found that a removal of 94 per cent of algae cells can be achieved at pH 7. At higher pH, the removal was lower and the improvement was achieved by lowering pH to pH 6.5.

TABLE 5.54

Experimental conditions of water filtration by magnetic seeding [D16]

Magnetic induction	1 T
Flow velocity	0.2 m s^{-1}
Matrix	Steel wool
Magnetite	1000 ppm
Al^{3+} ions	4 ppm

TABLE 5.55

Filtration of the river water (Charles River, Mass.) by HGMS using magnetic seeding [D16]

	Surface		Bottom	
	Influent	Effluent	Influent	Effluent
Coliform bacteria (per litre)	160 x 10^3	0	160 x 10^3	3 x 10^3
Turbidity (JTU units)	20	2	1700	1
Colour (colour units)	105	3	3700	1
Suspended solids (ppm)	7	5	690	5

The development of an industrial-scale process was carried out at Sala Magnetics. The applicability of the process to sewage and combined storm and sewage overflow has been studied in laboratory and pilot-scale systems under partial sponsorship of the United States Environmental Protection Agency [A19, A22]. A mobile trailer pilot plant, capable of processing 1.4 m^3/h at 250 m^3/hm^2, was constructed and used to test the parameters of water quality. Comparison between HGMS and other sewage treatment processes is given in Table 5.56.

The demonstration-scale 45 m^3/h system for magnetite-seeded water treatment was developed and built by Sala Magnetics and operated in Sweden by Boliden-Kemi AB at various sewage treatment plants. The purpose of these tests was to determine the effectiveness of this technique as tertiary treatment for the removal of phosphorus. A summary of the results is given in Table 5.57. In all cases the feed was the effluent from biological treatment.

TABLE 5.56

Comparison between HGMS and other sewage treatment processes [015]

Process	Removal (%)		
	Solids	COD	Bacteria
HGMS	88 - 95	60 - 75	99 - 99.9
Chemical clarification	60	-	-
Activated sludge treatment	55 - 95	70 - 80	90 - 98
Physical chemical treatment	99	80	99

TABLE 5.57

Sewage treatment by HGMS in Sweden [015]

Location	Parameter	Feed	HGMS effluent	Reduction (%)
Helsingborg	P (ppm)	10 - 15	0.5	95 - 97
	Turbidity(JTU)	20 - 30	5	75 - 83
Joenkoeping	P (ppm)	10 - 15	0.5	95 - 97
	Turbidity(JTU)	10	5	50
Stockholm	P (ppm)	1	0.13	87
	Turbidity(JTU)	15 - 20	3	80 - 85

V.4.3.2 Separation of dissolved phosphorus

In order to reduce the eutrophication of natural water, phosphorus must be removed from sewage or from receiving waters. Various sedimentation techniques, depending on the coagulation of the phosphorus with lime or alum, are available.

In contrast to these techniques, HGMS was found [B48] to be very rapid and capable of removing even very finely-suspended phosphate precipitates. Only 250 ppm of magnetite were necessary to remove coagulated ortho-phosphate by HGMS. When montmorillonite clay was used as a flocculant, removal efficiency was higher than 90 per cent at phosphate levels as low as 250 ppb.

V.4.3.3 Separation of heavy metals

Okuda et al. [014] developed a technique for the removal of heavy metals from waste water. They found that when divalent iron ions coexisted with non-ferrous divalent metal ions M in an aqueous solution, the addition of alkali, followed by oxidation, resulted in the formation of the ferrite $M_xFe_{3-x}O_4$; a flowchart of the process is shown in Fig. 5.50. Furthermore, magnetic separation of the precipitated ferrites resulted in an almost complete removal of heavy metals. The results of the pilot-plant tests are summarized in Table 5.58.

The virtually complete removal of heavy metals by their adsorption onto ferric hydroxide flocs, aided by a magnetite seed and followed by HGMS, was reported by Anand et al. [A29].

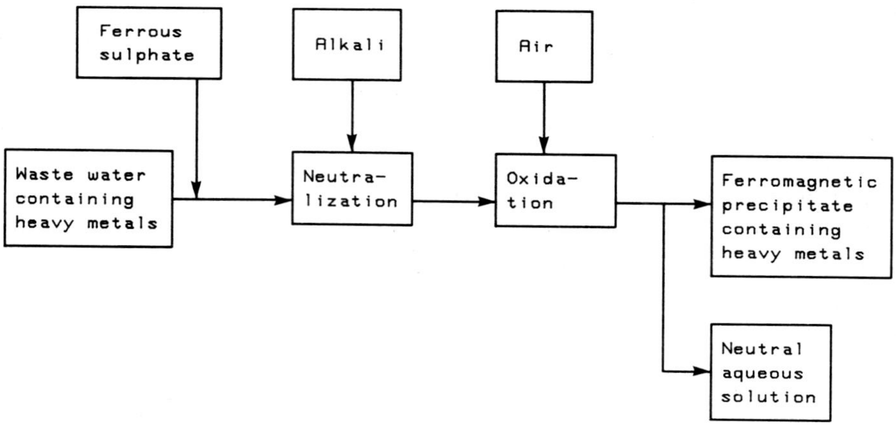

Fig. 5.50 Flowsheet showing the removal of heavy metals from waste water (after Okuda et al. [013]).

TABLE 5.58
Removal of heavy metals from waste water in a pilot plant [O14]

Metal	Concentration (ppm)	
	Influent	Effluent
Hg	7.4	0.001
Cd	240	0.008
Cu	10	0.01
Zn	18	0.016
Cr	10	< 0.01
Ni	1000	0.2
Mn	12	0.007
Fe	600	0.06
Bi	240	0.1
Pb	475	0.01

V.4.3.4 Separation of oil

The magnetic seeding technique can also be used to separate oil and other hydrocarbons from water. The experiments of Kaiser et al. [K13] and Petrakis et al. [P19, P26] demonstrated the applicability of the method to the treatment of waste water from oil refineries. Refinery process water effluent is treated first by an American Petroleum Institute separator (API), followed by a dissolved air flotation unit (AFU) and finally by a series of ponds. Table 5.59 compares the efficiency of AFU with that of HGMS. The effluent from API was used as feed water for both AFU and HGMS and a marked improvement was obtained with HGMS, even without magnetic seeding. However, the best results were shown when magnetite (3000 ppm) and a flocculant (1.7 ppm) were used.

TABLE 5.59
Purification of the refinery process water effluent [P26]

Process	Oil (ppm)	Solids (ppm)	Phenol (ppm)
API feed	1500	500	100
API effluent	190 - 240	140 - 200	1 - 4
AFU effluent	50 - 120	43	1 - 4
HGMS (no magnetite, no flocculant)	23	6	2
HGMS (magnetite, no flocculant)	19 - 23	3 - 5	2
HGMS (magnetite, flocculant)	5 - 20	1 - 5	0.3

V.5 THE APPLICATION OF HGMS IN NUCLEAR INDUSTRY

V.5.1 Removal of solids from fuel dissolver liquor

The reprocessing of the spent irradiated fuel from a nuclear reactor consists in separating the fuel into three basic ingredients: unburnt uranium, highly radioactive fission products and plutonium. The reprocessing cycle comprises three main operations:

1) Mechanical breakdown, in which the fuel is broken down by a massive shear to a size suitable for dissolving in acid
2) Dissolution in hot nitric acid
3) Separation of products by solvent extraction.

In the dissolution of oxide fuels, a significant amount of insoluble matter arises (0.1 to 0.3 per cent by mass for thermal fuel) which contains the fission products Ru, Rh, Mo, Te, Zr, Pd; particle size is smaller than 10 μm. In addition to fission products, fine particles of stainless steel and zircalloy are produced from the cladding when oxide fuel is sheared. All these insoluble components have to be removed from the liquor before it is passed to the separation sections.

Although several alternatives for the clarification process have been considered, the stream is processed at present by a centrifuge. The efficiency of removal is higher than 99 per cent for particles greater than 1.5 μm, and 94 to 99 per cent for particles between 0.5 and 1.5 μm [W34]. Owing to the high rotational speed in a centrifuge, problems relating to mechanical stability can arise, as well as heating problems in the caked radioactive materials. High-gradient magnetic separation has therefore been proposed [W18] as an alternative method for removing insoluble fission products (IFP) from the reprocessing stream.

Magnetic susceptibilities of insoluble fission products were measured by Mortimer [M42]. Most samples were found to possess an appreciable ferromagnetic component and to obey the Owen-Honda equation (eq. 1.54). The selected values are summarized in Table 5.60.

TABLE 5.60
Magnetic properties of insoluble fission products [M42]

Product	χ_∞ ($m^3kg^{-1}x10^{-8}$)	χ at 1.2 T ($m^3kg^{-1}x10^{-8}$)	σ_m ($Tm^3kg^{-1}x10^{-10}$)
UO_2 fuel	1.29	6.43	616
15 % PuO_2 fuel	-	10	1200
18 % PuO_2 fuel	3.85	14.9	1330
30 % PuO_2 fuel	5.28	10.3	600
	5.0	6.6	200
40 % PuO_2 fuel	6.13	6.39	31.4
	6.06	7.5	170

Since IFPs are too highly radioactive to be handled in normal laboratory facilities, experimental work was carried out on a simulant that closely duplicated their magnetic properties. Finely powdered chromium, with the specific magnetic susceptibility of 4.9 x 10^{-8} m^3 kg^{-1} and a size range from 1 to 8 μm, was selected as a simulant and its magnetic separation was investigated by Potts et al. [P21]. Table 5.61 shows that the removal efficiencies obtained are comparable with those from centrifugation.

V.5.2 Removal of ferric hydroxide floc

Waste-water streams from nuclear plant often contain significant portions of iron. If the pH is adjusted to a greater value than 3, the iron precipitates out as ferric hydroxide floc. It has been found that about 60 per cent of total radioactivity and approximately 99 per cent of the α - emitting radioactive contaminants are precipitated with ferric floc and can therefore be removed with it [W18].

Precipitates of this type are difficult to separate with conventional equipment, such as filters or centrifuges, and the mechanical complexity of such equipment does not make it suitable for operating in a radioactive environment. An experimental programme was carried out by Harding and Baxter [H36] to test the HGMS method for filtering ferric hydroxide floc. As follows from Table 5.62, iron can be removed with an efficiency of 93 to 94 per cent. The similarity between the removal efficiency for iron and for the other components suggests that the latter are associated with, or adsorbed on, the ferric hydroxide precipitate, rather than existing independently.

TABLE 5.61

Magnetic separation of insoluble fission products [W23]
Matrix: 3 mm balls, influent concentration 100 to 1000 ppm, mean particle size 2 μm

Magnetic induction (T)	Matrix height (m)	Superficial velocity ($ms^{-1} x10^{-2}$)	Removal efficiency (%)
1.7	0.5	1.5	95
1.7	0.5	3.0	91
1.7	0.5	4.5	88
0.8	0.5	3.0	85
1.2	0.5	3.0	89
1.7	0.5	3.0	91
1.7	0.17	3.0	86
1.7	0.5	3.0	91
1.7	1.0	3.0	98

TABLE 5.62

Removal of ferric hydroxide flocs by HGMS [W34]
Magnetic induction 1.8 T, flow velocity 7.5 mm s^{-1}, mean particle size 3 μm, specific magnetic susceptibility 3 to 5 x 10^{-6} m^3 kg^{-1}

	Concentration in influent	Solids removal (%)	
		Steel balls	Steel wool
Insoluble solids (iron, zinc as hydroxides)	22 ppm	93	93
Insoluble nuclides			
^{95}Nb	2.2 $\mu Ci/l$	92	93
^{106}Ru	15.5	94	94
^{239}Pu and ^{240}Pu	0.2	93	91
Total radionuclides		93	95

V.6 MAGNETIC FILTRATION OF GASES

Many steelmaking processes produce large emissions of fumes containing substantial amounts of iron-bearing dust particles that must be removed before the stack gases are discharged into the atmosphere. Gooding [G14, G15] built a mobile pilot-plant to evaluate the effectiveness and economics of applying HGMS to particulate emission control. Fig. 5.51 shows the laboratory pilot-plant that was used to investigate magnetic filtration of basic oxygen furnace dust (BOF) and electric arc furnace dust (EAF). The results of the investigation, as shown in Figs. 5.52 and 5.53 indicate that HGMS may be an efficient and economical method for particulate emission control in those applications where dust of high magnetic susceptibility (e.g. BOF), must be collected. The on-site testing of the pilot plant was carried out in a sinter plant and results have shown that a satisfactory level of filtration efficiency could not have been achieved because of the low magnetization value of sinter dust.

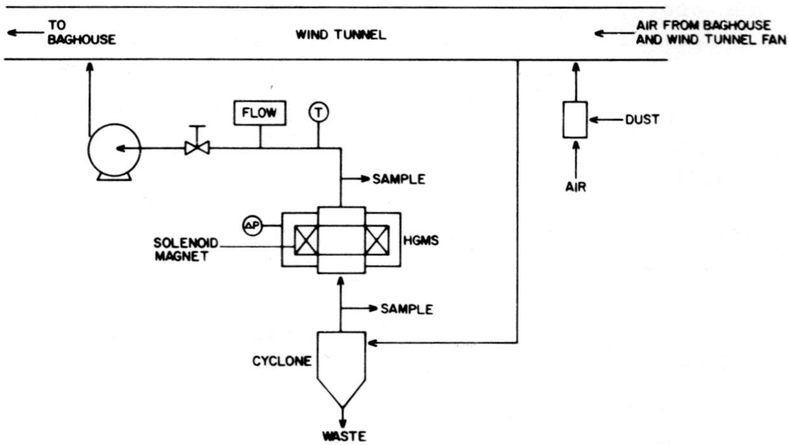

Fig. 5.51 Schematic diagram of a pilot plant for gas magnetic filtration [G14, L30].

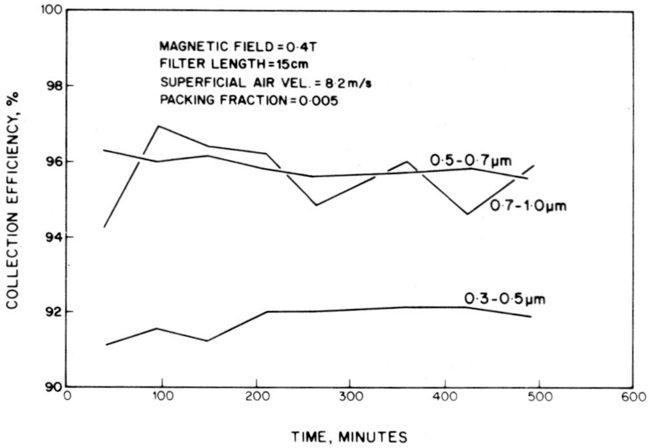

Fig. 5.52 Effect of matrix loading on capture efficiency of BOF dust [G14, L30]. (Reprinted from [L30], by courtesy of Gordon and Breach Science Publishers.)

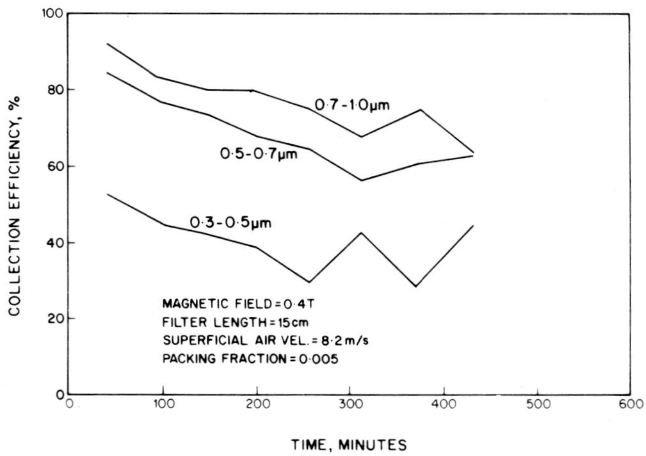

Fig. 5.53 Effect of matrix loading on capture efficiency of EAF dust [G14, L30]. (Reprinted from [L30] by courtesy of Gordon and Breach Science Publishers.)

A detailed investigation into the magnetic filtration of BOF dust was carried out by Boucher and Lua [L30, L31, B49]. Filtration efficiencies of 99 per cent and greater were achieved for submicron particles down to an optically measurable 0.24 μm diameter. For particles 1 μm and above, filtration efficiency was indistinguishable from 100 per cent. The steel wool matrix exhibits a very high loading capacity, up to 5×10^4 kg m^{-3}, which is equivalent to six times the mass of the matrix. The effect of matrix loading on the filtration efficiency is depicted in Fig. 5.54. Fig. 5.55 illustrates the effect of matrix height on filtration efficiency.

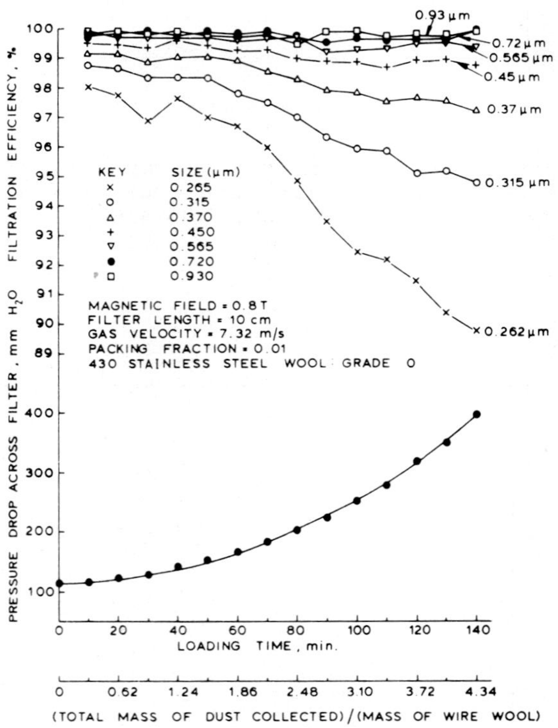

Fig. 5.54 Effect of matrix loading on the efficiency of magnetic filtration of BOF dust. (Reprinted from [L30], by courtesy of Gordon and Breach Science Publishers.)

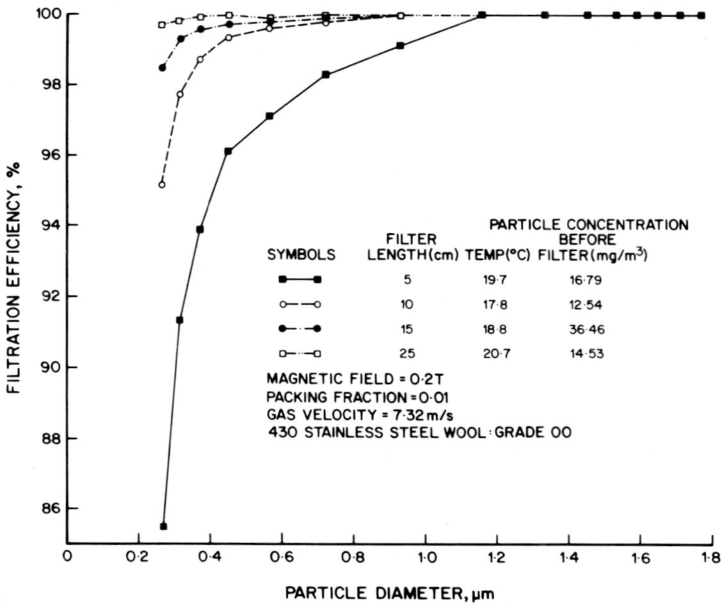

Fig. 5.55 Effect of matrix height on filtration efficiency of BOF dust. (Reprinted from [L30], by courtesy of Gordon and Breach Science Publishers).

Boucher and Lua compared the filtration efficiency of BOF dust with that of cupric oxide dust [B50]. Although the magnetic susceptiblity of cupric oxides is almost four orders of magnitude smaller than that of the BOF dust, efficiency of removal of cupric oxide particles was as high as 75 per cent, while for the smaller particles it dropped to about 50 per cent. On the other hand, matrix loading for cupric oxide dropped by a factor of 50 compared to BOF dust.

CHAPTER VI

THE ECONOMICS OF MAGNETIC SEPARATION

Although high-intensity and high-gradient separation has been used in the mining and other industries for a number of years, few data are available to evaluate the economical efficiency of the process. Like any beneficiation process, the economics of magnetic separation is strongly affected by general factors such as the cost of mining, the transport of ore and its subsequent treatment. Moreover, the cost of magnetic separation is difficult to evaluate because a wide spectrum of machines with widely varying degree of sophistication is available on the market. Factors like energy consumption, availability, mass of machine per unit mass of ore treated, and the capability for a scale-up, differ so substantially for different separators that a wrong choice can severely affect the economical analysis of magnetic separation.

VI.1 ENERGY COSTS

Although energy usually represents a mere 20 to 30 per cent of the running costs of a continuous high-gradient magnetic separator, the impact of energy costs on magnetic separation has been significant. At the present time, most developmental work on magnetic separators is carried out for the purpose of reducing energy consumption of existing machines or for developing new separators with lower energy demands. Recent progress in the development of powerful permanent magnets, and a consistent effort to introduce superconductivity into the mining industry, illustrate a serious impact of energy costs on magnetic separation.

It has been pointed out by Dahlstrom [D27] that energy cost indices have increased more rapidly, and to a far greater extent than other major factors affecting costs, as indicated in Table 6.1.

TABLE 6.1
Cost indices 1970 - 1983: percentage increase [D27]

	Increase (%)
Labour - construction	227
- mining and processing	293
Machinery	300
Rail transport	360
Energy - coal	354
- electricity	453
- diesel fuel	720
- natural gas	1337

The effect of energy costs on mineral processing can be visualised in Table 6.2, which gives the energy consumption for the treatment of 0.55 per cent copper ore. It can be seen that concentration consumes more energy than each of the other two operations. If the energy consumption is broken into four categories, as shown in Table 6.3, comminution of ore is the major consumer of energy in mineral processing. Although magnetic separation is more energy-demanding than flotation, the predominant role of comminution in primary energy consumption is preserved.

TABLE 6.2
Average energy consumption in the processing of copper ore (head 0.55 per cent Cu, cut-off grade 0.29 per cent Cu) [D27]

Operation	Energy consumption (kJ/t)x10^6	Distribution (%)
Mining	23.4	19.3
Comminution, concentration	49.5	40.9
Smelting, refining	48.2	39.8
Total	121.1	100

However, since magnetic separation is often used to re-treat tailings from other beneficiation processes or to filter solids from suspensions, where comminution costs are either low or absent, the energy consumption of a magnetic separator becomes an important issue. Consequently, the potential for savings in operating costs is very significant.

TABLE 6.3
Distribution of energy consumption in the concentration of copper sulphide [D27]

Operation	Distribution (%)	*
Crushing, grinding	73.5	71.8
Flotation	7.8	13.6
Filtration	1.6	1.3
Water	17.1	13.3

* includes energy for production of liners, grinding balls and flotation reagents

VI.2 THE COST OF MAGNETIC SEPARATORS AND THEIR ECONOMIC COMPARISON

VI.2.1 Cost of conventional magnetic separators

The average price of magnetic separators in South Africa, as at the last quarter of 1984, are given in Figs. 6.1 and 6.2 [R19]. The prices, as a function of capacity, mass and power, are given in South African rands and are f.o.r. Witwatersrand.

VI.2.2 The economic comparison of continuous high-gradient magnetic separators

Forssberg and Kostkevicius [F19] carried out a comparative study of various high-gradient magnetic separators. The metallurgical results are described in Section IV.14.1, while the economic comparison is given in Table 6.4. The prices shown are in Swedish krone (SEK) and are based on the price of energy of 0.2 SEK/kWh, cost of fresh water 0.14 SEK/m^3 and of recycled water 0.07 SEK/m^3. Wear and tear costs refer principally to the cost of the matrix.

It can be seen that the Readings separator is by far the cheapest alternative, on account of its very low capital cost. Its capacity however, is small and a large number of units are needed for a given capacity. The cost of one tonne of capacity is 30 000 krone for all separators, while the separation cost is between 1.25 and 1.90 SEK per tonne. The cost analysis disregards the fact that some separators are well established in the mining industry and have a proven record of successful operation while other separators are still in the development stage.

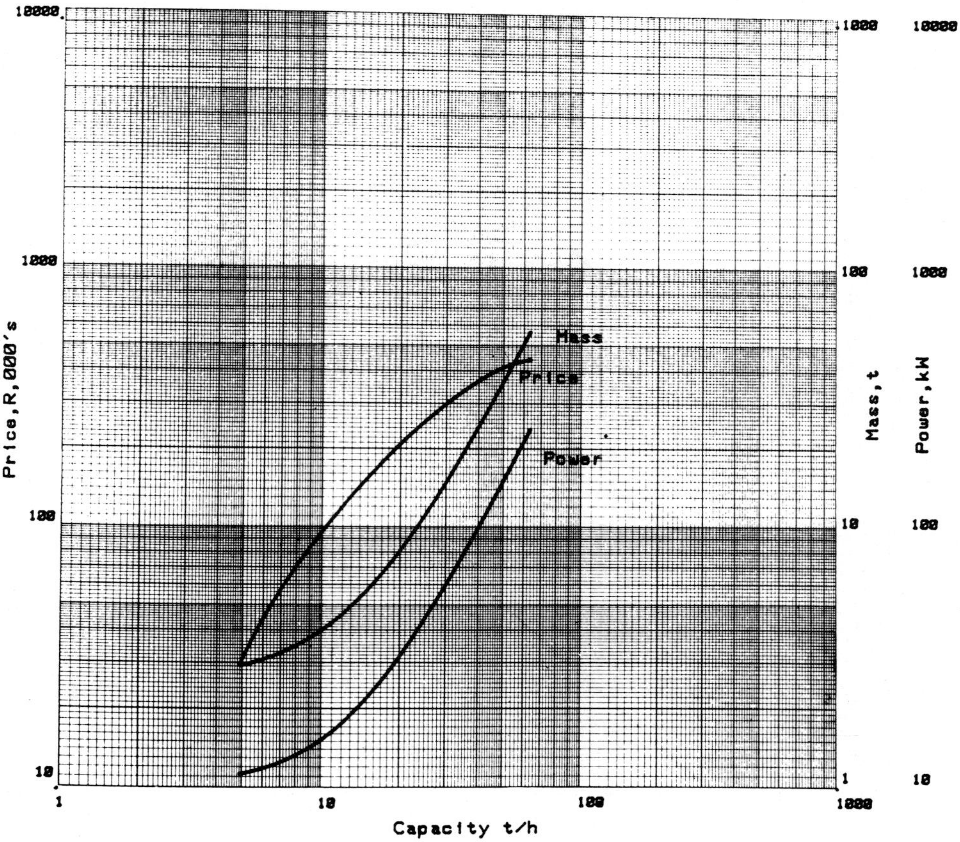

Fig. 6.1 High-gradient magnetic separators [R19].

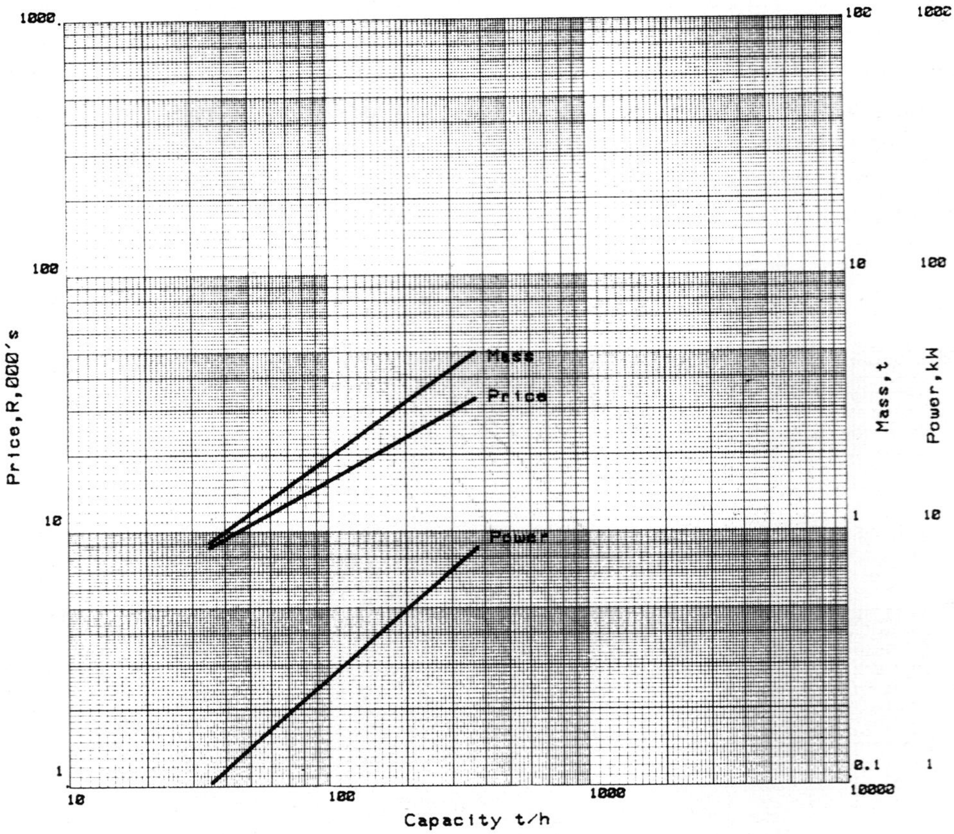

Fig. 6.2 Wet drum magnetic separators [R19].

VI.2.3 Cost of solenoid magnets

The capital cost of a solenoid magnet is determined by the area and height of the magnet and by the applied magnetic field. Total cost thus depends on the mass of the coil, the conductor material, the coil configuration, conductor shape and cost of the power supply.

In considering the mass and volume of the solenoid, one must make a choice between various suitable materials. For low and medium-field magnets, two metals are of interest: copper and

TABLE 6.4

Economic comparison of carousel high-gradient magnetic separators
Prices in Swedish krone in 1980 [F19]

		Jones PP 317	Boxmag-Rapid H1W 8	Readings WHIMS 16-pole	Krupp-Sol 24/14	Sala HGMS 480
Capacity	t/h	120	85	25	30	360
Price of high-intensity separator (1980)	MSEK[1]	3.8	2.5	0.47	1.0	12.0
Price of low-intensity separator	MSEK	—	0.2	0.06	—	—
Capital cost	SEK/t[2]	**1.09**	**0.97**	**0.66**	**1.15**	**1.15**
Total energy:						
(Consumption, kWh/t)		(0.49)	(0.62)	(1.60)	(0.94)	(0.70)
Cost	SEK/t	0.06	0.08	0.19	0.11	0.08
Dilution water:						
(Consumption, m^3/t)		(1.0)	(1.0)	(1.0)	(1.0)	(1.5)
Cost	SEK/t	0.07	0.07	0.07	0.07	0.10
Total flushing water:						
(Consumption, m^3/t)		(1.22)	(0.8)	(1.22)	(1.32)	(2.0)
Cost	SEK/t	0.18	0.12	0.18	0.20	0.30
Wear & tear	SEK/t	0.20	0.20	0.20	0.20	0.20
Running costs	SEK/t	**0.51**	**0.47**	**0.65**	**0.58**	**0.70**
Total cost	SEK/t	1.60	1.44	1.80	1.24	1.85

[1] Inclusive cost of installation, 15%. – [2] 15%, 10 years, running time 6.624 hours.

aluminium. Copper has a density of 8.89 x 10^3 kg m^{-3} at 300 K, aluminium 2.699 x 10^3 kg m^{-3}. Since copper is 3.3 times as dense as aluminium, the aluminium magnet is lighter. However, the resistivity of aluminium (2.55 x 10^{-8} Ωm at 300 K) is about 150 per cent that of copper (1.72 x 10^{-8} Ωm at 300 K) which, for the same magnet size and background field, requires 50 per cent more power.

For larger solenoids and higher applied fields, water-cooled hollow-square conductors are preferred. Coils in solenoids are composed of double pancakes with one or more hydraulic passages connected in parallel. Coil-winding is more expensive for copper solenoids than for aluminium coils, for which the aluminium conductor may have a continuous length of several hundred metres. Table 6.5 compares the prices of the copper and aluminium magnets, with and without steel cladding.

Prices are based on the price of extruded copper conductor at $ 2.20/kg, of extruded aluminium conductor $ 1.65/kg and of machined low-carbon steel at $ 1.10/kg.

TABLE 6.5

Cost comparison (1973) between various coil configuration with and without iron return yokes [B4]

Cost (US $ x 10^5)	Cu coil no iron	Al coil no iron	Cu coil iron	Al coil iron
Coil	3.82	1.87	2.11	1.51
Support structure	0.3	0.3	0.3	0.35
Iron	-	-	1.5	2.2
Power supply	3.6	4.8	2.4	3.0
Interlocks, instrumentation	0.5	0.5	0.5	0.5
Design, installation	0.7	0.7	0.7	0.7
Miscellaneous	0.6	0.6	0.5	0.6
Total	9.52	8.8	8.0	8.85

Collan et al. [C39] developed the equation for the price of a cyclic magnetic separator (in US dollars, 1981) as a function of the matrix area **S**, solenoid height **L** and background magnetic induction **B**.

$$K_m = 350\left[109.15 + 427.4S^{\frac{1}{2}} + 27.97S\right]\cdot\left[1 + \frac{S(L-0.15)}{10.098 + 0.7565}\right]\cdot\left[1 + 0.6(B-0.5)\right] \qquad (6.1)$$

The price as a function of **S** and **L** for an 0.5 T separator is shown in Fig. 6.3.

The price of pumps (in US dollars, 1981) is assumed to be proportional to the square root of the pressure head ΔP (kPa) and the capacity **Q** ($m^3\ h^{-1}$) [C39]

$$K_p = 900(100\Delta PQ)^{\frac{1}{2}} \qquad (6.2)$$

The installation costs of a magnet and pumps are assumed to be 25 per cent of the price of the magnet, and 50 per cent of the price of the pumps [C39]. The price of the buildings (in US dollars) is assumed to depend on the separator size as

$$K_b = 11 \times 10^3 + 4.5 \times 10^3 S \qquad (6.3)$$

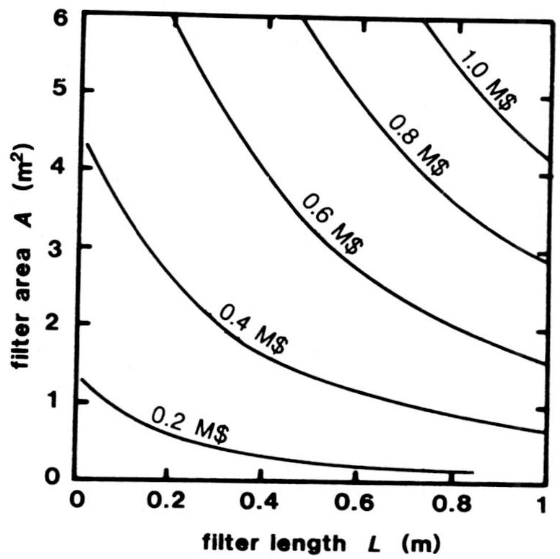

Fig. 6.3 Price of cyclic magnetic separators of different sizes with a background field 0.5 T, (1981). [K34].

The total capital costs, including pipings, electrical equipment, and contingency, are

$$K_{capital} = 1.39(1.25\,K_m + 1.5\,K_p + K_b) \tag{6.4}$$

Fig. 6.4 gives the prices of power suppplies per kW for various power ranges [B4]. We see that at high power levels, the power supply cost is appreciable, and magnet optimization must be performed accordingly.

VI.2.4 Running costs of a cyclic magnetic separator

Kokkala [K34] calculated the running costs of a cyclic magnetic separator as applied to the filtration of steel mill waste water. He assumed the annual operating time was 7700 h/y corresponding to a labour cost of $ 22000/year.

The cost of electricity (1982) was about $ 0.044/kWh or $ 340/kW/year. For magnetic induction smaller than 0.5 T, the input power (kW) is

$$P_s = 550 L S^{0.5} B^{1.25} \qquad (6.5)$$

If the temperature rise of the cooling water is 20° C, the cost of cooling will be approximately 3 per cent of the energy cost. The pumps are assumed to work at 80 per cent efficiency level. The pumping costs (in US dollars) are thus [K34]

$$k_p = 1200 Q \Delta P \qquad (6.6)$$

It was found that the matrix must be replaced every 4000 hours or 10 000 filtration cycles, whichever comes first. Since the price of a steel wool matrix is $ 28/kg (1982) the cost ot the matrix becomes

$$k_f = 8900 \max\left\{\frac{46}{t_D + t_N}, 1.9\right\} LS \qquad (6.7)$$

where t_N (in minutes) is the effective filtration time and t_D the dead time.

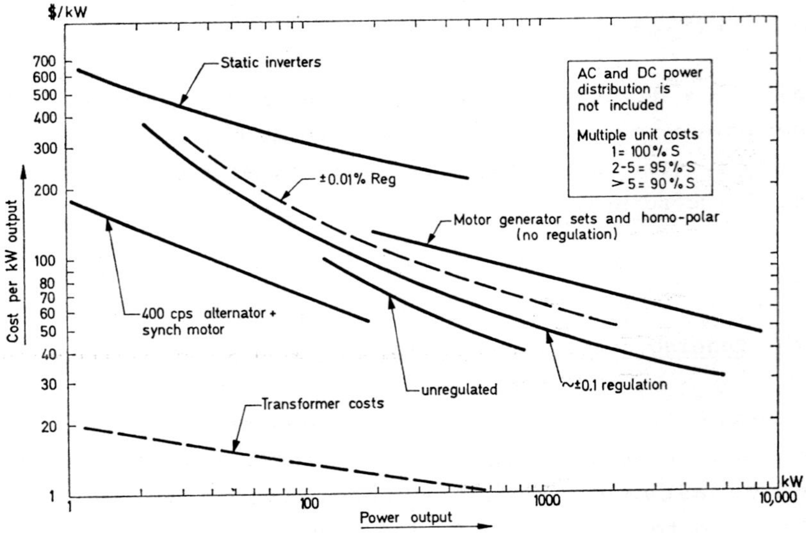

Fig. 6.4 Cost of power supplies (1973), including switches and voltage control (after Brechna [B4]).

VI.3 THE ECONOMICS OF MINERAL TREATMENT BY MAGNETIC SEPARATION

VI.3.1 Iron ore

VI.3.1.1 The production of superconcentrate by HGMS

In order to obtain an iron ore superconcentrate containing less than 2 per cent silica, the Jones separator was used to re-treat the preconcentrate produced by sizing, gravity separation or low-intensity magnetic separation. Table 6.6 shows the capital and operating costs of HGMS. The estimated total capital and operating costs to produce a tonne of superconcentrate by HGMS in 1974 were $ 0.43 for re-treatment of specular hematite spiral concentrate, $ 0.47 for Brazilian hematite fines and $ 0.49 for magnetite; cost analysis is given in Table 6.7.

Tables 6.8 and 6.9 summarize the costs of magnetic separation of the itabirite ore with the Boxmag Rapid high-gradient magnetic separators SHWI(single-pole model with a capacity of 6.5 t/h) and SHWX8 (8-pole model with a capacity of 104 t/h). The data are based on the annual throughput of 57 000 tonnes for the SHWI model and 911 000 tonnes for the SHWX8 model. The magnetic product would amount to 28 470 tonnes and 455 500 tonnes per year, respectively.

In comparison with the costs for the Jones machine, the Boxmag Rapid figures refer to operation in Great Britain where the capital charges were based on interest at 15 per cent and depreciation over 10 years.

VI.3.1.2 Beneficiation of semitaconite ore, Mesabi

Lawver and Hayes [L22] carried out a comparative study of beneficiation of taconite and semitaconite ores from the Mesabi deposit (USA). Four beneficiation processes were investigated, namely:

a) wet low-intensity magnetic separation
b) magnetizing roasting + LIMS + cationic flotation
c) anionic flotation
d) HGMS + flotation.

TABLE 6.6

Capital and operating costs for the production of iron ore superconcentrate by HGMS (1974) [B32]

Cost description	Superconcentrate production from spiral concentrate (400,000 tpy feed 339,200 tpy conc.)		Superconcentrate production from Brazilian hematite ore (600,000 tpy feed 315,000 tpy conc.)		Superconcentrate production from magnetite ore (650,000 tpy feed 297,700 tpy conc.)		Comments
	¢/ton feed	¢/ton conc.	¢/ton feed	¢/ton conc.	¢/ton feed	¢/ton conc.	
Equipment fob Montreal	14.1	16.7	9.4	17.9	8.7	19.0	taking into account service life of 15 years, interest rate of 10%;
Building and installation	3.8	4.4	2.5	4.8	2.3	5.1	
Capital cost	17.9	21.1	11.9	22.7	11.0	24.1	
Labor	2.3	2.7	1.5	2.9	1.4	3.0	
Power	4.5	5.3	3.0	5.7	2.8	6.1	
Water	4.3	5.0	2.8	5.4	2.6	5.7	based on costs prevailing in Canada;
Supplies	5.0	5.9	3.4	6.3	3.1	6.7	
Maintenance	2.7	3.2	1.8	3.5	1.7	3.7	
Operating cost	18.8	22.1	12.5	23.8	11.6	25.2	
Capital and operating cost	36.7*	43.2*	24.4	46.5	22.6	49.3	*cost based on double pass; sometimes single pass sufficient with corresponding lower costs; e.g. feed/spiral concentrate (800,000 tpy feed; 678,400 tpy conc.):

Cost description	¢/ton *feed*	¢/ton *conc.*
capital	8.9	10.6
operating	9.4	11.0
TOTAL	18.3	21.6

TABLE 6.7

Capital and operating costs for the DP 317 Jones separator (valid for Quebec, Canada, in US dollars, 1974, and Sweden, in Swedish krone, 1977) [B32]

	US dollars (1974)	Swedish krone (1977)
Capital costs		
A. Equipment		
1 Jones separator DP 317	386 000	3 150 000
6 Centrifugal pumps	24 000	105 000
Freight charge	20 000	84 000
Total equipment	430 000	3 339 000
B. Building and installation		
Ground area	20 000	100 000
Construction work	95 000	526 000
Total building and installation	115 000	626 000
C. Total capital investment	545 000	3 965 000
D. Annual depreciation and interest (interest rate 10 %, service life 15 years)	71 653	521 292
Annual operating costs		
A. Labour (20 % man-day)	9 000	44 100
B. Power (1 800 MWh at \$0.01/kWh or Skr 0.06/kWh)	18 000	113 400
C. Water (4 x 10^6 m^3/y)	17 000	84 000
D. Grooved plates	20 000	147 000
E. Maintenance	11 000	63 000
F. Total operating cost per year	75 000	451 500
Total capital and operating cost per year	146 653	972 292

The cost analysis of the four processes is given in Table 6.10.

The cost analysis assumes the annual production of 2.03 x 10^6 tonnes of pellets, 20 shifts per months, 52 weeks per year, wages \$ 4.1 per hour, energy cost \$ 0.01 per kWh, price of diesel fuel \$ 0.04 per litre and the cost of technological fuel \$ 0.1 per 1 GJ.

The metallurgical investigation of a semitaconite ore has shown that neither magnetizing-roasting followed by low-intensity magnetic separation, nor anionic flotation of quartz was able to give a high-grade concentrate comparable to that from magnetic taconites. Magnetizing roasting, followed by low-intensity magnetic

TABLE 6.8
Capital and operating costs for SHW Boxmag Rapid magnetic separators (valid for Great Britain, in British pounds (£), 1976) [E8]

	British pounds (£)	
	SHWI	SHWX 8
Capital costs		
Magnetic separator	17 250	100 000
Pumps	600	9 600
Buildings	3 600	11 700
Installation and start-up	900	2 925
Total	22 350	124 225
Depreciation and interest cost (interest rate 15 %, service life 10 years)	4 080	22 589
Annual operating costs		
Labour	365	730
Electrical power (at £ 0.017/kWh)	1 485	15 145
Maintenance and spares	1 295	7 500
Water requirement (m^3/y)	0.22	3.4
Water (at £ 0.06/m^3)	6 811	108 960
Total operating costs	9 956	132 335

TABLE 6.9
Capital and operating costs for beneficiation of itabirite iron ore by SHW magnetic separators (in British pounds, 1976) [E8]

	SHWI, cost (£)		SHWX 8, cost (£)	
	Tonne feed	Tonne conc.	Tonne feed	Tonne conc.
Capital	0.07	0.14	0.02	0.04
Operating	0.18	0.36	0.15	0.30
Total	0.25	0.50	0.17	0.34

separation and flotation, provided a high-grade concentrate at a recovery equal to 92 per cent.

Wet high-gradient magnetic separation, together with flotation of quartz, resulted in a high-grade concentrate at a recovery of iron of 75 per cent. Capital and running costs were lower than for magnetizing-roasting, followed by LIMS and flotation. On the other hand, recovery of iron was much lower than in the process comprising magnetizing-roasting, LIMS and flotation.

TABLE 6.10

Cost analysis of beneficiation of Mesabi taconite ores (1967) [L22]

	Taconite ore	Semitaconite ore		
	LIMS	Magnetizing roasting + flotation	Anionic flotation	HGMS + flotation
Capital investment (in US M$)				
General investment	6.27	4.72	5.18	5.28
Mining	8.20	4.77	5.50	5.68
Comminution	7.80	4.77	5.27	5.39
Magnetizing roasting	-	24.90	-	-
Beneficiation	11.70	10.77	14.49	19.37
Pelletizing	15.61	17.45	19.28	19.28
Transport of pellets	0.95	0.95	0.95	0.95
Tailings plant	1.79	1.25	1.40	1.39
Total capital investment	52.31	69.57	52.07	57.34
Operating costs (US $ per tonne of pellets)				
Mining	2.12	1.14	1.31	1.32
Comminution	0.96	0.65	0.70	0.71
Magnetizing roasting	-	2.37	-	-
Beneficiation	1.83	1.48	3.28	2.81
Pelletizing	1.61	1.78	1.96	1.96
Depreciation	2.57	3.26	2.48	2.73
Total operating cost	9.09	10.68	9.73	9.56

VI.3.2 Recovery of uranium and gold from leach residues

The application of HGMS for the concentration of gold and uranium was tested on West Rand cyanidation residues. The result indicated that recoveries of over 60 per cent of gold and uranium can be obtained [C7, C23]. The economic analysis of a HGMS plant treating cyanide residue, at a rate of 180 t/h, was developed by Corrans and Levin [C7].

The total capital cost of the HGMS plant was 4 x 10^6 rand (1979), which included the cost of two HGMS machines to treat 180 t/h of material (60 t/h per machine, 60 t/h slimes), together with two thickeners. A breakdown of the costs is given in Table 6.11.

Operating costs were obtained from the manufacturers of HGMS machines and are shown in Table 6.12.

TABLE 6.11

Capital costs (1979) for HGMS uranium-gold plant (180 t/h) [C7]

Item	Cost, kR
2 HGMS machines, 60 t/h, each with ball cleaning facilities	1200
2 thickeners	908
Other capital and installation costs	1065
Project personnel	373
Contractor's fee	220
Project contingency	200
Insurance	12
Final contract price	4000

TABLE 6.12

Operating costs (1979) of HGMS machines [C7]

HGMS machine	Total operating cost c/t
Eriez CF-60	15.0
Jones	13.8
SALA HGMS	13.7

Depreciation at 12 per cent over 15 years yields a cost of 47 South African cents per tonne treated. If allowance is made for non-availability of plant, this figure can be increased to 54 cents per tonne. Total cost of treatment by HGMS is thus 69 cents per tonne.

A production-size Eriez CF-60 machine was installed at a gold mine; details of the operating costs over a 12-month period during 1981-82 are given in Table 6.13.

An example of the economics of a HGMS installation at a gold mine is given in Table 6.14.

In the re-treatment of the tailings dumps, the flowsheet shown in Fig. 5.20 is considered to be optimum. It can be seen that the final tailings from such a circuit would be 0.14 g/t, while the extra gold recovery, due to the incorporation of the HGMS-regrinding operation, is 0.08 g/t. The economics for the recovery of this extra gold has been estimated in Table 6.15.

TABLE 6.13
Operating costs of HGMS at a gold mine (1981/82), in S.A. cents [C23]

Item	kRand	% of total	Cost c/t *)
Energy (at 1.64 cents/kWh)	37.6	16.9	11.5
Balls	23.3	10.6	7.2
Water	22.8	10.3	7.0
Wages	50.0	22.6	15.4
Matrix belting	3.6	1.6	1.1
Maintenance	9.9	4.4	3.0
Screens and pumps	60.4	27.4	18.6
Consumables	13.7	6.2	4.2
Total (desliming + HGMS)	221.2	100.0	68.0
Total (HGMS only)			44.0

*) per tonne of plant feed to the desliming cyclones
- cost of energy for HGMS machine itself 2.4 c per tonne

TABLE 6.14
An example of potential profitability of HGMS on a gold mine [C23]. (All costs (1981/82) are per tonne of plant feed, based on an operation of 400 kt per month)

Item	Au		U_3O_8
Head grade, g/t	0.5		175
Recovery by HGMS, %	43.2		32.2
Recovery by existing leach, %	90		90
Recovery by HGMS, g/t	0.19		51
Gross revenue, R/t	2.85		4.10
Operating cost of HGMS and desliming, R/t		0.54	
Operating cost of regrinding, R/t		0.11	
Leaching cost, R/t	0.30		0.50
Total operating cost, R/t		1.45	
Net operating profit, R/t		5.50	
Estimated amortization, R/t		0.60	
Profit, R/t		4.90	

TABLE 6.15
The economics of gold recovery by HGMS at Ergo (1984) [C36]

Unit operation	Treated feed (% by mass)	Cost of treatment (c/t)	Cost based on feed (c/t)
Desliming	100	30	30
HGMS	50	30	15
Milling	5	250	13
Total cost			58
Income from 0.08 g of gold per tonne at R15 per gram			120
Expected profit			62

An ideal HGMS machine for use in a large operation for the re-treatment of gold tailings should have the following features [C36]:

(1) The capacity of each individual machine must be large, i.e. of the order of 250 t/h. For Ergo, five such machines would be needed

(2) The capital cost of the machine must not exceed R3000 per tonne of capacity, i.e. for 250 t/h it must not exceed R 750 000 (1984 price)

(3) The operating costs of the machine must not exceed 30 cents per tonne of ore treated. As power constitutes about one-half of this figure, the energy consumption of the magnet must be optimized

(4) The machine must be so designed that blockage of its matrix by ferromagnetic matter and foreign material is avoided.

VI.3.3 The beneficiation of fluorspar

The operating costs of a carousel Eriez CF-60 magnetic separator, installed at Buffalo Fluorspar (S.A.) to reduce phosphate in the final concentrate, are summarized in Table 6.16. The consumption of balls used for the matrix is given in Table 6.17, while Table 6.18 shows the availability of the machine. In 1981 and 1982, new 6.3 mm balls were added to maintain the effective ball load in the separator. Since more than 60 per cent of the original mass was discarded, the operators of the HGMS changed to 7 mm balls and consumption dropped from approximately 460 kg/month to 260 kg/month. The higher consumption during 1985 is due to a change of supplier but cost per tonne is lower [J8].

Table 6.18 indicates that a slight increase in availability occurred during the years and a figure above 96 per cent can be expected. The magnetic conveyor belt and pumps are the major maintenance items.

TABLE 6.16

Operating costs of HGMS at a fluorspar mine (South African cents, 1982 and 1985) [J8]

Item	1982		1985	
	c/t	%	c/t	%
Power	17.3	41.4	26.3	34.1
Balls	9.6	22.9	26.1	33.8
Water	5.3	12.8	3.0	3.9
Wages	4.1	9.8	12.5	16.2
Belting	2.7	6.4	2.7	3.5
Maintenance	2.8	6.7	6.6	8.5
Total	41.8	100	77.2	100

TABLE 6.17

Consumption of ball matrix in HGMS operating at a fluorspar mine [J8]

Year	kg/month
1981	450
1982	470
1983	251
1984	270
1985	360

TABLE 6.18

Availability of HGMS machine at the fluorspar mine [J8]

Year	Hours	%
1980	7 888	89.8
1981	8 192	93.5
1982	8 164	93.2
1983	8 410	96.0
1984	8 491	96.4
1985	8 494	96.7

VI.3.4 The beneficiation of pyroxenite ore

The Phosphate Development Corporation (Foskor) has been producing apatite concentrate from the Phalaborwa complex by flotation for the past 30 years. Currently, 23 million tonnes of ore are treated annually to produce 3.4 million tonnes of apatite concentrate. From the data gathered during laboratory and pilot-plant tests, as well as from available in-house data regarding ore flotation costs, a feasibility study was generated in which the flotation route, together with the four dry high-intensity magnetic separation routes, was compared. A hypothetical new mine and plant, based on the southern pyroxenite ore-body, and utilizing the latest technology, such as in-pit mobile crushers and large-diameter autogenous mills, were costed [R3]. The results of the study are given in Table 6.19.

All the above cases refer to a plant processing 1000 tonnes of ore per hour at an availability of 88.5 per cent, to produce 1 million tonnes of apatite concentrate per annum. In this study, 20 per cent of the cost of flotation was allocated to each dry process option, thus reflecting indications that up to 20 per cent of the milled ore may be too fine for dry high-intensity magnetic separation and that it must be treated by flotation. Cost estimates based on dry magnetic separation, with one based on conventional flotation, indicate a similar level of capital investment, with savings in the region of 50 per cent in the cost of beneficiation and 20 per cent in total cost. With power being the only directly variable cost of beneficiation, the exposure to variations in the direct cost of consumables will be less than in the case of flotation.

VI.3.5 The beneficiation of kaolin

A high-gradient magnetic separator has been successfully integrated into the production process of high-grade ceramic kaolins. The estimated operating costs including amortization, energy, labour and maintenance of a 2.1 m-diameter unit are given in Table 6.20.

It is apparent from these figures that amortization accounts for over 50 per cent of the total cost and magnet energy consumption for only 22 per cent of the cost.

Magnetic separation is a nominal cost unit operation in wet processing that compares favourably with alternative beneficiation methods such as chemical leaching (reagent cost $1.20 to $ 6.00 per tonne [I7]), selective flocculation and froth flotation. Magnetic separation is less costly or of the same magnitude as other accessory unit operations in a wet flowsheet, including makedown, classification, filtration and drying. Besides the nominal cost of magnetic separation, in many cases there is simply no other viable substitute for this process other than selective mining.

TABLE 6.19

Projected economics for various routes of beneficiation of pyroxenite ore (1984) [R3]

Index basis: flotation route = 100

	Flotation	D.H.I.M.S. options			
		Sala H.G.M.S.	ELB-Yaniv	I.M.R.	Cryofos GLF
Fixed cost					
Salaries	3,51	3,51	3,51	3,51	3,51
Overheads	9,89	9,89	9,89	9,89	9,89
Depreciation	23,40	24,49	22,83	23,60	21,41
Total fixed cost	36,80	37,89	36,23	37,00	34,81
Variable cost					
Mining	24,12	24,12	24,12	24,12	24,12
Beneficiation	28,77	15,21	20,84	22,98	13,18
Filtration	1,46	0,28	0,28	0,28	0,28
Drying	4,67	0,94	0,94	0,94	0,94
Tailings disposal	3,17	0,65	0,65	0,65	0,65
Dispatch	1,01	1,01	1,01	1,01	1,01
Total variable cost	63,20	42,21	47,84	49,98	40,18
Total production cost	100,00	80,10	84,07	86,98	74,99
Capital investment	100	105	97	100	91

TABLE 6.20

Operating costs of a 2.1 m HGMS separator, in US dollars, 1982 [I7]

Item	Cost per hour	Cost per tonne	
		General use 60 t/h	Special use 15 t/h
Amortization (separator M$ 1.7 installation M$ 1.7 over 10 years 80 000 hours)	34.00	0.52	2.26
Energy consumption			
Magnet (270 kW at 5c per kWh)	13.50	0.20	0.90
Pumping and flushing (200 kW at 5c per kWh)	1.60	0.02	0.11
Labour	10.00	0.15	0.67
Maintenance	3.20	0.05	0.21
Total	62.30	0.94	4.15

APPENDIX I

LIST OF SYMBOLS

A

$\mathbf{a}$	Radius of matrix element (m)
$\mathbf{a}_c$	Centrifugal acceleration $(m\ s^{-2})$
$\mathbf{a}_d$	Radius of matrix element with deposit (m)
$\mathbf{a}_m$	Magnetic acceleration $(m\ s^{-2})$
$\mathbf{a}_p$	Acceleration $(m\ s^{-2})$
$\mathbf{A}$	Hamaker constant (J)

B

$\mathbf{b}$	Particle radius (m)
$\mathbf{b}_a$	Normalized particle radius
$\mathbf{B}$	Magnetic induction (T)
$\mathbf{B}_o$	Background magnetic induction (T)
$\mathbf{B}_c$	Critical magnetic induction (T)
$\mathbf{B}_F^p$	Threshold magnetic induction for flocculation into the primary minimum (T)
$\mathbf{B}_F^s$	Threshold magnetic induction for flocculation into the secondary minimum (T)

C

$\mathbf{c}$	Concentration of particles (m^{-3})
$\mathbf{c}_i$	Concentration of particles of type **i** (m^{-3})
$\mathbf{c}_{esc}$	Concentration of particles that did not collide with matrix collectors (m^{-3})
$\mathbf{c}_{in}$	Concentration of particles entering the filter (m^{-3})
$\mathbf{c}_{out}$	Concentration of particles leaving the filter (m^{-3})
$\mathbf{C}$	Curie constant (K^{-1})

D

$\mathbf{d}$	Diameter of a particle (m)
$\mathbf{d}_m$	Magnetic displacement (m)
$\mathbf{d}_s$	Distance of a splitter from the cryostat wall (m)

$\mathbf{D}$	Demagnetization factor
$\mathbf{D}_a$	Demagnetization factor in the direction **a**
$\mathbf{D}_i$	Internal demagnetizing factor
$\mathbf{D}_p$	Diffusion coefficient ($m^2\ s^{-1}$)

E

$\mathbf{e}$	Electronic charge (C)
$\mathbf{E}$	Intensity of electric field ($V\ m^{-1}$)
$\mathbf{E}_c$	Concentration ratio

F

$\mathbf{f}$	Filling factor (%)
$\mathbf{f}_c$	Critical revolution rate (r/min)
$\mathbf{f}_s$	Coefficient of friction
$\mathbf{F}(\alpha, \beta)$	Field factor
$\mathbf{F}_d$	Hydrodynamic drag (N)
$\mathbf{F}_g$	Force of gravity (N)
$\mathbf{F}_m$	Magnetic force (N)

G

$\mathbf{g}$	Acceleration of gravity ($m\ s^{-2}$)
$\mathbf{G}$	Grade (%)

H

$\mathbf{h}$	Height of a matrix (m)
$\mathbf{H}$	Magnetic field strength ($A\ m^{-1}$)
$\mathbf{H}_c$	Coercive force ($A\ m^{-1}$)
$\mathbf{H}_d$	Demagnetizing magnetic field ($A\ m^{-1}$)
$\mathbf{H}_i$	Internal magnetic field ($A\ m^{-1}$)
$\mathbf{H}_o$	Applied magnetic field strength ($A\ m^{-1}$)
$\mathbf{H}_r$	Radial component of **H** ($A\ m^{-1}$)
$\mathbf{H}_s$	Saturation magnetic field strength ($A\ m^{-1}$)
$\mathbf{H}_\theta$	Azimuthal component of **H** ($A\ m^{-1}$)

I

$\mathbf{I}$	Electric current (A)

J

j	Current density	$(A\ m^{-2})$
J	Magnetic polarization	(T)
J_D	Flux of particles as a result of diffusion	$(m^{-2}\ s^{-1})$
J_e	Exchange integral	(J)
J_F	Flux of particles as a result of external forces	$(m^{-2}\ s^{-1})$
J_p	Flux of particles	$(m^{-2}\ s^{-1})$
J_R	Remanent polarization	(T)
J_S	Saturation polarization	(T)

K

k	Boltzmann constant	$(J\ K^{-1})$
K_a	Constant of anisotropy	

L

l	Length	(m)
l_f	Mean free path	(m)
l_g	Effective length of field path	(m)
l_m	Length of magnetic zone of separator	(m)
l_p	Length of pole	(m)
l_t	Length of tortuous path	(m)
L	Length of coil	(m)
L_F	Matrix loading	$(kg\ m^{-3})$
L_M	Matrix loading	$(kg\ m^{-3})$
L(x)	Langevin function	

M

m	Mass	(kg)
m_i	Atomic mass of ion	
m_o	Rest mass of electron	(kg)
M	Magnetization	$(A\ m^{-1})$
M_s	Saturation magnetization	$(A\ m^{-1})$

N

n	Volume of slurry in units of canister volume	
n(V)	Fractional alignment	

$\mathbf{n_c}$	Number of particle collisions
$\mathbf{n_1}$	Volume of fluid (in units of canister volume) used to displace slurry from the matrix
$\mathbf{N}$	Number of turns of a coil
$\mathbf{N_a}$	Number of atoms per unit volume
$\mathbf{N_L}$	Loading number
$\mathbf{N_o}$	Initial concentration of particles (m^{-3})
$\mathbf{N_s}$	Saturation concentration of captured particles (m^{-3})
$\mathbf{N}(x,t)$	Number of captured particles between $\mathbf{x}$ and $\mathbf{x} + \Delta \mathbf{x}$, at time $\mathbf{t}$

P

$\mathbf{p}$	Packing density
$\Delta \mathbf{p}$	Pressure drop (Pa)
$\Delta \mathbf{p_m}$	Pressure drop in loaded matrix (Pa)
$\Delta \mathbf{P}$	Pressure difference (Pa)

Q

$\mathbf{q}$	Leakage factor
$\mathbf{Q}$	Capacity of a separator (t/h)
$\mathbf{Q_v}$	Volumetric flowrate ($m^3\ s^{-1}$)

R

$\mathbf{r}$	Radius (m)
$\mathbf{r_a}$	Normalized radius
$\mathbf{r_{aq}}$	Normalized radius where the fluid velocity attains its maximum value
$\mathbf{R}$	Gas constant ($J\ K^{-1}\ mol^{-1}$)
	Resistance (Ω)
	Radius (m)
	Recovery (%)
$\mathbf{R_c}$	Capture radius (m)
$\mathbf{R_{ca}}$	Normalized capture radius
$\mathbf{R_h}$	Hydraulic radius (m)
$\mathbf{R_m}$	Mass yield into magnetic product (%)
$\mathbf{Re}$	Reynolds number

S

$\mathbf{s}$	Active cross-sectional area of the winding of a coil (m^2)
$\mathbf{S}$	Cross-sectional area (m^2)
$\mathbf{S}_v$	Coefficient of magnetic viscosity

T

$\mathbf{t}$	Time (s)
$\mathbf{t}_d$	Thickness of a deposit (m)
$\mathbf{t}_D$	Dead time (s)
$\mathbf{t}_N$	Duration of separation in one cycle (s)
$\mathbf{t}_R$	Retention time (s)
$\mathbf{t}_S$	Thickness of a stream (s)
$\mathbf{t}_T$	Duration of one cycle (s)
$\mathbf{t}_{1/2}$	Half-time of flocculation (s)
$\mathbf{T}$	Absolute temperature (K)
$\mathbf{T}_c$	Curie temperature (K)
	Critical temperature (K)

U

$\mathbf{u}$	Particle mobility (kg^{-1} s^{-1})

V

$\mathbf{v}$	Interstitial velocity ($m\ s^{-1}$)
$\mathbf{v}_c$	Velocity of rotation of carousel ($m\ s^{-1}$)
$\mathbf{v}_d$	Drift velocity ($m\ s^{-1}$)
$\mathbf{v}_m$	Magnetic velocity ($m\ s^{-1}$)
$\mathbf{v}_s$	Flow velocity of slurry ($m\ s^{-1}$)
$\mathbf{v}_t$	Terminal velocity ($m\ s^{-1}$)
$\mathbf{v}_o$	Superficial velocity ($m\ s^{-1}$)
$\mathbf{V}$	Volume (m^3)
$\mathbf{V}_A$	London-van der Waals interaction energy (J)
$\mathbf{V}_M$	Energy of magnetic interaction (J)
$\mathbf{V}_{max}$	Height of a potential barrier (J)
$\mathbf{V}_{min}$	Depth of secondary minimum (J)
$\mathbf{V}_o$	Volumetric flow ($m^3\ s^{-1}$)
$\mathbf{V}_R$	Energy of electric double layer interaction (J)
$\mathbf{V}_T$	Total interaction energy (J)

W

w	Width of the matrix	(m)
W_{ij}	Stability ratio for flocculation of particle species i and j	
W_p	Stability ratio for primary minimum	
W_s	Stability ratio for secondary minimum	

X

x	Debye-Hückel factor	(m^{-1})

Z

z	Number of electrons in atom	
z_a	Normalized z-coordinate	

γ_c	Correction factor	
γ_m	Mass-normalized volume of build-up	
γ_V	Volume-normalized volume of build-up	

δ	Thickness of boundary layer	(m)
δ_d	Duty cycle	

ϵ	Porosity	(%)
ϵ_b	Porosity of a deposit	(%)
ϵ_d	Dielectric constant	$(F\ m^{-1})$
ϵ_i	Porosity of clean matrix	(%)
ϵ_o	Dielectric constant of vacuum	$(F\ m^{-1})$
ϵ_r	Relative dielectric constant	

η	Dynamic viscosity	(Pa·s)
η_c	Collection efficiency of a single collector	

θ_c	Critical angle
θ_o	Angular spacing between magnets in a drum separator
θ_s	Slip angle
κ	Volume magnetic susceptibility
κ_d	Diamagnetic susceptibility
κ_f	Magnetic susceptibility of fluid
κ_i	Intrinsic magnetic susceptibility
κ_M	Magnetic susceptibility of a mixture
κ_p	Paramagnetic susceptibility
κ_∞	Magnetic susceptibility extrapolated to infinite field strength
λ	Filling factor of the coil winding (%)
λ	Filter coefficient
λ_a	Aspect ratio
Λ	Reluctance (H^{-1})
μ	Magnetic permeability ($H\ m^{-1}$)
μ_B	Bohr magneton ($J\ T^{-1}$)
μ_f	Permeability of fluid ($H\ m^{-1}$)
μ_m	Magnetic moment ($T\ m^3$)
μ_o	Permeability of free space ($H\ m^{-1}$)
μ_r	Relative permeability
μ_W	Permeability of matrix material ($H\ m^{-1}$)
ν	Kinematic viscosity ($m^2\ s^{-1}$)
ν_m	Fraction of particle volume consisting of magnetite
ν_r	Frequency of rotation (s^{-1})

ρ	Density	$(kg\ m^{-3})$
ρ_b	Bulk density of feed	(%)
ρ_f	Density of fluid	$(kg\ m^{-3})$
ρ_p	Density of particle	$(kg\ m^{-3})$
ρ_R	Resistivity	$(\Omega\ m)$
ρ_w	Density of matrix material	$(kg\ m^{-3})$
ρ_s	Density of a suspension	$(kg\ m^{-3})$
σ	Surface charge density	$(C\ m^{-2})$
σ_d	Specific deposit (volume of deposited matter per unit volume of matrix)	
σ_m	Specific polarization	$(T\ m^3\ kg^{-1})$
σ_t	Surface tension	$(N\ m^{-1})$
τ_o	Shear stress at the wall	$(m^{-1}\ kg\ s^{-2})$
τ_R	Relaxation time	(s)
ϕ	Magnetic flux	(Wb)
χ	Specific (mass) magnetic susceptibility	$(m^3\ kg^{-1})$
χ_M	Magnetic susceptibility of a mixture	$(m^3\ kg^{-1})$
χ_∞	Magnetic susceptibility extrapolated to infinite field strength	$(m^3\ kg^{-1})$
ψ	Surface electrostatic potential	(V)
ω	Angular velocity	$(rad \cdot s^{-1})$

APPENDIX II

LIST OF ABBREVIATIONS

BOF	Basic oxygen furnace
BWR	Boiling water reactor
DHIMS	Dry high-intensity magnetic separation
DLVO	Derjaguin-Landau-Verwey-Overbeek theory
EAF	Electric arc furnace
HGMF	High-gradient magnetic filtration
HGMS	High-gradient magnetic separation
HIMS	High-intensity magnetic separation
HIWMS	High-intensity wet magnetic separation
IFP	Insoluble fission product
IMR	Induced magnetic roll separator
LIMS	Low-intensity magnetic separator
MS	Magnetic separation
OGMS	Open-gradient magnetic separation
PGM	Platinum-group metals
PM	Primary minimum
PWR	Pressurized water reactor
PZC	Point of zero charge
SEK	Swedish krone
SM	Secondary minimum
SRC	Solvent-refined coal
WHIMS	Wet high-intensity magnetic separation

APPENDIX III

VALUES OF PHYSICAL CONSTANTS

Magnetic permeability of a vacuum	μ_o	4 x 10^{-7} H m^{-1}
Dielectric constant (permittivity of a vacuum)	ϵ_o	8.854 x 10^{-12} F m^{-1}
Avogadro constant	N_a	6.022 x 10^{23} mol^{-1}
Rest mass of electron	m_o	9.1096 x 10^{-31} kg
Bohr magneton	μ_B	9.274 x 10^{-24} J T^{-1}
Gas constant	R	8.314 J K^{-1} mol^{-1}
Boltzmann constant	k	1.3806 x 10^{-23} J K^{-1}
Elementary charge	e	1.6022 x 10^{-19} C
Gravitational acceleration	g	9.807 m s^{-2}
Faraday constant	F	9.649 x 10^{4} C mol^{-1}

APPENDIX IV

CONVERSION OF UNITS

Length		
1 Å	=	10^{-10} m
1 μm	=	10^{-6} m
Area		
1 ft^2	=	0.0929 m^2
Density		
1 lb/ft^3	=	16.019 kg m^{-3}
Flowrate		
1 ft^3/min	=	1.699 m^3 h^{-1}
1 Imp. gpm	=	0.2728 m^3 h^{-1}
1 US gpm	=	0.2273 m^3 h^{-1}
Mass		
1 tonne	=	1000 kg
1 long ton	=	1016 kg
1 short ton	=	907 kg
Pressure		
1 psi	=	0.0681 atm
	=	0.0689 bar
	=	6.895 x 10^3 Pa
	=	51.71 mm Hg
Viscosity		
1 centipoise (cP)	=	1 g m^{-1} s^{-1}
	=	10^{-3} N s m^{-2}
Energy		
1 erg	=	0.1 μJ
1 cal	=	4.187 J
1 BTU	=	1.055 kJ
Power		
1 hp	=	0.746 kW

APPENDIX V

DEFINITIONS OF DERIVED UNITS

Physical quantity	SI unit		cgs unit	
Energy	joule	$m^2kg\ s^{-2}$	erg	$cm^2g\ s^{-2}$
Force	newton	$m\ kg\ s^{-2}$	dyne	$cm\ g\ s^{-2}$
Pressure	pascal	$m^{-1}kg\ s^{-3}$		$g\ cm^{-1}s^{-2}$
Power	watt	$m^2kg\ s^{-3}$		$g\ cm^2s^{-3}$
Electric charge	coulomb	s A		$g^{1/2}cm^{1/2}$
Electric potential difference	volt	$m^2kg\ s^{-3}A^{-1}$		$g^{1/2}cm^{3/2}s^{-2}$
Electric field strength		$kg\ m\ s^{-3}A^{-1}$		$g^{1/2}cm^{1/2}s^{-2}$
Electric resistance	ohm	$m^2kg\ s^{-3}A^{-2}$		$cm\ s^{-1}$
Electric capacitance	farad	$m^{-2}kg^{-1}s^4A^2$		$cm^{-1}\ s^2$
Magnetic flux	weber	$m^2kg\ s^{-2}A^{-1}$	maxwell	$g^{1/2}cm^{3/2}s^{-1}$
Magnetic induction	tesla	$kg\ s^{-2}A^{-1}$	gauss	$g^{1/2}cm^{-1/2}s^{-1}$
Magnetic field strength		$A\ m^{-1}$	oersted	$g^{1/2}cm^{-1/2}s^{-1}$
Permeability		$kg\ m\ s^{-2}A^{-2}$		1
Inductance	henry	$m^2kg\ s^{-2}A^{-2}$		cm
Reluctance		$kg^{-1}m^{-2}s^2A^2$		cm^{-1}

APPENDIX VI

LIST OF SELECTED COMPANIES MANUFACTURING MAGNETS, MAGNETIC SEPARATORS, MAGNETIC MEASURING INSTRUMENTS AND MATRICES

1. Magnets and magnetic separators

Applied Magnetic Systems, Inc.
P.O. Box 20911
GREENFIELD
WI. 53220
U.S.A

Aquafine Corp., Inc.
157 Darien Highway
BRUNSWICK
GA. 31520
U.S.A.

Bartles (Carn Brea) Ltd.
North Street
REDRUTH TR15 1HJ
Cornwall
England

Boxmag-Rapid Ltd.
Chester Street
Aston
BIRMINGHAM B6 4AJ
England

Carpco, Inc.
4120 Haines St.
JACKSONVILLE
Fla. 32206
U.S.A.

Cryogenics Consultants, Ltd.
Metrostore Building
231 The Vale
LONDON W3 7QS
England

E.L. Bateman, Ltd.
P.O. Box 565
BOKSBURG 1460
South Africa

Eriez Magnetics, Inc.
Asbury Road at Airport
ERIE
PA. 16514
U.S.A.

Intermagnetics General Corp.
P.O. Box. 566
GUILDERLAND
N.Y. 12084
U.S.A

International Research and Development Co., Ltd.
Fossway
NEWCASTLE upon Tyne NE6 2YD
England

Jones Ferro-Magnetics, Inc.
Oberneuhofstrasse 7
CH-6340 BAAR
Switzerland

KHD Humboldt Wedag AG
Herner Str. 299
D-4630 BOCHUM 1
West Germany

Kone OY
SF-15870 SALPAKANGAS
Finland

Krupp Polysius AG
P.O. Box 2340
D-4720 BECKUM
West Germany

Magnetic Corporation of America
179 Bear Hill Rd.
WALTHAM
Mass. 02154
U.S.A.

Ore Research Institute
252 10 MNÍŠEK pod Brdy
Czechoslovakia

Ore Sorters (Pty.) Ltd.
P.O. Box 565
BOKSBURG 1460
South Africa

Outokumpu OY
P.O. Box 27
02101 ESPOO 10
Finland

Oxford Instruments Ltd.
Osney Mead
OXFORD OX2 ODX
England

Pacific Electric Motor Co.,Inc.
1009 - 66th Avenue
OAKLAND
CA. 94621
U.S.A.

PROMEFEX
Z.I. VILLERS - LA MONTAGNE
54920 France

Readings of Lismore (Pty.) Ltd.
P.O. Box. 161
LISMORE
N.S.W. 2480
Australia

SALA International AB
S-73300 SALA
Sweden

SALA Magnetics, Inc.
247 Third St.
CAMBRIDGE
Mass. 02142
U.S.A.

S.G. Frantz, Inc.
P.O. Box 1138
TRENTON
N.J. 08606
U.S.A.

STROJEXPORT
Václavské nám. 56
113 26 PRAGUE 1
Czechoslovakia

Vacuumschmelze GmbH
Grüner Weg 37
D-6450 HANAU 1
West Germany

Železnorudné bane n.p.
ŠPIŠSKÁ NOVÁ VES
Czechoslovakia

2. Measuring instruments

Biomagnetic Technologies, Inc.
P.O. Box 210079
SAN DIEGO
CA 92 121
U.S.A.

Cahn Instruments, Inc.
16 207 S. Carmenita Rd.
CERRITOS
CA 90701
U.S.A.

Dowty RFL Industries, Inc.
Powerville Rd.
BOONTON
N.J. 07005
U.S.A.

GEOFYZIKA n.p.
Ječná 29a
612 46 BRNO
Czechoslovakia

LDJ, Inc.
P.O. Box 219
TROY
Mich. 48099
U.S.A.

List-Magnetik GmbH
Max-Lang-Strasse 56
D-7022 LEINFELDEN-ECHTERDINGEN
West Germany

EG & G Princeton Applied Research
P.O. Box 2565
PRINCETON
N.J. 08540
U.S.A.

F.W. Bell, Inc.
6120 Hanging Moss Rd.
ORLANDO
FLA 32807
U.S.A.

Magnet-Physik GmbH
Friedrich-Karl-Str. 243
D-5000 KÖLN 60
West Germany

Walker Scientific Inc.
Rockdale St.
WORCESTER
Mass. 01606
U.S.A.

3. Matrices

Bedfords & Co., Ltd.
74-84 Long Lane
LONDON SE1 4AZ
England

DINXPERLO
P.O. Box 2
7090 AA DINXPERLO
Holland

GANTOIS
B.P. 307
88105 ST. DIE CEDEX
France

G. Bopp & Co. AG
Bachmannweg 20
CH-8046 ZÜRICH
Switzerland

Haver & Boecker
P.O. Box 3320
D-4740 OELDE
West Germany

International Steel Wool Corp.
P.O. Box 1767
SPRINGFIELD
OH 45501
U.S.A.

Knitmesh Ltd.
Sanderstead Station Approach
SOUTH CROYDON CR2 0YY
Surrey
England

BIBLIOGRAPHY

A

A1 A. Aharoni: Traction force on paramagnetic particles in magnetic separators. IEEE Trans.Mag.,MAG-12 (1976), p.234.

A2 U. Andres: Magnetohydrodynamic and magnetohydrostatic methods of mineral separation. John Wiley & Sons, New York and Israel Universities Press, Jerusalem, 1976.

A3 C.C. Alexander, A.N. Thorpe, F.E. Senftle: Basic magnetic properties of bituminous coal. Fuel, 58 (1979), p.857.

A4 B.R. Arvidson, E. Barnea: Recent advances in dry high-intensity permanent magnet separator technology. Proc. 14th Internatl. Min. Proc. Congress, Toronto, 1982.

A5 B.R. Arvidson: New high-intensity roll separator using permanent magnets. CIM Bull., 78 (1985), p.88.

A6 B.R. Arvidson, A.J. Fritz: New inexpensive high-gradient magnetic separator. Proc. 15th Internatl. Min. Proc. Congress, Cannes, 1985, p.317.

A7 B.R. Arvidson, J.A. Oberteuffer, I. Wechsler: Continuous HGMS pilot plant: machine description and mineral processing results. AIME Preprint No. 76B-341 (1976).

A8 Anon: Superconducting high-gradient magnetic separator for kaolin. Mining Journal, 26th April 1985, p.289.

A9 Z. Adamczyk, T. Dabros, J.Czarnecki, T.G.M. van de Ven: Particle transfer to solid surfaces. Adv. Coll. Interface Sci., 19 (1983), p.183.

A10 I.Y. Akoto: Mathematical modelling of HGMS devices. IEEE Trans. Mag., MAG-13 (1977), p.1486.

A11 R.C. Arnolds, R.S. Kopish, D.W. Koch , G.T. Zirps: Electromagnetic filter purification of paper mill condensate. TAPPI, 60 (1977), p.107.

A12 N.A. Aleynikov, P.A. Usacheva, P.I. Zelenov: New method and apparatus for separation of magnetite from the slurry (in Russian). Obogascheniye bednykh rud, Nauka, Moscow, 1973, p.111.

A13 W.M. Aubrey, R.M. Funk: The quadrupole magnetic separator - a new concept in magnetic separation equipment. Proc. 32nd Ann. Min. Symp., Duluth, Minn., 1971, p.55.

A14 Anon: Magnetic drum separator. Min. Mag., September 1985, p.238.

A15 Anon: New superconducting magnet takes a bite out of energy costs, improves efficiency. E & MJ, August 1985, p.49.

A16 M.E. Arellano: pH contribution to HGMS recovery of tin ores. Digest of Intermag Conf. Canada, 1983.

A17 Yu.I. Azbel, A.I. Akatov, I.S. Kreymerzak: The fundamental factors of dry magnetic separation of feebly magnetic ores (in Russian). Proc. 8th Internatl. Min. Proc. Congress, Leningrad, 1968, p.208

A18 S. Arai, T. Shibata: Highly heat-resistant Nd-Fe-Co-B system permanent magnets. IEEE Trans. Mag., MAG-21 (1985), p.1952.

A19 D.M. Allen, R.L. Sargent, J.A. Oberteuffer: Treatment of combined sewer overflows by HGMS. EPA Report EPA-600/2-77-015, 1977.

A20 N.O. Aristaksesyan, V.V. Karmazin: Theory of separative mass transfer of particles in a magnetic field (in Russian). Obogashch. tonkovkraplennykh rud (Ed. P.A. Usachev), Akad. Nauk SSSR, Kol'sk. Fil., Apatity, USSR, 1985, p.88.

A21 M.E. Arellano, G.Z. Zambrana: HGMS applied to tin minerals. IEEE Trans. Mag., MAG-14 (1978), p.488.

A22 D.M. Allen: Treatment of combined sewer overflows by HGMS. On-site testing with mobile pilot plant trailer. EPA Report, EPA-600/2-78-209, 1978.

A23 B. Ande'n, E. Lindgren, D. Moore, J. Nolan: Beneficiation of a Swedish iron ore tailings with a SALA-HGMS. Min. Eng. 34 (1982), p.694.

A24 B.R. Arvidson: Metallurgical and economical advantages of HGMS. Proc. 12th Internatl. Min. Proc. Congress, Sao Paulo, Brazil, 1977, p.1.

A25 M.M.K. Abdelsalam: A centrifugal channel-straight wire superconducting magnetic ore separator. Ph.D. Thesis, University of Wisconsin-Madison, 1982.

A26 M. Abdelsalam, R.W. Boom: Superconductive helical flow magnetic ore separator. IEEE Trans. Mag., MAG-20 (1984), p.1207.

A27 Anon: Mexican iron mine to use Jones wet high intensity magnetic separators. E & MJ, July 1985, p.21.

A28 Anon: The selection and application of magnetic separation equipment. Publ. 60186 (1986), Applied Magnetic Systems, Inc., Wisconsin.

A29 P. Anaud, J.E. Etzel, F.J. Friedlaender: Heavy metals removal by HGMS. IEEE Trans. Mag., MAG-21 (1985), p.2062.

B

B1 C.P. Bean: Hysteresis loops of mixtures of ferromagnetic micropowders. J. Appl. Phys., 26 (1955), p.1381.

B2 S. Blackburn, T.J. Veasey: Effect of heat treatment in air at 820 - 960° C on magnetic properties of wolframite. Trans. Instn. Min. Metall. (Sect. C: Mineral Process. Extr. Metall.), 93 (1984), p.C55.

B3 F.Bitter: The design of powerful electromagnets. Rev. Sci. Instr., 7 (1936), p.479.

B4 H. Brechna: Superconducting magnet systems. Springer-Verlag, Berlin, 1973.

B5 P. Burgardt, M.S. Seehra: Magnetic susceptibility of iron pyrite between 4.2 & 620 K. Solid State Commun., 22 (1977), p.153.

B6 C.P. Bean, J.D. Livingston: Superparamagnetism. J. Appl. Phys. Suppl., 30 (1959), p.120S.

B7 R.M. Bozorth: Ferromagnetism. D. van Nostrand Co., Princeton, 1951.

B8 Magnetism and metallurgy (Ed: A.E. Berkowitz, E. Kneller), Vol.1. Academic Press, New York, 1969.

B9 A. Barone, G. Paterno: Physics and applications of the Josephson effects. J. Wiley & Sons, New York, 1982.

B10 W.F. Brown, Jr.: Analysis of the field of a magnetic separator. Univ. of Minnesota Mines Experimental Station Progress Report, No. 19 (1970), p.48.

B11 R.R. Birss, M.R. Parker, M.K. Wong: Modelling of fields in magnetic drum separators. IEEE Trans. Mag., MAG-15 (1979), p.1305.

B12 J.A. Bartnik, H.D. Wasmuth, W.H. Zabel: Production of coarse grained iron ore sinter feed using the Jones WHIMS. Aufbereitungs-technik, Sept. 1982, p.490.

B13 R.R. Birss, M.R. Parker: High-intensity magnetic separation. Progress in Filtration and Separation. (Ed.: R.J. Wakeman), Vol.2., Elsevier Sci. Publ. Co., Amsterdam, 1981, p.171.

B14 G.K. Batchelor: An introduction to fluid dynamics. Cambridge University Press, Cambridge, 1967.

B15 R.R. Birss, R. Gerber, M.R. Parker: Theory and design of axially ordered filters for HIMS. IEEE Trans. Mag., MAG-12 (1976), p.892.

B16 R.R. Birrs, R. Gerber, M.R. Parker, T.J. Sheerer: Laminar flow model of particle capture of axial magnetic filters. IEEE Trans. Mag., MAG-14 (1978), p.1165.

B17 R.R. Birss, R. Gerber, M.P. Parker, T.J. Sheerer: Theory and performance of axial magnetic filters in laminar flow conditions. IEEE Trans. Mag., MAG-14 (1978), p.389.

B18 R.B. Bird, W.E. Stewart, E.N. Lightfoot: Transport phenomena. J. Wiley & Sons, New York, 1960.

B19 R.R. Birss, M.R. Parker: Magnetic separation of diamagnetic particles. IEEE Trans. Mag., MAG-15 (1979), p.1523.

B20 R.R. Birss, M.R. Parker, T.J. Sheerer: Statistics of particle capture in HGMS. IEEE Trans. Mag., MAG-16 (1980), p.830.

B21 R.I. Berdichevskiy, V.V. Karmazin: Flocculation of magnetite concentrates during flotation in magnetic field (in Russian). Obogascheniye rud, 73 (1) (1968), p.41.

B22 J.A. Bartnik, R.W. Schively: Separation of solids and fluids by magnetic flocculation. Proc. 1971 AIME Cent. Ann. Meeting, New York (1971), Preprint No. 71-B-44.

B23 W.H. Benson, J.A. Bartnik, G.D. Rose: Demagnetizing coils and magnetic flocculators used in the magnetite beneficiation industry. Proc. 29th Ann. Min Symp. Duluth, Minn. (1968), p.139.

B24 J.A. Bartnik: Magnetic-chemical flocculation improves operation. Iron & Steel Eng., March 1969, p.106.

B25 J.A.Bartnik, C.F. Giermak: Chemical-magnetic flocculation process. CIM Bull., March 1969, p.263.

B26 C.N. Bensley, R.J. Hunter: The coagulation of concentrated colloidal dispersions at low electrolyte concentrations. J. Coll. Interface Sci., 92 (1983), p.436.

B27 A. Bullet, H. Laneau: Beneficiation of iron ores by high intensity magnetic separation (in Russian). Proc. 8th Internatl. Min. Proc. Congress, Leningrad (USSR), 1968, p.230.

B28 A.G. Bailey: Electrostatic phenomena during powder handling. Powder Tech., 37 (1984), p.71.

B29 V.I. Bagin, T.S. Gendler, R.N. Kuzmin, R.S. Rybak: Study of the magnetic properties and Mössbauer effect at the transformation temperature of siderite (in Russian). Fiz. Zemli, No.11 (1971), p.71.

B30 J.C. Banerjee, S.P. Banerjee, N.R. Sircar: Chemical and magnetic properties of magnesites. Ind. Cer. Soc. Trans., 26 (1967), p.93.

B31 J. Boehm, D. Fletcher, M.R. Parker: Analysis of wet OGMS. J. Phys. D.: Appl. Phys., 19 (1986), p.L97.

B32 J.A. Bartnik, W.H. Zabel, D.M. Hopstock: On the production of iron ore superconcentrates by WHIMS. Intl. J. Min. Proc., 2 (1975), p.117.

B33 R.F. Boucher, K.J. Bulmer: Cleaning of HGMF matrices. IEEE Trans. Mag., MAG-18 (1982), p.1665.

B34 M. Beech-Jones, T. Eby, W.H. Muloin: The development and operation of a WHIMS separator circuit for hematite recovery at Wabush mines. CIM Bull., Sept. 1979, p.109.

B35 L.F. Bates: Modern Magnetism. Cambridge University Press, Cambridge, 1963.

B36 P. Bolsaitis, D.R. Kelland, G. Schmitz, E. Maxwell: Particle size effects on capture and matrix loading in HGMS. IEEE Trans. Mag., MAG-18 (1982), p.1677.

B37 J.A. Bartnik, H.D. Wasmuth: Dressing of magnetite ore converted to martite with the aid of Jones WHIMS. Erzmetal, 38 (1985), p.243.

B38 J.A. Bartnik, H.G. Schnabel, H.D. Wasmuth: The Jones WHIMS using wide gaps for beneficiation of martised iron ores and production of coarse-grained sinter feed from hematite iron ores. Presented at the Colloquium on Magnetic Separation, Council for Mineral Technology, Randburg, 1986.

B39 G. Barzev, G. Klisuranov, I. Ivanov: Processing of fine-grained fractions of manganese ore from Obrochishte deposit by HGMS. Fizykochem Probl. Mineralugii (Warsaw), 17 (1985), p.185.

B40 R.K. Ballhorn, D. Rassi, J.H.P. Watson: Concentration of a complex ore by HGMS. IEEE Trans. Mag., MAG-20 (1984), p.1201.

B41 R.G. Boothroyd: Flowing gas-solids suspensions. Chapman and Hall Ltd., London, 1971.

B42 J. Babůrek: The preparation of kaolin by means of HIMS. Interceram, 21 (1972), p.54.

B43 J. Babůrek, V. Hanus, V. Kádě, Z. Štěpánek: The operation of HGMS for paper kaolin beneficiation (in Czech). Stavivo (Prague), 4 (1979), p.150.

B44 P. Bolsaitis, V. Chang, H. Schorin, R. Aranguren: Beneficiation of ferruginous bauxites by HGMS. Intl. J. Min. Proc., 8 (1981), p.249.

B45 D.O. Bluhm, G.E. Fanslow, S.R. Bech-Montgomery: Selective magnetic enhancement of pyrite in coal by dielectric heating. Ames Laboratory, Iowa State University, US/DOE/IS-4766, 1982.

B46 G. Bitton, R. Mitchell: The removal of Escherichia Coli-bacteriophage T_7 by magnetic filtration. Water Res., 8 (1974), p.549.

B47 G. Bitton, J.L. Fox, H.G. Strictland: Removal of algae from Florida lakes by magnetic filtration. App. Microbiol., 30 (1975), p.905.

B48 G. Bitton, R. Mitchell, C. de Latour, E. Maxwell: Phosphate removal by magnetic filtration. Water Res., 8 (1974), p.107.

B49 R.F. Boucher, A.C. Lua: Loadability of high-gradient magnetic gas filter matrix. IEEE Trans. Mag., MAG-18 (1982), p.1662.

B50 R. Boucher, A. Lua: High-gradient magnetic filtration of gas flows. Proc. "Electrical and Magnetic Separation and Filtration Technology", Antwerp 1984, Koninklijke Vlaamse Ingenieursvereniging Tech. Inst. Antwerpen (Ed. R. Vanbrabant and R. Weiler), 1984, p.131.

C

C1 M.R. Chevallier, S. Mathieu: Magnetic properties of the hematite powder. Ann. des Phys., 18 (1943), p.258.

C2 P.P. Cioffi: High-power solenoid with external magnetic circuit. Proc. Conference on High Magnetic Fields, M.I.T., Cambridge, Mass., USA, (1961), p.202.

C3 V.I. Chichernikov: Magnetic measurements (in Russian). Izd. MGU, Moscow, 1969.

C4 J. Cibulka: The development of machines for the beneficiation of ores (in Czech). Rudy, 30 (1982), p.281.

C5 J.H. Carpenter: Carpco-Amax high-intensity wet magnetic separator. Proc. 7th Intl. Min. Proc. Congress, New York, (1964), p.399.

C6 I.J. Corrans: Magnetic separators. US Patent 4,260,477 (1981).

C7 I.J. Corrans, J.Levin: Wet high-intensity magnetic separation for the concentration of Witwatersrand gold-uranium ores and residues. J. South Afr. Min. Metall., 79 (1979), p. 210.

C8 J. Cibulka, F. Žůrek, O. Kolář, M. Horáček, V. Hencl, G.F. Suslikov, L.A. Lomovtsev, V.A. Gramm, Yu. A. Davydov: A new concept of high-gradient magnetic separators. Proc. 15th Internatl. Miner. Proc. Congress, Cannes (1985), p.363.

C9 V. Chrz, J. Fojtek, J. Máša: A core of a magnetic separator. Czechoslovak Patent Appl. PV 5191/82 (1982).

C10 H.E. Cohen, J.A. Good: Principles, design and performance of a superconducting magnet system for mineral separation in magnetic fields of high intensity. Proc. 11th Internatl. Min. Proc. Congress, Cagliari (1975), Paper 28.

C11 H.E. Cohen, J.A. Good: The principles and operation of a very high intensity magnetic mineral separator. IEEE Trans. Mag., MAG-12 (1976), p.552.

C12 H.E. Cohen: Open-channel magnetic separators with use of superconducting magnets. Proc. Intl. Conf. "Mineral processing and extractive metallurgy", Kunming, China (1984), p.427.

C13 H.K. Collan et al.: Superconducting open-gradient magnetic separator. Trans. Instn. Min. Metall (Sect. C: Mineral process. Extr. Metall.), 91 (1982), p.C5.

C14 P.A. Cheremnykh, V.K. Fyodorov, A.N. Piskunov: An analysis of the axial disk VGMS characteristics. J. Physique Cl, Suppl. No. 1, 45 (1984), C1-793.

C15 I.J. Corrans, J. Svoboda: Magnetic separation in South Africa. Mag. Sep. News, 1 (1985), p.205.

C16 E. Cohen, S. Roberts, N. Kerley: A new superconducting magnet system for high intensity magnetic separation. Proc. "Electrical and Magnetic Separation and Filtration Technology", Antwerp 1984. Koninklijke Vlaamse Ingenieursvereniging Tech. Inst. Antwerpen (Ed. R. Van Brabant and R. Weiler), 1984, p.85.

C17 D.L. Cummings, D.C. Prieve, G.J. Powers: The motion of small paramagnetic particles in HGMS. IEEE Trans Mag., MAG-12 (1976), p.471.

C18 D.L. Cummings, D.A. Himmelblau, J.A. Oberteuffer, G.J. Powers: Capture of small paramagnetic particles by magnetic forces from low speed fluid flows. AICHE J., 22 (1976), p.569.

C19 C.J. Clarkson, D. Kelland, T.B. King: Model for calculation of capture radii of a HGMS at moderate Reynolds numbers. IEEE Trans. Mag., MAG-12 (1978), p.901.

C20 C. Cowen, F.J. Friedlaender, R. Jaluria: Single wire model of HGMS process. IEEE Trans. Mag., MAG-12 (1976), p.466 and 898.

C21 H.K. Collan, J. Jantunen, M. Kokkala, A. Ritvos: Inversion of the breakthrough curve of HGMF: Theory and experiment. Industrial Applications of Magnetic Separation (Ed. Y.A. Liu), IEEE Publ. No. 78CH144-2 MAG, 1978, p.175.

C22 H.K. Collan, M. Kokkala, A. Ritvos, V. Pohjola: Rapid evaluation of magnetic filter performance. IEEE Trans. Mag., MAG-13 (1977), p.1480.

C23 I.J. Corrans, W.A. Gilbert, K.S. Liddell, R.C. Dunne: The performance of an industrial wet high-intensity separator for the recovery of gold and uranium. J. South Afr. Min. Metall., 84 (1984), p.57.

C24 A.F. Colombo: Selective flocculation and flotation of iron-bearing materials. Proc. Internatl. Symp. on Fine Particles Processing (Ed. P. Somasundaran), Las Vegas, Nevada, AIME, New York, 1980, p.1034.

C25 D.Y. Chan, D. Henderson: The orientation-averaged interaction of a pair of magnetic dipolar spheres in an external magnetic field. J. Coll. Interface Sci., 101 (1984), p.419.

C26 D.Y.C. Chan, D. Henderson, J. Barojas, A.M. Homola: The stability of a colloidal suspension of coated magnetic particles in an aqueous solution. IBM J. Res. Develop., 29 (1985), p.11.

C27 D.Y.C. Chan, B. Halle: Dissociation kinetics of secondary minimum flocculated colloidal particles. J. Coll. Interface Sci., 102 (1984), p.400.

C28 F. Chovanec, I. Hlásnik, J. Pitel, V. Hencl: The removal of sulphur and ash contamination from lignite coal by superconducting OGMS. Proc. 10th ICEC Conf., Otaniemi, Finland, 1984.

C29 K.M. Creer, E. Schmidbauer: The influence of crystallographic phase transitions on the magnetization of some iron-chromium spinels. J. Phys. Chem. Solids, 33 (1972), p.461.

C30 H.K. Collan, J. Jantunen, M. Kokkala, A. Ritvos: Experimental determination of the separation performance of various types of magnetic filter matrix. IEEE Trans. Mag., MAG-14 (1978), p.398.

C31 C.J. Clarkson D.R. Kelland: Theory and experimental verification of a model for HGMS. IEEE Trans. Mag., MAG-14 (1978), p.97.

C32 I.J. Corrans, J. Svoboda: Magnetic separator. SA Pat. 83/3323 (1983).

C33 W.A. Colburn: Method and apparatus for separating material. US Patent 3,994,801 (1976).

C34 H. Cordes: Examples of hematite ore beneficiation with Jones high-intensity wet magnetic separators. Erzmetall, No.5 (1981), p.269.

C35 I.J. Corrans: A development in the application of wet high-intensity magnetic separation. "International Symp. on Fine Particles Processing" (Ed. P. Somasundaran), Vol.2, Las Vegas, Nevada, AIME, New York 1980, p. 1294.

C36 I.J. Corrans, R.C. Dunne: Optimization of the recovery of gold and uranium from Witwatersrand residues. 15th Internatl. Min. Proc. Congress, Cannes, France, 1985.

C37 T.M. Conway, R.C. Dunne: The influence on gold recovery of desliming before WHIMS. Proc. Intl. Conf. Gold 100, Johannesburg (1986), p.419.

C38 J.L. Clements: Recovery of metals from coal fly ash: technological challenge or uneconomical dream? Conf. Recycle and Secondary Recovery of Metals, Ft. Lauderdale, Florida, 1985, p.747.

C39 H.K. Collan, M.A. Kokkala, O.E. Toikka: Magnetic filtration of steel mill effluent waters: principles, techniques and economics. Acta Polytech. Scand., Appl. Physics Series. No. 133 (1982).

C40 J. Czarnecki: The effect of surface inhomogenities on the interactions in colloidal systems and colloidal stability. Adv. Coll. Interface Sci., 24 (1986), p.283.

D

D1 V.G. Derkach: Special methods of the beneficiation of mineral (in Russian). Nedra, Moscow, 1966.

D2 R.S. Dean, C.W. Davis: Magnetic separation of ores. Bulletin No. 425, US Bureau of Mines (1941).

D3 I.E. Dzyaloshinski: Theory of weak ferromagnetism of antiferromagnetics (in Russian). ZhETF, 32 (1957), p.1547.

D4 D.J. Dunlop: Magnetic properties of fine-particle hematite. Ann. Geophys., 27 (1971), p.269.

D5 D. de Klerk: The construction of high field electromagnets. Newport Instruments Ltd., Newport Pagnell, 1965.

D6 K. Dwight: Experimental techniques with general applicability for the study of magnetic phenomena. J. Appl. Phys., 38 (1967), p.1505.

D7 G. Dobby, J.A. Finch: Capture of mineral particles in a high gradient magnetic field. Powder Technol, 17 (1977), p.73.

D8 F.D. Devaney: New developments in the magnetic concentration of iron ores. Proc. Internatl. Min. Proc. Congress, London, 1960, p.675.

D9 H. Dörr, F.D. Monostory, M. Seidel: Solid fuel desulphurization in power stations by superconducting magnets. Glückauf, 117 (1981), p.11.

D10 G. Dobby, J. Nesset, J.A. Finch: Mineral recovery by HGMS. Canad. Metall. Quart., 18 (1979), p.293.

D11 P.G. de Gennes, P.A. Pincus: Pair correlations in a ferromagnetic colloid. Phys. Kondens. Materie, 11 (1970), p.189.

D12 B.V. Derjaguin, L.D. Landau: Theory of molecular interaction. Acta Physicokhim. (USSR), 14 (1941), p.633.

D13 N.V. Dobrovolskaya, V.A. Maximova, Yu.A. Voevodin: The effect of heat treatment on the magnetic properties of stannite and its oxidation products (in Russian). Izv. Vyssh. Ucheb. Zaved. Tsvetn. Metall., No.1 (1985), p.23.

D14 D.J. Dunlop: Theory of magnetic viscosity of lunar and terrestial rocks. Rev. Geophys. Space Phys., 11 (1973), p.855.

D15 C. de Latour, G. Schmitz, E. Maxwell, D. Kelland: Designing HGMS matrix arrays for selective filtration. IEEE Trans. Mag., MAG-19 (1983), p.2127.

D16 C. de Latour: Magnetic separation in water pollution control. IEEE Trans. Mag., MAG-9 (1973), p.314.

D17 S. Dayton: Brazil '75.: For mining, the best is yet to come. E/MJ, 176 (1975), p.89.

D18 M.F. Dawson: Testwork on samples of oxidized ore from the Potgietersrus prospect. Report M 59D, Council for Mineral Technology, Randburg (RSA), 1982.

D19 J.M. Doulin, I. Wechsler: HGMS and its adaptability to iron ore processing. Ann. Min. Symp. and Ann. 42nd Meeting Minn. Sect. AIME, 1981, 16-1.

D20 D. De Klerk, C.J. Gorter: An efficiency criterion for iron-core electromagnets. High Magnetic Fields (Ed. H. Kolm et al.), M.I.T., Cambridge, 1962, p.194.

D21 L. Dobrescu, N. Muresan, A. Bogdan: Effects of the dispersing reagents in the magnetic cleaning of some clay raw materials by HGMS. 15th Internatl. Min. Proc. Congress, Cannes (France), 1985, p.343.

D22 J.I. Dijkhuis, C.B.W. Kerkdijk: Upgrading of coal using cryogenic HGMS. IEEE Trans. Mag., MAG-17 (1981), p.1503.

D23 G.S. Dobby, D.R. Kelland: Magnetite-coal separation by continuous HGMS. IEEE Trans. Mag., MAG-18 (1982), p.1698.

D24 M.S. Dobbins, G. Burnet: Production of an iron ore concentrate from iron-rich fraction of power plant fly ash. Resources and Conservation, 9 (1982), p.231.

D25 L. Dolle, J. Chenouard: How can electromagnetic filtration of corrosion products work with efficiency in electronuclear power plant? Water Chemistry 3. BNES, London, 1983, p.171.

D26 C. de Latour, H. Kolm: Magnetic separation in water pollution control II. IEEE Trans. Mag., MAG-11 (1975), p.1570.

D27 D.A. Dahlstrom: Impact of changing energy economics on mineral processing. Min. Eng., January 1986, p.45.

D28 R. Day, M. Fuller, V.A. Schmidt: Hysteresis properties of titano magnetites: grain size and compositional dependence. Phys. Earth Planet. Int., 13 (1977), p.260.

D29 D.J. Dunlop: Superparamagnetic and single domain threshold sizes in magnetite. J. Geophys. Res., 78 (1973), p.1780.

E

E1 I. Eisenstein: Magnetic separators: Traction force between ferromagnetic and paramagnetic spheres. IEEE Trans. Mag., MAG-13 (1977), p.1646.

E2 I. Eisenstein: Magnetic traction force in an HGMS with an ordered array of wires. IEEE Trans. Mag., MAG-14 (1978), p.1148 and 1155.

E3 S. Endoh, K. Yamaguchi: Charactetistics of pressure drop across spongy nickel filter in magnetic separation process. Powder Tech., 45 (1985), p.35.

E4 Y.M. Eyssa, R.W. Boom: Flow-through magnetic separators for weakly magnetic ores designed for superconducting magnets. IEEE Trans. Mag., MAG-11 (1975), p.1585.

E5 M.M. El Tawil, M.M. Morales: Application of WHIMS to sulfide mineral beneficiation. Intl. Symp. Complex Sulphides, (Ed. A.E. Zunkel), San Diego, Metall. Soc. Warrendale, Pa., 1985, p.507.

E6 S. Ergun, E.H. Bean: Magnetic separation of pyrite from coals. US Dept. of Interior, Bureau of Mines, Rept. No. 7181, 1968.

E7 B.B. Emory: Nuclear power reactor application of HGMS. IEEE Trans. Mag., MAG-18 (1982), p.1686.

E8 Economics of Mineral Engineering. Mining Journal Books Ltd., London, 1976, p.157.

F

F1 S.G. Frantz, G.W. Jarman: Magnetic beneficiation of nonmetallics. Trans. AIME, 102 (1932), p.122.

F2 S.G. Frantz: US Patent 2,074,085 (1937).

F3 M.D. Fuller: Geophysical aspects of paleomagnetism. CRC Crit. Rev. Solid State Sci., 1 (1970), p.137.

F4 S. Foner: Special magnetic measurement techniques. J. Appl. Phys., 38 (1967), p.1570.

F5 S. Foner: Review of magnetometry. IEEE Trans. Mag., MAG-17 (1981), p.3358.

F6 S. Foner: Vibrating sample magnetometer. Rev. Sci. Instr., 27 (1956), p.548.

F7 J. Fojtek, Z. Kaiser: Superconducting magnetic separator for kaolin brightening (in Czech). Proc. Conf. Kryogenika, Prague, 1984, p.27.

F8 B.H. Flinter: The magnetic separation of some alluvial minerals in Malaya. Amer. Mineral., 44 (1959), p.738.

F9 F.J. Friedlaender, M. Takayasu, J.B. Rettig, C.P. Kentzer: Particle flow in single wire HGMS studies. IEEE Trans. Mag., MAG-14 (1978), p.1158.

F10 F.J. Friedlaender, R. Gerber, W. Kurz, R.R. Birss: Particle motion near and capture on single spheres in HGMS. IEEE Trans. Mag., MAG-17 (1981), p.2801.

F11 F.J. Friedlaender, M. Takayasu, J.B. Rettig, C.P. Kentzer: Particle flow and collective process in single wire HGMS studies. IEEE Trans. Mag., MAG-14 (1978), p.1158.

F12 F.J. Friedlaender, M. Takayasu, T. Nakano, W.H. McNeese: Diamagnetic capture in single wire HGMS. IEEE Trans. Mag., MAG-15 (1979), p.1526.

F13 H.V. Fairbanks, W. Morton, J. Wallis: Separation processes aided by ultrasound. Filtration and Separation, July/August 1986, p.236.

F14 J.E. Forciea, R.W. Salmi: Primary magnetic separator specifications. Trans. Soc. Min. Eng., December 1965, p.339.

F15 F.J. Friedlaender, R. Gerber, H.P. Henkel, R.R. Birss: Particle build-up on single spheres in HGMS. IEEE Trans. Mag., MAG-17 (1981), p.2804.

F16 F.J. Friedlaender, M. Takayasu: A study of the mechanisms of particle build-up on single ferromagnetic wires and spheres. IEEE Trans. Mag., MAG-18 (1982), p.817.

F17 R. Flükiger: Advanced superconducting materials. J. Physique, 45, Coll. C1, Suppl. no. 1, (1984), C1-365.

F18 N. Fuchs: On the stability and charging of aerosol (in German). Z. Physik, 89 (1934), p.736.

F19 K.S.E. Forssberg, N.R. Kostkevicius: Comparative pilot scale tests with WHIMS separators. Erzmetall, 35 (1982), p.285.

F20 F. Fraas: Magnetization delay in the separation of minerals. US Bureau of Mines Report RI 6411, 1964.

F21 L.V. Favorskaya, A.N. Naumchik: The possibility of extracting chromium-containing minerals from North-Onega bauxites by magnetic separation. Sov. Non-Ferrous Met. Res., 3 (1984), p.186.

G

G1 R. Gerber, R.R. Birss: High gradient magnetic separation. John Wiley & Sons Ltd., Chichester, 1983.

G2 G.V. Gubin, P.A. Sanzharovskiy, N.S. Vitkovskaya: Magnetic properties of artificial magnetite (in Russian). Electrophysical methods of the treatment of rare metal ores (Ed. U. Andres). AN SSSR, Moscow, 1972.

G3 S.R. Guha Thakurta: Magnetic studies on natural crystals of wolframite. Indian J. Phys., 44 (1970), p.117.

G4 C.G. Gnehm: Rare-earth cobalt permanent magnets: market and technology. Met. Powder Rep., 32 (1984), p.576.

G5 A.M. Gaudin: Progress in magnetic separation using high intensity, high-gradient separators. Min. Congr. J., 60 (1974), p.18.

G6 R. Guillen, J.R. Regnard: Magnetic properties of natural and synthetic wolframites $Fe_xMn_{1-x}WO_4$. Phys. Chem. Minerals, 12 (1985), p.246.

G7 J.C. Gallop, B.W. Petley: SQUIDS and their applications. J. Phys. E: Sci. Inst., 9 (1976), p.417.

G8 G.M. Greene, L.E. Cornitius: A technique for magnetically separating minerals in a liquid mode. J. Sed. Petr., 41 (1971), p.310.

G9 J.A. Good, L. Lam: 0.3 m bore mobile reciprocating superconducting separator. Digests of Symposium on Superconducting Magnetic Separation, London, 1985.

G10 J.A. Good, K.White: Superconducting magnet for 60 tonne/hour mineral separator with closed cycle 4 K refrigeration. Journal de Physique, Coll. C1, Suppl. 1, 45 (1984), C1-759.

G11 J.A. Good, J. Kopp: Superconducting magnet for 60 tonne/ hour mineral separator with closed cycle 4 K refrigeration. Proc. Internatl. Conf. MINTEK 50, Johannesburg, 1984, Paper A6/1.

G12 S.J. Gill, C.P. Malone: Force on a small particle in the vicinity of a cylinder in a homogeneous magnetic field. Rev. Sci. Instr., 34 (1963), p.788.

G13 C.H. Gooding, T.W. Sigmon, L.K. Monteith: Application of HGMS to fine particle control. US EPA Research Report No. EPA-600/2-77-230, 1977.

G14 C.H. Gooding: Pilot-scale field tests of HGMF. US EPA Research Report No. EPA-600/7-80-037, 1980.

G15 C.H. Gooding, T.W. Sigmon, L.K. Monteith, D.C. Drehmel: Application and modelling of HGMS in particulate/gas system. IEEE Trans. Mag., MAG-14 (1978), p.407.

G16 R. Gerber: Theory of particle capture in axial filters for HGMS. J. Phys. D: Appl. Phys., 11 (1978), p.2119.

G17 R. Gerber, M. Takayasu, F.J. Friedlaender: Generalization of HGMS theory: the capture of ultrafine particles. IEEE Trans. Mag., MAG-19 (1983), p.2115.

G18 R. Gerber: Magnetic filtration of ultrafine particles. IEEE Trans. Mag., MAG-20 (1984), p.1159.

G19 J.P. Glew, M.K. Wong, M.R. Parker, R.R. Birss: Single-wire build-up of diamagnetic particles at low Reynolds numbers. IEEE Trans. Mag., MAG-18 (1982), p.1656.

G20 R. Gerber, M.H. Watmough: Magnetic separator equations. IEEE Trans. Mag., MAG-18 (1982), p.1671.

G21 H. Greiner, H. Hoffmann: Efficiencies of axially ordered HGMS filters in a multiwire model. J. de Physique, Coll. Cl, Suppl. 1, 45 (1984), p.Cl-789.

G22 H. Greiner, H. Hoffmann: The magnetic field and force in a perpendicularly magnetized matrix consisting of parallel ordered ferromagnetic wires. J. Mag. Mag. Mater., 38 (1983), p.187.

G23 H. Greiner, H. Hoffmann: Efficiencies of axially ordered HGMS filters calculated from a multiwire model. J. Mag. Mag. Mater., 38 (1983), p.194.

G24 H. Greiner, G. Reger, H. Hoffmann: The comparative performance of HGMS filters with a strictly ordered wire matrix in the axial and transverse configurations. IEEE Trans. Mag., MAG-20 (1984), p.1171.

G25 H. Greiner, H. Hoffmann: Ordered axial HGMS filters: theory and experiment. "Electric and Magnetic Separation and Filtration Technology", Antwerp 1984, (Ed. R. Vanbrabant, R. Weiler). Koninklijke Vlaamse Ingenieursvereniging Tech. Inst., Antwerpen, 1984, p. 67.

G26 R. Gerber, M.H. Watmough: A design of a superconducting split-coil open-gradient magnetic separator. IEEE Trans. Mag., MAG-21 (1985), p.2053.

G27 R.E. Gomez, J.A. Finch: HGMS of Fe from Pb and Ag in residue of sulphation roast-leach process. Trans. Instn. Min. Metall., (Sect. C: Mineral Process. Extr. Metall.), 92 (1983), p.C100.

G28 T.L. Gak, A.A. Bayshulakov, T.G. Gucalyuk: Recovery of wolframite from the tailings (in Russian). Izv. Vyssh. Ucheb. Zaved. Cvet. Metal., No. 2 (1983), p.16.

G29 T.M. Gilliam et al.: Economic metal recovery from fly ash. Resources and Conservation, 9 (1982), p.155.

H

H1 V. Hencl: Wet magnetic separation of weakly magnetic minerals (in Czech). Ore Research Institute, Prague, Vol. 19 (1974).

H2 K. Honda: The thermomagnetic properties of elements (in German). Ann. Phys. (Lpz.), 32 (1910), p.1048.

H3 M. Homma, M. Imakawa, M. Okada, S. Tsutsumi: Au-Fe-Ni permanent magnet alloys. IEEE Trans. Mag., MAG-21 (1985), p.1245.

H4 J.F. Herbst, J.J. Croat, F.E. Pinkerton, W.B. Yelon: Relationship between crystal structure and magnetic properties in $Nd_2Fe_{14}B$. Phys. Rev., B29 (1984), p.4176.

H5 W. Heinz: Status and trends of superconducting magnet development in Europe. IEEE Trans. Mag., MAG-19 (1983), p.167.

H6 K.N. Henrichsen: Large-scale magnetic field measurements and mapping. J. Physique, Coll. C1, Suppl. 1, 45 (1984), C1-937.

H7 F.B. Humphrey: Magnetic measurement techniques for thin films and small particles. J.Appl. Phys., 38 (1967), p.1520.

H8 D.M. Hopstock: Fundamental aspects of design and performance of low-intensity dry magnetic separators. Trans. AIME/SME, 258 (1975), p.222.

H9 W.E. Horst, W.P. Dyrenforth: Wet high-intensity magnetic separation of industrial minerals. Min. Eng., March (1971), p.57.

H10 V. Hencl: New HIMS with vertical rotor weighs less and separates finer particles. World Mining, January 1970, p.41.

H11 V. Hencl, M. Horáček, O. Kolář: New high-intensity wet magnetic separator for treatment of fine fraction of weakly magnetic materials. Proc. 9th Internatl. Min. Proc. Congress, Prague, 1970, p.69.

H12 M. Horáček, O. Kolář, V. Hencl, J. Hudík, F. Neupauer: Wet disk rotating magnetic separator (in Czech). Czechoslovak Patent 136403 (1970).

H13 V. Hencl, M. Horáček, O. Kolář, J. Hudík, J. Cibulka: High gradient drum magnetic separator for wet beneficiation of weakly magnetic minerals (in Czech). Czechoslovak Patent 194464 (1982).

H14 V. Hencl: Current developments in the application of high gradient magnetic separation (in Czech). Rudy, 30 (1982), p.41.

H15 R. Houot, G. Gillet, C. Leschevin: Superconducting solenoid magnetic separator (in French). Industrie Minerale-Les Techniques, February 1982, p.1.

H16 I. Hlásnik: Simplified theory of multipole superconducting magnetic separator. IEEE Trans. Mag., MAG-17 (1981), p.1990.

H17 A.S. Holman, E.C. Hise, J.E. Jones: Initial investigation of open-gradient magnetic separation. Oak Ridge National Laboratory, Report ORNL-5764. Oak Ridge, Tenn., USA, 1982.

H18 E.C. Hise, A.S. Holman, F.J. Friedlaender: Development of high-gradient and open-gradient magnetic separation of dry fine coal. IEEE Trans. Mag., MAG-17 (1981), p.3314.

H19 M. Hollingworth, J.A. Finch: Downstream capture in longitudinal HGMS. IEEE Trans. Mag., MAG-18 (1982), p.1674.

H20 K. Hayashi, S. Uchiyama: On particle trajectory and capture efficiency around many wires. IEEE Trans. Mag., MAG-16 (1980), p.827.

H21 H.G. Heitmann: Use of electromagnetic filters in nuclear power plants: experimental and operating experience. Intl. Conf. "Water Chemistry of Nuclear Reactor Systems", Bournemouth, England, 1977, Paper no. 42.

H22 J. Happel: Viscous flow in multiparticle systems: slow motion of fluid relative to beds of spherical particles. Am. Inst. Chem. Eng. J., 4 (1958), p.197.

H23 V. Hencl, J. Svoboda: The possibility of magnetic flocculation of wekly magnetic minerals. Proc. 13th Internatl. Min. Proc. Congress, Warsaw, 1979, p.209.

H24 H. Hirabayashi: Present status of A15 compounds. J. Physique, 45, Coll.1, Suppl. no. 1 (1984), p.C1-359.

H25 R. Hogg, T.W. Healy, D.W. Fuerstenau: Mutual coagulation of colloidal dispersions. Trans. Farad. Soc., 62 (1966), p.1638.

H26 V. Hencl, J. Svoboda, J. Cibulka: A method and a device for a separation of weakly magnetic materials by selective flocculation (in Czech). Czechoslovak Patent AO 200060 (1978).

H27 E.C. Hise: Development of high-gradient and open-gradient magnetic separation. IEEE Trans. Mag., MAG-18 (1982), p.847.

H28 S.T. Hall, J.A. Finch, N. Rowlands: Enhancement of magnetic susceptibility of pyrite by oxidative pressure leaching. Trans. Instn. Min. Metall., Sect. C: Mineral Process. Extr. Metall., 94 (1985), p.C35.

H29 H.E. Hawkes: U.S. Geol. Sur. Bull., A973 (1951).

H30 M. Hollingworth, J.E. Nesset, J.A. Finch: The build-up model of HGMS modified for field-dependent susceptibility minerals. Can. Met. Qart., 23 (1984), p.479.

H31 G.C. Hadjipanayis, Y.F. Tao: Magnetic properties of melt-spun and sintered Fe-Nd-B magnets at elevated temperatures. J. Physique Coll. C6, Suppl. 9, 46 (1985), C6-237.

H32 E.P. Hyland, S.J. Dale. M.D. Croft: Magnetic separators. South African Patent 84/10068 (1984).

H33 E.P. Hyland: The development of a submerged roll permanent magnetic separator. Paper presented at the Magnetic separation Colloquium, Council for Mineral Technology , Randburg, S.A., January 1986.

H34 J.R. Harland, L. Nilsson, M. Wallin: Pilot-scale high-gradient magnetic filtration of steel mill waste water. IEEE Trans. Mag., MAG-12 (1976), p.904.

H35 V. Hencl, G. Šebor: Optimization of the parameters of HGMS of complex siderite ore. Proc. 15th Intl. Min. Proc. Congress, Cannes, 1985, p.354.

H36 K. Harding, W. Baxter: Application of HGMS to ferric hydroxide filtration. IEEE Trans. Mag.,MAG-17 (1981), p.2795.

H37 H.G. Heitmann: Studies of the application of electromagnetic filters in power plants. "Industrial Applications of Magnetic separation." (Ed. Y.A. Liu). IEEE Publication No. 78 CH 1447-2MAG, New York, 1979, p.115.

H38 D.M. Hopstock: Wet high-intensity magnetic beneficiation of oxidized taconites. US Bureau of Mines, Report RI 8363, 1983.

H39 D.M. Hopstock, A.F. Colombo: Processing finely ground oxidized taconite by WHIMS. "International Symp. on Fine Particles Processing." (Ed. P. Somasundaran). Las Vegas, Nevada, AIME, New York, 1980.

H40 J.Y. Hwang, M. Takayasu, F.J. Friedlaender, G. Kullerud: Application of magnetic susceptibility gradients to magnetic separation. J. Appl. Phys., 55 (1984), p.2592.

H41 V. Hencl: Complex treatment of Slovakian siderites. Proc. 6th Internatl. Min. Proc. Congress, Cannes, 1963, p.483.

H42 E.C. Hise, I. Wechsler, J.M. Doulin: Separation of dry crushed coals by HGMS. Oak Ridge National Laboratory Report ORNL-5571, Oak Ridge, Tenn., USA, 1979.

H43 E.C. Hise, I. Wechsler, J.M. Doulin: Continuous separation of dry crushed coal at one ton per hour by HGMS. Oak Ridge National Laboratory Report ORNL-5763, Oak Ridge, Tenn., USA, 1981.

H44 S.T. Hall, J.A. Finch: Enhanced magnetic desulphurization of coal. Min. Metal. Proc., November 1984, p.179.

H45 H.G. Heitmann, V. Schneider, K. Fröhlich, G. Rupp: Investigations of feed water purification in nuclear power plants by electromagnetic filtration. Proc. "Electrical and Magnetic Separation and Filtration Technology" Antwerp 1984. (Ed. R. Vanbrabant, R. Weiler). Koninklijke Vlaamse Ingenieursvereniging Tech. Inst., Antwerpen, 1984, p.73.

H46 J.Y. Hwang et al.: Selective seeding for magnetic separation. IEEE Trans. Mag., MAG-18 (1982), p.1689.

H47 V. Hencl, K. Jahoda, E. Madai: Analysis of magnetic and hydraulic forces in an oriented real matrix of HGMS. Mag. Sep. News, 1 (1985), p.173.

H48 J.P. Herzig, D.M. Leclerc, P. De Goff: Flow of suspensions through porous media - application to deep filtration. Ind. Eng. Chem, 62 (1970), p.8.

I

I1 J. Iannicelli: New development in magnetic separation. IEEE Trans. Mag., MAG-12 (1976), p.436.

I2 J. Iannicelli: High extraction magnetic filter separator. West German Patent 2,111,986 (1971).

I3 J. Iannicelli, N. Millman, W.J.D. Stone: Process for improving the brightness of clay. US Patent 3,471,011 (1969).

I4 J. Iannicelli: New developments in high extraction magnetic filtration of kaolin clay. Proc. 13th Intl. Min. Proc. Congress, Beneficiation of Clay Raw Minerals, Warsaw, 1979, p. 105.

I5 J. Iannicelli: High extraction magnetic filtration of kaolin clay. Clays and Clay Min. 24 (1976), p.64.

I6 J. Iannicelli: High intensity magnetic separation used to process bauxite. Min. Eng., October 1984, p.1470.

I7 J. Iannicelli: Paramagnetic separation in ultrafine industrial minerals and coal. AIME 1983, Chapter 11, p.105.

I8 K.J. Ives: Capture mechanisms in filtration. The Scientific Basis of Filtration (Ed. K.J. Ives). Noordhoof, Leiden, 1975, p.183.

J

J1 H.P. Johnson, W. Lowrie, D.V. Kent: Stability of anhysteretic remanent magnetization in fine and coarse magnetite and maghemite particles. Geophys. J. Roy. Ast. Soc., 41 (1975), p.1.

J2 K.P. Jüngst, G. Ries, S. Förster, F. Graf, G. Obermaier, W. Lehmann: Magnet system for a superconducting magnetic separator. Cryogenics, 24 (1984), p.648.

J3 L.S. Jurng, T. Fujita, J. Shimoiizaka: HGMS of fine low grade iron ore. J. Inst. Min. Metall. Japan, 101 (1985), p.525.

J4 G.H.Jones: Wet magnetic separator for feebly magnetic minerals. I. Description and theory. Proc. 5th Internatl. Min. Proc. Congress, London, 1960, p.717.

J5 J.D. Jackson: Classical electrodynamics. John Wiley & Sons, New York, 1975.

J6 P.G. Jordan: Association phenomena in a ferromagnetic colloid. Mol. Phys., 25 (1973), p.961.

J7 D.M. Jacobson, D.S. Evans: Possible takeover by neodymium alloy supermagnets. Metals and Materials, 1 (1985), p.667.

J8 P.J. Jonker: The operational efficiency of an Eriez WHIMS at Buffalo Fluorspar Mine. Presented at Colloquium on Magnetic Separation, Council for Mineral Technology, Randburg, RSA, 1986.

J9 I.S. Jacobs, L.M. Levinson: Solvent refining of Wyodak, Illinois No. 6 and Black Mesa Coals. EPRI Report RP389, Vol. 2, 1976.

K

K1 V.I. Karmazin: Modern methods of magnetic beneficiation of ores (in Russian). Gosgortekhizdat, Moscow, 1962.

K2 H.H. Kolm: The future of superconducting technology. Cryogenics, 15 (1975), p.63.

K3 H.H. Kolm: Process for magnetic separation. US Patent 3,676,337 (1972).

K4 J. Kulda: Field in a magnetic filter (in Czech). Elektrotechn. obzor (Prague), 70 (1981), p.67.

K5 Kappa bridge KLY-2. The instruction manual. Geofyzika Brno, Czechoslovakia, 1983.

K6 P.G. Kihlstedt, B. Sköld: Concentration of magnetite ores with dry magnetic separator of the Mörtsell-Sala type. Proc. Internatl. Min. Proc. Congress, London, 1960, p.691.

K7 O. Kolář, M. Horáček, J. Cibulka: Magnetic separators with permanent magnets (in Czech). Proc. 12th National Mineral Processing Conference, Prague, 1977, p.49.

K8 J. Kopp: Permanent magnet disk separator. IEEE Trans. Mag., MAG-20 (1984), p.1204.

K9 H. Kolm: Magnetic device. US Patent 3,567,026 (1971).

K10 H. Kolm: The large-scale manipulation of small particles. IEEE Trans. Mag., MAG-11 (1975), p.1567.

K11 H. Kolm, J.A. Oberteuffer, D.R. Kelland: High-gradient magnetic separation. Scientific American, November 1975, p.47.

K12 D.R. Kelland, Y. Hiresaki, F.J. Friedlaender, M. Takayasu: Diamagnetic particle capture and mineral separation. IEEE Trans. Mag., MAG-17 (1981), p.2813.

K13 R. Kaiser, C.K. Colton, C. Miskolczy, L. Mir: Magnetically induced separation of stable emulsions. AICHE Symp. Series, 68 (1971), 124. Water 115.

K14 S. Kuwabara: The forces experienced by randomly distributed parallel circular cylinders or spheres in a viscous flow at small Reynolds numbers. J. Phys. Soc. Japan, 14 (1959), p.527.

K15 V.V. Karmazin, V.I. Karmazin, V.A. Binkevich: Magnetic regeneration and separation in the beneficiation of ores and coal (in Russian). Nedra, Moscow, 1968.

K16 J. Kopp, J.A. Good: The physics of high-intensity dry magnetic separation. IEEE Trans. Mag., MAG-18 (1982), p.833.

K17 J. Kopp: The physics of "falling curtain" dry magnetic separation. Intl. J. Min. Proc., 10 (1983), p.297.

K18 R.N. Kuzmin, T.S. Gendler: A study of the structural and magnetic phase transitions in siderite (in Russian). Kristallografiya, 15 (1971), p.736.

K19 A. Kumar, M.S. Bhalla: Source of stable remanence in chromite ores. Geophys. Res. Lett. 11 (1984), p.177.

K20 C.B.W. Kerkdijk, J.I. Dijkhuis, H.R. Segal, R. van Kleef, H.A. van der Sloot: High-gradient magnetic separation of fly ash. IEEE Trans. Mags., MAG-18 (1982), p.853.

K21 V.I. Karmazin et al.: Finishing tungsten-tin slimes on the 2VK-3 high-intensity magnetic separator (in Russian). Tsvetn. Metal., (10) (1969), p.35.

K22 K.H. Kukuck: The concept of new SOL separator. Presented at the Colloquium on Magnetic Separation in the Mineral Industry, Council for Mineral Technology, Randburg (RSA), 1986.

K23 Y.S. Kim, T. Suzuki. I. Matsuoka: Magnetic susceptibility of some sulphide minerals. J. Min. Metal. Inst. Japan, 102 (1986), p.85.

K24 V.I. Karmazin, L.F. Mostipan. A.I. Urvantsev, Yu.S. Mostika: Upgrading an oxidized quartzite magnetic concentrate from Krivoi Rog (in Russian). Proc. Conf. "Pererab. okislennykh rud" (Ed. V.A. Chanturiya, E.A. Trofimova). Nauka, Moscow, 1985, p.97.

K25 D.R. Kelland, E. Maxwell: Oxidized taconite beneficiation by continuous HGMS. IEEE Trans. Mag., MAG-11 (1975), p.1582.

K26 A. Károli: A new approach to the beneficiation of the siderite ore from the Rudňany deposit (in Slovak). Rudy (Prague), 30 (1982), p.125.

K27 Y.S. Kim, T. Fujita, S. Hashimoto, J. Shimoiizaka: The removal of Cu sulphide minerals from Pb flotation concentrate of black ore by HGMS. Proc. 15th Internatl. Min. Proc. Congress, Cannes (France), 1985, p.381.

K28 W.M. Kester: Magnetic demineralization of pulverized coal. Min. Eng., 17 (1965), p.72.

K29 J.K. Kindig: The Magnex process: review and current status. "Industrial Applications of Magnetic Separation" (Ed. Y.A. Liu). IEEE Publication No. 78CH1447-2MAG, New York, 1979, p.99.

K30 D.R. Kelland, E. Maxwell: Improved magnetite recovery in coal cleaning by HGMS. IEEE Trans. Mag., MAG-14 (1978), p.401.

K31 D.R. Kelland, G.S. Dobby, E. Maxwell: Efficient HGMS for highly magnetic material. IEEE Trans. Mag., MAG-17 (1981), p.3308.

K32 D.R. Kelland: A review of HGMS methods of coal cleaning. IEEE Trans. Mag., MAG-18 (1982), p.841.

K33 D.R. Kelland: Continuous heavy medium recovery,by HGMS. IEEE Trans. Mag., MAG-20 (1984), p.1180.

K34 M.A. Kokkala: Water treatment by means of magnetic separation. Ph.D. Thesis, Helsinki University of Technology, 1982.

K35 Z. Kaiser, J. Fojtek, M. Kouba, J. Kotva, J Šůma: Magnetic separator with superconducting magnet and reciprocating matrix. Proc. Conf. CryoPrague 86, Prague (1986), p.192.

K36 J. Kopp: Linear superconducting separator magnets. Proc. Conference "Magnet Technology 9", Zürich, 1985.

K37 J. Kopp: The performance of 'racetrack' superconducting magnetic separators. Intl. J. Min. Proc., 18 (1986), p.33.

L

L1 E. Laurila: On the magnetic permeability of mixtures containing ferromagnetic particles and porous ferromagnetic materials. Ann. Acad. Scient. Fenn. A, Physica 70 (1961), p.1.

L2 K. Lark-Horovitz, V.A. Johnson (Ed.): Solid State Physics, Vol. 6, Part B. Academic press, New York, 1959.

L3 K.K. Likharev, V.T. Ulrikh: Systems with Josephson contacts (in Russian). Moscow University Press, Moscow, 1978.

L4 E. Laurila: On the fields of permanent magnets in magnetic pulley and separators. Acta Polytechnica Scand. (1962), p.312.

L5 E. Laurila: On the design principles of a high-intensity separator with permanent magnets. Acta Polytechnica Scand., Ph102 (1974).

L6 E. Laurila, T. Meinander: HIMS with permanent magnets works well in lab tests. World Mining, November 1986, p.56.

L7 L.D. Landau, E.M. Lifshitz: Fluid Mechanics. Pergamon Press, Oxford, 1959.

L8 E.E. Luborsky, B.J. Drummond: HGMS: theory versus experiment. IEEE Trans. Mag., MAG-11 (1975), p.1696.

L9 W.F. Lawson, W.H. Simons, R.P. Treat: The dynamics of a particle attracted by a magnetized wire. J. Appl. Phys., 48 (1977), p.3213.

L10 W.F. Lawson, R.P. Treat: Particle trajectory observations in dry HGMS. IEEE Trans. Mag., MAG-18 (1982), p.1668.

L11 F.E. Luborsky, B.J. Drummond: HGMS: Theory versus experiment. IEEE Trans. Mag., MAG-11 (1975), p.1696.

L12 F.E. Luborsky, B.J. Drummond: Build-up of particles on fibers in HGMS. IEEE Trans. Mag.,MAG-12 (1976), p.463.

L13 J. Levin: Wet high-intensity magnetic separation. Laboratory tests conducted in 1978 and 1979. Report 2076. National Institute for Metallurgy, Randburg, RSA, 1980.

L14 L.A. Lomovtsev, L.S. Shugol: The effect of magnetic separation process (in Russian). Gorn. Zh., 8 (1967), p.56.

L15 H. Lantto: The effect of magnetic flocculation in the beneficiation of magnetic materials. Acta Polytech. Scand., Ch133 (1977).

L16 E. Laurila: On the behaviour of magnetic powder in a rotating field. Ann. Acad. Sci. Fenn., AVI34 (1959).

L17 F.E. Luborsky: HGMS for removal of sulphur from coal. Report EPA-600/7-78-208, 1978.

L18 P. Le Goff, D. Leclerc, J. Dodds: The structure of packed beds: continuity of research in Nancy and some new results. Powder Tech., 42 (1985), p.47.

L19 C.H. Lofthouse, C.W. Scobie: Beneficiation of kaolin using commercial high-intensity magnetic separator. Proc. 13th Internatl. Min. Proc. Congress, Round Table Seminar: "Beneficiation of Clay Raw Materials", Warsaw, Poland, 1979, p.163.

L20 T. Lindberg: Experience with the Jones separator at Buchanan. Min. Mag., Sept. 1979, p.199.

L21 J.E. Lawver, D.M. Hopstock: Wet magnetic separation of weakly magnetic minerals. Min. Sci. Eng., 6 (1974), p.154.

L22 J.W. Lawver, R.M. Hays: A technical and economical appraisal of methods of beneficiating Mesabi Range iron ore (in Russian). Proc. 8th Internatl. Min. Proc. Congress, Leningrad, USSR, 1968. p.148.

L23 J.E. Lawver, J.L. Wright, H.R. Kokal: The behaviour of Mesabi iron and silicate minerals in 20 kG magnetic field. Trans AIME, 241 (1968), p.194.

L24 J. Levin: Concentration tests on the gold uranium ores of the Witwatersrand for the recovery of uranium. J. South Afr. Inst. Min. Metall., 57 (1956/57), p.209.

L25 D. Larbalestier, G. Fisk, B. Montgomery, D. Hawksworth: High-field superconductivity. Physics Today, March 1986, p.24.

L26 S.L. Lefond (Ed.): Industrial minerals and rocks. AIMPE, New York, 1975.

L27 C.J. Lin, Y.A. Liu, D.L. Vives, M.J. Oak, G.E. Crow, E.L. Huffman: Pilot-scale studies of sulphur and ash removal from coals by HGMS. IEEE Trans. Mag., MAG-12 (1976), p.513.

L28 Y.A. Liu, C.J. Lin: Assessment of sulphur and ash removal from coals by magnetic separation. IEEE Trans. Mag., MAG-12 (1976), p.538.

L29 Y.A. Liu, G.E. Crow: A pilot-scale study of HGMS desulphurization of SRC. Fuel, 58 (1979), p.345.

L30 A.C. Lua, R.F. Boucher: High-gradient magnetic gas filtration. Mag. Sep. News, 1 (1984), p.67.

L31 A.C. Lua, R.F. Boucher: An investigation of efficiency of high gradient magnetic gas filtration. IEEE Trans. Mag., MAG-18 (1982), p.1659.

M

M1 M. McCaig: Permanent magnets in theory and practice. Pentech Press, London, 1977.

M2 F. J. Morin: Magnetic susceptibility of α-Fe_2O_3 and α-Fe_2O_3 with added titanium. Phys. Rev., 78 (1950), p.819.

M3 P.A. Montano, M.S. Seehra: Magnetism of iron pyrite - a Mössbauer study in an external magnetic field. Solid State Commun., 20 (1976), p.897.

M4 P.G. Marston: Magnetic design and its effects on the economic of HGMS process. Proc. of High-Gradient Magnetic Separation Symposium, M.I.T., 1973, p.25.

M5 D.B. Montgomery: Solenoid magnet design. Wiley-Interscience, New York, 1969.

M6 P.G. Marston: The application of superconductivity to magnetic separation. IEEE Trans. Mag., MAG-11 (1975), p.602.

M7 D.B. Montgomery: An assessment of superconducting magnets for the next decade of applications. IEEE Trans. Mag., MAG-12 (1976), p.908.

M8 The magnetic handbook. RFL Industries, Inc., Boonton, N.J., 1976.

M9 J. McAndrew: Calibration of a Frantz isodynamic separator and its application to mineral separation. Proc. Aust. I.M.M., 181 (1957), p.59.

M10 G.B. McCabe: unpublished results. Council for Mineral Technology, Randburg, RSA.

M11 T.E. Masoner: High-intensity wet magnetic separation research at Climax. Soc. Min. Eng. AIME, Preprint no. 68-B-327.

M12 P.G. Marston, J.J. Nolan, L.M. Lontai: Magnetic separation and magnetic separation methods. US Patent 3,627,678 (1971).

M13 P.G. Marston, J.J. Nolan: Moving matrix magnetic separator. U.S. Patent 3,920,543 (1975).

M14 C. Moyer, M. Natenapit, S. Arajs: Particle capture by an assemblage of spheres in HGMS. J. Appl. Phys., 55 (1984), p.2589.

M15 C. Moyer, M. Natenapit, S. Arajs: Magnetic filtration of particles in laminar flow through a bed of spheres. J. Mag. Mag. Mater., 44 (1984), p.99.

M16 C. Mills: Process design, scale-up and plant design for gravity separation. Mineral Processing Plant Design, AIME, New York, 1978, p.404.

M17 P. Mucha, V. Hencl, E. Madai: Magnetic forces in an ordered matrix of the HGMS. Intl. J. Min. Proc., 9 (1982), p.259.

M18 J.J. Miles, R.W. Chantrell, M.R. Parker: Model of magnetic field-induced ordering in dispersions of fine paramagnetic particles. J. Appl. Phys., 57 (1985), p.4271.

M19 J.M.W. MacKenzie, V.M. Lovell: The application of double layer properties to the upgrading of uranium in Witwatersrand ores by selective flocculation. Report 1413, National Institute of Metallurgy, Johannesburg, 1972.

M20 V.F. Moroz, S.B. Ol'khovik: Influence of comminution on the structure and properties of magnetite (in Russian). Izv. Akad. Nauk SSSR, Metally No.5 (1984), p.9.

M21 T.R. McGuire, L.N. Howard, J.S. Smart: Magnetic properties of the chromites. Ceramic Age, July 1952, p.22.

M22 I.V.R. Murthy, G. Gopalakrishna: Remanence hysteresis properties of chromites. Proc. Indian Acad. Sci., 91 (1982), p.159.

M23 H.J. Mauritsch: The paleomagnetic investigation of some magnesites (in German). Mitt. Österr. Geol. Ges., 73 (1980), p.1.

M24 A. Mookherji, S.C. Mathur: Magnetic studies of siderite. J. Phys. Soc. Japan, 20 (1965), p.1336.

M25 J. Murkes: Separation - an art or a science? Filtration & Separation, Sept./Oct. 1969, p.491.

M26 E. Madai: On the problem of matrix adaptation in HGMS operation. Aufbereitungs-Technik, 24 (1983), p.99.

M27 V.M. Malyi, T.B. Ganzenko, E.A. Titlyanov: Magnetic beneficiation of oxidized ferruginous quartzites (in Russian). Proc. Conf. "Pererab. okisl. rud." (Ed. V.A. Chanturiya, E.A. Trofimova). Nauka, Moscow, 1985.

M28 E. Maxwell, A. Shalom, D.R. Kelland: Particle size dependence in HGMS. IEEE Trans. Mag., MAG-17 (1981), p.1293.

M29 O.I. Martynova, A.S. Kopylov: The use of electromagnetic filters for removing ferromagnetic impurities from water (in Russian). Teploenergetika, 19 (3) (1972), p.67.

M30 M. Mamula, J. Křepelka, V.Vejskal: A method for separating solid materials dispersed in a liquid (in Czech). Czechoslovak Patent 132624 (1969).

M31 I. Makarow, T. Agrafiotis, J.A. Finch: Magnetizing roast for chalcopyrite: effect of particle size and dilution. Mag. Sep. News, 1 (1984), p.115.

M32 S.G. Malghan, J.P. Van Dillen: Influence of process variables in the HGMS of uranium ores. Min. Eng., December 1983, p.1650.

M33 D.B. Montgomery: Iron magnet design. High Magnetic Fields (Ed. H. Kolm et al.). M.I.T., Cambridge, 1961, p.180.

M34 E. Martinez: Magnetic concentration of asbestos ore. CIM Bull., May 1979, p.100.

M35 E. Maxwell, D.R. Kelland: HGMS in coal desulphurization. IEEE Trans. Mag., MAG-14 (1978), p.482.

M36 S.E. Male: The magnetic beneficiation of coal. Mag. Sep. News, 2 (1985), p.1.

M37 H.H. Murray: Beneficiation of selected industrial minerals and coal by HGMS. IEEE Trans. Mag., MAG-12 (1976), p.498.

M38 E. Maxwell, D.R. Kelland, I.Y. Akoto: HGMS of mineral particulates from solvent refined coal. IEEE Trans. Mag., MAG-12 (1976), p.507.

M39 S.E. Male: Magnetic susceptibility and separation of inorganic material from UK coals. J. Phys. D: Appl. Phys., 17 (1984), p.155.

M40 F. Michalíková, F. Špaldon: The beneficiation of the power plant fly ash (in Slovak). Rudy (Prague), 30 (1982), p.179.

M41 M.J. McNallan, K.C. Russel, J.A. Oberteuffer: High-gradient magnetic filtration of steel plant waste water. Iron and Steel Eng., 53 (1976), p.40.

M42 M.J. Mortimer: Magnetic susceptibility measurements on fission product insolubles. IEEE Trans. Mag., MAG-17 (1981), p.2819.

M43 E. Maxwell: Magnetic separation - the prospects for superconductivity. Cryogenics, April 1975, p.179.

M44 T. Mizoguchi, I. Sakai, K. Inomata: Nd-Fe-B-Co-Al based permanent magnets with improved magnetic properties and temperature characteristics. Appl. Phys. Lett., 48 (1986), p.1309.

N

N1 T. Nagata: Rock magnetism. Maruzen Co. Ltd., Tokyo, 1961.

N2 J.E. Nesset, J.A. Finch: A loading equation for HGMS and application in identifying the fine size limit of recovery. Fine Particles Processing (Ed. P. Somasundaran), Vol. 2., AIME, New York, 1980, p.1217.

N3 J.E. Nesset, J.A. Finch: Determination of magnetic parameters for field-dependent susceptibility minerals using the Frantz isodynamic magnetic separator. Trans. IMM, Sect. C, 89 (1980), p.C161.

N4 J.E. Nesset, J.A. Finch: A static model of HGMS based on force within the fluid boundary layer. Proc. Intl. Conf. "Industrial Applications of Magnetic Separation", (Ed. Y.A. Liu). IEEE Publication No. 78 CH 1447-2 MAG, New York, 1979, p. 188.

N5 J.E. Nesset, J.A. Finch: The static build-up model of particle accumulation on single wires in HGMS: experimental confirmation. IEEE Trans. Mag., MAG-17 (1981), p.1506.

N6 K.S.V.L. Narasimhan: Iron-based rare-earth magnets. J. Appl. Phys., 57 (1985), p.4081.

N7 R. Nebolsine: Steel plant waste water treatment and reuse. Iron and Steel Eng., 44 (1967), p.122.

O

O1 J.A. Oberteuffer: Magnetic separation: a review of principles, devices and applications. IEEE Trans. Mag., MAG-10 (1974), p.223.

O2 W. O'Reilly: Rock and mineral magnetism. Blackie & Son Ltd., Glasgow, 1984.

O3 M. Owen: Magnetochemical investigations. The thermomagnetic properties of elements II. Ann. Phys. (Lpz), 37 (1912), p.657.

O4 J.A. Oberteuffer: HGMS: basic principles, devices and applications. Proc. Intl. Conf. "Industrial Applications of Magnetic Separation", (Ed. Y.A. Liu). IEEE Publication No. 78 CH 1447-2 MAG, New York, 1979. p.3.

O5 M. Odehnal: Some non-standard applications of SQUIDS (in Russian). Fiz. Nizkikh Temp.(USSR), 11 (1985), p.5.

O6 J.A. Oberteurffer: Engineering development of HGMS. IEEE Trans. Mag., MAG-12 (1976), p.444.

O7 J. Ormerod: The physical metallurgy and processing of sintered rare earth permanent magnets. J. Less Common Metals, 111 (1985), p.49.

O8 S. Owada, T. Harada: Grindability and magnetic property of chromites from various localities in relation to their mineralogical properties. J. Min. Metal. Inst. Japan, 101 (1985), p.781.

O9 P.E. Ostapenko, V.A. Arsent'ev, A.M. Komlev: Basic trends in dressing of oxidized iron, manganese and chromium ores (in Russian). Proc. Conf. "Pererab. Okisl. Rud." (Ed. V.A. Chanturiya, E.A. Trofimova). Nauka, Moscow, 1985, p.83.

O10 P.W. Overbeek: Personal communication. Council for Mineral Technology, Randburg, RSA, 1986.

O11 R.R. Oder: Magnetic desulphurization of liquified coals: conceptual process design and cost estimation. IEEE Trans. Mag., MAG-12 (1976), p.532.

O12 R.R. Oder: The application of high-field and high-gradient methods to the magnetic separation of mineral matter from micronized coal. Sep. Sci. Technol., 19 (1984-85), p.761.

O13 J.A. Oberteuffer, I. Wechsler, P.G. Marston, M.J. McNallan: High-gradient magnetic filtration of steel mill process and waste waters. IEEE Trans. Mag., MAG-11 (1975), p.1591.

O14 T. Okuda, I. Sugano, T. Tsuji: Removal of heavy metals from waste water by ferrite co-precipitation. Filt. Separ., 12 (1975), p.472.

O15 J.A. Oberteuffer, I. Wechsler: Recent advances in HGMS. "International Symp. on Fine Particles Processing (Ed. P. Somasundaran). Las Vegas, Nevada, AIME, New York, 1980, p.1178.

O16 Ö. Özdemir, W. O'Reilly: High temperature hysteresis and other magnetic properties of synthetic monodomain titanomagnetites. Phys. Earth Planet. Int., 25 (1981), p.406.

O17 J.B. O'Donovan, W. O'Reilly: An experimental study of TRM in pseudo-single domain titanomagnetite. Geophys. J.R. Astr. Soc., 69 (1982), p.290.

P

P1 N.G. Petrova: Investigation of the pulverized magnetite (in Russian). Izv. AN SSSR, Ser. geogr. i geophys., 12 (1948).

P2 J.M. Pastrana, D.M. Hopstock: Magnetic properties of natural hematite and goethite. Trans. Soc. Min. Eng. AIME, 262 (1977), p.1.

P3 D.E. Polk: Permanent magnet materials: an opportunity to reduce cobalt consumption. Naval Res. Rev., 34 (1982), p.24.

P4 D.H. Parkinson. B.E. Mulhall: The generation of high magnetic fields. Plenum Press, New York, 1967.

P5 R.J. Parker, R.J. Studders: Permanent magnets and their application. John Wiley & Sons, New York, 1962.

P6 P. Parsonage: Extension of range and sensitivity of laboratory isodynamic magnetic separator to fine sizes. Trans. IMM, Sect. C (1979), p.C182.

P7 M.R. Parker: Use of superconducting magnets in magnetic separation. J. Physique C1, Suppl. No. 1, 45 (1984), C1-753.

P8 F. Paul: HGMS of erythrocytes. Ph.D. Thesis, University of Southampton (1981).

P9 F. Paul, D. Mellvile, S. Roath: Inviscid approximation trajectories in HGMS. IEEE Trans. Mag., MAG-18 (1982), p.792.

P10 M.R. Parker: Evaluation of magnetic filter performance. IEEE Trans. Mag., MAG-18 (1982), p.822.

P11 M.R. Parker, R.P.A.R. van Kleef, H.W. Myron, P. Wyder: Particle aggregation in colloids in high magnetic fields. J. Mag. Mag. Mater., 27 (1982), p.250.

P12 M.R. Parker et al.: Kinetics of magnetic flocculation in colloidal dispersions. J. Coll. Interface Sci., 101 (1984), p.314.

P13 M.R. Parker, R.P.A.R. van Kleef, H.W. Myron, P. Wyder: Study of particle behaviour in high field magnetic flocculation. IEEE Trans. Mag., MAG-18 (1982), p.1647.

P14 J. Parsonage: Selective magnetic coating for mineral separation. Trans. Inst. Min. Metall.,(Sec. C: Mineral Process. Extr. Metall.),93 (1984), p.C37.

P15 J. Pitel, L. Cesnak, I. Hlásnik: Winding geometry optimization of superconducting multipole magnets for a linear OGMS. Proc. of Magnet Technology Conf. MT-9, Zürich, 1985.

P16 M.R.Parker: Recent developments in high field magnetic separation. Proc. "Electrical and Magnetic Separation and Filtration Technology" Antwerp 1984, (Ed. R. Van Brabant, R. Weiler). Koninklijke Vlaamse Ingenieursvereniging Tech. Inst., Antwerpen, 1984, p.1.

P17 H.E. Powel, C.K. Miller: Magnetic susceptibility of siderite. US Bureau of Mines Rept. Inv. 6224, 1963.

P18 J.W. Peoples, G.P. Eaton: Magnetic susceptibility of chromite from Montana and its relation to other physical and chemical properties. Amer. Mineralogist, 37 (1952), p.298.

P19 L. Petrakis, P.F. Ahner: HGMS in water effluents. IEEE Trans. Mag., MAG-14 (1978), p.491.

P20 D.H. Peregrine: A note on the steady high-Reynolds-number flow about a circular cylinder. J. Fluid Mech., 157 (1985), p.493.

P21 R. Potts, D.H. Prothero, W. Baxter: Magnetic separation trials for clarification of radioactive liquor using chromium powder simulant. IEEE Trans. Mag., MAG-17 (1981), p.2798.

P22 C.R. Price, W.F. Abercombie: Practical aspects of HGMS. "Industrial Applications of Magnetic Separation", (Ed. Y.A. Liu). IEEE Publication No. 78 CH 1447-2MAG, New York, 1979, p.14.

P23 M.R. Parker: Magnetic filtration by a current-carrying wire matrix. IEEE Trans. Mag., MAG-17 (1981), p.2816.

P24 C.R. Porter, D.N. Goens: Coal preparation using magnetic separation. Volume 1: Magnetic separation study of Magnex processed coal. EPRI CS-1517, 1980.

P25 L. Petrakis, P.F. Ahner, F.E. Kiviat: Desulphurization and deashing of SRC by HGMS techniques. Sep. Sci. Technol., 16 (1981), p.745.

P26 L.Petrakis, P.F. Ahner, F.E. Kiviat: HGMS of fine particles from industrial streams. "International Symp. on Fine Particles Processing", (Ed. P. Somasundaran), Las Vegas, Nevada, AIME, New York, 1980, p.1261.

R

R1 P.W. Riley, D. Hocking: A reciprocating canister superconducting magnetic separator. IEEE Trans. Mag., MAG-17 (1981), p.3299.

R2 P.A. Reeve: Magnetic separation in quadrupole fields. Proc. 6th Internatl. Conf. Magnet Technology. Bratislava, 1977, p.319.

R3 E.H. Roux, J.G. Goodey, E.F. Wepener, K.R. Hodierne: Industrial-scale dry beneficiation of phosphate-bearing pyroxenite ore. Proc. Internatl. Conf. MINTEK 50, Johannesburg, 1984, Paper A6/2.

R4 G. Rowlands: Magnetostatic energy of paramagnetic particles in magnetic separators. IEEE Trans. Mag., MAG-13 (1977), p.992.

R5 G. Rowlands: Magnetostatic energy of paramagnetic particles in magnetic separators. IEEE Trans. Mag., MAG-15 (1979), p.991.

R6 G. Rowlands: On the invariance of the magnetic fields strength due to an uniformly magnetized body. IEEE Trans. Mag., MAG-16 (1980), p.480.

R7 J. Reffle, M. Duschl, H. Hoffmann, F.J. Friedlaender: Critical capture radius in single wire HGMS. J. Appl. Phys., 24 (1981), p.225.

R8 G. Reger, R. Gerber, F.J. Friedlaender, H. Hoffmann: The efficiency of particle capture by an infinite array of ferromagnetic wires at low Reynolds numbers. J. Mag. Mag. Mater, 49 (1985), p.291.

R9 R. Rajagopalan, Chi Tien: The theory of deep bed filtration. Progress in filtration and separation (Ed. R.J. Wakeman), Vol. 1., Elsevier Sci. Publishing Co., Amsterdam, 1979, p.179.

R10 Ya. Rabinovich, N.V. Tchukaryev: Effect of electromagnetic retardation on the force of molecular attraction (in Russian). Kolloid. Zh., 41 (1979), p.468.

R11 M. Robbins, G.K. Wertheim, R.C. Sherwood, D.N.E. Buchanan: Magnetic properties and site distribution in the system $FeCr_2O_4$-Fe_3O_4($Fe^{2+}Cr_{2-x}Fe_x^{3+}O_4$). J. Phys. Chem. Solids, 32 (1971), p.717.

R12 A.T. Rao: Magnetic chromites from Kondapalli, Andhra Pradesh, India. Miner. Mag., 42 (1978), p.406.

R13 W. Rodewald: Magnetization and aging of sintered Nd-Fe-B magnets. J. Less Common Metals, 111 (1985), p.77.

R14 P.W. Riley, J.H.P. Watson: The use of paramagnetic matrices for magnetic separations. Filtration ad Separation, July/August 1977, p.1.

R15 J. Reffle, V. Krautwurm, H. Schewe, H. Hoffmann: Investigation of the matrix cleaning process in HGMS. J. Mag. Mag. Matter, 13 (1979), p.11.

R16 R.J. Rots: Chernaya Metallurgiya, No.4 (936) (1983), p.29.

R17 W.T. Ruhmer: Down in the dumps-cheerfully. Nuclear Active, June 1984, p.2.

R18 N.K. Roy, M.J. Murtha, G. Burnet: Recovery of iron oxide from power plant fly ash by magnetic separation. "Industrial Applications of Magnetic Separation", (Ed. Y.A. Liu). IEEE Publication No. 78 CH 1447-2MAG, New York, 1979, p.32.

R19 W.T. Ruhmer, O. Svoboda, B.M. Wilson: South African costs of equipment for the metallurgical industry (1984). Council for Mineral Technology Report M212, 1985.

S

S1 J.A. Stratton: Electromagnetic theory. McGraw-Hill, Inc., New York, 1941.

S2 M.M. Schieber: Experimental magnetochemistry. North-Holland Publ. Co., Amsterdam, 1967.

S3 F.D. Stacey, S.K. Banerjee: The physical principles of rock magnetism. Elsevier Scientific Publishing Co., Amsterdam, 1974.

S4 J. Svoboda: unpublished results.

S5 G.A. Smelyanskaya: The application of magnetic and other physical properties of minerals to geological and mineralogical problems (in Russian). Electrophysical methods of the treatment of rare metal ores, (Ed. U. Ts. Andres). AN SSSR, Moscow, 1972.

S6 E.M. Spokes, D.R. Mitchell: Relation of magnetic susceptibility to mineral composition. Min. Engng., New York, 10 (1958), p.373.

S7 M. Sagawa, S. Fujimura, H. Yamamoto, Y. Matsura, K. Hiraga: Permanent magnet materials based on the rare-earth-iron-boron tetragonal compounds. IEEE Trans. Mag., MAG-20 (1984), p.1584.

S8 Sagawa, S. Fujimura, H. Yamamoto, Y. Matsuura, S. Hirosawa: Magnetic properties of rare-earth-iron-boron permanent magnet materials. J. Appl. Phys., 57 (1985), p.4094.

S9 M. Sagawa, S. Fujimura, N. Togawa, H. Yamamoto, Y. Matsura: J. Appl. Phys., 55 (1984), p.2083.

S10 B.G. Street: Future development trends in the production of sintered permanent magnets. Met. Powder Rep., 32 (1984), p.581.

S11 R.G. Sharma, M.M. Krishna: High-field industrial superconductors - recent developments. Indian J. Cryog., 8 (1983), p.213.

S12 Z.J.J. Stekly: A superconducting high-intensity magnetic separator. IEEE Trans. Mag., MAG-11 (1975), p.1594.

S13 E. Steingroever: Magnetic measurements with fluxmeter. Magnet Physik GmbH, Köln, 1984, p. 1218.

S14 J. Suleski: New magnets and tank designs for wet magnetic drum separators. World Mining, April 1972, p.60.

S15 W.J.D. Stone: Application of the Jones wet magnetic separator to the beneficiation of iron ore. Canad. Min. Metall. Bull., Sept. 1961, p.677.

S16 W.J.D. Stone: Effect of variable adjustments on separation in the Jones magnetic separator. Mine & Quarry Eng., May 1964, p.202.

S17 SMI Bulletin No. S79-0001. Sala Magnetics, Inc., Cambridge, Mass., USA.

S18 W. Süsse, H. Alter: Strong field magnetic separator. German Patent P 2222003.7 (1973).

S19 W. Süsse: SOL-separator - a new development in the field of high-intensity wet magnetic separation. Erzmetall, 32 (1979), p.321.

S20 Z.J.J. Stekly, G.Y. Robinson, G.J. Powers: Magnetic processing: commercial supercon? Cryogenics and Ind. Gases, 8/4 (1973), p.26.

S21 K. Schönert, A. Supp, H. Dörr: Solenoid-pile separator, a new high intensity magnetic separator with super conductive coils. Proc. 12th Internatl. Miner. Process. Congress, Sao Paulo, 1977, Paper 4/1.

S22 N.F. Schultz: Determination of the magnetic separation characteristics with the Davis magnetic tube. Trans. Soc. Min. Eng. AIME, June 1964, p.211.

S23 H. Schewe, M. Takayasu, F.J. Friedlaender: Observation of particle trajectories in an HGMS single-wire system. IEEE Trans. Mag., MAG-16 (1980), p.149.

S24 Z.J.J. Stekly, J.V. Minervini: Shape effect of the matrix on the capture cross-section of particles in HGMS. IEEE Trans. Mag., MAG-12 (1976), p.474.

S25 H. Schlichting: Boundary-layer theory. McGraw-Hill Co., New York, 1979.

S26 V. Strouhal: Über eine besondere Art der Tonerregung (in German). Ann. Phys. und Chemie, New Series, 5 (1878), p.216.

S27 T.J. Sheerer, M.R. Parker, F.J. Friedlaender, R.R. Birss: Theory of capture of weakly magnetic particles in random matrices in the longitudinal configuration in HGMS. IEEE Trans. Mag., MAG-17 (1981), p.2807.

S28 W.H. Simons, R.P. Treat: Results on particle trajectories in a lattice of magnetized fibres in HGMS. "Industrial Applications of Magnetic Separation" (Ed. Y.A. Liu). IEEE Publication No. 78 CH 1447-2MAG, New York, 1979, p.145.

S29 J. Svoboda: The selection of a matrix for the recovery of uranium by WHIMS. Council for Mineral Technology Report M204, Randburg, S.A., 1985.

S30 J. Svoboda, R.N. Guest, W.J.C. Venter: Recovery of copper and lead minerals from Tsumeb flotation tailings. J. South Afr. Min. Metall. (1987).

S31 J. Svoboda: HGMS: a search for a matrix material. Intl. J. Min. Proc., 8 (1981), p.165.

S32 J. Svoboda: Current problems of practical applications of HGMS. Mag. Sep. News., 2 (1986), p.51.

S33 P. Somasundaran: Fine particles treatment. Research Needs in Mineral Processing. Columbia University School of Mines, New York, 1976, p.125.

S34 J. Svoboda: A theoretical approach to the magnetic flocculation of weakly magnetic minerals. Intl. J. Min. Proc., 8 (1981), p.377.

S35 J. Svoboda: Magnetic flocculation and treatment of fine weakly magnetic minerals. IEEE Trans. Mag., MAG-18 (1982), p.796.

S36 J. Svoboda, J. Žofka: Magnetic flocculation in secondary minimum. J. Coll. Interface Sci., 94 (1983), p.37.

S37 J. Svoboda: The selective magnetic flocculation of fine weakly magnetic minerals. Proc. Internatl. Conf. MINTEK 50, Sandton, 1984. Council for Mineral Technology, Randburg(RSA), 1984, p.329.

S38 P.C. Scholten, D.L.A. Tjaden: Mutual attraction of superparamagnetic particles. J. Coll. Interface Sci., 73 (1980), p.254.

S39 J.H. Schenkel, J.A. Kitchener: A test of the DLVO theory with colloidal suspensions. Trans. Farad. Soc., 56 (1960), p.161.

S40 J. Svoboda: The influence of surface forces on magnetic separation. IEEE Trans. Mag., MAG-18 (1982), p.862.

S41 J. Svoboda: The investigation of the effect of colloid stability in magnetic separation. Aufbereitungs-Technik, 24 (1983), p.520.

S42 J. Svoboda, I.J. Corrans, M.H.E. Spitze: The effect of pH on the recovery of uranium and gold by HGMS. Intl. J. Min. Proc., 17 (1986), p.83.

S43 J. Svoboda, M. Lazer, W.A.M. te Riele: Selective magnetic separation of uranium and gold. IEEE Trans. Mag., MAG-23 (1987),

S44 V.D. Samygin, L.A. Barsky, S.M. Angelova: Mechanisms of mutual flocculation of particles differing in size (in Russian). Kol. Zh. (USSR), 30 (1968), p.581.

S45 J. Svoboda, J. Cibulka: A method of magnetic separation of suspended minerals. Czechoslovak Patent PV1730 (1982).

S46 J. Svoboda: Magnetic flocculation a treatment of fine feebly magnetic minerals (in Czech). Rudy (Prague), 29 (1981), p.321.

S47 F.C. Schwerer, W. Gundaker: Magnetic properties of natural chromites: mechanical and thermal effects. Trans. Soc. Min. Eng. AIME, 258 (1975), p.88.

S48 I.C. Slater: One-electron atoms, molecules and solids. Phys. Rev., 98 (1955), p.4.

S49 F.E. Senftle, T. Pankey, F.A. Grant: Magnetic susceptibility of tetragonal titanium dioxide. Phys. Rev., 120 (1960), p.820.

S50 J.J.N. Sun, D. Lewis: Method and apparatus for magnetic separation of particles in a fluid carrier. US Patent 4,102,780 (1978).

S51 J. Svoboda: The particle capture in a matrix of a magnetic separator. Technical Memorandum No. 15227. Council for Minerals Technology, Randburg (RSA), 1986.

S52 J. Svoboda, V. Ross: Particle capture in a matrix of a magnetic separator. J. Mag. Mag. Mater., (1987),

S53 F.T. Smith: Laminar flow of an incompressible fluid past a bluff body: The separation, reattachment, eddy properties and drag. J. Fluid Mech., 92 (1979), p.171.

S54 J. Svoboda: The effect of magnetic field strength on the efficiency of magnetic separation. Report. Council for Mineral Technology, Randburg (RSA), 1987.

S55 F. Soda, T. Yukawa, H. Ishibe: Diffusion-bonded matrix of HGMF. IEEE Trans. Mag., MAG-21 (1985), p.1181.

S56 J. Svoboda, V. Hencl, J. Cibulka: Elimination of the effect of extraneous bodies in magnetic separation (in Czech). Proc. Conf. "Hornická Příbram", Prague, 1979.

S57 J. Suchánek, M. Vocel, J. Bakula: The selection of material for machine parts subjected to the erosion effect of hard particles (in Czech). Strojirenství (Prague), 30 (1980), p.294.

S58 L.A. Spielman: Flow through porous media and fluid - particle hydrodynamics. The Scientific Basis of Filtration. (Ed. K.J. Ives). Noordhoff, Leyden, 1975. p.129.

S59 J. Svoboda, I.J. Corrans: The removal of particles from the matrix of a high-gradient magnetic separator. IEEE Trans. Mag., MAG-21 (1985), p.53.

S60 A. Shalom, C.R. Price: Effect of matrix fiber size on performance of HGMS. IEEE Trans. Mag., MAG-17 (1981), p.3305.

S61 P. Somasundaran: Principles of flocculation, dispersion and selective flocculation. Fine Particles Processing, Vol. 2. (Ed. P. Somasundaran). AIME, New York, 1980, p.947.

S62 J. Svoboda: The effect of particle size and colloid stability on the WHIMS of uranium from cyanidation residues. Report M236, Council for Mineral Technology, Randburg (RSA), 1986.

S63 A.L. Shalom, J.M. Doulin: Study of particle size selectivity in WHGMS with coarse particles in industrial matrices. IEEE Trans. Mag., MAG-19 (1983), p.2130.

S64 J. Svoboda: The limiting particle size in HGMS (in Czech). Rudy (Prague), 29 (1981), p.133.

S65 F. Špaček, M. Dočkal: The application of pulse magnetic field in iron ore treatment (in Czech). Rudy (Prague), 34 (1986), p.45.

S66 J. Svoboda, P. Manganya, S. Tsilevhulevhu: The effect of magnetic viscosity on magnetic separation. Powder Technology (to be published).

S67 Z. Sun, Z. Li: A sudy of vibration HGMS. J. Central South Inst. Min. Metal (China), No. 2 (1985), p.1.

S68 Z. Sun, Z. Li: A study of vibration HGMS. Proc. 15th Internatl. Min. Proc. Congress, Cannes, 1985, p.410.

S69 T. Sabelin, A.S. Malicsi, I. Iwasaki: Cuyuna manganese resources and their concentratability by HGMS. 58th Ann. Meeting Minn. Section, AIME, and 46th Ann. Min. Symp., Duluth, 1985, paper 13.

S70 Sun Shanlun, Chen Qiren, Yang Yue: New techniques in processing tungsten ore slimes. Proc. Miner. Proc. Extr. Metal. Conference (Ed. M.J. Jones, P. Gill), Inst. Min. Metal. and Chinese Soc. Metals, Kunming, China, 1984, p.521.

S71 W.J.D. Stone: Uranium beneficiation by magnetic means and Jones separator development. Ann. Meeting: Canadian Uranium Producers Marketing Committee, 1964.

S72 Z.J.J. Stekly, M.L. Mallary, H.R. Segal: Coal preparation using magnetic separation. Volume 2. EPRI CS-1517, 1980.

S73 V.D. Semin, G.P. Medvedev, Z.F. Semina: Production methods of using low-grade aluminosilicate ores. Sov. Non-Ferrous Met. Res., 11 (4) (1983), p.279.

T

T1 R.K. Tenzer: Estimating leakage factors for permanent magnets from geometry of magnetic circuit. Electrical Manuf., Feb. 1957, p.94.

T2 A.F. Taggart: Handbook of mineral dressing. John Wiley & Sons, Inc., New York, 1927.

T3 R. Tille, W. Kirkpatrick: Investigation of the magnetic separation of ores. Z. Erzbergbau Metallhüttenwessen, 8 (1955), p.B117.

T4 R.P. Treat, W.F. Lawson, J.T. Johnson: Observation of particle trajectories near a magnetized fibre. J. Appl. Phys., 50 (1979), p.3596.

T5 M. Takayasu, R. Gerber, F.J. Friedlaender: Magnetic separation of submicron particles. IEEE Trans. Mag., MAG-19 (1983), p.2112.

T6 M. Takayasu, F.J. Friedlaender, Y. Hiresaki, D.R. Kelland: Matrices for selective diamagnetic HGMS. IEEE Trans. Mag., MAG-17 (1981), p.2810.

T7 I. Todd, J. Finch: Measurement of susceptibility using a wet modification to the Frantz isodynamic separator. Can. Met. Qart., 23 (1984), p.475.

T8 S. Takeno, K. Masumoto, T. Kamigaichi: Electric and magnetic properties of bornite with a review of the physical properties of Cu-Fe-S minerals. J. Sci. Hiroshima Univ., Ser. C, 5 (1968), p.321.

T9 K. Takino, T. Tankaka, T. Shichiri: Operation of a full-scale device of high gradient magnetic filter for treatment of gas scrubber water from vacuum degassing process in a steel mill. "Industrial Applications of Magnetic Separation. (Ed. Y.A. Liu). IEEE Publication No. 78 CH 1447-2MAG, New York, 1979, p. 137

T10 V. Timbrell: Magnetic separation of respirable asbestos fibres. Filtration & Separation, May/June 1977, p.241.

T11 S. Taneda: Experimental investigation of the wakes behind cylindres and plates at low Reynolds numbers. J. Phys. Soc. Japan, 11 (1956), p.302.

T12 M. Troy: Study of magnetic filtration applications to the primary and secondary systems of PWR plants. EPRI Report NP-514, 1978.

T13 M. Takayasu, E. Maxwell, D.R. Kelland: Continuous selective HGMS in the repulsive force mode. IEEE Trans. Mag., MAG-20 (1984), p.1186.

T14 M. Takayasu, J.Y. Hwang, F.J. Friedlaender, L. Petrakis, R. Gerber: Magnetic separation utilizing a magnetic susceptibility gradient. IEEE Trans. Mag., MAG-20 (1984), p.155.

T15 M. Takayasu, D.R. Kelland: Dynamic magnetic separation by a magnetic susceptibility distribution. IEEE Trans. Mag., MAG-20 (1984), p.1183.

T16 K.I. Tishschenko, N.K. Voroniev, M.A. Fedorov: Improvement of a technology for dressing manganese ore slurries (in Russian). Conf. Pererab. Okisl. Rud (Ed. V.A. Chanturiya, E.A. Trofimova). Nauka, Moscow, 1985, p.135.

T17 D.M. Thayer, P.B. Linkson: The beneficiation of specular hematite by high-intensity magnetic separation. Ann. Meeting Soc. Min. Eng. AIME, New Orleans, La., 1978. SME Preprint No. 79-16.

T18 E.M. Taits, I.A. Andreeva: The reduction of sulphur content in coals from the Moscow deposit (in Russian). Khim. Tverd. Topl., No. 5 (1972), p.44.

T19 S. Trindade, H.H. Kolm: Magnetic desulphurization of coal. IEEE Trans. Mag., MAG-9 (1973), p.310.

T20 S.C. Trindade, M. Saddy, J.L.F. Monteiro: Magnetic recovery of sulphur from coal in Brazil. IEEE Trans. Mag., MAG-12 (1976), p.355.

T21 P.A. Tatsienko et at.: Obogascheniye Rud, 27/3 (1981), p.3.

T22 P. Tucker, W. O'Reilly: A magnetic study of single crystal titanomagnetite. Phys. Earth Planet. Int., 16 (1978), p.183.

U

U1 S. Uchiyama, S. Kondo, M. Takayasu, I. Eguchi: Performance of parallel stream type magnetic filter for HGMS. IEEE Trans. Mag., MAG-12 (1976), p.895.

U2 S. Uchiyama, K. Hayashi: Analytical theory of magnetic particle capture process and capture radius in HGMS. Proc. Intl. Conf. " Industrial Applications of Magnetic Separation" (Ed. Y.A. Liu). IEEE Publication no. 78 CH 1447-2MAG, New York, 1979, p.169.

U3 S. Uchiyama, S. Kurinobu, M. Takayasu, T. Fujii: Particle build-up effect on recovery in parallel stream type HGMS filter. IEEE Trans. Mag., MAG-13 (1977), p.1493.

U4 S. Uchiyama, S. Kurinobu, M. Kumazawa, M. Takayasu: Magnetic particle build-up process in parallel stream type HGMS filter. IEEE Trans. Mag., MAG-13 (1977), p.1490.

U5 S. Uchiyama, S. Kurinobu: Steady-state solution of filter equations of HGMS. "Industrial Applications of Magnetic Separation" (Ed. Y.A. Liu). IEEE Publ. No. 78 CH 1447-2MAG, 1979, p.166.

V

V1 Y.A. Voevodin, L.A. Grekulova, N.Yu. Yakubovskaya: Utilization of the magnetic characteristics of stannite and its oxidation products for the concentration of tin ores. Sov. Non-Ferrous Metals Research, 11 (1983), p.423.

V2 R.P.A.R. van Kleef et al.: Limits of magnetic flocculation in colloidal dispersions. IEEE Trans. Mag., MAG-19 (1983), p.2118.

V3 R.P.A.R. van Kleef, H.W. Myron, M.R. Parker, P. Wyder: Magnetic aggregation of particles in very high fields. Proc. 3rd World Filtration Congress. Downington, Po. USA, 1982, p.769.

V4 E.J.W. Verwey, J.T.G. Overbeek: Theory of the stability of lyophobic colloids. Elsevier, Amsterdam, 1948.

V5 J. Visser: The concept of negative Hamaker coefficient. Adv. Coll. Interface Sci., 15 (1981), p.157.

V6 R.P.A.R van Kleef, H.W. Myron, P. Wyder, M.R. Parker: Continuous flow separation, an application of selective magnetosedimentation. J. Physique, 45, Cl, Suppl. no.l, (1984), Cl-763.

V7 M. Velicescu: High-energy product magnets from (Sm, Pr, Nd)-Co alloys. J. Less Common Metals, 111 (1985), p.71.

V8 C.P. van Driel, C.B.W. Kerkdijk, H.R. Segal, J. Sikkenga: Coal cleaning by HGMS. FDO Technische Adviseurs Report, 1983.

W

W1 M.N. Wilson: Superconducting magnets. Clarendon Press, Oxford, 1983.

W2 T.D. Wellington: Comments on "The large-scale manipulation of small particles". IEEE Trans. Mag., MAG-12 (1976), p.586.

W3 J.H.P. Watson: Superconducting high-gradient magnetic separation. Mining Magazine, August 1983, p.121.

W4 J.H.P. Watson, N.O. Clark, W. Windle: A superconducting magnetic separator and its application in improving ceramic raw materials. Proc. 11th Internatl. Min. Proc. Congress, Cagliari, 1975, Paper no. 29.

W5 J.H.P. Watson, D. Hocking: The beneficiation of clay using a superconducting magnetic separator. IEEE Trans. Mag., MAG-11 (1975), p.1588.

W6 W. Windle: A method of, and apparatus for, separating magnetisable particles from a fluid in which they are suspended. Brit. Patent 1,469,765 (1977).

W7 J.H.P. Watson, R.G. Scurlock, A.W. Swales, T.A. Holden: New methods of force compensation in reciprocating superconducting magnetic separators. Proc. 9th Internatl. Cryog. Eng. Confer. Kobe, Japan, 1982, p.120.

W8 J.H.P. Watson, R.G. Scurlock, A.W. Swales: Magnetic separators. British Patent Appl. 8127084 (1981).

W9 J.H.P. Watson: Magnetic filtration. J. Appl. Phys., 44 (1973), p.4209.

W10 J.H.P. Watson: Theory of capture of particles in magnetic high-intensity filters. IEEE Trans. Mag., MAG-11 (1975), p.1597.

W11 J.H.P. Watson: Theory of high-intensity magnetic separation. Filtration & Separation, May/June 1977, p.242.

W12 J.H.P. Watson, S.J.P. Watson: The ball matrix magnetic separator. IEEE Trans. Mag., MAG-19 (1983), 2698.

W13 J.H.P. Watson: Particle retention in HIMS. Proc. Symp. on Deposition and Filtration of Particles from Gases and Liquids. Loughborough University, 1978.

W14 J.H.P. Watson: HGMS at moderate Reynolds number. IEEE Trans. Mag., MAG-15 (1979), p.1538.

W15 J.H.P. Watson: Superconducting magnetic separation at moderate Reynolds numbers. Proc. XVth Internatl. Congress of Refrigeration, Venezia, 1979, Paper A1/2.

W16 J.H.P. Watson: Theory of high-intensity magnetic filter performance. Digests of INTERMAG Conf. IEEE Publ. No. 75CH0932-4MAG, 1975, Paper 24.3.

W17 J.H.P. Watson: Approximate solutions of the magnetic separator equations. IEEE Trans. Mag., MAG-14 (1978), p.240.

W18 J.A. Williams, C.M. Leslie: HGMS in the nuclear fuel cycle. IEEE Trans. Mag, MAG-17 (1981), p.2790.

W19 J.H.P. Watson: Magnetic separation at high magnetic fields. Proc. 6th Intl. Conf. on Cryog. Eng., Grenoble, 1976, p. 223.

W20 J.H.P. Watson, K.B.M. Yusof: Magnetic flocculation of paramagnetic particles in high magnetic field. Digest of INTERMAG 1981. Grenoble, 1981, p.14-3.

W21 G.R. Wiese, T.W. Healy: Effect of particle size on colloid stability. Trans. Farad. Soc., 66 (1970), p.490.

W22 M. Wada, M. Majima, A. Ohba: Laboratory hot high-intensity IMR separator and its applications. Proc. 7th Internatl. Min. Proc. Congress, New York, 1964, p.391.

W23 J.A. Williams et al.: The application of HGMS in nuclear fuel reprocessing. Nuclear Technology, 52 (1981), p.284.

W24 M.F. Williams, L.G. Hendrickson: Depolarizing magnetite pulps. Min. Trans. AIME, February 1956, p.201.

W25 I. Wechsler, J. Doulin, R. Eddy: Coal preparation using magnetic separation, Vol.3. EPRI Report CS-1517, 1980.

W26 J.H.P. Watson, D. Rassi, A.S. Bahaj: The recovery of gold and uranium from gold leached residues by HGMS. IEEE Trans. Mag. MAG-19 (1983), p.2136.

W27 C.W. Wegst: Stahlschlüssel. Verlag Stahlschlüssel Wegst GMBH, Marbach (BRD), 1983.

W28 J.H.P. Watson, D. Rassi, D. Prothero, R. Potts: HGMS of uranium oxide and gold from mine residues. Proc. Conf. "Electrical and Magnetic Separation adn Filtration Technology" Antwerp 1984 (Ed. R. Vanbrabant, R. Weiler). Koninklijke Vlaamse Ingenieursvereniging Tech. Inst., Antwerpen, 1984, p.93.

W29 D. Walton: Viscous magnetization. Nature, 305 (1983), p.616.

W30 J.H.P. Watson: Personal communication.

W31 J.H.P. Watson, A.S. Bahaj, C.H. Boorman, D. Rassi: A superconducting HGMS separator with a current carrying matrix. 15th Internatl. Min. Proc. Congress, Cannes, 1985, p.330.

W32 J.H.P. Watson, A.S. Bahaj, C.H. Boorman, D. Rassi: Superconducting HGMS separator incorporating a current carrying wire matrix. IEEE Trans. Mag., MAG-21 (1985), p.2056.

W33 J.R. Walklate: Magnetic separation of heavy minerals at Richards Bay Minerals. Presented at the Colloquium on Magnetic Separation in Mineral Industry, Council for Mineral Technology, Randburg (RSA), 1986.

W34 W.L. Wilkinson: Filtration techniques in the nuclear fuel industry. Filt. Separ., Jan./Feb. 1982, p.17.

W35 R.P. Walker et al.: Aspects of magnetic filtration of gas/particle mixtures. IEEE Trans. Mag., MAG-21 (1985), p.2050.

W36 I.S. Wells: Operational experience with Boxmag-Rapid high intensity wet magnetic separator. Presented at Colloquium on Magnetic Separation in Mineral Industry. Council for Mineal Technology, Randburg (RSA), 1986.

W37 M.E. Weber, D. Paddock: Interceptional and gravitational collision efficiencies for single collectors at intermediate Reynolds numbers. J. Coll. Interface Sci., 94 (1983), p.328.

Y

Y1 Ying-Shan Yang et al.: Magnetic separation of kaolin clay with amorphous ferromagnetic fibre. IEEE Trans. Mag., MAG-16 (1980), p.934.

Y2 E.A. Yanchuk: Coordination of manganese; magnetic properties and stability of its natural oxides (in Russian). Miner. Sb. L'vov Gos. Univ. No. 24, 1970, p.30.

Y3 F.S. Yoong, D. Fletcher, M.R. Parker: Magnetic particle capture and build-up on a current-carrying wire. IEEE Trans. Mag., MAG-19 (1983), p.2109.

Y4 A.Z. Yurovskii, I.D. Remesnikov: Thermomagnetic methods of concentrating and desulphurizing coal (in Russian). Koks i khim., No. 12 (1958), p.8.

Y5 Jun-ichi Yauo, I. Eguchi: Application of HGMS for water treatment in steel industry. "Industrial Applications of Magnetic Separation" (Ed. Y.A. Liu). IEEE Publication No. 78 CH 1447-2MAG, New York, 1979, p.134.

Z

Z1 H. Zijlstra: Experimental methods in magnetism. Vol.2: Measurement of magnetic quantities. North-Holland Publ. Co., Amsterdam, 1967.

Z2 G. Zebel: Deposition of aerosol flowing past a cylindrical fiber in a uniform electric field. J. Coll. Sci., 20 (1965), p.522.

SUBJECT INDEX

C

N

O

P

Q

R

S

W

Z